METHODS OF MODERN MATHEMATICAL PHYSICS

IV: ANALYSIS OF OPERATORS

METHODS OF MODERN MATHEMATICAL PHYSICS

IV: ANALYSIS OF OPERATORS

MICHAEL REED

Department of Mathematics
Duke University

BARRY SIMON

Departments of Mathematics
and Physics
Princeton University

ACADEMIC PRESS, INC.
Harcourt Brace Jovanovich, Publishers
Boston   San Diego   New York
London   Sydney   Tokyo   Toronto

This book is printed on acid-free paper. ∞

Copyright © 1978, by Academic Press, Inc.

All rights reserved.
No part of this publication may be reproduced or
transmitted in any form or by any means, electronic
or mechanical, including photocopy, recording, or
any information storage and retrieval system, without
permission in writing from the publisher.

ACADEMIC PRESS, Inc.
1250 Sixth Avenue, San Diego, CA 92101-4311

United Kingdom edition published by
ACADEMIC PRESS LIMITED
24-28 Oval Road, London NW1 7DX

Library of Congress Cataloging in Publication Data

Reed, Michael.
    Methods of modern mathematical physics.

    Vol. 4. Analysis of Operators.
    Includes bibliographical references.
    CONTENTS: v. 1. Functional analysis.-v. 2 Fourier
analysis, self-adjointness. -v. 3. Scattering theory.-v. 4.
Analysis of operators.
    1. Mathematical physics. 1. Simon, Barry, joint
author. II. Title.
QC20.R37   1972   530.1'5   75-182650
ISBN 0-12-585004-2 (v. 4)
AMS (MOS) 1970 Subject Classifications: 47-02, 81-02

Printed in the United States of America
93 94 95 96 97 QW 9 8 7 6 5

To David
To Rivka and Benny

Preface

... of making books there is no end, and much study is a weariness of the flesh
Koheleth (*Ecclesiastes*) *12:12*

With the publication of Volumes III and IV we have completed our presentation of the material which we originally planned as "Volume II" at the time of publication of Volume I. We originally promised the publisher that the entire series would be completed nine months after we submitted Volume I. Well! We have listed the contents of future volumes below. We are not foolhardy enough to make any predictions.

We were very fortunate to have had T. Kato and R. Lavine read and criticize Chapters XII and XIII, respectively. In addition, we received valuable comments from J. Avron, P. Deift, H. Epstein, J. Ginibre, I. Herbst, and E. Trubowitz. We are grateful to these individuals and others whose comments made this book better.

We would also like to thank:

J. Avron, G. Battle, C. Berning, P. Deift, G. Hagedorn, E. Harrell, II, L. Smith, and A. Sokol for proofreading the galley and/or page proofs.

G. Anderson, F. Armstrong, and B. Farrell for excellent typing.

The National Science Foundation, the Duke Research Council, and the Alfred P. Sloan Foundation for financial support.

Academic Press, without whose care and assistance these volumes would have been impossible.

Martha and Jackie for their encouragement and understanding.

Introduction

Il libro della natura é scritto in lingua matematica.   Galileo Galilei

---

The first step in the mathematical elucidation of a physical theory must be the solution of the existence problem for the basic dynamical and kinematical equations of the theory. Once that is accomplished, one would like to find general qualitative features of these solutions and also to study in detail specific special systems of physical interest.

Having discussed the general question of the existence of dynamics in Chapter X, we present methods for the study of general qualitative features of solutions in this volume and its companion (Volume III) on scattering theory. We concentrate on the Hamiltonians of nonrelativistic quantum mechanics although other systems are also treated. In Volume III, the main theme is the long-time behavior of dynamics, especially of solutions which are "asymptotically free." In this volume, the main theme involves the five kinds of spectra defined in Sections VII.2 and VII.3: the essential spectrum, σ_{ess}; the discrete spectrum, σ_{disc}; the absolutely continuous spectrum, σ_{ac}; the pure point spectrum, σ_{pp}; and the singular continuous spectrum, σ_{sing}. It turns out that the study of the absolutely continuous spectrum as well as the problem of showing that the continuous singular spectrum is empty are intimately connected with scattering theory. Thus, the separation of the material in Volumes III and IV is somewhat artificial. For this reason, we preprinted in Volume III three sections from Volume IV.

These are not the only sections in which the themes of the two volumes overlap.

In these volumes specific systems are usually presented to illustrate the application of general mathematical methods, but the detailed analysis of the specific systems is not carried very far. Mathematical physicists have to some extent neglected the detailed study of specific systems; we believe that this neglect is unfortunate, for there are many interesting unsolved problems in specific systems, even in the purely Coulombic model of atomic physics. For example, it has not been shown that H^{--} has no bound states even though the analogous classical system of one positive and three negative charges has the property that its energy is lowered by moving a suitable electron to infinity. And it is not known rigorously that the energy needed to remove the first electron from an atom is less than the energy needed to remove the second, even though this is "physically obvious." We hope that by collecting the general mathematical methods in Volumes II, III, and IV, we have made the analysis of specific systems easier and more attractive.

Nonrelativistic quantum mechanics is often viewed by physicists as an area whose qualitative structure, especially on the level treated here, is completely known. It is for this reason that a substantial fraction of the theoretical physics community would regard these volumes as exercises in pure mathematics. On the contrary, it seems to us that much of this material is an integral part of modern quantum theory. To take a specific example, consider the question of showing the absence of the singular continuous spectrum and the question of proving asymptotic completeness for the purely Coulombic model of atomic physics. The former problem was solved affirmatively by Balslev and Combes in 1970, the latter is still open. Many physicists would approach these questions with Goldberger's method: "The proof is by the method of reductio ad absurdum. Suppose asymptotic completeness is false. Why that's absurd! Q.E.D." Put more precisely: If asymptotic completeness is not valid, would we not have discovered this by observing some bizarre phenomena in atomic or molecular physics? Since physics is primarily an experimental science, this attitude should not be dismissed out of hand and, in fact, we agree that it is extremely unlikely that asymptotic completeness fails in atomic systems. But, in our opinion, theoretical physics should be a science and not an art and, furthermore, one does not fully understand a physical fact until one can derive it from first principles. Moreover, the solution of such mathematical problems can introduce new methods of calculational interest (for example, Faddeev's treatment of completeness in three-body systems and the application of his ideas in nuclear physics) and can provide important elements of clarity (for example, the physical artificiality of "adiabatic switching" in nonrigorous scattering theory and the clarifying work of Cook, Jauch, and Kato).

The general remarks about notes and problems in earlier introductions are applicable here with one addition: the bulk of the material presented in this volume is from advanced research literature, so many of the "problems" are quite substantial. Some of the starred problems summarize the contents of research papers!

Contents

Preface	*vii*
Introduction	*ix*
Contents of Other Volumes	*xv*

XII: PERTURBATION OF POINT SPECTRA

1. Finite-dimensional perturbation theory	1
Appendix Algebraic and geometric multiplicity of eigenvalues of finite matrices	9
2. Regular perturbation theory	10
3. Asymptotic perturbation theory	25
4. Summability methods in perturbation theory	38
5. Spectral concentration	45
6. Resonances and the Fermi golden rule	51
Notes	60
Problems	69

XIII: SPECTRAL ANALYSIS

1. The min–max principle	75
2. Bound states of Schrödinger operators I: Quantitative methods	79
3. Bound states of Schrödinger operators II: Qualitative theory	86

A.	Is $\sigma_{\text{disc}}(H)$ finite or infinite?	86
B.	Bounds on $N(V)$ in the central case	90
C.	Bounds on $N(V)$ in the general two-body case	98
4.	Locating the essential spectrum I: Weyl's theorem	106
5.	Locating the essential spectrum III: The HVZ theorem	120
6.	The absence of singular continuous spectrum I: General theory	136
7.	The absence of singular continuous spectrum II: Smooth perturbations	141
A.	Weakly coupled quantum systems	151
B.	Positive commutators and repulsive potentials	157
C.	Local smoothness and wave operators for repulsive potentials	163
8.	The absence of singular continuous spectrum III: Weighted L^2 spaces	168
9.	The spectrum of tensor products	177
10.	The absence of singular continuous spectrum IV: Dilation analytic potentials	183
11.	Properties of eigenfunctions	191
12.	Nondegeneracy of the ground state	201
Appendix 1	The Beurling–Deny criteria	209
Appendix 2	The Levy–Khintchine formula	212
13.	Absence of positive eigenvalues	222
Appendix	Unique continuation theorems for Schrödinger operators	239
14.	Compactness criteria and operators with compact resolvent	244
15.	The asymptotic distribution of eigenvalues	260
16.	Schrödinger operators with periodic potentials	279
17.	An introduction to the spectral theory of non-self-adjoint operators	316
Notes		338
Problems		364
List of Symbols		387
Index		389

Contents of Other Volumes

Volume I: Functional Analysis

- I   *Preliminaries*
- II *Hilbert Spaces*
- III *Banach Spaces*
- IV *Topological Spaces*
- V   *Locally Convex Spaces*
- VI *Bounded Operators*
- VII *The Spectral Theorem*
- VIII *Unbounded Operators*

Volume II: Fourier Analysis, Self-Adjointness

- IX *The Fourier Transform*
- X *Self-Adjointness and the Existence of Dynamics*

Volume III: Scattering Theory

- XI *Scattering Theory*

Contents of Future Volumes: *Convex Sets and Functions, Commutative Banach Algebras, Introduction to Group Representations, Operator Algebras, Applications of Operator Algebras to Quantum Field Theory and Statistical Mechanics, Probabilistic Methods.*

XII: Perturbation of Point Spectra

In the thirties, under the demoralizing influence of quantum-theoretic perturbation theory, the mathematics required of a theoretical physicist was reduced to a rudimentary knowledge of the Latin and Greek alphabets.

Res Jost

In this chapter we shall examine the following general situation: An operator H_0 has an eigenvalue E_0, which we usually assume is in the discrete spectrum. Suppose that H_0 is perturbed a little; that is, consider $H_0 + \beta V$ where V is some other operator and $|\beta|$ is small. What eigenvalues of $H_0 + \beta V$ lie near E_0 and how are they related to V? What are their properties as functions of β? Such a situation is familiar in quantum mechanics where there are *formal* series for the perturbed eigenvalues. These **Rayleigh–Schrödinger series** are not special to quantum-mechanical operators but exist for many perturbations of the form $H_0 + \beta V$. The heart of this chapter is the second section where we shall discuss the beautiful Kato–Rellich theory of regular perturbations; this theory gives simple criteria under which one can prove that these formal series have a nonzero radius of convergence. We then discuss what the perturbation series means in cases where it is divergent or not directly related to eigenvalues.

XII.1 Finite-dimensional perturbation theory

We first discuss finite-dimensional matrices. Not only will this allow us to present explicit formulas in the simplest case, but we shall eventually treat degenerate perturbation theory by reducing it to an essentially finite-dimensional problem. Furthermore, an important difficulty already occurs in the finite-dimensional case, namely proving analyticity in β when there is a degenerate eigenvalue. Recall that E_0 is called a **degenerate eigenvalue** when the characteristic equation for H_0, $\det(H_0 - \lambda) = 0$, has a multiple root at $\lambda = E_0$. In an appendix to this section we review the theory of matrices with degenerate eigenvalues and, in particular, we discuss the Jordan normal form.

First consider the elementary example

$$T(\beta) = \begin{bmatrix} 1 & \beta \\ \beta & -1 \end{bmatrix}$$

By our definition of operator-valued analytic function in Section VI.3, $T(\beta)$ is a matrix-valued analytic function. To find its eigenvalues, we need only solve $\det(T(\beta) - \lambda) = 0$ (the **secular** or **characteristic equation**). Thus

$$\lambda_\pm(\beta) = \pm\sqrt{\beta^2 + 1}$$

are the eigenvalues. This problem has several characteristic features:

(i) Even though $T(\beta)$ is entire in β, the eigenvalues are not entire but have singularities as functions of β.

(ii) The singularities are not on the real β axis where $T(\beta)$ is self-adjoint but occur at nonreal β, namely at $\beta = \pm i$. Thus, while there are no singularities at "physical" values, the **perturbation series**, i.e., the Taylor series for $\lambda_\pm(\beta)$ at $\beta = 0$, have a finite radius of convergence due to complex singularities.

(iii) "Level crossing" takes place at the singular values of β; that is, at $\beta = \pm i$ there are fewer distinct eigenvalues, namely one, than at other points, where there are two.

(iv) At the singular values of β the matrix $T(\beta)$ is *not* diagonalizable. Explicitly

$$T(i)\begin{bmatrix} 2 \\ 2i \end{bmatrix} = \begin{bmatrix} 0 \\ 0 \end{bmatrix}, \quad T(i)\begin{bmatrix} 1 \\ -i \end{bmatrix} = \begin{bmatrix} 2 \\ 2i \end{bmatrix}$$

so the matrix of $T(i)$ in the basis $\langle 2, 2i \rangle$, $\langle 1, -i \rangle$, is

$$\begin{bmatrix} 0 & 1 \\ 0 & 0 \end{bmatrix}$$

While this "Jordan anomaly" is typical, we leave a discussion of it to the Notes; see also Problem 23.

(v) The analytic continuation of an eigenvalue is an eigenvalue.

For the remainder of this section, we shall suppose that $T(\beta)$ is a matrix-valued analytic function in a connected region R of the complex plane. Notice that we do not require $T(\beta)$ to be linear in β. Later, we shall be able to reduce the infinite-dimensional, linear, finitely degenerate perturbation problem to a finite-dimensional problem, but one that is no longer *linear* in β. Thus, greater generality at this point will be crucial.

To find the eigenvalues of $T(\beta)$ we must solve a secular equation

$$\det(T(\beta) - \lambda) = (-1)^n[\lambda^n + a_1(\beta)\lambda^{n-1} + \cdots + a_n(\beta)] = 0$$

The basic theorem about such functions is:

Theorem XII.1 Let $F(\beta, \lambda) = \lambda^n + a_1(\beta)\lambda^{n-1} + \cdots + a_n(\beta)$ be a polynomial of degree n in λ whose leading coefficient is one and whose coefficients are all analytic functions of β. Suppose that $\lambda = \lambda_0$ is a simple root of $F(\beta_0, \lambda)$. Then for β near β_0, there is exactly one root $\lambda(\beta)$ of $F(\beta, \lambda)$ near λ_0, and $\lambda(\beta)$ is analytic in β near $\beta = \beta_0$.

Proof This is a special case of the implicit function theorem. Since $F(\beta, \lambda)$ is analytic near β_0 and λ_0, we can write $F(\beta, \lambda) = \sum_{m=0}^{n} (\lambda - \lambda_0)^m f_m(\beta)$ with $f_0(\beta_0) \equiv F(\beta_0, \lambda_0) = 0$, and $f_1(\beta_0) \equiv (\partial F/\partial \lambda)(\beta_0, \lambda_0) \neq 0$ since λ_0 is a simple root. Thus to find solutions of $F(\beta, \lambda) = 0$, we need only solve the equivalent equation

$$\lambda = \lambda_0 - \frac{f_0(\beta)}{f_1(\beta)} - \sum_{m=2}^{n} (\lambda - \lambda_0)^m \frac{f_m(\beta)}{f_1(\beta)} \tag{1}$$

Because $f_1(\beta_0) \neq 0$, all the coefficients $f_k(\beta)/f_1(\beta)$ are analytic near $\beta = \beta_0$. We try to solve this last equation with a solution of the form $\lambda(\beta) = \lambda_0 + \sum_{k=1}^{\infty} \alpha_k(\beta - \beta_0)^k$. The α_k can be computed by recursive substitution into (1); for example,

$$\alpha_1 = -\left[\frac{f_0(\beta)}{f_1(\beta)}\right]'\bigg|_{\beta=\beta_0}$$

and

$$\alpha_2 = -\frac{1}{2}\left[\frac{f_0(\beta)}{f_1(\beta)}\right]''\bigg|_{\beta=\beta_0} - \alpha_1^2 \frac{f_2(\beta_0)}{f_1(\beta_0)}$$

It is not very hard to prove that the α's determined recursively yield a power series with a nonzero radius of convergence (Problem 1a). Uniqueness is also fairly easy (Problem 1b). ∎

Corollary Let $T(\beta)$ be a matrix-valued analytic function near β_0 and suppose λ_0 is a simple eigenvalue of $T(\beta_0)$. Then:

(a) For β near β_0, $T(\beta)$ has exactly one eigenvalue, $\lambda_0(\beta)$, near λ_0.

(b) $\lambda_0(\beta)$ is a simple eigenvalue if β is near β_0.
(c) $\lambda_0(\beta)$ is analytic near $\beta = \beta_0$.

For multiple roots, a more complicated but still straightforward analysis is necessary. We do not prove the following basic theorem for this case (proofs can be found in the references in the Notes).

Theorem XII.2 Let $F(\beta, \lambda) = \lambda^n + a_1(\beta)\lambda^{n-1} + \cdots + a_n(\beta)$ be an nth degree polynomial in λ whose leading coefficient is one and whose coefficients are all analytic functions of β. Suppose $\lambda = \lambda_0$ is a root of multiplicity m of $F(\beta_0, \lambda)$. Then for β near β_0, there are exactly m roots (counting multiplicity) of $F(\beta, \lambda)$ near λ_0 and these roots are the branches of one or more multivalued analytic functions with at worst algebraic branch points at $\beta = \beta_0$. Explicitly, there are positive integers $p_1, \ldots, p_k$ with $\sum_{i=1}^{k} p_i = m$ and multivalued analytic functions $\lambda_1, \ldots, \lambda_k$ (not necessarily distinct) with convergent **Puiseux series** (Taylor series in $(\beta - \beta_0)^{1/p}$)

$$\lambda_i(\beta) = \lambda_0 + \sum_{j=1}^{\infty} \alpha_j^{(i)}(\beta - \beta_0)^{j/p_i}$$

so that the m roots near λ_0 are given by the p_1 values of λ_1, the p_2 values of λ_2, etc.

Corollary If $T(\beta)$ is a matrix-valued analytic function near β_0 and if λ_0 is an eigenvalue of $T(\beta_0)$ of algebraic multiplicity m, then for β near β_0, $T(\beta)$ has exactly m eigenvalues (counting multiplicity) near λ_0. These eigenvalues are all the branches of one or more multivalued functions analytic near β_0 with at worst algebraic singularities at β_0.

If A and B are self-adjoint, the perturbed eigenvalues of $A + \beta B$ are analytic at $\beta = 0$ even if A has degenerate eigenvalues. That the branch points allowed by the last theorem do not occur in this case is a theorem of Rellich. This theorem and its sister theorem on the analyticity of the eigenvectors in this case are the really deep results of finite-dimensional perturbation theory. The example at the beginning of this section shows that branch points can occur for *nonreal* β even in the "self-adjoint case," $T(\beta)^* = T(\bar{\beta})$.

Theorem XII.3 (Rellich's theorem) Suppose that $T(\beta)$ is a matrix-valued analytic function in a region R containing a section of the real axis, and that $T(\beta)$ is self-adjoint for β on the real axis. Let λ_0 be an eigenvalue of $T(\beta_0)$ of multiplicity m. If β_0 is *real*, there are $p \leq m$ distinct functions $\lambda_1(\beta)$, $\ldots, \lambda_p(\beta)$, *single-valued* and analytic in a neighborhood of β_0, which are *all* the eigenvalues.

Proof Consider one of the functions $\lambda_i(\beta)$ given in Theorem XII.2:

$$\lambda(\beta) = \lambda_0 + \sum_{j=1}^{\infty} \alpha_j(\beta - \beta_0)^{j/p}$$

The crucial fact that we shall use is that each branch of $\lambda(\beta)$ is an eigenvalue so that, in particular, *each branch is real for β real and near β_0*. Thus

$$\alpha_1 = \lim_{\beta \downarrow \beta_0} (\lambda(\beta) - \lambda_0)|\beta - \beta_0|^{-p^{-1}}$$

is real and

$$e^{i\pi/p}\alpha_1 = \lim_{\beta \uparrow \beta_0} (\lambda(\beta) - \lambda_0)|\beta - \beta_0|^{-p^{-1}}$$

is real. So, if $p \neq 1$, then $\alpha_1 = 0$. By induction, one shows that $\alpha_j = 0$ if j/p is not an integer. Therefore $\lambda(\beta)$ is actually analytic at $\beta = \beta_0$. ∎

We now want to consider the special case $H(\beta) = H_0 + \beta V$. Suppose that E_0 is a nondegenerate eigenvalue of H_0. From Theorem XII.1 we know that, for β small, $H_0 + \beta V$ has a unique eigenvalue $E(\beta)$ near E_0 and that $E(\beta)$ is analytic near $\beta = 0$. The coefficients of its Taylor series are called **Rayleigh–Schrödinger coefficients** and the Taylor series is called the **Rayleigh–Schrödinger series**. We can use the results described in the appendix to find formulas for the coefficients. The formulas are simpler when H_0 is self-adjoint, so we restrict ourselves to that case. $E(\beta)$ is the only eigenvalue of $H_0 + \beta V$ near E_0, so if $|E - E_0| < \varepsilon$, and ε is small, $E(\beta)$ is the only eigenvalue of $H_0 + \beta V$ in the circle $\{E\,|\,|E - E_0| < \varepsilon\}$. By the functional calculus,

$$P(\beta) = -\frac{1}{2\pi i} \oint_{|E - E_0| = \varepsilon} (H_0 + \beta V - E)^{-1}\, dE$$

is the projection onto the eigenvector with eigenvalue $E(\beta)$. We shall show in Theorem XII.9 that $(H_0 + \beta V - E)^{-1}$ is analytic in β near $\beta = 0$. Thus $P(\beta)$ is analytic in β at $\beta = 0$. In particular, if Ω_0 is the unperturbed eigenvector, then $P(\beta)\Omega_0 \neq 0$ for β small since $P(\beta)\Omega_0 \to \Omega_0$ as $\beta \to 0$. Since $P(\beta)\Omega_0$ is an unnormalized eigenvector for $H(\beta)$,

$$E(\beta) = \frac{(\Omega_0, H(\beta)P(\beta)\Omega_0)}{(\Omega_0, P(\beta)\Omega_0)} = E_0 + \beta\frac{(\Omega_0, VP(\beta)\Omega_0)}{(\Omega_0, P(\beta)\Omega_0)}$$

This formula is very important in the development of perturbation theory and plays a critical role in the discussions in Sections 2–4. For it says that to find the Taylor series for $E(\beta)$, we need only find the Taylor series for $P(\beta)$.

To do this, we need only find a Taylor series for $(H_0 + \beta V - E)^{-1}$ and integrate it. But the Taylor series for $(H_0 + \beta V - E)^{-1}$ is just a geometric series:

$$(H_0 + \beta V - E)^{-1} = (H_0 - E)^{-1} - \beta(H_0 - E)^{-1}V(H_0 - E)^{-1}$$
$$+ \cdots + (-1)^n \beta^n (H_0 - E)^{-1}[V(H_0 - E)^{-1}]^n + \cdots$$

Not only is this series simple, but there is a simple form for the error term when the series is truncated.

Thus, the Rayleigh–Schrödinger series for $E(\beta)$ is given by

$$E(\beta) = E_0 + \beta \frac{\sum_{n=0}^{\infty} a_n \beta^n}{\sum_{n=0}^{\infty} b_n \beta^n}$$

with

$$a_n = \frac{(-1)^{n+1}}{2\pi i} \oint_{|E-E_0|=\varepsilon} (\Omega_0, [V(H_0 - E)^{-1}]^{n+1}\Omega_0)\, dE$$

$$b_n = \frac{(-1)^{n+1}}{2\pi i} \oint_{|E-E_0|=\varepsilon} (\Omega_0, (H_0 - E)^{-1}[V(H_0 - E)^{-1}]^n\Omega_0)\, dE$$

Because of the contour integration and the division of power series, the formulas for the Rayleigh–Schrödinger coefficients are complicated. To illustrate this, let us compute $E(\beta)$ up to order β^4. Since H_0 is self-adjoint, we can choose a basis of eigenvectors, $\Omega_0, \ldots, \Omega_{n-1}$, with $H\Omega_i = E_i \Omega_i$. Let $V_{ij} = (\Omega_i, V\Omega_j)$. Then

$$b_0 = -\frac{1}{2\pi i} \oint_{|E-E_0|=\varepsilon} (\Omega_0, (H_0 - E)^{-1}\Omega_0)\, dE$$

$$= -\frac{1}{2\pi i} \oint_{|E-E_0|=\varepsilon} (E_0 - E)^{-1}\, dE = 1$$

$$b_1 = \frac{1}{2\pi i} \oint_{|E-E_0|=\varepsilon} V_{00}(E_0 - E)^{-2}\, dE = 0$$

$$b_2 = -\frac{1}{2\pi i} \oint_{|E-E_0|=\varepsilon} (E_0 - E)^{-2} \sum_{i=0}^{n} (E_i - E)^{-1} V_{0i} V_{i0}\, dE$$

The $i = 0$ term in this last sum has a very different status from the $i \neq 0$ terms. For,

$$\frac{1}{2\pi i} \oint_{|E-E_0|=\varepsilon} (E_0 - E)^{-3}\, dE = 0$$

while
$$\frac{1}{2\pi i} \oint_{|E-E_0|=\varepsilon} (E_0 - E)^{-2}(E_i - E)^{-1} \, dE = (E_i - E_0)^{-2}$$
Thus,
$$b_2 = -\sum_{i\neq 0} (E_i - E_0)^{-2} V_{0i} V_{i0}$$
Similarly,
$$b_3 = \sum_{i\neq 0\neq j} [(E_i - E_0)^{-1}(E_j - E_0)^{-2} + (E_i - E_0)^{-2}(E_j - E_0)^{-1}] V_{0i} V_{ij} V_{j0}$$
$$- 2\sum_{i\neq 0} (E_i - E_0)^{-3} V_{0i} V_{i0} V_{00}$$

$$a_0 = V_{00}$$
$$a_1 = -\sum_{i\neq 0} (E_i - E_0)^{-1} V_{0i} V_{i0}$$
$$a_2 = \sum_{i\neq 0\neq j} (E_i - E_0)^{-1}(E_j - E_0)^{-1} V_{0i} V_{ij} V_{j0}$$
$$- 2\sum_{i\neq 0} (E_i - E_0)^{-2} V_{0i} V_{i0} V_{00}$$
$$a_3 = -\sum_{\substack{i\neq 0\neq j \\ k\neq 0}} (E_i - E_0)^{-1}(E_j - E_0)^{-1}(E_k - E_0)^{-1} V_{0i} V_{ij} V_{jk} V_{k0}$$
$$+ 2\sum_{i\neq 0\neq j} [(E_i - E_0)^{-1}(E_j - E_0)^{-2}$$
$$+ (E_j - E_0)^{-1}(E_i - E_0)^{-2}] V_{00} V_{0i} V_{ij} V_{j0}$$
$$+ 2\sum_{i\neq 0\neq j} (E_i - E_0)^{-2}(E_j - E_0)^{-1} V_{0i} V_{i0} V_{0j} V_{j0}$$
$$- 3\sum_{i\neq 0} (E_i - E_0)^{-3} V_{0i} V_{i0} V_{00}^2$$

Thus, if we write $E(\beta) = E_0 + \sum_{n=1}^{\infty} \alpha_n \beta^n$, we have computed:
$$\alpha_1 = a_0 = V_{00}$$
$$\alpha_2 = a_1$$
$$= -\sum_{i\neq 0} (E_i - E_0)^{-1} V_{0i} V_{i0}$$
$$\alpha_3 = a_2 - b_2 a_0$$
$$= \sum_{i\neq 0\neq j} (E_i - E_0)^{-1}(E_j - E_0)^{-1} V_{0i} V_{ij} V_{j0} - \sum_{i\neq 0} (E_i - E_0)^{-2} V_{0i} V_{i0} V_{00}$$

$$\alpha_4 = a_3 - b_3 a_0 - b_2 a_1$$
$$= - \sum_{\substack{i \neq 0, j \neq 0 \\ k \neq 0}} (E_i - E_0)^{-1}(E_j - E_0)^{-1}(E_k - E_0)^{-1} V_{0i} V_{ij} V_{jk} V_{k0}$$
$$+ \sum_{i \neq 0, j \neq 0} [(E_i - E_0)^{-1}(E_j - E_0)^{-2}$$
$$+ (E_i - E_0)^{-2}(E_j - E_0)^{-1}] V_{00} V_{0i} V_{ij} V_{j0}$$
$$+ \sum_{i \neq 0, j \neq 0} (E_i - E_0)^{-2}(E_j - E_0)^{-1} V_{0i} V_{i0} V_{0j} V_{j0}$$
$$- \sum_{i \neq 0} (E_i - E_0)^{-3} V_{0i} V_{i0} V_{00}^2$$

We can draw several conclusions from these elementary but tedious computations:

(i) The nth Rayleigh–Schrödinger coefficient α_n is considerably more complicated than the leading term

$$(-1)^{n+1} \sum_{i_1 \neq 0, i_2 \neq 0, \ldots, i_{n-1} \neq 0} \prod_{j=1}^{n-1} (E_{i_j} - E_0)^{-1} V_{0 i_1} V_{i_1 i_2} \cdots V_{i_{n-1} 0}$$

which one might guess from the familiar second-order term found in quantum-mechanics books.

(ii) The denominator in $(\Omega_0, VP(\beta)\Omega_0)/(\Omega_0, P(\beta)\Omega_0)$ does not add new complications to the Taylor series but actually provides cancellations with terms already present in the numerator.

(iii) Most importantly, the terms in the Taylor series are quite complicated, although they arise from a simple geometric series. This suggests that the simplest object to study is the resolvent: To deduce rigorous theorems about $E(\beta)$ in the infinite-dimensional case, we shall generally first prove results about the resolvent and then obtain information about the eigenvalues by formulas that give the eigenvalue as a ratio of contour integrals of matrix elements of the resolvent.

As a final result in finite-dimensional perturbation theory, we mention:

Theorem XII.4 Let Ω_0 be a nondegenerate eigenvector for T_0 with $T_0 \Omega_0 = E_0 \Omega_0$, and let $T(\beta)$ be a matrix-valued analytic function with $T(0) = T_0$. Then, for β small, there is a vector-valued analytic function $\Omega(\beta)$ that obeys $T(\beta)\Omega(\beta) = E(\beta)\Omega(\beta)$, where $E(\beta)$ is the eigenvalue of $T(\beta)$ near E_0. Moreover, if $T(\beta)$ is self-adjoint for β real, $\Omega(\beta)$ can be chosen so that $\|\Omega(\beta)\| = 1$ for β real.

Proof Take

$$\psi(\beta) = -\frac{1}{2\pi i} \oint_{|E-E_0|=\varepsilon} (T(\beta) - E)^{-1} \Omega_0 \, dE \equiv P(\beta)\Omega_0$$

Then $\psi(\beta)$ is analytic and an eigenvector. Since $\psi(\beta) \to \Omega_0$ as $\beta \to 0$, $(\Omega_0, \psi(\beta)) \neq 0$ for small β. Let $\Omega(\beta) = (\Omega_0, \psi(\beta))^{-1/2} \psi(\beta)$ Then $\Omega(\beta)$ is normalized when $T(\beta)$ is self-adjoint for β real since then $(\Omega_0, \psi(\beta)) = (\Omega_0, P(\beta)\Omega_0) = \|\psi(\beta)\|^2$. ∎

One can also construct analytic eigenvectors in the situation covered by Rellich's theorem; see Problems 16 and 17.

Appendix to XII.1 Algebraic and geometric multiplicity of eigenvalues of finite matrices

We first recall some elementary definitions about roots of algebraic equations:

Definition A root λ_0 of an algebraic equation $F(\lambda) = \lambda^n + a_1 \lambda^{n-1} + \cdots a_n$ is called **nondegenerate** or **simple** if $F'(\lambda_0) \neq 0$. Equivalently λ_0 is simple if the decomposition $F(\lambda) = \prod_{i=1}^n (\lambda - \lambda_i)$ has $\lambda_i = \lambda_0$ for exactly one value of i. λ_0 is said to have **multiplicity** m if $F'(\lambda_0) = \cdots = F^{(m-1)}(\lambda_0) = 0$, $F^{(m)}(\lambda_0) \neq 0$ or equivalently if exactly m of the λ_i equal λ_0. An eigenvalue of a matrix is called **simple** or **nondegenerate** if it is a nondegenerate root of the secular equation. In general, the **algebraic multiplicity** of an eigenvalue is its multiplicity as a root of the secular equation.

The connection between algebraic multiplicity and geometric multiplicity is explained by the following series of remarks:

(i) Let $\mu(\lambda)$ be the algebraic multiplicity of λ. The fundamental theorem of algebra immediately implies that $\sum_{\lambda \in \mathbb{C}} \mu(\lambda) = n$ if T is an $n \times n$ matrix.

(ii) Let $m(\lambda) = \dim\{v \mid Tv = \lambda v\}$ be the **geometric multiplicity**. Then $m(\lambda) \leq \mu(\lambda)$.

(iii) If T is self-adjoint, $m(\lambda) = \mu(\lambda)$.

(iv) In general $\mu(\lambda) = \dim\{v \mid (T - \lambda)^k v = 0 \text{ for some } k\}$. This space is called the **generalized** or **geometric eigenspace** for λ.

The statements (ii) and (iv) become transparent once it is known that T can be put in Jordan normal form, i.e., there is a basis in which T has the block form

$$T = \begin{bmatrix} T_{\lambda_1} & 0 & 0 & 0 \\ 0 & \ddots & 0 & 0 \\ 0 & 0 & \ddots & 0 \\ 0 & 0 & 0 & T_{\lambda_n} \end{bmatrix}$$

$$T_{\lambda_i} = \begin{bmatrix} \lambda_i & x & 0 & \cdots & 0 \\ 0 & \lambda_i & x & 0 & \cdots & 0 \\ 0 & 0 & \lambda_i & x & & 0 \\ & \cdots & & & & \\ & \cdots & & & & x \\ & \cdots & & & & \lambda_i \end{bmatrix}$$

where each $x = 0$ or 1. In this case, the generalized eigenspace $\{v \mid (T - \lambda_i)^k v = 0\}$ is spanned by the $\mu(\lambda_i)$ basis elements associated with the block T_{λ_i} and clearly $\mu(\lambda_i)$ is the number of times λ_i appears as a root of $\det(T - \lambda) = 0$.

From the fact that any matrix can be put in Jordan normal form, it is also easy to see (Problem 2) that if ε is chosen sufficiently small, then

$$P_{\lambda_i} = -\frac{1}{2\pi i} \oint_{|\lambda - \lambda_i| = \varepsilon} (T - \lambda)^{-1} \, d\lambda$$

is the projection onto the generalized eigenspace associated with λ_i and $P_{\lambda_i} P_{\lambda_j} = \delta_{ij} P_{\lambda_i}$. In fact, one of the ways of establishing properties (i)–(iv) is through the use of these P_{λ_i} (see Problems 3 and 4).

XII.2 Regular perturbation theory

We now turn to the main result of this chapter and prove that under very general circumstances the Rayleigh–Schrödinger series has a nonzero radius of convergence for perturbations of unbounded operators in infinite-dimensional Hilbert spaces. An example where such results are applicable is $H(\beta) = -\Delta + \beta V$ on $\mathbb{R}^3$ where $V \in L^2$ is real-valued and β is real and positive. We shall see in Section XIII.4 that $\sigma_{\text{ess}}(H(\beta)) = [0, \infty)$ and in Section

XIII.1 that inf $\sigma(H(\beta)) \equiv E(\beta)$ is a monotonic decreasing function of β. If V is negative in some region of $\mathbb{R}^3$, $E(\beta)$ will be negative for β larger than some β_0 and thus, by the result on $\sigma_{\text{ess}}(H(\beta))$, an eigenvalue. It is reasonable to ask if this "ground state energy" $E(\beta)$ is analytic in β, at least in a neighborhood of the interval (β_0, ∞).

This section is divided into four parts: (1) A brief discussion of the discrete spectra of not necessarily self-adjoint operators. (2) A proof of the analyticity of discrete eigenvalues in the nondegenerate case for "analytic families of operators." This is the general theory of regular perturbations. This theory has many applications in quantum mechanics where eigenvalues are possible values of the energy. For this reason, we shall sometimes use the words **energy level** in place of eigenvalue. Another term we borrow from quantum mechanics is **coupling constant**, which we shall use for the variable β. (3) Two simple criteria (type (A) and type (B)) for $H_0 + \beta V$ to be an analytic family; these techniques enable one to apply the general theory to specific cases. (4) A brief discussion of degenerate perturbation theory.

We defined the discrete spectrum of a self-adjoint operator A in Section VII.3. For such operators, $\lambda \in \sigma_{\text{disc}}(A)$ means that λ is an isolated point of $\sigma(A)$ and dim $P_{\{\lambda\}} < \infty$ where P_Ω is the projection-valued measure associated with A. In the case of a general operator, we obviously should keep the requirement that λ be an isolated point of $\sigma(A)$. To replace the spectral projection, we use the projection which we introduced in Section XII.1:

Theorem XII.5 Suppose that A is a closed operator and let λ be an isolated point of $\sigma(A)$. Explicitly, suppose that $\{\mu \mid |\mu - \lambda| < \varepsilon\} \cap \sigma(A) = \{\lambda\}$. Then,

(a) For any r with $0 < r < \varepsilon$,

$$P_\lambda = -\frac{1}{2\pi i} \oint_{|\mu - \lambda| = r} (A - \mu)^{-1} \, d\mu$$

exists and is independent of r.
(b) $P_\lambda^2 = P_\lambda$. Thus P_λ is a (*not necessarily orthogonal*) projection.
(c) If $G_\lambda = \text{Ran } P_\lambda$ and $F_\lambda = \text{Ker } P_\lambda$, then G_λ and F_λ are complementary (not necessarily orthogonal) closed subspaces; that is, $G_\lambda + F_\lambda = \mathcal{H}$ and $G_\lambda \cap F_\lambda = \{0\}$. Moreover, A leaves G_λ and F_λ invariant in the following precise sense: $G_\lambda \subset D(A)$, $AG_\lambda \subset G_\lambda$, $F_\lambda \cap D(A)$ is dense in F_λ, and $A[F_\lambda \cap D(A)] \subset F_\lambda$.
(d) If $\psi \in G_\lambda$ and G_λ is finite dimensional, then $(A - \lambda)^n \psi = 0$ for some n. If $B \equiv A \upharpoonright F_\lambda$, then $\lambda \notin \sigma(B)$.

12 XII: PERTURBATION OF POINT SPECTRA

Proof (a) We already know that $(A - \mu)^{-1}$ is an analytic function on $\mathbb{C}\backslash\sigma(A) \equiv \rho(A)$. Thus the integral exists as a Banach-space-valued Riemann integral. That it is independent of r is a consequence of the Cauchy integral theorem.

(b) Let $r < R < \varepsilon$. Then, using the resolvent equation,

$$P_\lambda^2 = (2\pi i)^{-2} \oint_{|\mu - \lambda| = r} \oint_{|v - \lambda| = R} (A - \mu)^{-1}(A - v)^{-1} \, dv \, d\mu$$

$$= (2\pi i)^{-2} \oint_{|\mu - \lambda| = r} \oint_{|v - \lambda| = R} (v - \mu)^{-1}[(A - v)^{-1} - (A - \mu)^{-1}] \, d\mu \, dv$$

$$= (2\pi i)^{-2} \left[\oint_{|v - \lambda| = R} dv (A - v)^{-1} \oint_{|\mu - \lambda| = r} d\mu (v - \mu)^{-1} \right.$$

$$\left. - \oint_{|\mu - \lambda| = r} d\mu (A - \mu)^{-1} \oint_{|v - \lambda| = R} dv (v - \mu)^{-1} \right]$$

$$= (2\pi i)^{-2} \left[\oint_{|v - \lambda| = R} (A - v)^{-1} 0 \, dv - \oint_{|\mu - \lambda| = r} (2\pi i)(A - \mu)^{-1} \, d\mu \right] = P_\lambda$$

(c) That $G_\lambda = \text{Ker}(1 - P_\lambda)$ and $F_\lambda = \text{Ker } P_\lambda$ are closed complementary subspaces is elementary algebra (see Problem 6). Let $\psi = P_\lambda \psi \in G_\lambda$. Since P_λ is given by a Riemann integral, $\psi = \lim_{n \to \infty} \psi_n$ where

$$\psi_n = \sum_{i=1}^{n} c_i^{(n)}(A - \mu_i^{(n)})^{-1}\psi$$

and the $c_i^{(n)}$ and $\mu_i^{(n)}$ are chosen so that the sums converge to $-(2\pi i)^{-1} \oint (A - \mu)^{-1}\psi \, d\mu$. A simple computation using the formula $A(A - \mu)^{-1} = 1 + \mu(A - \mu)^{-1}$ proves that $\psi_n \to \psi$ and that $\{A\psi_n\}$ is Cauchy. Since A is closed, we conclude that $\psi \in D(A)$ and the above approximation procedure proves that $A\psi = AP_\lambda\psi = P_\lambda(A\psi)$. Thus $A\psi \in G_\lambda$. The statements about $D(A)$ and F_λ are left to the reader.

(d) Suppose that $A\psi = v\psi$. Then

$$P_\lambda \psi = (-2\pi i)^{-1} \oint_{|\mu - \lambda| = r} (v - \mu)^{-1}\psi \, d\mu = \begin{cases} \psi & \text{if } v = \lambda \\ 0 & \text{if } v \neq \lambda \end{cases}$$

It follows that the only eigenvalue of $A \upharpoonright G_\lambda$ is λ. If G_λ is finite dimensional, the Jordan normal form of $C \equiv A \upharpoonright G_\lambda$ has only λ along the diagonal and some 1's above the diagonal. Thus $(C - \lambda)^{(\dim G_\lambda)} = 0$, i.e., $(A - \lambda)^n \psi = 0$ for all $\psi \in G_\lambda$.

Finally, let

$$R_\lambda = (-2\pi i)^{-1} \oint_{|\mu - \lambda| = r} (\lambda - \mu)^{-1}(A - \mu)^{-1} d\mu$$

By doing computations similar to those in (b), one finds that $R_\lambda P_\lambda = P_\lambda R_\lambda$ and that $(A - \lambda)R_\lambda = R_\lambda(A - \lambda) = 1 - P_\lambda$. $R_\lambda(A - \lambda) = 1 - P_\lambda$ indicates an operator equality applied to vectors in $D(A)$. Thus R_λ takes F_λ into itself and $(B - \lambda)R_\lambda = R_\lambda(B - \lambda) = I \upharpoonright F_\lambda$. ∎

We are now in a position to define discrete spectrum:

Definition A point $\lambda \in \sigma(A)$ is called **discrete** if λ is isolated and P_λ (given by Theorem XII.5) is finite dimensional; if P_λ is one dimensional, we say λ is a **nondegenerate** eigenvalue.

The reader should check that this definition of discrete spectrum agrees with the definition given in Chapters VII and VIII when A is self-adjoint. Note that if λ is a nondegenerate eigenvalue, any $\psi \in \text{Ran } P_\lambda$ obeys $A\psi = \lambda\psi$. To complete our discussion of the discrete spectrum, we prove a converse to Theorem XII.5.

Theorem XII.6 Let A be an operator with $\{\mu \mid |\mu - \lambda| = r\} \subset \rho(A)$. Then $P = (-2\pi i)^{-1} \oint_{|\mu - \lambda| = r} (A - \mu)^{-1} d\mu$ is a projection. If P has dimension $n < \infty$, then A has at most n points of its spectrum in $\{\mu \mid |\mu - \lambda| < r\}$ and each is discrete. If $n = 1$, there is exactly one spectral point in $\{\mu \mid |\mu - \lambda| < r\}$ and it is nondegenerate.

Proof The proof of Theorem XII.5b carries through without change to prove that P is a projection and the proof of (c) implies that $G = \text{Ran } P$ and $F = \text{Ker } P$ are closed complementary invariant subspaces. Let $A_1 = A \upharpoonright G$ and $A_2 = A \upharpoonright F$. As in the proof of Theorem XII.5d, $v \notin \sigma(A_2)$ if $|v - \lambda| < r$. Thus $(A - v)^{-1}$ exists for such v if and only if $(A_1 - v)^{-1}$ exists. If G is finite dimensional, A_1 has eigenvalues $v_1 \ldots, v_k$ ($k \leq n$), so $\sigma(A) \cap \{v \mid |v - \lambda| < r\}$ is a finite set. To see that each spectral point in the circle is discrete, we note that if P_v is the spectral projection of Theorem XII.5 and if v is in the circle, then $P_v P = P P_v = P_v$. Thus $\text{Ran } P_v \subset \text{Ran } P$, which completes the proof. ∎

Having completed our brief discussion of discrete spectra, we can get down to the real object of study:

Definition A (possibly unbounded) operator-valued function $T(\beta)$ on a complex domain R is called an **analytic family** or an **analytic family in the sense of Kato** if and only if:

(i) For each $\beta \in R$, $T(\beta)$ is closed and has a nonempty resolvent set.
(ii) For every $\beta_0 \in R$, there is a $\lambda_0 \in \rho(T(\beta_0))$ so that $\lambda_0 \in \rho(T(\beta))$ for β near β_0 and $(T(\beta) - \lambda_0)^{-1}$ is an analytic operator-valued function of β near β_0.

If $T(\beta)$ is a family of bounded operators, this definition is equivalent to the definition of bounded operator-valued analytic function (Problem 8). The number λ_0 in the above definition does not play a special role:

Theorem XII.7 Let $T(\beta)$ be an analytic family on a domain R. Then

$$\Gamma = \{\langle \beta, \lambda \rangle \mid \beta \in R, \lambda \in \rho(T(\beta))\}$$

is open and the function $(T(\beta) - \lambda)^{-1}$ defined on Γ is an analytic function of two variables.

Proof Let $\langle \beta_0, \lambda_1 \rangle \in \Gamma$ and suppose that $(T(\beta) - \lambda_0)^{-1}$ exists and is analytic in β for β near β_0. By the first resolvent identity, $1 - (\lambda_1 - \lambda_0) \times (T(\beta_0) - \lambda_0)^{-1}$ has an inverse equal to $(T(\beta_0) - \lambda_0)(T(\beta_0) - \lambda_1)^{-1}$. Since the set of invertible operators in $\mathscr{L}(\mathscr{H})$ is open, $[1 - (\lambda - \lambda_0)(T(\beta) - \lambda_0)^{-1}]$ is invertible if λ is near λ_1 and β is near β_0. For such $\langle \beta, \lambda \rangle$, $T(\beta) - \lambda$ has an inverse equal to

$$(T(\beta) - \lambda_0)^{-1}[1 - (\lambda - \lambda_0)(T(\beta) - \lambda_0)^{-1}]^{-1}$$

so $\langle \beta, \lambda \rangle \in \Gamma$. Thus Γ is open. To prove the analyticity of $(T(\beta) - \lambda)^{-1}$, we note that $1 - (\lambda - \lambda_0)(T(\beta) - \lambda_0)^{-1}$ is analytic for λ near λ_0 and β near β_0 with values in the invertible operators. By a general theorem (Problem 9), it follows that $(1 - (\lambda - \lambda_0)(T(\beta) - \lambda_0)^{-1})^{-1}$ and therefore $(T(\beta) - \lambda)^{-1}$ is analytic. ∎

Only a simple technical lemma remains to complete the machinery for an effortless proof of the Kato–Rellich theorem:

Lemma If P and Q are two (not necessarily orthogonal) projections and $\dim(\operatorname{Ran} P) \neq \dim(\operatorname{Ran} Q)$, then $\|P - Q\| \geq 1$. In particular, if $P(x)$ is a continuous projection-valued function of x on a connected topological space, then $\dim(\operatorname{Ran} P(x))$ is a constant.

Proof Without loss of generality suppose dim(Ran P) < dim(Ran Q). Let $F = \text{Ker } P$ and let $E = \text{Ran } Q$. Then $\dim(F^\perp) = \dim(\text{Ran } P) < \dim E$. As a result, $F \cap E \neq \{0\}$ (see Problem 4 of Chapter X). Let $\psi \neq 0$, $\psi \in F \cap E$. Then $P\psi = 0$, $Q\psi = \psi$, so $\|(P - Q)\psi\| = \|\psi\|$. This implies that $\|P - Q\| \geq 1$. The final statement follows from an elementary connectedness argument. ∎

Theorem XII.8 (Kato–Rellich theorem) Let $T(\beta)$ be an analytic family in the sense of Kato. Let E_0 be a nondegenerate discrete eigenvalue of $T(\beta_0)$. Then, for β near β_0, there is exactly one point $E(\beta)$ of $\sigma(T(\beta))$ near E_0 and this point is isolated and nondegenerate. $E(\beta)$ is an analytic function of β for β near β_0, and there is an analytic eigenvector $\Omega(\beta)$ for β near β_0. If $T(\beta)$ is self-adjoint for $\beta - \beta_0$ real, then $\Omega(\beta)$ can be chosen to be normalized for $\beta - \beta_0$ real.

Proof Pick ε so that the only point of $\sigma(T(\beta_0))$ within $\{E \mid |E - E_0| \leq \varepsilon\}$ is E_0. Since the circle $\{E \mid |E - E_0| = \varepsilon\}$ is compact and the set Γ of the last theorem is open, we can find δ so that $E \notin \sigma(T(\beta))$ if $|E - E_0| = \varepsilon$ and $|\beta - \beta_0| \leq \delta$. Let $N = \{\beta \mid |\beta - \beta_0| \leq \delta\}$. Then

$$P(\beta) = -(2\pi i)^{-1} \oint_{|E - E_0| = \varepsilon} (T(\beta) - E)^{-1} \, dE$$

exists and is analytic for $\beta \in N$. The nondegeneracy of E_0 as an eigenvalue of $T(\beta_0)$ implies that $P(\beta_0)$ is one dimensional. The last lemma then implies $P(\beta)$ is one dimensional for all $\beta \in N$. Thus, by Theorem XII.6, there is exactly one eigenvalue $E(\beta)$ of $T(\beta)$ with $|E(\beta) - E_0| < \varepsilon$ when $\beta \in N$ and this eigenvalue is nondegenerate. The analyticity of $E(\beta)$ follows from the formula

$$(E(\beta) - E_0 - \varepsilon)^{-1} = \frac{(\Omega_0, (T(\beta) - E_0 - \varepsilon)^{-1} P(\beta)\Omega_0)}{(\Omega_0, P(\beta)\Omega_0)}$$

We obtain an analytic eigenvector by choosing $\Omega(\beta) = P(\beta)\Omega_0$ or

$$\Omega(\beta) = (\Omega_0, P(\beta)\Omega_0)^{-1/2} P(\beta)\Omega_0$$

in the real case, where Ω_0 is the unperturbed eigenvector. ∎

We thus see how easy it is to prove that energy levels are analytic in the coupling constant *if* we know that $T(\beta)$ is an analytic family. This would not be very useful if we did not have convenient criteria for $T(\beta)$ to be analytic. Fortunately, there are two simple ones reflecting the usual operator/form dualism. We shall discuss the operator criterion in detail and the form criterion briefly.

Definition Let R be a connected domain in the complex plane and let $T(\beta)$, a closed operator with nonempty resolvent set, be given for each $\beta \in R$. We say that $T(\beta)$ is an **analytic family of type (A)** if and only if

(i) The operator domain of $T(\beta)$ is some set D independent of β.
(ii) For each $\psi \in D$, $T(\beta)\psi$ is a vector-valued analytic function of β.

Of course, every family of type (A) is an analytic family in the sense of Kato. We leave the general case of this theorem to the problems and consider only the linear case $T(\beta) = H_0 + \beta V$. We first prove a lemma that is of interest in itself since it is a convenient criterion for a family to be type (A).

Lemma Let H_0 be a closed operator with nonempty resolvent set. Define $H_0 + \beta V$ on $D(H_0) \cap D(V)$. Then $H_0 + \beta V$ is an analytic family of type (A) near $\beta = 0$ if and only if:

(a) $D(V) \supset D(H_0)$.
(b) For some a and b and for all $\psi \in D(H_0)$,
$$\|V\psi\| \leq a\|H_0\psi\| + b\|\psi\|$$

That is, $H_0 + \beta V$ is type (A) if and only if V is H_0-bounded in the sense of Section X.2.

Proof Suppose first that $H_0 + \beta V$ is an analytic family of type (A). Then $D(H_0) = D(H_0 + \beta V) = D(H_0) \cap D(V)$ so (a) holds. Since H_0 is closed, $D(H_0)$ with the norm $\||\psi\|| = \|H_0\psi\| + \|\psi\|$ is a Banach space $\hat{D}$. Fix β small and positive so that β and $-\beta$ are both in the domain of analyticity. $H_0 + \beta V: \hat{D} \to \mathcal{H}$ is everywhere defined and has a closed graph in $\hat{D} \times \mathcal{H}$ since the graph is closed in $\mathcal{H} \times \mathcal{H}$ with a weaker topology. Thus, by the closed graph theorem,
$$\|(H_0 + \beta V)\psi\| \leq a_1 \||\psi\||$$
and
$$\|(H_0 - \beta V)\psi\| \leq a_2 \||\psi\||$$
Thus,
$$\|V\psi\| \leq (2\beta)^{-1}[\|(H_0 + \beta V)\psi\| + \|(H_0 - \beta V)\psi\|]$$
$$\leq (2\beta)^{-1}(a_1 + a_2)\||\psi\||$$
so that condition (b) holds.

Conversely, let (a) and (b) hold. Then, for $\psi \in D(H_0)$,
$$\|H_0\psi\| \leq \|(H_0 + \beta V)\psi\| + |\beta|\|V\psi\|$$
$$\leq \|(H_0 + \beta V)\psi\| + |\beta|a\|H_0\psi\| + |\beta|b\|\psi\|$$
Thus, if $|\beta| < a^{-1}$, we have
$$\|H_0\psi\| \leq (1 - |\beta|a)^{-1}\|(H_0 + \beta V)\psi\| + (1 - |\beta|a)^{-1}b|\beta|\|\psi\|$$
Therefore, $H_0 + \beta V$ is closed on $D(H_0)$ for if $\psi_n \to \psi$ in $\mathcal{H}$ with $\psi_n \in D(H_0)$ and $(H_0 + \beta V)\psi_n$ is Cauchy, then $H_0\psi_n$ is Cauchy by the above inequality and thus $\psi \in D(H_0)$. That $(H_0 + \beta V)\psi$ is analytic for $\psi \in D(H_0)$ is obvious. ∎

It is a corollary of the above proof that if V is infinitesimally small with respect to H_0, then $H_0 + \beta V$ is an *entire* family of type (A).

Example 1 Let $V \in L^2(\mathbb{R}^3) + L^\infty(\mathbb{R}^3)$ and let $H_0 = -\Delta$ on $L^2(\mathbb{R}^3)$. More generally, let $V = \sum V_{ij}$ with $V_{ij} \in L^2 + L^\infty$ and $H_0 = -\Delta$ on $\mathbb{R}^{3n}$. Then $H_0 + \beta V$ is an entire analytic family of type (A).

Example 2 It can be shown that if $V << H_0$ and $W << H_0$, then $W << H_0 + V$ (Problem 11). Thus, letting $H_0 = -\Delta_1 - \Delta_2 - 2/r_1 - 2/r_2$ on $L^2(\mathbb{R}^6)$ and $V = |r_1 - r_2|^{-1}$, we see $H_0 + \beta V$ is an analytic family of type (A). In the approximation of infinite nuclear mass, $H + V$ is the helium atom Hamiltonian (see Section XI.5 for the kinematics).

Theorem XII.9 Let $H_0 + \beta V$ be an analytic family of type (A) in a region R. Then $H_0 + \beta V$ is an analytic family in the sense of Kato. In particular, if $0 \in R$ and if E_0 is an isolated nondegenerate eigenvalue of H_0, then there is a unique point $E(\beta)$ of $\sigma(H_0 + \beta V)$ near E_0 when $|\beta|$ is small which is an isolated nondegenerate eigenvalue. Moreover, $E(\beta)$ is analytic near $\beta = 0$.

Proof Since analyticity is a local property, we suppose that $0 \in R$ and prove analyticity in the sense of Kato near $\beta = 0$. Choose $\lambda \notin \sigma(H_0)$. Then $(H_0 - \lambda)^{-1}$ and $H_0(H_0 - \lambda)^{-1} = 1 + \lambda(H_0 - \lambda)^{-1}$ are bounded. Thus, for any $\varphi \in \mathcal{H}$,
$$\|V(H_0 - \lambda)^{-1}\varphi\| \leq a\|H_0(H_0 - \lambda)^{-1}\varphi\| + b\|(H_0 - \lambda)^{-1}\varphi\|$$
$$\leq (a\|H_0(H_0 - \lambda)^{-1}\| + b\|(H_0 - \lambda)^{-1}\|)\|\varphi\|$$

Thus $V(H_0 - \lambda)^{-1}$ is bounded; so for β small, $[1 + \beta V(H_0 - \lambda)^{-1}]^{-1}$ exists and is analytic in β (being given by a geometric series). Direct computation (Problem 12) shows that $(H_0 - \lambda)^{-1}[1 + \beta V(H_0 - \lambda)^{-1}]^{-1}$ is an inverse for $(H_0 + \beta V - \lambda)$, so for β small, $\lambda \notin \sigma(H_0 + \beta V)$ and $(H_0 + \beta V - \lambda)^{-1}$ is analytic in β. This proves that $H_0 + \beta V$ is an analytic family in the sense of Kato near $\beta = 0$. By writing $H_0 + \beta V = (H_0 + \beta_0 V) + (\beta - \beta_0)V$, we prove analyticity at $\beta = \beta_0$. ∎

Example 1, revisited By Theorem X.15 and Theorem XII.9, it follows that $E_0(\beta)$, the lowest eigenvalue of $-\Delta + \beta V$ is an analytic function of β in a neighborhood of (β_0, ∞) where $\beta_0 = \inf\{\beta > 0 \,|\, E_0(\beta) < 0\}$. In applying Theorem XII.9, we are assuming the nondegeneracy of the ground state which we shall prove in Section XIII.12.

Example 2, revisited $h \equiv -\Delta_1 - 2/r_1$ is an operator with an exactly solvable eigenvalue problem. Its lowest eigenvalue is $E = -1$. H_0 is of the form $h \otimes 1 + 1 \otimes h$ on $L^2(\mathbb{R}^3) \otimes L^2(\mathbb{R}^3) = L^2(\mathbb{R}^6)$, so its ground state energy is -2. For $|\beta|$ small, the ground state energy $E(\beta)$ is analytic with the Taylor coefficients at $\beta = 0$ given by the Rayleigh–Schrödinger formula discussed in Section 1.

Physically, one is interested in the ground state energy $E(1)$ of the helium atom. The question immediately arises as to whether the Taylor series for $E(\beta)$ about $\beta = 0$ has a radius of convergence bigger than 1. In Theorem XII.11 below, we shall obtain explicit lower bounds on the radius of convergence of the Rayleigh–Schrödinger series, but our bounds will be crude and we shall not be able to use them directly to prove that $\beta = 1$ is within the circle of convergence. By hard work one might be able to show $\beta = 1$ is actually in the circle of convergence (we expect it is true), but the question is really academic! For when $\beta = 1$, even if the series is convergent, a large number of terms of the Taylor series are necessary to approximate $E(1)$ well, and the higher order Rayleigh–Schrödinger coefficients are hard to compute. For example, the first-order approximate value $(\Omega_0, V\Omega_0)$ for $E(1) - E(0)$ disagrees with experiment by about 15%. It turns out that other methods, which we shall discuss in Section XIII.2, can be used to obtain an accuracy of better than 1% with experiment (and if various relativistic corrections are taken into account, of one part in 10^6). However, if β is small, perturbation theory is more accurate. It turns out that the ground state energy of Li^+ is directly related to $E(\frac{4}{9})$ and it is given by the first-order approximant to within 5%. $E(\frac{1}{4})$, which is related to the ground state energy of Be^{++}, is given within 2%.

Example 3 (hyperfine structure in hydrogen) Perturbation theory is connected with one of the more spectacular agreements between theory and experiment in quantum physics. In the usual model for the hydrogen atom there is one energy level near -13 eV, the ground state energy. The physical atom has two levels; this splitting is due to interactions between the magnetic moments of the electron and proton. It is the transition between these levels that radio astronomers observe when looking for intergalactic gas clouds, and it is this transition that is the dominant one in a hydrogen maser. For the latter reason, the energy difference is very well measured; in fact, in units with $\hbar = 1$, so that ΔE has units of hertz (Hz) $\equiv$ cycles per second,

$$\Delta E(1s_{1/2}) = 1{,}420{,}405{,}751.800 \text{ Hz}$$

There is an old theory of the magnetic interaction due to Fermi and Segrè which is suggested by classical models of interacting magnets. The Fermi-Segrè potential has a coupling constant β made up of fundamental constants (magnetic moments of the electron and proton, the electric charge), a spin-spin interaction and multiplication by $\rho(r)$, the effective charge distribution of the proton. In practice $\rho(r)$ is approximated by a δ function, which means that it is technically outside the mathematical theory we have discussed, but a peaked smooth function $\rho(r)$ is within our theory, and leads to approximately the same lowest order contributions to perturbation theory.

In comparing theory and experiment an interesting problem arises. The physical constants needed to compute β are known to only about one part in 10^5 or 10^6 and $\Delta E(1s_{1/2}) = \beta a_1 + \beta^2 a_2 + \cdots$, where $\beta \simeq 10^{-4}$ and a_1, a_2, measured in units of the ground state energy of hydrogen, are about 1. For a truly accurate comparison with experiment, one also looks at the hyperfine splitting in the first excited state $\Delta E(2s_{1/2}) = \beta b_1 + \beta^2 b_2$. If we look at the ratio $\Delta E(2s_{1/2})/\Delta E(1s_{1/2})$, then since β is already approximately 10^{-4}, an error of β in its sixth place only affects $(a_1 + \beta a_2)/(b_1 + \beta b_2)$ in its tenth place! Experimentally:

$$\frac{\Delta E(2s_{1/2})}{\Delta E(1s_{1/2})} = \frac{1}{8}(1.000034495)$$

The Fermi–Segrè theory (with relativistic corrections) and lower order perturbation theory predicts

$$\frac{\Delta E(2s_{1/2})}{\Delta E(1s_{1/2})} = \frac{1}{8}(1.00003445)$$

Better agreement than this would actually be embarrassing since the above calculations ignore the finite size of the nucleus, corrections due to strong interactions, etc.

Let us return to general criteria for a linear function $H_0 + \beta V$ to be an analytic family in the sense of Kato. There is a form notion analogous to the operator notion of family of type (A). An **analytic family of type (b)** is a family of closed, strictly m-sectorial forms, $q(\beta)$, one for each β in a region R of the complex plane, so that:

(i) The form domain of $q(\beta)$ is some subspace F independent of β.
(ii) $(\psi, q(\beta)\psi)$ is an analytic function in R for each $\psi \in F$.

If $q(\beta)$ is an analytic family of type (b), then, for each $\beta \in R$, there is associated a unique closed operator $T(\beta)$ by Theorem VIII.16. $T(\beta)$ is called an **analytic family of type (B)**. As in the type (A) case, any analytic family of type (B) is an analytic family in the sense of Kato, and $H_0 + \beta V$ defined as a form on $Q(H_0) \cap Q(V)$ is an analytic family of type (B) near $\beta = 0$ if and only if V is H_0 form bounded.

Type (B) methods can be used to extend the results discussed under Example 1 above to potentials in the Rollnik class $R + L^\infty$. Type (B) techniques imply strong analyticity properties for $H_0 + \beta V$ if H_0 and V are positive:

Theorem XII.10 Let H_0 be positive and self-adjoint and let V be self-adjoint. Let $V_+ = \frac{1}{2}(V + |V|)$; $V_- = \frac{1}{2}(|V| - V)$. Suppose that:

(i) $Q(V_+) \cap Q(H_0)$ is dense.
(ii) V_- is H_0 form bounded with relative bound zero.

Then $H_0 + \beta V$ is an analytic family of type (B) in the cut plane $\{\beta \,|\, \beta \notin (-\infty, 0]\}$.

A reference for the proof of this theorem can be found in the Notes.

Example 4 From our discussion in Section XIII.12, it will follow that the ground state of $-d^2/dx^2 + x^2 + \beta x^4$ is nondegenerate if $\beta > 0$. Thus, Theorem XII.10 says that its ground state energy $E(\beta)$ is analytic in a neighborhood of the positive real axis.

There are examples of analytic families that are neither type (A) nor type (B). For example, let $T(\beta)$ be an analytic family of type (A) and let C be any bounded self-adjoint operator. Then $U(\beta) = \exp(i\beta C)$ is an entire analytic function. It is not hard to see that $\tilde{T}(\beta) = U(\beta)T(\beta)U(\beta)^{-1}$, defined on $U(\beta)D$, is an analytic family. However, C and T can be chosen so that neither $D(T(\beta))$ nor $Q(T(\beta))$ is constant.

We would like to make a few remarks, some of which are warnings about pitfalls. First, we note that as in Section 1 one has explicit formulas for the coefficients of the Taylor series for $E(\beta)$ given as contour integrals of resolvents. If H_0 has purely discrete spectrum, we can do the integrals explicitly and obtain formulas identical to those of the preceding section. If H_0 is self-adjoint, we can still do these contour integrals, obtaining spectral integrals in place of sums; for example, if E_0 is an isolated nondegenerate eigenvalue of H_0 so that $\operatorname{dist}(E_0, \sigma(H_0)\setminus E_0) > \varepsilon$, then

$$\alpha_2 = -\int_{|\lambda - E_0| > \varepsilon} (\lambda - E_0)^{-1} \, d(V\Omega_0, P_\lambda V\Omega_0)$$

Secondly, we warn the reader that it may happen that the power series for $E(\beta)$ has a circle of convergence larger than the circle in which $H(\beta)$ has $E(\beta)$ as an eigenvalue.

Example 5 Let $H_0 = -\Delta - 1/r$ and $V = 1/r$. Then, the eigenvalues of $H_0 + \beta V$ for β small are $-\frac{1}{4}n^{-2}(1-\beta)^2$, $n = 1, 2, \ldots$. In particular, the ground state energy $(n = 1)$ $E_0(\beta) = -\frac{1}{4} + \frac{1}{2}\beta - \frac{1}{4}\beta^2$ is given by a function with an analytic continuation to the entire complex plane. But for $\beta > 1$, H_0 has no eigenvalues at all!

Thus, one vestige of the finite-dimensional theory is not present: In general, the analytic continuation of an eigenvalue need not be an eigenvalue. However, in one important special case it can be proven that the analytic continuation of an eigenvalue is an eigenvalue (see Problem 13).

Finally, we note that one can obtain explicit lower bounds on the radius of convergence of the Taylor series:

Theorem XII.11 Suppose that $\|V\varphi\| \le a\|H_0\varphi\| + b\|\varphi\|$. Let H_0 be self-adjoint with an unperturbed isolated, nondegenerate eigenvalue E_0, and let $\varepsilon = \frac{1}{2}\operatorname{dist}(E_0, \sigma(H_0)\setminus\{E_0\})$. Define

$$r(a, b, E_0, \varepsilon) = [a + \varepsilon^{-1}[b + a(|E_0| + \varepsilon)]]^{-1}$$

Then the eigenvalue $E(\beta)$ of $H_0 + \beta V$ near E_0 is analytic in the circle of radius $r(a, b, E_0, \varepsilon)$.

The reader is asked to provide a proof in Problem 14.

* * *

As a final subject in regular perturbation theory, we shall discuss the case where E_0 is an isolated degenerate eigenvalue of finite multiplicity of $T(\beta_0)$. Because of our experience with the finite-dimensional case, we shall suppose that $T(\beta)$ is self-adjoint for β real. If $T(\beta)$ is a Kato family, we have no trouble in proving that $P(\beta) = (-2\pi i)^{-1} \oint (T(\beta) - E)^{-1} dE$ is analytic in β for β near β_0. We are thus faced with finding the eigenvalues of $H(\beta)$ restricted to the *variable* finite-dimensional subspace Ran $P(\beta)$. To reduce this to a truly finite-dimensional problem, we need the following technical result of Kato, which has several other applications (see Problems 15 and 17).

Theorem XII.12 Let R be a connected, simply connected region of the complex plane containing 0. Let $P(\beta)$ be a projection-valued analytic function in R. Then, there is an analytic family $U(\beta)$ of invertible operators with

$$U(\beta)P(0)U(\beta)^{-1} = P(\beta)$$

Moreover, if $P(\beta)$ is self-adjoint for β real and in R, then we can choose $U(\beta)$ unitary for β real.

We defer the proof to the conclusion of this section.

Theorem XII.13 Let $T(\beta)$ be an analytic family in the sense of Kato for β near 0 that is self-adjoint for β real. Let E_0 be a discrete eigenvalue of multiplicity m. Then, there are m not necessarily distinct single-valued functions, analytic near $\beta = 0$, $E^{(1)}(\beta), \ldots, E^{(m)}(\beta)$, with $E^{(k)}(0) = E_0$, so that $E^{(1)}(\beta), \ldots, E^{(m)}(\beta)$ are eigenvalues of $T(\beta)$ for β near 0 (with a repeated entry in $E^{(1)}, \ldots, E^{(m)}$ indicating a degenerate eigenvalue). Further, these are the only eigenvalues near E_0.

Proof Since E_0 is an isolated point of $\sigma(T(0))$, and $T(\beta)$ is an analytic family, $P(\beta) = (-2\pi i)^{-1} \oint (T(\beta) - E)^{-1} dE$ exists and is analytic for β small. By the proof of Theorem XII.6,

$$\sigma(T(\beta) \restriction \text{Ran } P(\beta)) = \sigma(T(\beta)) \cap \{E \mid |E - E_0| < \varepsilon\}$$

From Theorem XII.12 we know that there exists a family $U(\beta)$, analytic near $\beta = 0$, unitary for β real, so that $U(\beta)P(0)U(\beta)^{-1} = P(\beta)$. Let $\tilde{T}(\beta) = U(\beta)^{-1}T(\beta)U(\beta)$. Then Ran $P(0)$ is an invariant subspace for all the $\tilde{T}(\beta)$. Thus $S(\beta) \equiv \tilde{T}(\beta) \restriction \text{Ran } P(0)$ is a finite-dimensional analytic family, self-adjoint for β real. The theorem now follows from Rellich's theorem (Theorem XII.3). ∎

XII.2 Regular perturbation theory

Since we have reduced the infinite-dimensional problem to a finite-dimensional one, the existence of analytic eigenvectors in the finite-dimensional case implies their existence in the infinite-dimensional case.

Example 1, revisited If $H_0 + \beta_0 V$ has n eigenvalues in $(-\infty, 0)$, then $H_0 + \beta V$ has at least n eigenvalues for $|\beta - \beta_0|$ small, and the ones near those of $H_0 + \beta_0 V$ are analytic in β near β_0.

Finally, we shall prove Theorem XII.12. The idea of the proof comes from differentiating $U(\beta)P(0)U(\beta)^{-1} = P(\beta)$, finding $P'(\beta) = [U'(\beta)U(\beta)^{-1}, P(\beta)]$ where $[A, B] = AB - BA$. Thus, one seeks an operator $Q(\beta)$ that satisfies $P'(\beta) = [Q(\beta), P(\beta)]$ and then solves the differential equation $U'(\beta) = Q(\beta)U(\beta)$.

Lemma Let R be a connected, simply connected subset of $\mathbb{C}$ with $0 \in R$ and let $A(\beta)$ be an analytic function on R with values in the bounded operators on some Banach space X. Then for any $x_0 \in X$, there is a unique function $f(\beta)$, analytic in R, with values in X obeying

$$\frac{d}{d\beta} f(\beta) = A(\beta) f(\beta), \qquad f(0) = x_0 \tag{2}$$

Proof By standard methods of analytic continuation, it is enough to suppose that R is a circle of radius r_0 and to show that there is an analytic solution in the circle of radius $r_0 - 2\varepsilon$ for any ε. We first note that uniqueness follows from (2): By supposing $f(\beta) = \sum_{n=0}^{\infty} f_n \beta^n$ and knowing $A(\beta) = \sum_{n=0}^{\infty} A_n \beta^n$, one finds that

$$f_0 = x_0 \tag{3a}$$

$$f_n = n^{-1} \left[\sum_{k=0}^{n-1} A_k f_{n-1-k} \right] \tag{3b}$$

We now show that if the f_n are defined by (3), then $\sum_{n=0}^{\infty} f_n \beta^n$ converges if $|\beta| < r_0 - 2\varepsilon$. Let $M = \max\{1 + \|A(\beta)\| \mid |\beta| \leq r_0 - \varepsilon\}$. By the Cauchy integral formula, $\|A_n\| < M(r_0 - \varepsilon)^{-n}$. A simple inductive argument using (3) shows that $\|f_n\| < (M(r_0 - \varepsilon)^{-1})^n \|x_0\|$. It follows that $f(\beta)$ is analytic in a circle of radius $(r_0 - \varepsilon)/M$. By repeating this argument at points β with $|\beta|$ near $(r_0 - \varepsilon)/M$, we can prove that f is analytic in a circle of radius $(r_0 - \varepsilon)M^{-1} + M^{-1}(r_0 - \varepsilon)(1 - M^{-1})$. See Figure XII.1. After a finite number of repetitions we get analyticity in a circle of radius $r_0 - 2\varepsilon$. ∎

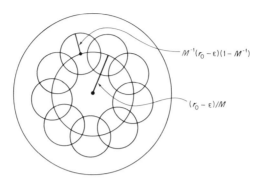

FIGURE XII.1

Proof of Theorem XII.12 We divide the proof into four parts.

(i) Let $Q(\beta) = [P'(\beta), P(\beta)]$. $P^2(\beta) = P(\beta)$, so

$$P'(\beta) = P'(\beta)P(\beta) + P(\beta)P'(\beta) \qquad (4)$$

Thus $P(\beta)P'(\beta)P(\beta) = 2P(\beta)P'(\beta)P(\beta)$ or $P(\beta)P'(\beta)P(\beta) = 0$. As a result

$$[Q(\beta), P(\beta)] = P'(\beta)P(\beta) + P(\beta)P'(\beta) - 2P(\beta)P'(\beta)P(\beta) = P'(\beta)$$

by (4).

(ii) Using the lemma with $X = \mathscr{L}(\mathscr{H})$, solve $dU/d\beta = Q(\beta)U(\beta)$ and $dV/d\beta = -V(\beta)Q(\beta)$ with initial conditions $U(0) = 1 = V(0)$. Then $U(\beta)V(\beta) = V(\beta)U(\beta) = 1$; in particular, $U(\beta)$ is invertible. For

$$\frac{d}{d\beta}(V(\beta)U(\beta)) = \frac{dV}{d\beta}U(\beta) + V(\beta)\frac{dU}{d\beta} = 0$$

and thus $VU \equiv 1$. On the other hand, if $F(\beta) = U(\beta)V(\beta)$, then $F(\beta)$ solves the differential equation $dF/d\beta = Q(\beta)F(\beta) - F(\beta)Q(\beta)$; $F(0) = 1$. Since $F(\beta) \equiv 1$ solves this equation with initial condition, we conclude that $F(\beta) = 1$ by the uniqueness of solutions proven in the lemma.

(iii) $U(\beta)P(0)V(\beta) = P(\beta)$. For let $\tilde{P}(\beta) = U(\beta)P(0)V(\beta)$. Then $d\tilde{P}(\beta)/d\beta = [Q(\beta), \tilde{P}(\beta)]$ with initial condition $\tilde{P}(0) = P(0)$. On the other hand, by step (i), $P(\beta)$ also solves $dP(\beta)/d\beta = [Q(\beta), P(\beta)]$ with $P(\beta)|_{\beta=0} = P(0)$. By the uniqueness of solutions of differential equations, $\tilde{P}(\beta) = P(\beta)$.

(iv) Finally, we must prove that $U(\beta)$ is unitary for β real if $P(\beta)$ is self-adjoint for β real. Thus, let us suppose that $P(\beta)^* = P(\beta)$ if $\beta = \bar{\beta}$. By the Schwarz reflection principle, it follows that $P(\beta)^* = P(\bar{\beta})$ for all β. By the definition of Q, $Q(\beta)^* = -Q(\bar{\beta})$. Let $\tilde{V}(\beta) = U(\bar{\beta})^*$. Then $\tilde{V}$ obeys

$d\tilde{V}/d\beta = -\tilde{V}(\beta)Q(\beta)$; $\tilde{V}(0) = I$. By the uniqueness of solutions of differential equations, $\tilde{V}(\beta) = V(\beta)$. Thus, for β real, $U(\beta)^* = \tilde{V}(\beta) = V(\beta) = U(\beta)^{-1}$, so U is unitary. ∎

XII.3 Asymptotic perturbation theory

The elegant regular perturbation theory developed in the preceding section is not always applicable, even to simple-looking examples. Consider the family of Hamiltonians $H(\beta) = H_0 + \beta V$ where $\mathcal{H} = L^2(\mathbb{R})$, $H_0 = -d^2/dx^2 + x^2$, and $V = x^4$. We discussed the self-adjointness of this family from several points of view in Chapter X. For any $\beta > 0$, $H(\beta)$ is self-adjoint on $D(H_0) \cap D(V) = D(p^2) \cap D(x^4)$; see Problem 23 of Chapter X. Since $D(H_0) = D(p^2) \cap D(x^2)$, we see that the domain changes as soon as the perturbation is turned on. Thus the analyticity criterion of Theorem XII.9 is not applicable. A similar change occurs in the form domain $Q(H(\beta))$. In fact, no analyticity criterion can hold since the perturbation series about $\beta = 0$ *diverges*.

The following argument has been made by various authors to predict this divergence of the perturbation series for the eigenvalues of $H(\beta)$ for $\beta \neq 0$: If β is negative, then $x^2 + \beta x^4 \to -\infty$ as $x \to \pm\infty$, so $H_0 + \beta V$ is qualitatively very different from H_0 (it is actually not even essentially self-adjoint). For this reason, one expects that the perturbation series should diverge for β negative. Since power series converge in circles, the series should not converge for any β. Whether or not one wants to accept this heuristic argument, its conclusion is correct. A detailed analysis allows one to prove that the Rayleigh–Schrödinger coefficients a_n for $E_0(\beta)$, the ground state energy, obey $|a_n| \geq AB^n \Gamma(n/2)$ for suitable constants A and B.

Thus, one is faced with deciding whether the perturbation series makes any sense in this case. It is precisely this question that we consider in this section and the next. The meaning of divergent perturbation series is of interest in a wider area than the realm of nonrelativistic quantum mechanics. The most useful calculational tool in certain (at present ill-defined) quantum field theories is another perturbation series known as the Gell'Mann–Low series or Feynman series. In some cases these series have been proven to diverge and they are believed to be divergent in others. For this reason and especially since there are formal similarities between some field theory Hamiltonians and $p^2 + x^2 + \beta x^4$ (see Section X.7), the problems we study in these two sections are relevant to quantum field theory.

The simplest interpretation of a formal series is as an asymptotic series:

Definition Let f be a function defined on the positive real axis. We say that $\sum_{n=0}^{\infty} a_n z^n$ is **asymptotic** to f as $z \downarrow 0$ if and only if, for each fixed N,

$$\lim_{z \downarrow 0} \left(f(z) - \sum_{n=0}^{N} a_n z^n \right) \bigg/ z^N = 0$$

If f is defined in a sectorial region of the complex plane $\{z \mid 0 < |z| < B;\ |\arg z| \leq \theta\}$, we say that $\sum a_n z^n$ is **asymptotic** to f as $|z| \to 0$ **uniformly in the sector** if, for each N,

$$\lim_{\substack{|z| \to 0 \\ |\arg z| \leq \theta}} \left(f(z) - \sum_{n=0}^{N} a_n z^n \right) \bigg/ z^N = 0$$

If $\sum a_n z^n$ is asymptotic to f, we shall sometimes write

$$f \underset{z \downarrow 0}{\sim} \sum a_n z^n$$

Notice that if $f \sim \sum a_n z^n$ and $f \sim \sum b_n z^n$ as $z \downarrow 0$, then

$$\lim_{z \downarrow 0} \left(\sum_{n=0}^{N} a_n z^n - \sum_{n=0}^{N} b_n z^n \right) \bigg/ z^N = 0$$

for all N, so $a_n = b_n$ for all n. Thus, *any function f has at most one asymptotic series as $z \downarrow 0$.*

Example 1 Consider the function $f(z) = \exp(-z^{-1})$ for $z > 0$. Then $z^{-n} f(z) \to 0$ as $z \downarrow 0$, so f has zero asymptotic series. In fact, the zero series is asymptotic uniformly in any sector $|\arg z| \leq \theta$ with $\theta < \pi/2$.

This example illustrates an important fact about asymptotic series: *Two different functions may have the same asymptotic series.* Saying that f has a certain asymptotic series gives us no information about the value of $f(z)$ for some fixed nonzero value of z. We know that $f(z)$ is well approximated by $a_0 + a_1 z$ as z gets "small," but the definition says nothing about how small is "small." If an asymptotic series $\sum_{n=0}^{\infty} a_n z^n$ is not convergent, the typical behavior is the following: If z is "small," the first few partial sums are a fairly good approximation to $f(z)$, but as $N \to \infty$, the sums oscillate wildly and no longer approximate f very well. For example, we shall presently show that the Rayleigh–Schrödinger series for $E_0(\beta)$, the ground state energy of the Hamiltonian, $p^2 + x^2 + \beta x^4$ $(\beta > 0)$, is asymptotic to $E_0(\beta)$ as $\beta \downarrow 0$. For

$\beta = 0.2$, variational methods (see Section XIII.2) show that $E_0(\beta) = 1.118292\ldots$. The first 15 partial sums are given in the accompanying tabulation.

N	$\sum_{n=0}^{N} a_n(0.2)^n$	N	$\sum_{n=0}^{N} a_n(0.2)^n$
1	1.150000	9	2.353090
2	1.097500	10	−2.442698
3	1.153750	11	13.253968
4	1.105372	12	−42.333586
5	1.176999	13	168.895730
6	1.049024	14	−796.466406
7	1.314970	15	3005.179546
8	0.686006		

Thus, we see the typical behavior of wandering near the right answer for a while (and not even that near!) and then going wild. And as N gets larger, things get worse: The 50th partial sum is about 10^{45} in magnitude and the 1000th about 10^{2000} in magnitude.

Example 2 Let f be C^∞ on $[-1, 1]$. Then $\sum_{n=0}^{\infty} (f^{(n)}(0)/n!)x^n$ is asymptotic as $x \downarrow 0$ or $x \uparrow 0$ to f. For Taylor's theorem with a remainder says that

$$\left| f(x) - \sum_{n=0}^{N} \frac{f^{(n)}(0)}{n!} x^n \right| \leq \frac{|x|^{N+1}}{(N+1)!} \sup_{|a| \leq |x|} [|f^{(n+1)}(a)|]$$

Since C^∞ functions can be nonanalytic, this example shows that asymptotic series may not converge; and even if they do, the sum may not have anything to do with the function f (see Example 1).

We defined analytic families on open sets. We will abuse this terminology by saying that a family is analytic on a closed set if it is norm continuous on the set and analytic in the interior.

Theorem XII.14 Let H_0 be a self-adjoint operator. Suppose that $H(\beta)$ is an analytic family in the region $\{\beta \,|\, 0 < |\beta| < B; \, |\arg \beta| \leq \theta\}$ and that the following conditions are obeyed:

(a) $\lim_{\substack{|\beta| \to 0 \\ |\arg \beta| \leq \theta}} \|(H(\beta) - \lambda)^{-1} - (H_0 - \lambda)^{-1}\| = 0$ for some $\lambda \notin \sigma(H_0)$.

(b) There is a closed symmetric operator V so that $C^\infty(H_0) \subset D(V)$ and $V[C^\infty(H_0)] \subset C^\infty(H_0)$.

(c) $C^\infty(H_0) \subset D(H(\beta))$ for all β in the sector and for $\psi \in C^\infty(H_0)$, $H(\beta)\psi = H_0\psi + \beta V\psi$.

Let E_0 be an isolated nondegenerate eigenvalue of H_0. Then if $|\beta|$ is small and $|\arg \beta| \leq \theta$, there is exactly one eigenvalue $E(\beta)$ of $H(\beta)$ near E_0. Moreover, the formal Rayleigh–Schrödinger series $\sum a_n \beta^n$ for the eigenvalue of $H_0 + \beta V$ is finite term by term and is an asymptotic series for $E(\beta)$ uniformly in the sector. Explicitly,

$$\lim_{\substack{|\beta| \to 0 \\ |\arg \beta| \leq \theta}} \left| E(\beta) - \sum_{n=0}^{N} a_n \beta^n \right| \Big/ \beta^N = 0$$

for all N.

This is the main result of this section. The conclusion makes several distinct statements, and for this reason we divide the proof into several lemmas. The first conclusion states that there is an eigenvalue $E(\beta)$ of $H_0 + \beta V$ near E_0 if β is small. We shall emphasize this property by giving it a name (stability). In Sections 5 and 6 we discuss situations where such stability does not hold. The second part of the conclusion concerns the asymptotic nature of the perturbation series for $E(\beta)$. A similar result holds for the eigenvector associated with $E(\beta)$ (see Problem 24). The main tool used in establishing the asymptotic property is familiar from Section 2: We use the formulas

$$E(\beta) = \frac{(\Omega_0, (H_0 + \beta V)P(\beta)\Omega_0)}{(\Omega_0, P(\beta)\Omega_0)}$$

and

$$P(\beta) = -\frac{1}{2\pi i} \oint (H(\beta) - E)^{-1} dE$$

As we have seen, these formulas allow one to unravel the complicated structure of the Rayleigh–Schrödinger coefficients into several simple operations on a geometric series. Our main tool will be the well-known error term for this series.

Definition Let $A(\beta)$ be a family of operators in the set $\{\beta \mid 0 < |\beta| < B, |\arg \beta| \leq \theta\}$. Suppose that an operator A_0 exists so that for some $\lambda \notin \sigma(A_0)$,

$$\text{s-lim}_{\substack{|\beta| \to 0 \\ |\arg \beta| \leq \theta}} (A(\beta) - \lambda)^{-1} = (A_0 - \lambda)^{-1}$$

XII.3 Asymptotic perturbation theory

An isolated nondegenerate eigenvalue E_0 of A_0 is called **stable** if $A(\beta)$ has exactly one eigenvalue near E_0 for β small and that eigenvalue is isolated and nondegenerate. Thus, E_0 is stable if, for all sufficiently small ε, there is a δ so that $|\beta| < \delta$ and $|\arg \beta| \leq \theta$ imply that there is exactly one point of $\sigma(A(\beta))$ in $\{E \mid |E - E_0| \leq \varepsilon\}$ and that point is a nondegenerate eigenvalue.

Lemma 1 If hypothesis (a) of Theorem XII.14 holds, then any isolated nondegenerate eigenvalue of H_0 is stable.

Proof An application of the first resolvent formula shows that if $(H(\beta) - \lambda)^{-1} \to (H_0 - \lambda)^{-1}$ in norm for some $\lambda \notin \sigma(H_0)$, it converges for all $\lambda \notin \sigma(H_0)$ and the convergence is uniform on compact sets of $\rho(H_0)$. Thus if ε is given so that H_0 has only one eigenvalue in $\{E \mid |E - E_0| \leq \varepsilon\}$, then for some δ, we conclude that $E \notin \sigma(H(\beta))$ if $|\beta| \leq \delta$, $|\arg \beta| \leq \theta$, and $|E - E_0| = \varepsilon$. Therefore, for $|\beta|$ small,

$$P(\beta) = -(2\pi i)^{-1} \oint_{|E-E_0|=\varepsilon} (H(\beta) - E)^{-1} \, dE$$

exists and converges in norm to

$$P_0 = -(2\pi i)^{-1} \oint_{|E-E_0|=\varepsilon} (H_0 - E)^{-1} \, dE$$

as $|\beta| \to 0$. By Theorem XII.5, P_0 is the projection onto the eigenvector of H_0 with eigenvalue E_0. By the lemma preceding Theorem XII.8, $P(\beta)$ is one dimensional if $|\beta|$ is small. Thus Lemma 1 follows from Theorem XII.6. ∎

Next we develop an asymptotic series for $P(\beta)\Omega_0$:

Lemma 2 Suppose that the hypotheses of Theorem XII.14 hold and that Ω_0 is the eigenvector of H_0 with eigenvalue E_0. For $|\beta|$ small, let $\Omega(\beta) = P(\beta)\Omega_0$ with $P(\beta)$ given as above. Then, for all N,

$$\lim_{\substack{|\beta| \to 0 \\ |\arg \beta| \leq \theta}} \left\| \Omega(\beta) - \sum_{n=0}^{N} \varphi_n \beta^n \right\| \Big/ |\beta|^N = 0$$

where

$$\varphi_n = (-1)^{n+1}(2\pi i)^{-1} \oint_{|E-E_0|=\varepsilon} (H_0 - E)^{-1}[V(H_0 - E)^{-1}]^n \Omega_0 \, dE$$

Proof Let $|E - E_0| = \varepsilon$. Formally,

$$(H(\beta) - E)^{-1} = \sum_{n=0}^{N} (-\beta)^n (H_0 - E)^{-1} [V(H_0 - E)^{-1}]^n$$
$$+ (-\beta)^{N+1} (H(\beta) - E)^{-1} [V(H_0 - E)^{-1}]^{N+1}$$

Consider applying both sides of this formal equation to Ω_0. Since $\Omega_0 \in C^\infty(H_0)$, both sides are well defined by condition (b). By condition (c), both sides give the same result when $H(\beta) - E$ is applied. Since $E \notin \sigma(H(\beta))$, the two sides are equal. By using the closed graph theorem, one can prove that $[V(H_0 - E)^{-1}]^N \Omega_0$ is continuous in E on $\{E | |E - E_0| = \varepsilon\}$ (see Problem 25). Thus,

$$\left\| \Omega(\beta) - \sum_{n=0}^{N} \varphi_n \beta^n \right\| / |\beta|^N$$

$$= \frac{|\beta|}{2\pi} \left\| \oint_{|E-E_0|=\varepsilon} (H(\beta) - E)^{-1} [V(H_0 - E)^{-1}]^{N+1} \Omega_0 \, dE \right\|$$

$$\leq \varepsilon |\beta| \sup_{\substack{|E-E_0|=\varepsilon \\ |\beta| < \delta \\ |\arg \beta| \leq \theta}} \|(H(\beta) - E)^{-1}\| \, \|[V(H_0 - E)^{-1}]^{N+1} \Omega_0\|$$

$$\to 0 \quad \text{as} \quad |\beta| \to 0 \quad \blacksquare$$

To conclude the proof of the theorem, we need the following lemma whose proof is left to Problem 26c.

Lemma 3 Let f and g be two functions defined in the sector $\{\beta | 0 < |\beta| < B; |\arg \beta| \leq \theta\}$. Suppose that $f \sim_{\beta \to 0} \sum a_n \beta^n$ and $g \sim_{\beta \to 0} \sum b_n \beta^n$ with $b_0 \neq 0$. Let $\sum c_n \beta^n$ be the formal series for $\sum a_n \beta^n / \sum b_n \beta^n$. Then the function $h(\beta) = f(\beta)/g(\beta)$ is asymptotic to $\sum c_n \beta^n$ as $\beta \to 0$.

Proof of Theorem XII.14 Since Lemma 1 has been proven, all that remains is to prove that the Rayleigh–Schrödinger series is finite term by term and asymptotic. By Lemma 2 and hypothesis (b), the series for $(H(\beta)\Omega_0, P(\beta)\Omega_0)$ and $(\Omega_0, P(\beta)\Omega_0)$ are finite term by term and asymptotic. Since $\lim_{|\beta| \to 0} (\Omega_0, P(\beta)\Omega_0) = 1 \neq 0$, Lemma 3 completes the proof. ∎

Finally we must ask when condition (a) can be proven. Let us state two general theorems. We shall only sketch the proof of one of them (for references, see the Notes). As in Chapter XIII, we will say that two self-adjoint operators A and B which are bounded from below satisfy $A \leq B$ if $Q(B) \subset Q(A)$ and $(\varphi, A\varphi) \leq (\varphi, B\varphi)$ for all φ in $Q(B)$.

Theorem XII.15 Suppose that H_0 and V are self-adjoint. Suppose, moreover, that:

(i) $H_0 \geq 0$.
(ii) For each $a > 0$, there is a b so that
$$V_- \leq aH_0 + b$$
(V_- is the negative part of V given by the spectral theorem; in particular, if $V \geq 0$, $V_- = 0$ so (ii) holds.)
(iii) For some c and d
$$|V| \leq cH_0^2 + d$$

Then $H_0 + \beta V$ is an analytic family of type (B) in the cut plane $\{\beta \,|\, |\arg \beta| < \pi;\ |\beta| > 0\}$ and
$$\lim_{\substack{|\beta| \to 0 \\ |\arg \beta| \leq \theta}} \|(H(\beta) - \lambda)^{-1} - (H_0 - \lambda)^{-1}\| = 0$$
for any $\theta < \pi$.

Theorem XII.16 Suppose that H_0 and V are self-adjoint operators. Suppose, moreover, that:

(i) $H_0 \geq 0$.
(ii) For each $a > 0$, there is a b so $V_- \leq aH_0 + b$.
(iii) For some d, some $c < 1$, and some B,
$$\pm \beta[H_0^{1/2}, [H_0^{1/2}, V]] \leq c(H_0^2 + \beta^2 V^2) + d \quad \text{if} \quad 0 < \beta < B$$
(iv) For all $e > 0$, there is an f so that
$$\pm \beta i[H_0, V] \leq e(H_0^2 + \beta^2 V^2) + f \quad \text{if} \quad 0 < \beta < B$$
(v) For some $p > 1$, g, and h,
$$|V|^{2/p} \leq gH_0^2 + h$$
(vi) $H_0 + \beta V$ is essentially self-adjoint on $D(H_0) \cap D(V)$ if $0 < \beta < B$.

Then $H_0 + \beta V$ is an analytic family of type (B) in the cut plane and
$$\lim_{\substack{|\beta| \to 0 \\ |\arg \beta| \leq \theta}} \|(H(\beta) - \lambda)^{-1} - (H_0 - \lambda)^{-1}\| = 0$$
for any $\theta < \pi$.

If in addition:

(iii'), (iv') For any B, we can choose d, f suitably so that the basic estimates in (iii), (iv) hold; and

(vi') $H_0 + \beta V$ is essentially self-adjoint on $D(H_0) \cap D(V)$ for all $\beta > 0$,

then $H(\beta)$ is an analytic family of type (A) in the cut plane.

The main idea of the proof of Theorem XII.16 is to use (ii), (iii), and (iv) to prove "quadratic" estimates. Specifically, given $\theta < \pi$, one can find, α, $\gamma > 0$ so that

$$H_0^2 + |\beta|^2 V^2 \leq \alpha(H_0 + \beta V)^*(H_0 + \beta V) + \gamma$$

for all β with $0 < |\beta| < B$; $|\arg \beta| < \theta$. It then follows that $H_0 + \beta V$ is closed on $D(H_0) \cap D(V)$ (Problem 27). Now write

$$\|(H_0 + \beta V - \lambda)^{-1} - (H_0 - \lambda)^{-1}\|$$
$$\leq |\beta|^{1/p} \|(H(\beta) - \lambda)^{-1} |\beta V|^{1-(1/p)} |V|^{1/p}(H_0 - \lambda)^{-1}\|$$

By the quadratic estimate, $(H(\beta) - \lambda)^{-1} |\beta V|^{1-(1/p)}$ is bounded uniformly in the sector $\{\beta | 0 < |\beta| < B; |\arg \beta| < \theta\}$. By (v), $|V|^{1/p}(H_0 - \lambda)^{-1}$ is bounded. Thus $\|(H_0 + \beta V - \lambda)^{-1} - (H_0 - \lambda)^{-1}\|$ goes to zero at least as fast as $|\beta|^{1/p}$.

Example 3 (the anharmonic oscillator) Let $H_0 = -d^2/dx^2 + x^2$ on $L^2(\mathbb{R})$ and $V = x^{2m}$. Then all the conditions of Theorem XII.16 hold. (i) and (ii) are immediate. (vi) was proven in Example 2 of Section X.4 and Example 2 of Section X.9. To prove (iii) and (iv), one does explicit computations with the creation and annihilation operators A, $A^\dagger$, introduced in the Appendix to Section V.3 (see Problem 28). (v) holds with $p = m$ by again using A and $A^\dagger$ (Problem 28). Since V leaves $C^\infty(H_0)$ invariant and Theorem XII.16 is applicable, all the conditions of Theorem XII.14 hold. Thus the eigenvalues of $p^2 + x^2 + \beta x^{2m}$ have Rayleigh–Schrödinger series as asymptotic series, uniformly in sectors $|\arg \beta| \leq \theta < \pi$. It can also be proven that the eigenvalues have analytic continuations onto many-sheeted surfaces, $\arg |\beta| \leq \theta$, $0 < \beta < B_\theta$, for any $\theta < ((m+1)/2)\pi$ where the series is asymptotic on the many-sheeted surface.

Example 4 ($(\varphi^4)_2$ field theory with a spatial cut-off) Let H_0 be the free boson field of mass $m_0 > 0$ in two-dimensional space–time. Let $V = \int g(x) {:}\varphi^4(x){:}\, dx$ with $g \in L^1 \cap L^2$ and $g \geq 0$ (as discussed in Section

X.7). Then, one can prove all the conditions of Theorem XII.16 (or Theorem XII.15). The only isolated eigenvalue of H_0 is the Fock vacuum. Theorem XII.14 is not applicable in the stated form because $C^\infty(H_0)$ is not invariant under V. However, if N is the number operator, V and $(H_0 - E)^{-1}$ leave $C^\infty(N)$ invariant and $\Omega_0 \in C^\infty(N)$. It is easy to extend the method used in Theorem XII.14 to prove that the Rayleigh–Schrödinger series is asymptotic when the invariance of $C^\infty(H_0)$ is replaced by invariance of $C^\infty(N)$. In this way one can conclude that the ground state energy of $H_0 + \beta V$ has an asymptotic series. The coefficients are given by sums of Feynman type diagrams (see the Notes).

When condition (a) of Theorem XII.14 holds, it is a convenient tool for proving stability, but there are cases where it fails in which stability can be proven by the following method:

Theorem XII.16.5 Let H_0 be a closed operator and P a finite dimensional *orthogonal* projection so that $\operatorname{Ran} P \subset D(H_0)$, and $H_0 P = P H_0$. Suppose that the operator $H_0 \restriction \operatorname{Ran}(1 - P)$ is sectorial with sector $S_0 \equiv \{z \mid |\arg z| \leq \theta_0 < \pi/2\}$ and that $H_0 \restriction \operatorname{Ran} P$ has spectrum outside S_0. Let V be a closed sectorial operator with sector S_0. Suppose that $\operatorname{Ran} P \subset D(V) \cup D(V^*)$. Let E_0 be an isolated nondegenerate eigenvalue of $H_0 \restriction \operatorname{Ran} P$. Then, for any $\varepsilon, \delta > 0$, there is a B so that the form sum $H_0 + \beta V$ has exactly one eigenvalue, $E(\beta)$, in $\{E \mid |E - E_0| < \varepsilon\}$ for β in the region $Q \equiv \{\beta \mid |\arg \beta| \leq \pi/2 - \theta_0 - \delta; |\beta| \leq B\}$. The eigenvalue is nondegenerate, it is the only spectral point of $H_0 + \beta V$ near E_0, and

$$\sup\{\|(H_0 + \beta V - E)^{-1}\| \mid |E - E_0| = \varepsilon; \beta \in Q\} < \infty$$

Moreover, suppose that $H_0 \Omega_0 = E_0 \Omega_0$ and that for $|E - E_0| = \varepsilon$, we have that $((H_0 - E)^{-1} V)^j (H_0 - E)^{-1} \Omega_0 \in D(V)$ for $j = 0, 1, \ldots, k$. Then the Rayleigh–Schrödinger coefficients for $E(\beta)$ are finite up to order at least β^{k+1} and the Rayleigh–Schrödinger series is asymptotic to at least that order in the region Q.

Proof It suffices to prove stability since the asymptotic series result then follows as in the proofs of Lemmas 2 and 3. Write $V = V_1 + V_2$ where $V_1 = (1 - P)V(1 - P)$, $V_2 = V - V_1$. V_2 is a bounded finite rank operator since $\operatorname{Ran} P \subset D(V) \cup D(V^*)$. Clearly, $H_0 + \beta V_1$ is sectorial on $\operatorname{Ran}(1 - P)$ for $\beta \in Q$ no matter what we take for B and δ and it equals H_0 on $\operatorname{Ran} P$. Since βV_2 is bounded, we can study its effect on $\sigma(H_0 + \beta V_1)$ by regular perturbation theory. The result is that for β small the sup in question is finite so the projection $-(2\pi i)^{-1} \int (H_0 + \beta V - E)^{-1} dE$ has constant dimension. ∎

Example 5 Let $H_0 = -\Delta + W$ where W is real-valued and $W(x) \to 0$ as $x \to \infty$. Let $V = |x|$ (we could take $V(x) = |x|^n$). In this case $H_0 + \beta V \to H_0$ in strong resolvent sense but not in norm resolvent sense. The convergence cannot be in norm since $(H_0 + \beta V - i)^{-1}$ is compact (see Section XIII.14) but $(H_0 - i)^{-1}$ is not. Thus Theorem XII.14 is not applicable.

If H_0 has negative eigenvalues, we can take P to be the projection onto the corresponding eigenvectors (or those corresponding to the m lowest eigenvectors in which case we add a constant to H_0 to make sure that $H_0 \upharpoonright (1 - P)$ is sectorial with sector S_0). Since the eigenvectors of H_0 fall off exponentially (see Section XIII.11), Ran $P \subset D(V) = D(V^*)$, so the hypotheses of the first part of Theorem XII.16.5 holds. Moreover, under mild hypotheses on W, one can show, using the method of Section XIII.11, that for E near E_0 and a small, $(H_0 - E)^{-1}$ is a bounded map from $\mathscr{H}_a$ to itself where

$$\mathscr{H}_a = \{\psi \mid \|\psi\|_a = \|e^{a|x|}\psi\|_2 < \infty\}$$

Since $\Omega_0 \in \mathscr{H}_a$ for a small and V is bounded from $\mathscr{H}_a$ to $\mathscr{H}_{a/2}$, we can verify the hypotheses of the second part of Theorem XII.16.5 for any k. Thus, the Rayleigh–Schrödinger series is asymptotic to all orders for any β satisfying Re $\beta > 0$. If $\beta < 0$, the eigenvalue is not stable, but the Rayleigh–Schrödinger series is still relevant as we shall see in Section 5.

Finally, we consider an example which shows that one can sometimes prove that a Rayleigh–Schrödinger series is asymptotic even though the eigenvalues are not stable. However, a detailed argument will be required. This example also illustrates, in a dramatic way, the fact that two distinct functions can have the same asymptotic series.

Example 6 (The double well potential)   Consider the family of operators on $L^2(\mathbb{R})$:

$$H(\beta) = -d^2/dx^2 + x^2 + 2\beta x^3 + \beta^2 x^4$$

Since the total potential can be written $x^2(1 + \beta x)^2$, we see that for each $\beta \geq 0$, $H(\beta)$ is bounded from below. By the methods of Chapter X, $H(\beta)$ is essentially self-adjoint on $\mathscr{S}(\mathbb{R})$ and, by the methods of Section XIII.4 or XIII.14, it has a purely discrete spectrum and a complete set of eigenvectors. Although $H(\beta)$ is not linear in β, it is not hard to extend the formulas we have given in Section 1 to obtain formal series for the eigenvalues: one need only use $2\beta x^3 + \beta^2 x^4$ for V in the formulas of Section 1 up to order n and then collect all terms of order β^n. Because $(\Omega_n, (2\beta x^3 + \beta^2 x^4)\Omega_m) = 0$ if $|n - m| > 4$ for the unperturbed eigenvectors Ω_n

of $H(0)$, the sums in the definition of the Rayleigh–Schrödinger series are all finite and so one has a formal series $\sum_{n=0}^{\infty} a_n \beta^n$ for the eigenvectors of $H(\beta)$. Since the problem is invariant under the changes $x \to -x$, $\beta \to -\beta$, we have that $a_{2n+1} = 0$ for all n.

So far, this problem does not seem so different from the anharmonic oscillator of Example 3, and indeed, if the 2 in the $2\beta x^3$ term were replaced by some θ with $-2 < \theta < 2$, then one could prove norm resolvent convergence as $\beta \downarrow 0$. But notice that the total potential vanishes also at $x = -\beta^{-1}$; indeed, if we translate by $-\tfrac{1}{2}\beta^{-1}$ units we obtain the unitarily equivalent operator

$$\tilde{H}(\beta) = -d^2/dx^2 + \tfrac{1}{16}\beta^{-2} - \tfrac{1}{2}x^2 + \beta^2 x^4$$

Since the potential term in $\tilde{H}(\beta)$ is symmetric under $x \to -x$, it is evident that $H(\beta)$ has two identical wells in its potential, one centered at $x = 0$ and one at $x = -\beta^{-1}$. As a result, for β very small, we expect two eigenvalues near E_0, the ground state energy of $H(0)$, since we can obtain two nearly orthogonal vectors with energy expectation near $E(0)$ by using the ground state $\Omega_0(x)$ for $H(0)$ and its translate $\Omega_0(x + \beta^{-1})$. For this reason, we will label the eigenvalues of $H(\beta)$ for $\beta > 0$ as $E_0(\beta) < E_0'(\beta) < E_1(\beta) < E_1'(\beta) < \cdots$ and corresponding vectors $\Omega_0(x, \beta)$, etc. To avoid confusion, we will only use prime below to distinguish these eigenvalues and *not* for derivatives. $E_0, E_1, \ldots$ are just the eigenvalues of $\tilde{H}(\beta)$ whose eigenvectors are invariant under $x \to -x$, and $E_0', E_1', \ldots$ those whose eigenvectors are odd under $x \to -x$. As $\beta \downarrow 0$, we expect both $E_0(\beta)$ *and* $E_0'(\beta)$ to go to E_0; i.e., the eigenvalue E_0 of $H(0)$ is *not* stable. Moreover, one expects that $E_0'(\beta) - E_0(\beta)$ goes to zero extremely fast, namely as $\exp(-a\beta^{-2})$ for suitable a, because of the following argument: The true eigenvectors of $\tilde{H}(\beta)$ are even and odd under $x \to -x$. If one starts at $t = 0$ with a state like $N(\beta)^{-1}[\Omega_0(x, \beta) + \Omega_0'(x, \beta)]$ which is concentrated near $x = 0$, then after a time $t = \pi/(E_0' - E_0)$, the state, moving under the dynamics given by $H(\beta)$, will be concentrated at $x = -\beta^{-1}$; that is, it will tunnel through the barrier between the wells. Thus $E_0'(\beta) - E_0(\beta)$ should be of the same order as the probability for tunnelling and this should be of order $\exp(-d\sqrt{h})$ where d is the width of the barrier, i.e., $1/\beta$, and h is the height of the barrier, i.e., $\tfrac{1}{16}\beta^{-2}$.

Therefore, our expectation is that the two eigenvalues $E_0(\beta)$ and $E_0'(\beta)$ approach the single eigenvalue E_0 of $H(0)$, and that both have the same Rayleigh–Schrödinger series as asymptotic series. We want to sketch a proof of these facts leaving the details to the reader (Problem 36). The arguments are rather involved and we will use freely tools from Chapter XIII. We will establish three facts: (i) As $\beta \downarrow 0$, both $E_i(\beta)$ and $E_i'(\beta)$

approach $E_i(0) \equiv E_i$ and in particular, there is a β-independent gap between $E'_0(\beta)$ and $E_1(\beta)$ as β goes to zero. (ii) We will prove that $\exp(-a\beta^{-2}) \leq E'_0(\beta) - E_0(\beta) \leq \exp(-b\beta^{-2})$ as $\beta \downarrow 0$ for suitable a and b. (iii) We will show that $E_0(\beta)$ (and thus by (ii), also $E'_0(\beta)$) has the Rayleigh–Schrödinger series as asymptotic series.

As a preliminary to step (i), we consider the operators $H^D(a)$ and $H^N(a)$ on $L^2(-a, a)$ given by $-d^2/dx^2 + x^2$ with Dirichlet (respectively, Neumann) boundary conditions at $x = \pm a$. Let $E_n^D(a)$ and $E_n^N(a)$ denote the eigenvalues of $H^D(a)$ and $H^N(a)$. Let E_n denote the eigenvalues of $-d^2/dx^2 + x^2$ on all of $L^2(\mathbb{R})$. By the method of Section XIII.15, we have

$$E_n^N(a) \leq E_n \leq E_n^D(a) \tag{5a}$$

so long as $a^2 \geq E_n = 2n + 1$. (This condition is needed so that after Neumann conditions are imposed at $x = \pm a$ on the problem for $L^2(\mathbb{R})$, the lowest eigenvalues for the intervals $|x| \geq a$ are above E_n.) We will show that

$$E_n^D(a) \leq E_n^N(a) + \exp(-ca^2) \tag{5b}$$

for a large, where how large will depend on n. Let $\psi_n^N(x; a)$ denote the nth normalized eigenvector of $H^N(a)$. We claim that for a sufficiently large,

$$|\psi_n^N(a; a)| \leq \exp(-\tfrac{1}{12}a^2) \tag{6}$$

To prove (6), let $\varphi(x)$ solve $-\varphi'' + x^2\varphi = E_n^N(a)\varphi$ with boundary conditions $\varphi'(a) = 0$, $\varphi(a) = 1$. Choose a so large that $a > 16$ and $(a/2)^2 \geq (a^2/9) + (2n + 1)$; then in the interval $[\tfrac{1}{2}a, a]$, we have $x^2 - E_n^N(a) \geq a^2/9$ since $E_n^N(a) \leq 2n + 1$. Thus, by a comparison argument of the type used in Section XI.2, $\varphi(x) \geq \cosh(\tfrac{1}{3}a)(x - a))$ for x in $[\tfrac{1}{2}a, a]$. In particular, $\varphi(x) \geq \tfrac{1}{2}\exp(\tfrac{1}{12}a^2)$ for $x \in [\tfrac{1}{2}a, \tfrac{3}{4}a]$, and thus, since $a > 16$,

$$\int_{a/2}^{3a/4} |\varphi(x)|^2 \, dx \geq \cosh(\tfrac{1}{6}a^2)$$

Since $\psi_n^N(x, a)$ is a normalized multiple of φ, (6) holds. For later purposes, we note that a similar argument shows that

$$\left|\frac{\partial \psi_n^D}{\partial x}(a; a)\right| \leq \exp(-ca^2) \tag{6'}$$

Now fix n_0 and pick a_0 so that (6) holds for $n = 0, \ldots, n_0$ and $a \geq a_0$. Let

$$\eta_i(x) = \begin{cases} \psi_i^N(x, a) - \psi_i^N(a, a) & i = 0, 2, \ldots \\ \psi_i^N(x, a) - a^{-1}x\psi_i^N(a, a) & i = 1, 3, \ldots \end{cases}$$

Then the η_i vanish at $x = \pm a$, so they are suitable trial functions for $H^D(a)$; i.e., they lie in $Q(H^D(a))$. On account of (6), $\Delta_{ij} = (\eta_i, \eta_j)$ and $E_{ij} = (\eta_i, H^D(a)\eta_j)$ obey, for $0 \le i, j \le n_0$:

$$|\Delta_{ij} - \delta_{ij}| \le d \exp(-\tfrac{1}{20} a^2)$$

$$|E_{ij} - E_i^N(a)\delta_{ij}| \le d \exp(-\tfrac{1}{20} a^2)$$

for d suitable. Let $E_{(i)}$ be the ith eigenvalue of $\Delta^{-1/2} E \Delta^{-1/2}$ where $\Delta = \{\Delta_{ij}\}$ and $E = \{E_{ij}\}$. By the Rayleigh–Ritz method of Section XIII.2, $E_{(i)} \ge E_i^D(a)$ (for $i = 0, \ldots, n_0$) and by the above,

$$\|\Delta^{-1/2} E \Delta^{-1/2} - E_i^N \delta_{ij}\| \le d_1 \exp(-\tfrac{1}{20} a^2)$$

It follows that $|E_{(i)} - E_i^N| \le d_1 \exp(-\tfrac{1}{20} a^2)$ so that (5b) holds.

We can now accomplish step (i). By the form of $\tilde{H}(\beta)$, E_i (respectively, E_i') is the $(i + 1)$st eigenvalue of $-d^2/dx^2 + x^2 + 2\beta x^3 + \beta^2 x^4$ on $L^2(-\tfrac{1}{2}\beta^{-1}, \infty)$ with Neumann (respectively, Dirichlet) boundary condition at $x = -\tfrac{1}{2}\beta^{-1}$. Fix $a > \sqrt{2n+1}$, and put additional Dirichlet or Neumann boundary conditions at $x = \pm a$. Let $E_{i,\,D}(\beta, a)$, $E_{i,\,N}(\beta, a)$, $E_{i,\,D}'(\beta, a)$, $E_{i,\,N}'(\beta, a)$ denote the corresponding eigenvalues. By Dirichlet–Neumann bracketing (Section XIII.15),

$$E_{i,\,N}(\beta, a) \le E_i(\beta) \le E_{i,\,D}(\beta, a)$$

and similarly for the primed quantities. Taking β to zero we see that

$$E_i^N(a) \le \underline{\lim}\, E_i(\beta) \le \overline{\lim}\, E_i(\beta) \le E_i^D(a)$$

where $E_i^N(a)$ is as defined in the last paragraph. Taking now $a \to \infty$ and using (5), we see that $|E_i(\beta) - E_i| \to 0$ as $\beta \to 0$.

As a preliminary to step (ii), we note that as $\beta \downarrow 0$, $\|\Omega_i(\beta) - \tilde{\Omega}_i(\beta)\| \to 0$ and $\|\Omega_i'(\beta) - \tilde{\Omega}_i'(\beta)\| \to 0$ where $\tilde{\Omega}_i(x, \beta) = 2^{-1/2}(\Omega_i(x) + \Omega_i(x + \beta^{-1}))$ and $\tilde{\Omega}_i'(x, \beta) = 2^{-1/2}(\Omega_i(x) - \Omega_i(x + \beta^{-1}))$. For $\tilde{\Omega}_i$ is even under reflection about $x = -\tfrac{1}{2}\beta^{-1}$ and

$$\|(H(\beta) - E_i(\beta))\tilde{\Omega}_i\| \to 0$$

as $\beta \downarrow 0$. Since $H(\beta)$ has only one eigenvalue with an even eigenvector near $E_i(\beta)$ we have that $\|\Omega_i(\beta) - \tilde{\Omega}_i(\beta)\| \to 0$. The argument with primes is similar. In particular,

$$\int_{-\tfrac{1}{2}\beta^{-1}}^{\infty} \Omega_i(x, \beta)\Omega_i'(x, \beta)\, dx \to \tfrac{1}{2} \qquad (7)$$

as $\beta \downarrow 0$.

Using (7), the boundary conditions at $x = -\tfrac{1}{2}\beta^{-1}$, and an integration by parts in

$$\int_{-\frac{1}{2}\beta^{-1}}^{\infty} [\Omega_i(H(\beta)\Omega_i') - \Omega_i'(H(\beta)\Omega_i)]\,dx$$

one finds that

$$\tfrac{1}{2}(E_i' - E_i) = \Omega_i(-\tfrac{1}{2}\beta^{-1}, \beta)\frac{d}{dx}\Omega_i'(-\tfrac{1}{2}\beta^{-1}, \beta) \tag{8}$$

By the method used in the proof of (6) and (6′) and the fact that on $(-\tfrac{1}{2}\beta^{-1}, 0)$, we have $\tfrac{1}{4}x^2 \leq x^2(1+\beta x)^2 \leq x^2$, it is easy to show that the right side of (8) is bounded below by $\exp(-a\beta^{-2})$ and above by $\exp(-b\beta^{-2})$. This completes the proof of step (ii).

Finally we turn to step (iii). For $\varphi \in \mathscr{S}$, it is easy to see that $(H(\beta) - H(0))\varphi \to 0$ as $\beta \downarrow 0$. Since the spectrum of $H(\beta)$ approaches that of $H(0)$ by step (i), we see that $(H(\beta) - z)^{-1} - (H(0) - z)^{-1} \to 0$ strongly as $\beta \downarrow 0$ for $z \neq 2n + 1$. In particular, if

$$P(\beta) = (2\pi i)^{-1} \int_{|z-1|=1/2} (z - H(\beta))^{-1}\,dz$$

then $P(\beta) \to P(0)$ strongly as $\beta \downarrow 0$. Of course $P(0)$ is a rank 1 projection and $P(\beta)$ for $\beta \neq 0$ is a rank 2 projection. As usual $e(\beta) \equiv (\Omega_0, P(\beta)\Omega_0)^{-1}(\Omega_0, H(\beta)P(\beta)\Omega_0)$ has the Rayleigh–Schrödinger series as asymptotic series. Moreover, $e(\beta) = \theta(\beta) E_0(\beta) + (1 - \theta(\beta))E_0'(\beta)$ for a suitable function $\theta(\beta)$. Since $E_0 - E_0' \to 0$ faster than any power, both $E_0(\beta)$ and $E_0'(\beta)$ have the same Rayleigh–Schrödinger series as asymptotic series.

XII.4 Summability methods in perturbation theory

The idea that a function could be determined by a divergent asymptotic series was a foreign one to the nineteenth century mind. Borel, when an unknown young man, discovered that his summability method yielded the "right" answer for many classical divergent series. He decided to make a pilgrimage to Stockholm to see Mittag-Leffler, who was the recognized lord of complex analysis. Mittag-Leffler listened politely to what Borel had to say and then, placing his hand upon the complete works of Weierstrass, his teacher, he said, in Latin, "The Master forbids it."

A tale of Mark Kac

We have just seen that a divergent perturbation series such as the one for energy levels of $p^2 + x^2 + \beta x^4$ can nonetheless give some information about the actual energy levels and that under fairly general circumstances, the

Rayleigh–Schrödinger series is an asymptotic series. We also saw that saying a series is an asymptotic series is a very weak statement; in particular, saying that $\sum_{n=0}^{\infty} a_n \beta^n$ is asymptotic for $E(\beta)$ gives one no information about $E(\beta_0)$ for fixed $\beta_0 \neq 0$. In this section, we shall study the question of whether, nevertheless, the divergent perturbation series might determine the energy level $E(\beta)$ by some method more subtle than direct summation.

We first seek some condition stronger than saying that $\sum_{n=0}^{\infty} a_n \beta^n$ is asymptotic to $E(\beta)$, and weaker than saying the sum is convergent, but that determines $E(\beta)$ uniquely. Recall that the nonuniqueness associated with asymptotic series came from the fact that nonzero functions like $\exp(-\beta^{-1})$ could have a zero asymptotic series. We thus seek a stronger condition that guarantees uniqueness. Such a condition is supplied by the following theorem of Carleman:

Theorem XII.17 (Carleman's theorem)   Let g be a function analytic in the interior of $S = \{z \mid 0 \leq |z| \leq B;\ |\arg z| \leq \pi/2\}$ and continuous on S. Suppose that for each n,

$$|g(z)| \leq b_n |z|^n$$

for all z in the sector. If moreover $\sum_{n=1}^{\infty} b_n^{-1/n} = \infty$, then g is identically zero.

In applications, we need only the special case of Carleman's theorem proven below. We remark that the condition $|g(z)| \leq b_n |z|^n$ says exactly that g has 0 asymptotic series. The additional condition $\sum_{n=1}^{\infty} b_n^{-1/n} = \infty$ is a bound on how quickly $b_n^{-1/n}$ can go to zero or equivalently how fast b_n can go to infinity. Since $\sum_{n=1}^{\infty} n^{-1}$ is barely divergent, $\sum_{n=1}^{\infty} b_n^{-1/n} = \infty$ is only slightly weaker than a bound $b_n \leq C_0 B_0^n n^n$ for suitable C_0, B_0 or equivalently $b_n \leq CB^n n!$, for suitable C, B. We prove the following special case of Carleman's theorem:

Theorem XII.18   Suppose that g is a function analytic in the interior of and continuous on $S = \{z \mid |\arg z| \leq \tfrac{1}{2}\pi + \varepsilon; 0 < |z| \leq R\}$ for some $\varepsilon > 0$. If there exist C and B so that

$$|g(z)| \leq CB^n n! |z|^n$$

for all $z \in S$ and all n, then g is identically zero.

Proof   For simplicity of notation, let $w = z^{-1}$ and $f(w) = g(w^{-1})$. Then f is analytic in

$$\tilde{S} = \{w \mid |\arg w| \leq \tfrac{1}{2}\pi + \varepsilon;\ |w| \geq R^{-1}\}$$

For all n, $|f(w)| \leq CB^n n!\,|w|^{-n}$ for all w in $\tilde{S}$. By Stirling's formula, for any $\delta > 0$ we can find D_δ so that $n! \leq D_\delta n^n e^{-n(1-\delta)}$. Taking n to be the largest integer less than $|w|/B$, we see $|f(w)| \leq C'e^{-\varepsilon|w|}$ for some C' and ε. Using a technique known as the Phragmén–Lindelöf principle, we shall show that $f(w_0) = 0$ if w_0 is real and $w_0 > R^{-1}$ and conclude that $f \equiv 0$ by analytic continuation. Pick α so that $\alpha(\frac{1}{2}\pi + \varepsilon) = \frac{1}{2}\pi$. Let $f_m(w) = f(w)\exp[m(w/w_0)^\alpha]$. Since $\alpha < 1$ and $|f(w)| \leq C'e^{-\varepsilon|w|}$, $f_m(w) \to 0$ as $|w| \to \infty$. Thus, by the maximum principle,

$$|f_m(w_0)| \leq \max\{|f_m(w)|\,|\,|w| = R^{-1} \text{ or } |\arg w| = \tfrac{1}{2}\pi + \varepsilon\}$$

Letting $M_1 = \max\{|f(w)|\,|\,|w| = R^{-1}\}$ and $M_2 = \max\{|f(w)|\,|\,|\arg w| = \frac{1}{2}\pi + \varepsilon\}$ we see that

$$|f_m(w_0)| \leq \max\{M_1 \exp[m(R^{-1}/w_0)^\alpha],\ M_2\}$$

so that

$$|f(w_0)| \leq \max\{M_1 \exp[m(-1 + (R^{-1}/w_0)^\alpha)],\ M_2 \exp[-m]\}$$

Since m is arbitrary, $f(w_0) = 0$. ∎

In the proof just given we established that if f is analytic in $\tilde{S}$ with exponential falloff, then f is zero. We shall prove and use an extension of this fact (Carlson's theorem) in Section XIII.13.

This last theorem suggests that we single out a stronger condition than the statement that $E(\beta)$ has $\sum_{n=0}^\infty a_n \beta^n$ as asymptotic series.

Definition We say that a function $E(\beta)$, analytic in a sectorial region $\{\beta\,|\,0 < |\beta| < B;\ |\arg \beta| < \frac{1}{2}\pi + \varepsilon\}$, obeys a **strong asymptotic condition** and has $\sum_{n=0}^\infty a_n \beta^n$ as **strong asymptotic series** if there are C and σ so that

$$\left| E(\beta) - \sum_{n=0}^N a_n \beta^n \right| \leq C\sigma^{N+1}(N+1)!\,|\beta|^{N+1}$$

for all N and all β in the sector.

Theorem XII.19 If $\sum_{n=0}^\infty a_n \beta^n$ is a strong asymptotic series for two analytic functions f and g, then $f = g$.

Proof By hypothesis,

$$|f(\beta) - g(\beta)| \leq 2C\sigma^{N+1}(N+1)!\,|\beta|^{N+1}$$

in $\{\beta\,|\,0 < |\beta| \leq \frac{1}{2}B;\ |\arg \beta| \leq \frac{1}{2}\pi + \frac{1}{2}\varepsilon\}$, so Theorem XII.18 implies that $f - g = 0$. ∎

XII.4 Summability methods in perturbation theory

To apply the idea of strong asymptotic conditions to eigenvalues it is useful to have the following simple technical lemma (Problem 26).

Lemma If f and g obey strong asymptotic conditions and if $\lim_{|\beta|\to 0} g(\beta) \neq 0$, then f/g obeys a strong asymptotic condition with asymptotic series given by the formal quotient of the asymptotic series for f and g.

Example 1 Consider the energy levels of the anharmonic oscillator $p^2 + x^2 + \beta x^4$. We have seen that $E(\beta) = (H(\bar{\beta})\Omega_0, P(\beta)\Omega_0)/(\Omega_0, P(\beta)\Omega_0)$. By the preceding lemma, it is sufficient to get a bound of the form $C\sigma^{N+1}(N+1)!|\beta|^{N+1}$ on the norm of the remainder of $P(\beta)\Omega_0$ after the first N terms of its asymptotic series have been subtracted. Since $P(\beta)$ is given as an integral of the resolvent, we need only establish a strong asymptotic condition for $(H(\beta) - E)^{-1}\Omega_0$ with bounds uniform for $|E - E_0| = \varepsilon$. The remainder term we have to bound is just the remainder term of a geometric series:

$$[H(\beta) - E]^{-1}\Omega_0 - \sum_{n=0}^{N} (-\beta)^n (H_0 - E)^{-1}[V(H_0 - E)^{-1}]^n \Omega_0$$

$$= (-\beta)^{N+1}(H(\beta) - E)^{-1}[V(H_0 - E)^{-1}]^{N+1}\Omega_0$$

Since $\|(H(\beta) - E)^{-1}\|$ is uniformly bounded by the norm convergence we proved in the preceding section, we need only prove

$$\|[x^4(p^2 + x^2 - E)^{-1}]^{N+1}\Omega_0\| \leq C\sigma^{N+1}(N+1)!$$

for $|E - \tfrac{1}{2}| = \tfrac{1}{2}$. In terms of the standard A, $A^\dagger$ operators discussed in the appendix to Section V.3, $x^4 = \tfrac{1}{4}(A + A^\dagger)^4$ so that the norm in question can be bounded by $(2^4)^{N+1}$ terms of the form

$$4^{-N-1}\|A_1^\# \cdots A_4^\# (H_0 - E)^{-1} A_5^\# \cdots A_{4N+4}^\#(H_0 - E)^{-1}\Omega_0\|$$

where each $A^\#$ is either A or $A^\dagger$. The bound $C\sigma^{N+1}(N+1)!$ is proven by using the formulas

$$A\Omega_n = \sqrt{n}\,\Omega_{n-1}$$
$$A^\dagger \Omega_n = \sqrt{n+1}\,\Omega_{n+1}$$
$$(H_0 - E)^{-1}\Omega_n = (2n + 1 - E)^{-1}\Omega_n.$$

Therefore, the energy levels of a $p^2 + x^2 + \beta x^4$ oscillator obey a strong asymptotic condition; in particular, the energy levels of the anharmonic oscillator are determined uniquely by the Rayleigh–Schrödinger series.

Generalizations of the above argument to an abstract setting have been made by Simon. Typical of the results obtained is:

Theorem XII.20 Let H_0, V, and M be self-adjoint operators obeying:
(i) $H_0 \geq 0$.
(ii) For each $a > 0$, there is a b so that $V_- \leq aH_0 + b$.
(iii) For some c and d
$$|V| \leq cH_0^2 + d$$
(iv) $0 \leq M \leq H_0$.
(v) M and H_0 commute.
(vi) $C^\infty(M) \subset D(V)$ and V takes $C^\infty(M)$ into itself.
(vii) There are constants e and f so that for all n and all $\psi \in C^\infty(M)$,
$$\|(M+1)^n V\psi\| \leq e\|(M+f)^{n+2}\psi\|$$

Let $H_0 + \beta V$ be defined as a sum of forms for $|\beta|$ small, $|\arg \beta| < \pi - \varepsilon$. Then, for any isolated nondegenerate eigenvalue of H_0 there is an eigenvalue of $H_0 + \beta V$ for $|\beta|$ small that has the Rayleigh–Schrödinger series as strong asymptotic series.

For a proof, see the reference discussed in the Notes.

This theorem allows one to extend the idea of Example 1 to n-dimensional anharmonic oscillators. Specifically, if $\mathscr{H} = L^2(\mathbb{R}^n)$,
$$H_0 = \sum_{i=1}^n -\frac{\partial^2}{\partial x_i^2} + \omega_i^2 x_i^2$$
and
$$V = \sum_{i,j,k,\ell=1}^n a_{ijk\ell} x_i x_j x_k x_\ell$$
where a is such that $V(x) \geq 0$ for all $x \in \mathbb{R}^n$, then for any nondegenerate eigenvalue E_0 of H_0, the Rayleigh–Schrödinger series is a strong asymptotic series for the eigenvalue $E(\beta)$ of $H(\beta) = H_0 + \beta V$ with $E(0) = E_0$.

Example 2 $((\varphi^4)_2$, spatially cut-off$)$ In Section X.7, we briefly discussed spatially cut-off field theories in two-dimensional space time.
$$H_0 = \int \sqrt{m^2 + k^2}\, a^\dagger(k)a(k)\, dk = d\Gamma(\sqrt{k^2+m^2})$$
$$V = \int g(x){:}\varphi^4(x){:}\, dx$$

with $g \in L^2 \cap L^1$, $g \geq 0$, and

$$M = \int a^\dagger(k)a(k)\,dk = d\Gamma(1)$$

the number operator. All the conditions of Theorem XII.20 can be verified. The hardest condition to prove is (ii), which we discussed in Section X.9. Therefore, the Rayleigh–Schrödinger series for the ground state energy is a strong asymptotic series for the exact ground state energy. As we have already mentioned, the Rayleigh–Schrödinger series in this case is given by sums of Feynman diagrams. This example suggests that in more general circumstances the Feynman series might be a strong asymptotic series.

The strong asymptotic condition implies that $|a_n| \leq C\sigma^n n!$. There are series associated with simple examples for which a_n behaves like $(kn)!$ with $k > 1$. Thus, a strong asymptotic condition cannot hold in such cases. But a simple scaling argument and Carleman's theorem imply that any function g, analytic in the many-sheeted region $\{z \mid 0 < |z| < B;\ |\arg z| < \tfrac{1}{2}k\pi + \varepsilon\}$ with $|g(z)| \leq C\sigma^N[k(N+1)]!\,|z|^{N+1}$ for all N and z in the region, is identically zero. This suggests that we define:

Definition We say that a function $E(\beta)$, analytic in a sectorial region $\{\beta \mid 0 < |\beta| < B;\ |\arg \beta| < \tfrac{1}{2}k\pi + \varepsilon\}$, obeys a **modified strong asymptotic condition of order k** and has $\sum_{n=0}^{\infty} a_n \beta^n$ as an **order k strong asymptotic series** if there exist C and σ so that

$$\left| E(\beta) - \sum_{n=0}^{N} a_n \beta^n \right| \leq C\sigma^{N+1}[k(N+1)]!\,|\beta|^{N+1}$$

for all N and all β in the sector.

Because of the extension of Carleman's theorem, Theorem XII.19 extends: If f and g both have $\sum_{n=0}^{\infty} a_n \beta^n$ as order k strong asymptotic series, then $f = g$.

Example 3 (x^{2m} oscillator) Let $H_0 = -d^2/dx^2 + x^2$ and $V = x^{2m}$ ($m \geq 3$). There is strong numerical evidence that the Rayleigh–Schrödinger coefficients a_n for the ground state energy behave as $(-1)^{n+1} C\sigma^n[(m-1)n]!\,[1 + O(1/n)]$ for n large, which would imply that an ordinary strong asymptotic condition cannot hold. However, one can show that $E(\beta)$ has a continuation to a multisheeted sector $\{\beta \mid |\arg \beta| \leq \tfrac{1}{2}(m+1)\pi - \varepsilon;\ |\beta| \leq B_\varepsilon\}$ for any $\varepsilon > 0$ and that a strong asymptotic condition of order $m - 1$ holds in such a sector. Thus, the asymptotic series determines the eigenvalues in these cases also.

Since a strong asymptotic series determines $E(\beta)$ uniquely, one might expect there is a way of constructing E from the a_n and the knowledge that E obeys a strong asymptotic condition. The content of the following theorem of Watson, whose proof can be found in the references, is a method of constructing E explicitly:

Theorem XII.21 (Watson's theorem) Suppose that $E(\beta)$ has $\sum_{n=0}^{\infty} a_n \beta^n$ as strong asymptotic series in $\{\beta \,|\, 0 < |\beta| < B;\, |\arg \beta| \leq \tfrac{1}{2}\pi + \varepsilon\}$. Form the function

$$g(z) = \sum_{n=0}^{\infty} \frac{a_n}{n!} z^n$$

which is analytic in some circle about $z = 0$ because of the strong asymptotic condition. Then:

(a) g has an analytic continuation to the region $\{z \,|\, |\arg z| < \varepsilon\}$.
(b) If $|\beta| < B$ and $|\arg \beta| < \varepsilon$, then $\int_0^\infty |g(x\beta)| e^{-x} \, dx < \infty$.
(c) If $|\beta| < B$ and $|\arg \beta| < \varepsilon$, then $E(\beta) = \int_0^\infty g(x\beta) e^{-x} \, dx$.

Notice that since $\int_0^\infty x^n e^{-x} \, dx = n!$, if we could interchange $\int_0^\infty$ and $\sum_{n=0}^{\infty}$, we would have

$$\int_0^\infty g(x\beta) e^{-x} \, dx = \sum_{n=0}^{\infty} a_n \beta^n \quad \text{(formally)}$$

The method we describe here for obtaining a sum for $\sum_{n=0}^{\infty} a_n \beta^n$ is a particular example of a **summability method**, that is, a method of obtaining a finite answer from a divergent series that is *formally* a sum for the series. This method is known as the **Borel summability method**; $g(z)$ is called the **Borel transform** of $\{a_n\}_{n=0}^{\infty}$; and the Laplace transform formula $\int_0^\infty g(x\beta) e^{-x} \, dx$ is sometimes called the inverse Borel transform.

Example 1, revisited We can take ε arbitrarily close to π so g is analytic in the cut plane $\mathbb{C}\backslash(-\infty, -R)$, where R is the radius of convergence of the Borel transform. In the special case of a one-dimensional oscillator, it is known that $E(\beta)$ is analytic in $\{\beta \,|\, |\arg \beta| < \pi\}$, so the inverse Borel transform converges to E for any positive β (and more generally for any β with $\operatorname{Re} \beta > 0$).

Example 2, revisited   In this case, we can recover $E(\beta)$ for $|\beta|$ and $|\arg \beta|$ sufficiently small from the Borel transform for the Rayleigh–Schrödinger series.

Example 3, revisited   The Borel method is not directly applicable, but a generalization works (Problem 29(b)). Explicitly, for the x^{2m} oscillator, the modified Borel transform

$$g(z) = \sum_{n=0}^{\infty} \frac{a_n}{[n(m-1)]!} z^n$$

is analytic in the cut plane $\mathbb{C}\setminus(-\infty, -R_m)$ and for β positive and small,

$$E(\beta) = \int_0^{\infty} g(\beta x^{m-1}) e^{-x} \, dx$$

We have just seen that a particular summability method can be used to recover the eigenvalues. In certain special cases, other methods that are often computationally more convenient have been shown to work; we discuss them briefly in the Notes.

XII.5 Spectral concentration

In the preceding two sections we analyzed perturbed systems that have an isolated eigenvalue near an isolated unperturbed eigenvalue and a perturbation series that is finite order by order but that does not converge. In this section we wish to consider the more complex situation where there is an isolated unperturbed eigenvalue and a perturbation series that is finite term by term but where there is no perturbed eigenvalue!

Consider the following simple example: Let $H_0 = -\Delta - 1/r$ and let $V = E_0 \, \mathbf{e} \cdot \mathbf{r}$ where $\mathbf{e}$ is a fixed unit vector in $\mathbb{R}^3$. For β real, $H_0 + \beta V$ is the Hamiltonian of a hydrogen atom in a constant electric field $-\beta E_0 \, \mathbf{e}$, the so-called Stark effect Hamiltonian. In Section X.5 we proved that $H_0 + \beta V$ is essentially self-adjoint on $C_0^{\infty}(\mathbb{R}^3)$. On the other hand, it is easy to see that if $\beta \neq 0$, $H_0 + \beta V$ is not bounded below. We will show (Section XIII.4) that $\sigma(H_0 + \beta V) = (-\infty, \infty)$ if $\beta \neq 0$. Thus, as soon as the perturbation is turned on, all the eigenvalues of H_0 are drowned in the continuous spectrum. Nevertheless, the perturbation series for the ground state energy, given by the formal expressions in Section 1, is finite term by term. What does it mean?

Before presenting the mathematical resolution of this question, let us consider an alternative interpretation. Let

$$V_n(r) = \begin{cases} E_0 \, \mathbf{e} \cdot \mathbf{r} & \text{if } |\mathbf{e} \cdot \mathbf{r}| < n \\ E_0 n & \text{if } \mathbf{e} \cdot \mathbf{r} \geq n \\ -E_0 n & \text{if } \mathbf{e} \cdot \mathbf{r} \leq -n \end{cases}$$

Each V_n is a perturbation of type (A) so, for some $B^{(n)}$, the perturbation series $\sum a_m^{(n)} \beta^m$ for the ground state energy $E^{(n)}(\beta)$ of $H_0 + \beta V_n$ converges to $E^{(n)}(\beta)$ if $|\beta| \leq B^{(n)}$. Suppose that $\sum_{m=0}^{\infty} a_m \beta^m$ is the formal perturbation series for the ground state energy of $H_0 + \beta V$. We notice that $a_m^{(n)}$ converges to a_m as $n \to \infty$. We thus interpret $\sum_{m=0}^{\infty} a_m \beta^m$ as a formal limit of the convergent series $\sum_{m=0}^{\infty} a_m^{(n)} \beta^m$. This interpretation is not very satisfying mathematically for it is likely that $B^{(n)} \to 0$ and that $\sum_{m=0}^{\infty} a_m \beta^m$ diverges for all β, but it must be emphasized that it is the right physical interpretation! For experiments done on real atoms in real fields are not done with an electric field extending through all of space—the potential V is an idealization. V_n is much closer to the experimentally obtained potential and $E^{(n)}(\beta)$ to the measured spectroscopic observables. But since $E^{(n)}(\beta) \simeq a_0 + a_1^{(n)} \beta$ for β small and $a_1^{(n)} \simeq a_1$ for n large, we can estimate $E^{(n)}(\beta)$ using a_1.

In spite of the above remark on physical relevance, one cannot help feeling that the perturbation series has something to do with $H_0 + \beta V$. What we shall see is that the spectrum of $H_0 + \beta V$ "bunches up" near the unperturbed eigenvalue and that the center of this spectral concentration is given by an asymptotic series $\sum a_m \beta^m$. The exact meaning of this notion of "bunching up" is given by:

Definition Let H_n be a family of self-adjoint operators and let $\{P_n(\Omega)\}$ be the family of spectral projections of H_n. Let $\{S_n\}_{n=1}^{\infty}$ and T be subsets of $\mathbb{R}$. We say **the part of the spectrum of H_n in T is asymptotically in S_n** if and only if

$$\operatorname*{s-lim}_{n \to \infty} P_n(T \setminus S_n) = 0$$

If $H_n \to H$ in strong resolvent sense, we say **the part of the spectrum of H_n in S_n is asymptotically the part of the spectrum of H in T** if

$$\operatorname*{s-lim}_{n \to \infty} P_n(S_n) = P(T)$$

where $\{P(\Omega)\}$ is the projection-valued measure of H.

Example 1 Let $H = I$, $H_n = I + n^{-1}X$ where X is the operator $(Xf)(x) = xf(x)$ on $L^2(-\infty, \infty)$. If $\chi_{(a,b)}$ is multiplication by the characteristic function of (a, b), then $P_n(1 - \alpha, 1 + \beta) = \chi_{(-n\alpha, n\beta)}$. In particular, the part of the spectrum of H_n in $(0, 2)$ is asymptotically in $(1 - n^{-1/2}, 1 + n^{-1/2})$, and the part of the spectrum of H_n in $(1 - n^{-1/2}, 1 + n^{-1/2})$ is asymptotically the part of the spectrum of H in $(0, 2)$. To see the connection between this asymptotic containment and the intuitive notion of "bunching up," consider the spectral measures associated with a fixed vector ψ. If $d\mu_n(\lambda) = d(\psi, P_n(\lambda)\psi)$, then $d\mu_n(\lambda) = d\mu_1(1 + (\lambda - 1)n^{-1})$. Thus $d\mu_n \to \delta(\lambda - 1)$.

In this example, we observe a characteristic symmetry:

Proposition Suppose that $H_n \to H$ in strong resolvent sense. Let $T = (a, b)$ with $a, b \notin \sigma_{pp}(H)$. Let $S_n \subset T$ for n sufficiently large. Then the part of the spectrum of H_n in T is asymptotically in S_n if and only if the part of the spectrum of H_n in S_n is asymptotically the part of the spectrum of H in T.

Proof By Theorem VIII.24, s-lim $P_n(T) = P(T)$. Thus s-lim $P_n(S_n) = P(T)$ if and only if s-lim $P_n(T\backslash S_n) = 0$. ∎

We also emphasize that the notion of asymptotic concentration is a property of a *sequence* of operators $\{H_n\}_{n=1}^\infty$. It makes no sense to say that a single operator H_n has spectrum concentrated in S_n. In Example 1 all the H_n are unitarily equivalent, so $\sigma(H_n) = \sigma(H_m)$.

We can now make a conjecture about the relation of $a_0 + a_1\beta$ to $H_0 + \beta V$ in our Stark effect example. Suppose that we can find an interval I about the unperturbed ground state energy a_0 with $\sigma(H_0) \cap I = \{a_0\}$ and a function $f(\beta)$ with $f(\beta)/\beta \to 0$ so that the part of the spectrum of $H(\beta)$ in $(a_0 + a_1\beta - f(\beta), a_0 + a_1\beta + f(\beta))$ is asymptotically the part of spectrum of H_0 in I. This property determines a_1 uniquely (Problem 31). Thus, while $H(\beta)$ does not have an eigenvalue near $a_0 + \beta a_1$, it does have spectrum concentrated near $a_0 + \beta a_1$ as $\beta \to 0$. Our tool for proving this spectral concentration is the notion of pseudo-eigenvalue:

Definition Suppose that $H(\beta)$ is a family of self-adjoint operators defined for β real and small so that $H(\beta) \to H_0$ as $\beta \to 0$ in strong resolvent sense. Let E_0 be an isolated nondegenerate eigenvalue of H_0 and ψ_0 be the corresponding normalized eigenvector. A family of vectors $\psi(\beta)$, β real, and numbers

$E_0 + \beta E_1$ are called a **first order pseudo-eigenvector** and a **first order pseudo-eigenvalue** respectively if and only if:

(i) $\lim_{\beta \to 0} \|\psi(\beta) - \psi_0\| = 0$;
(ii) $\lim_{\beta \to 0} \beta^{-1} \|(H(\beta) - E_0 - \beta E_1)\psi(\beta)\| = 0$.

The statement "$E_0 + \beta E_1$ is a first order pseudo-eigenvalue of $H(\beta)$" will be used as a shorthand for the statement that there exists a first order pseudo-eigenvector $\psi(\beta)$ with $E_0 + \beta E_1$ the associated pseudo-eigenvalue.

Theorem XII.22 Suppose that $H(\beta) \to H_0$ in strong resolvent sense as $\beta \to 0$ with all the $H(\beta)$ self-adjoint. Let E_0 be an isolated nondegenerate eigenvalue of H_0 and I an interval so that $\bar{I} \cap \sigma(H_0) = \{E_0\}$. Then there exists a function $f(\beta)$, obeying $f(\beta)/\beta \to 0$, so that the part of the spectrum of $H(\beta)$ in I is asymptotically in

$$I_\beta \equiv (E_0 + \beta E_1 - f(\beta), E_0 + \beta E_1 + f(\beta))$$

if and only if $E_0 + \beta E_1$ is a first order pseudo-eigenvalue for $H(\beta)$.

Proof Let $P^\beta(\Omega)$ be the spectral projections of $H(\beta)$. If the part of $\sigma(H(\beta))$ in I is asymptotically in I_β, then $P^\beta(I_\beta)$ converges strongly to $P^0(I) = P^0(\{E_0\})$. In particular, if ψ_0 is the unperturbed eigenvector for H and $\psi(\beta) = P^\beta(I_\beta)\psi_0$, then $\psi(\beta) \to \psi_0$ in norm. Moreover,

$$\beta^{-1}\|[H(\beta) - E_0 - E_1 \beta]\psi(\beta)\| \leq \beta^{-1} f(\beta) \|\psi(\beta)\| \to 0$$

as $\beta \to 0$, so $E_0 + E_1 \beta$ is a first order pseudo-eigenvalue.

Conversely, suppose that $E_0 + \beta E_1$ is a pseudo-eigenvalue with $\psi(\beta)$ as pseudo-eigenvector. Let

$$f(\beta) = \|[H(\beta) - E_0 - E_1 \beta]\psi(\beta)\|^{1/2} \beta^{1/2}$$

Then $\beta^{-1} f(\beta) \to 0$ as $\beta \to 0$. We shall show that $P^\beta(I_\beta)$ converges to $P^0(\{E_0\})$ strongly.

$$\beta^{-2} f(\beta)^4 = \|[H(\beta) - E_0 - E_1 \beta]\psi(\beta)\|^2$$

$$= \int_{-\infty}^{\infty} (\lambda - E_0 - E_1 \beta)^2 \, d(\psi(\beta), P^\beta(\lambda)\psi(\beta))$$

$$\geq f(\beta)^2 \int_{\lambda \notin I_\beta} d(\psi(\beta), P^\beta(\lambda)\psi(\beta))$$

$$= f(\beta)^2 \|(I - P^\beta(I_\beta))\psi(\beta)\|^2$$

Thus $\|(I - P^\beta(I_\beta))\psi(\beta)\| \leq \beta^{-1} f(\beta) \to 0$. Since
$$\|(I - P^\beta(I_\beta))\psi_0\| \leq \|(I - P^\beta(I_\beta))\psi(\beta)\| + \|\psi(\beta) - \psi_0\|$$
we conclude that
$$(I - P^\beta(I_\beta))P^0(\{E_0\}) \to 0$$
in norm. Therefore,
$$P^\beta(I \setminus I_\beta) P^0(\{E_0\}) = P^\beta(I \setminus I_\beta)(1 - P^\beta(I_\beta)) P^0(\{E_0\}) \to 0$$
In addition, the strong resolvent convergence implies that $P^\beta(I) \to P^0(\{E_0\})$, and thus that
$$P^\beta(I \setminus I_\beta)(1 - P^0(\{E_0\})) \xrightarrow{s} 0$$
We conclude that $P^\beta(I \setminus I_\beta) \xrightarrow{s} 0$. ∎

Theorem XII.23 Suppose that self-adjoint operators $H(\beta)$ are given for each β in some neighborhood N of $0 \in \mathbb{R}$. Let $H_0 \equiv H(0)$ and suppose that there is a symmetric operator V such that:

(i) H_0 is essentially self-adjoint on $D(H_0) \cap D(V)$.
(ii) If $\varphi \in D(H_0) \cap D(V)$, then for any $\beta \in N$, we have that $\varphi \in D(H(\beta))$ and $H(\beta)\varphi = H_0 \varphi + \beta V \varphi$.
(iii) E_0 is an isolated eigenvalue of H_0 of multiplicity 1 and the associated eigenvector ψ_0 is in $D(V)$.

Let I be an open interval with $\sigma(H_0) \cap \bar{I} = \{E_0\}$ and let $E_1 = (\psi_0, V\psi_0)$ be the first-order Rayleigh–Schrödinger coefficient for the perturbation of the eigenvalue E_0.

Then, there is a function $f(\beta)$ with $\beta^{-1} f(\beta) \to 0$ so that the part of the spectrum of $H_0 + \beta V$ in I is asymptotically in $(E_0 + E_1 \beta - f(\beta), E_0 + E_1 \beta + f(\beta))$.

Proof A simple argument using (i), (ii) and the first resolvent equation proves that $H(\beta) \to H_0$ in strong resolvent sense as $\beta \to 0$ (Problem 32). By Theorem XII.22, it is sufficient to construct a first order pseudo-eigenvector $\psi(\beta)$ with pseudo-eigenvalue $E_0 + E_1 \beta$. Let $\{P^\beta(\Omega)\}$ be the spectral projections for $H(\beta)$. Choose $\psi(\beta) = P^\beta(I)\psi_0$. By Theorem VIII.24, s-lim $P^\beta(I) =$

$P^0(I)$, so $\psi(\beta) \to \psi_0$ as $\beta \to 0$. Since $\psi_0 \in D(H(\beta))$ for any $\beta \in N$, $P^\beta(I)H(\beta)\psi_0 = H(\beta)P^\beta(I)\psi_0$. Thus

$$\begin{aligned}\|\beta^{-1}(H(\beta) - E_0 - \beta E_1)\psi(\beta)\| &= \|\beta^{-1}P^\beta(I)(H_0 + \beta V - E_0 - E_1\beta)\psi_0\| \\ &= \|P^\beta(I)(V - E_1)\psi_0\| \\ &\leq \|[P^\beta(I) - P^{(0)}(I)](V - E_1)\psi_0\| \\ &\quad + \|P^{(0)}(I)(V - E_1)\psi_0\|\end{aligned}$$

Since s-lim $P^\beta(I) = P^{(0)}(I)$, the first term in the final expression converges to 0. Since E_0 is nondegenerate, $P^{(0)}(I)\varphi = (\psi_0, \varphi)\psi_0$ and thus

$$P^0(I)(V\psi_0) = E_1\psi_0$$

Therefore the second term in the final expression is 0. This proves that $\psi(\beta)$ is a first-order pseudo-eigenvector. ∎

Example 2 (Stark effect for the hydrogen atom)   Let $H_0 = -\Delta - 1/r$ and let $V = E_0 \mathbf{e} \cdot \mathbf{r}$. Then $E_0 = -\frac{1}{4}$ is the ground state energy of H_0. By Theorem X.38, $H_0 + \beta V$ is essentially self-adjoint on $D(H_0) \cap D(V)$. Moreover, ψ_0 is known explicitly and $|\psi_0(x)| \leq C \exp(-|x|/2)$ for suitable C. In particular, $\psi_0 \in D(V)$. Thus, the hypotheses of Theorem XII.23 are satisfied. Since $E_1 \equiv (\psi_0, V\psi_0) = 0$, we conclude that, as $\beta \to 0$, the part of the spectrum of $H_0 + \beta V$ near E_0 is concentrated in $(E_0 - \lambda\beta, E_0 + \lambda\beta)$ for any $\lambda > 0$. In addition one can prove that all the coefficients of the Rayleigh–Schrödinger series are finite. Moreover, for any n, one can find $f^{(n)}(\beta)$ with $|\beta|^{-n}f^{(n)}(\beta) \to 0$ so that the part of the spectrum of $H_0 + \beta V$ near E_0 is concentrated in $(\sum_{m=0}^n E_m\beta^m - f^{(n)}(\beta), \sum_{m=0}^n E_m\beta^m + f^{(n)}(\beta))$ as $|\beta| \to 0$.

In addition to the generalization to order n mentioned in the preceding example, Theorem XII.23 has another important generalization: Discrete levels of finite but nontrivial degeneracy can be treated. For instance, in the Stark effect example above, the first eigenvalue above the ground state is fourfold degenerate and splits into one two-dimensional pseudo-eigenvalue and two one-dimensional pseudo-eigenvalues to first order. Explicitly, one can find four linearly independent first-order pseudo-eigenvectors converging to vectors ψ_i satisfying $H_0\psi_i = -\frac{1}{16}\psi_i$ and with pseudo-eigenvalues, $-\frac{1}{16}$ twice and $-\frac{1}{16} \pm a\beta$, where a can be computed using "degenerate perturbation theory."

XII.6 Resonances and the Fermi golden rule

> *Whenever calculus is brought in, or higher algebra, you could take it as a warning signal that the operator was trying to substitute theory for experience.*
>
> B. Graham, *The Intelligent Investor*

Thus far we have been considering successively more singular perturbations of isolated eigenvalues. In Section 2 we considered the case where a perturbed eigenvalue and a convergent perturbation series exist, and in Sections 3 and 4 the case where a perturbed eigenvalue exists but the associated perturbation series is divergent. In the examples of Section 5, there is not even a perturbed eigenvalue. In this section we consider a still more singular, although, as we shall see, physically relevant, case where it is not only true that there is no perturbed eigenvalue but where the unperturbed eigenvalue is not isolated.

A simple example of this phenomenon occurs if

$$H_0 = -\Delta_1 - \frac{2}{r_1} - \Delta_2 - \frac{2}{r_2}$$

and $V = |\mathbf{r}_1 - \mathbf{r}_2|^{-1}$ on $L^2(\mathbb{R}^6)$. We will write points of $\mathbb{R}^6$ as $\langle \mathbf{r}_1, \mathbf{r}_2 \rangle$ with $\mathbf{r}_i \in \mathbb{R}^3$. $H_0 + V$ is the Hamiltonian of the helium atom if we ignore relativistic corrections and go to the limit of infinite nuclear mass. If $h = -\Delta - 2/r$ on $L^2(\mathbb{R}^3)$, then $H_0 = 1 \otimes h + h \otimes 1$. Since h and H_0 are self-adjoint, we can use Theorem VIII.33 to compute the spectrum of H_0 in terms of the spectrum of h under the natural identification of $L^2(\mathbb{R}^6)$ with $L^2(\mathbb{R}^3) \otimes L^2(\mathbb{R}^3)$. The eigenvalues of h are explicitly known to be $\{-1/n^2\}_{n=1}^{\infty}$, so H_0 has eigenvalues

$$\{E_{n,m}\} \equiv \left\{ -\left(\frac{1}{n^2} + \frac{1}{m^2} \right) \right\}_{n,m=1}^{\infty}$$

Moreover, $\sigma_{\text{ess}}(h) = [0, \infty)$, so $\sigma_{\text{ess}}(H_0) = [-1, \infty)$. Thus

$$\sigma(H_0) = \{x + y \mid x, y \in [0, \infty)\} \cup \{-1/n^2\}_{n=1}^{\infty}$$

$$= \left\{ -1 - \frac{1}{n^2} \right\}_{n=1}^{\infty} \cup [-1, \infty)$$

Therefore, the eigenvalues $E_{n,m}$ are in the continuous spectrum if $n \geq 2$ and $m \geq 2$. See Figure XII.2. When the perturbation V is turned on, one expects from physical considerations that the eigenvalues in the continuous spectrum will "dissolve." This expectation is correct except for two qualifications

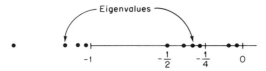

FIGURE XII.2 The spectrum of H_0.

discussed in the Notes. However, one observes a "memory" of these eigenvalues of H_0 in the physics of the helium atom. In the scattering of electrons off helium ions, one observes bumps in the scattering cross section for total $He^+ + e$ energies near the energies $E_{n,m}$ (in practice, only the bumps for small m and n can be distinguished from "background"). See Figure XII.3. Similar bumps are found in the absorption of light by helium, i.e., at frequencies of incident light for which the energy of a light quantum is

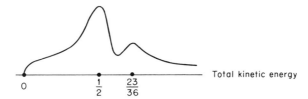

FIGURE XII.3 Schematic elastic scattering cross section for $e + He^+ \to e + He^+$.

near the difference of $E_{n,m}$ and $E_{1,1}$ (= ground state energy) light is absorbed strongly (Figure XII.4). We emphasize that our diagrams for scattering and absorption are schematic. Specifically, the energies $E_{n,m}$ are shifted due to the perturbation V.

Not only are these bumps, called **Auger** or **autoionizing** states, observed, but their widths are fairly well described by a calculational method known as Fermi's golden rule which has a purely heuristic derivation which we describe in the Notes. The derivation is based on a physical model for the mechanism that produces the bumps. In the remainder of this section we first isolate the mathematical quantity corresponding to the widths of the bumps and then show how the Fermi golden rule is a consequence of the

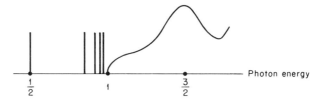

FIGURE XII.4 Schematic absorption cross section for $He + \gamma$.

XII.6 Resonances and the Fermi golden rule

Kato–Rellich theory of regular perturbations, at least in this helium atom situation. We emphasize that the Fermi golden rule has been applied successfully to many physical situations where it has not yet been rigorously justified.

Our interpretation of the Fermi golden rule will be in three steps, two of them "quasi-mathematical" and the last rigorous mathematics. The goal of the first two steps will be to isolate a quantity susceptible to exact mathematical analysis. While this quantity is not precisely the width of a bump, it will be related to the width by a set of nonrigorous arguments. This situation is not an uncommon one in mathematical physics: The mathematically precise quantities are not equal to experimental quantities but are related to them by some nonrigorous argument. There is thus an important element of extramathematical taste in certain problems of mathematical physics.

Our first step is to relate the width of bumps to the imaginary part of the position of poles of "the scattering amplitude." The second step is to look at poles in the analytic continuation of the resolvent. Thus the quasi-mathematical part of the analysis will be to replace the nonprecise "width of the bump" with the precise "imaginary part of the position of a pole on the second sheet of the analytic continuation of the expectation value of the resolvent on a dense set of states." Our study of this precise quantity will be accomplished by proving it equal to the imaginary part of an eigenvalue of a certain non-self-adjoint operator. The main technical device in this proof will be scaling techniques which we study further in Section XIII.10.

Scattering is described in quantum-mechanical situations as the square of the modulus of the "scattering amplitude" (see Section XI.6). The scattering amplitude, which is a complex-valued function of the energy and scattering angle, is typically an analytic function of energy (see Section XI.7) in a cut plane $\mathbb{C}\backslash\sigma(H)$, where H is the Hamiltonian of the interacting quantum system. Suppose that the scattering amplitude $f(E)$ has an analytic continuation onto a second sheet (Figure XII.5) and that there is a simple pole on the second sheet at a position $E_r - i\Gamma/2$ very near the real axis. Thus,

$$f(E) = \frac{C}{E - E_r + \tfrac{1}{2}i\Gamma} + f_b(E)$$

where the background f_b is analytic at $E = E_r - \tfrac{1}{2}i\Gamma$. If the pole is very near the real axis and if $f_b(E_r)$ is not too large, then

$$|f(E)|^2 = \frac{|C|^2}{(E - E_r)^2 + \tfrac{1}{4}\Gamma^2} + R$$

where the remainder R will be small near $E = E_r$. In Figure XII.6, we plot the **Breit–Wigner** resonance shape: $|C|^2((E - E_r)^2 + \tfrac{1}{4}\Gamma^2)^{-1}$. Notice that its

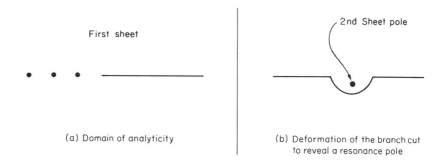

(a) Domain of analyticity

(b) Deformation of the branch cut to reveal a resonance pole

FIGURE XII.5 Continuation of the scattering amplitude.

width at half maximum is Γ. The pole in $f(E)$ is called a resonance pole and Γ is called the **width** of the resonance. If the pole term is much larger than the background term for $E = E_r$, Γ will approximate the width of a "bump" in $|f(E)|^2$. We have thus accomplished half of our quasi-mathematical task of "isolating" the width of a bump.

The second half of the task requires us to appeal to the fact that the scattering amplitude is related to boundary values of the resolvent $(H - E)^{-1}$ as E approaches the real axis from the upper half-plane (see Sections XI.6 and XI.7). For this reason, we seek a method of continuing $(\psi, (H - E)^{-1}\psi) = R_\psi(E)$ to the second sheet. If we find that such a continuation is possible for a dense set of $\psi \in \mathscr{H}$ and that for this dense set, the function $R_\psi(E)$ has a pole at $E = E_r - \tfrac{1}{2}i\Gamma$, we will associate E with a resonance pole, and Γ with a resonance width. Of course, we need a reason to think the pole is to be associated with H rather than the particular dense set of ψ, and so we shall consider only ψ for which $(\psi, (H_0 - z)^{-1}\psi)$ also has a continuation to the second sheet but without a pole at $E_r - \tfrac{1}{2}i\Gamma$.

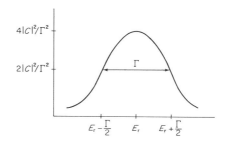

FIGURE XII.6 A Breit–Wigner resonance shape.

XII.6 Resonances and the Fermi golden rule

We have thus completed the task of finding a mathematical quantity to analyze:

Definition Suppose that there is a dense set of vectors $D \subset \mathcal{H}$ such that for all $\psi \in D$, both $(\psi, (H - z)^{-1}\psi) \equiv R_\psi(z)$ and $(\psi, (H_0 - z)^{-1}\psi) \equiv R_\psi^{(0)}(z)$ have analytic continuations onto the second sheet (across the real axis from the upper half-plane of the first sheet). If $R_\psi^{(0)}(z)$ is analytic at $z_0 = E_r - \frac{1}{2}i\Gamma$ and $R_\psi(z)$ has a pole at z_0 for some ψ, we say that z_0 is a **resonance pole**. Γ is called the **width** of the resonance.

We like to think of Γ as the width of a bump, but this step is not precise: First, the formula relating the scattering amplitude to the resolvent does not always involve expectation values of vectors $\psi \in D$; secondly, the "background" may not be negligible.

To study these resonance poles, we need the technical device of dilations. Here we just sketch the ideas involved and apply them to the operator $-\Delta - 2/r$ and the helium Hamiltonian. In Section XIII.10 we present a detailed development of dilation methods and apply them to a general class of operators. For θ real, we define $u(\theta)$ on $L^2(\mathbb{R}^3)$ by

$$(u(\theta)f)(r) = e^{3\theta/2}f(e^\theta r)$$

$u(\theta)$ is a one-parameter unitary group on $L^2(\mathbb{R}^3)$. It is easy to see that the $u(\theta)$ leave $D(h) = D(-\Delta)$ invariant and that

$$h(\theta) \equiv u(\theta)hu(\theta)^{-1} = -e^{-2\theta}\Delta - \frac{2e^{-\theta}}{r}$$

The first crucial fact, which follows from Theorems X.12 and XII.9, is that $h(\theta)$ has a continuation to a family of operators analytic in the sense of Kato in the whole plane.

Let us find the spectrum of $h(\theta)$. The spectrum of $-\Delta$ is $[0, \infty)$. By Theorem XIII.14 and the fact that $r^{-1}(-\Delta + 1)^{-1}$ is Hilbert–Schmidt, $-\Delta - 2e^\theta/r$ will have spectrum $[0, \infty)$ plus a possible set of isolated points outside of $[0, \infty)$ even when θ is nonreal. Moreover, each of the points outside $[0, \infty)$ is an eigenvalue of finite multiplicity. Thus $h(\theta)$ has spectrum $\{e^{-2\theta}\lambda \mid \lambda \in [0, \infty)\}$ plus some possible additional discrete spectrum. The points of the discrete spectrum will be given by analytic functions $f(\theta)$ of θ (with possible algebraic singularities) according to Theorems XII.8 and XII.13. But

$$u(\theta_0)h(\theta_1 + i\theta_2)u(\theta_0)^{-1} = h(\theta_1 + \theta_0 + i\theta_2)$$

for $\theta_1, \theta_2, \theta_0$ real. Thus $h(\theta_1 + i\theta_2)$ is unitarily equivalent to $h(\theta'_1 + i\theta_2)$. As a consequence, they have the same eigenvalues. Therefore, the functions $f(\theta)$

56   XII: PERTURBATION OF POINT SPECTRA

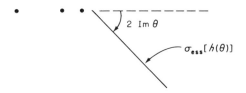

FIGURE XII.7 $\sigma(h(\theta))$, $0 \leq \text{Im } \theta \leq \pi/2$.

are constant if Im θ is constant. Since the $f(\theta)$ are analytic, they are constant for all θ. For $\theta = 0$, $\sigma_{pp}(h) = \{-1/n^2\}_{n=1}^{\infty}$ so

$$\sigma(h(\theta)) = \{e^{-2\theta}\lambda \,|\, \lambda \in [0, \infty)\} \cup \{-1/n^2\}_{n=1}^{\infty}$$

(Figure XII.7). In the above we have not proven that $h(\theta)$ could not have eigenvalues in the sector $0 > \arg \lambda > -2 \text{ Im } \theta$, but a detailed analysis using the explicit solutions of the Coulombic–Schrödinger equation shows that they do not occur.

Next we consider scaling for H_0 and $H_0 + \beta V$. Define $U(\theta)$ on $\mathbb{R}^6$ by $(U(\theta)F)(\mathbf{r}_1, \mathbf{r}_2) = e^{3\theta} F(e^\theta \mathbf{r}_1, e^\theta \mathbf{r}_2)$, for θ real. Also for θ real, let $V(\theta) \equiv U(\theta)VU(\theta)^{-1} = e^{-\theta}V$ and $H_0(\theta) \equiv U(\theta)H_0 U(\theta)^{-1} = h(\theta) \otimes 1 + 1 \otimes h(\theta)$. Both $H_0(\theta)$ and $V(\theta)$ have type (A) analytic continuations to the whole complex plane. For any θ, $H_0(\theta)$ is of the form $A \otimes 1 + 1 \otimes B$, although A and B are not self-adjoint. We shall prove that

$$\sigma(A \otimes 1 + 1 \otimes B) = \{x + y \,|\, x \in \sigma(A), y \in \sigma(B)\}$$

in Section XIII.9. Thus

$$\sigma(H_0(\theta)) = \{x + y \,|\, x, y \in \sigma(h_0(\theta))\} = \bigcup_{n=1}^{\infty} \left\{ -\frac{1}{n^2} + \lambda e^{-2\theta} \,\middle|\, \lambda \geq 0 \right\}$$

$$\cup \left\{ -\frac{1}{n^2} - \frac{1}{m^2} \,\middle|\, n, m \text{ integral} \right\} \cup \{\lambda e^{-2\theta} \,|\, \lambda \geq 0\}$$

(see Figure XII.8). In particular, the eigenvalues embedded in $\sigma(H_0)$ are isolated discrete eigenvalues of $H_0(\theta)$ for θ nonreal. The eigenvalues $E_{n,m}$

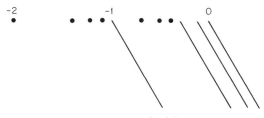

FIGURE XII.8 $\sigma(H_0(\theta))$.

($n, m \geq 2$) are all degenerate. It is possible to use symmetries to decompose $L^2(\mathbb{R}^6)$ into an infinite direct sum of spaces $\mathcal{H}_k$, each left invariant by all of the operators $H_0(\theta)$, $V(\theta)$. For each n and m, $E_{n,m}$ is an eigenvalue of only finitely many of the operators $H_0(\theta) \restriction \mathcal{H}_k$. For some of the $\mathcal{H}_k$, $E_{n,m}$ is nondegenerate. Those $\mathcal{H}_k$ for which $E_{n,m}$ is an eigenvalue of $H_0(\theta) \restriction \mathcal{H}_k$ and those for which it is nondegenerate depend on n and m. Below, we assume that we have restricted to an $\mathcal{H}_k$ on which $E_{n,m}$ is nondegenerate without introducing new notation. We discuss some of the details of the degenerate theory and the reduction by symmetries in the Notes.

Let $H(\theta; \beta) = H_0(\theta) + \beta V(\theta)$. By the theory of regular perturbations described in Section 2, if θ is fixed and $|\beta|$ is sufficiently small, then $H(\theta; \beta)$ has an eigenvalue near $E_{n,m}$ given by a convergent power series in β. See Figure XII.9a,b. Now, let us fix β and vary θ. Since $H(\theta; \beta)$ is an analytic family in θ for β fixed we have an eigenvalue $E^{(\theta)}(\beta)$ *as long as the eigenvalue stays isolated.* $E^{(\theta)}(\beta)$ is analytic in θ for β fixed. Since $H(\theta; \beta)$ and $H(\theta'; \beta)$ are unitarily equivalent if $\theta - \theta'$ is real, $E^{(\theta)}(\beta)$ is constant in θ. We can now prove that Im $E_{2,2}^{(\theta_0)}(\beta_0) \leq 0$ if β_0 is real and Im $\theta_0 > 0$. For suppose that Im $E_{2,2}^{(\theta_0)}(\beta_0) > 0$. Then the eigenvalue stays isolated if we continue θ to 0 keeping Im $\theta \geq 0$. This would imply $H(0; \beta_0)$ has a complex eigenvalue, which is impossible. We thus conclude that Im $E_{2,2}^{(\theta_0)}(\beta) \leq 0$. If Im $E_{2,2}^{(\theta_0)}(\beta) < 0$, and we try to continue θ to zero, the essential spectrum swings into the eigenvalue, invalidating discrete perturbation theory. There is thus no contradiction between the self-adjointness of $H_0 + \beta_0 V$ for β_0 real and the occurrence of complex eigenvalues, $E_{2,2}^{(\theta_0)}(\beta_0)$ in the lower half-plane when β_0 is real and Im $\theta_0 > 0$.

The second crucial remark is that these complex eigenvalues of $H(\theta; \beta)$ are related to second sheet poles of the resolvent of $H(\beta) = H_0 + \beta V$. Let $\psi \in \mathcal{H}$ be a vector so that $U(\theta)\psi$ has an analytic continuation to all of $\mathbb{C}$. Take ψ to be an entire vector for the infinitesimal generator of the group $U(\theta)$. Consider the function $R_\psi(E) = (\psi, (H_0 + \beta V - E)^{-1}\psi)$ originally defined in the cut plane $\mathbb{C}\setminus\sigma(H_0 + \beta V)$. For θ real,

$$R_\psi(E) = (U(\theta)\psi, U(\theta)(H_0 + \beta V - E)^{-1}U(\theta)^{-1}U(\theta)\psi)$$
$$= (U(\bar\theta)\psi, (H(\theta; \beta) - E)^{-1}U(\theta)\psi)$$

(a) $\sigma[H(\theta; 0)]$ $E_{2,2}(\beta = 0)$, Eigenvalue

(b) $\sigma[H(\theta; \beta)]$ $E_{2,2}(\beta > 0)$, Resonance of $H(\theta; \beta)$

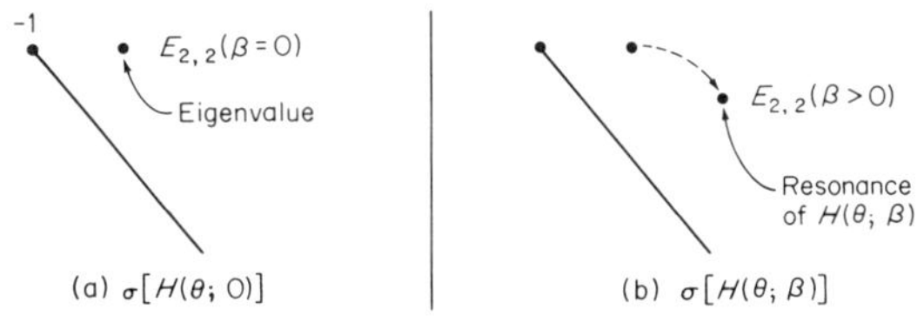

FIGURE XII.9 Trajectory of $E(\beta, \theta)$ for θ fixed; Im $\theta > 0$.

58 XII: PERTURBATION OF POINT SPECTRA

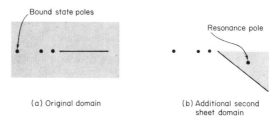

FIGURE XII.10   Domain of analyticity for $R_\psi(E)$.

Fix E in the upper half-plane. The last formula then holds by analytic continuation if $0 \leq \operatorname{Im} \theta < \tfrac{1}{2}\pi$. Thus we can continue $R_\psi(E)$ across the real E axis onto parts of the second sheet (Figure XII.10). Since $(H(\theta; \beta) - E)^{-1}$ has poles when E is an eigenvalue of $H(\theta; \beta)$, the analytic continuation of $R_\psi(E)$ will have poles at the complex eigenvalues of $H(\theta; \beta)$. In summary: *The complex eigenvalues of $H(\theta; \beta)$ are the positions of second sheet poles of the resolvent.*

The width of these resonance poles is, by definition, the imaginary part of $E^{(\theta)}(\beta)$. $E^{(\theta)}(\beta)$ is in turn given for θ fixed and β small by a convergent perturbation series $\sum_{n=0}^{\infty} a_n \beta^n$. By our basic constancy-in-θ argument, the a_n are independent of θ. Clearly a_0 is real and since $\operatorname{Im} E \leq 0$ for all real β, $\operatorname{Im} a_1 = 0$. a_2 is thus the first a_n for which it is possible that $\operatorname{Im} a_n \neq 0$. If the perturbation series is rapidly convergent, then $|\operatorname{Im} E(\beta)| = \tfrac{1}{2}\Gamma \approx (\operatorname{Im} a_2)\beta^2$. Thus

$$\Gamma \approx (\beta^2)(2 \operatorname{Im} a_2)$$

a_2 is a Rayleigh–Schrödinger coefficient and can be computed by methods we discussed in Section 1. Explicitly, if $\Omega(\theta; 0)$ is the unperturbed eigenvector of $H(\theta; 0) \equiv H_0(\theta)$ and $\overline{\Omega(\theta; 0)}$ is its complex conjugate as an element of $L^2(\mathbb{R}^6)$, then (Problem 33)

$$a_2 = (2\pi i)^{-1} \oint_{|E - E_0| = \varepsilon} (\overline{\Omega(\theta; 0)}, V[H_0(\theta) - E]^{-1} V \Omega(\theta; 0)) \frac{dE}{E - E_0}$$

where E_0 is the unperturbed eigenvalue. Now consider the function

$$f(\theta, E) = (\overline{\Omega(\theta, 0)}, V(\theta)(H_0(\theta) - E)^{-1} V(\theta) \Omega(\theta; 0))$$
$$- |(\overline{\Omega(\theta; 0)}, V(\theta) \Omega(\theta; 0))|^2 (E_0 - E)^{-1}$$

$f(\theta, E)$ is analytic at $E = E_0$ for θ fixed with $\operatorname{Im} \theta > 0$, for we have subtracted the pole term explicitly. Thus by the Cauchy integral theorem,

$$a_2 = f(\theta, E)\bigg|_{E = E_0} = \lim_{\varepsilon \downarrow 0} f(\theta, E_0 + i\varepsilon) = \lim_{\varepsilon \downarrow 0} f(\theta = 0, E_0 + i\varepsilon)$$

In the last step we have used the fact that for Im $E > 0$, $f(\theta, E)$ is defined for Im $\theta = 0$ and is independent of θ in the strip $0 \leq \text{Im } \theta < \frac{1}{2}\pi$. Thus

$$\text{Im } a_2 = \lim_{\varepsilon \downarrow 0} \frac{1}{2i} \Big[(\Omega_0, V\{(H_0 - E_0 - i\varepsilon)^{-1} - (H_0 - E_0 + i\varepsilon)^{-1}\} V\Omega_0)$$

$$- |(\Omega_0, V\Omega_0)|^2 \{(E - E_0 - i\varepsilon)^{-1} - (E - E_0 + i\varepsilon)^{-1}\} \Big]_{E = E_0}$$

By the Stone formula (Theorem VII.13), Im a_2 is a function of the spectral projections for H_0 except that the subtraction of the pole subtracts the projection onto the eigenvector Ω_0. We can now prove:

Theorem XII.24 (the Fermi golden rule for Auger states in helium) Let n, m be given with $n > 1$, $m > 1$, and restrict to a symmetry subspace on which $E_{n,m}$ is a nondegenerate eigenvalue. Then, for β real and small, the resolvent of

$$-\Delta_1 - \Delta_2 - \frac{2}{r_1} - \frac{2}{r_2} + \frac{\beta}{|r_1 - r_2|}$$

has a second sheet pole near $E_0 \equiv a_0 = -n^{-2} - m^{-2}$. The position of the pole is given by a function $E(\beta)$, which is analytic for β small,

$$E(\beta) = a_0 + a_1\beta + a_2\beta^2 + \cdots$$

Let $\{P_\Omega\}$ be the spectral family for $H_0 = -\Delta_1 - \Delta_2 - 2/r_1 - 2/r_2$ and define $\tilde{P}(E) = P_{(-\infty, E)\setminus\{E_0\}}$. Let Ω_0 be the unperturbed eigenvector of H_0. Then:

(a) $g(E) = (\Omega_0, V\tilde{P}(E)V\Omega_0)$ is analytic at $E = E_0$.

(b) $\text{Im } a_2 = \pi \dfrac{dg}{dE}\bigg|_{E = E_0}$

Proof We need only prove (a) and (b) because of our previous discussion. By Stone's formula, to prove (a) and (b) we need only show that

$$(\Omega_0, V\{(H_0 - E)^{-1} - (E_0 - E)^{-1} P_{\{E_0\}}\} V\Omega_0)$$

has an analytic continuation from the region Im $E > 0$ up to and below the real axis near $E = E_0$. For then $g(E)$ is the integral of an analytic function and its derivative is just (up to a factor of π) the expression we already have for Im a_2. To see that the required function is analytic at $E = E_0$, we need only show that $U(\theta)V\Omega_0$ has an analytic continuation to a strip $|\text{Im } \theta| < \varepsilon$. But

$$U(\theta)V\Omega_0 = [U(\theta)VU(\theta)^{-1}]U(\theta)\Omega_0 = (e^{-\theta}V)[U(\theta)\Omega_0]$$

That $U(\theta)\Omega_0$ has an analytic continuation is an argument using the Schwarz reflection principle (Problem 35). ∎

Thus we have shown that in lowest order perturbation theory,

$$\Gamma = 2 \text{ Im } a_2 = 2\pi \frac{d}{dE} (\Omega_0, V\tilde{P}(E)V\Omega_0)\bigg|_{E=E_0}$$

This formula is the Fermi golden rule, written in a slightly different form than is usual in the physics literature.

NOTES

For a wealth of information about the perturbation theory of discrete spectra, see the classic: T. Kato, *Perturbation Theory for Linear Operators*, Springer-Verlag, Berlin and New York, 1966 (2nd ed. 1976).

Section XII.1   For additional discussion of the material in this section, see Kato, Chapters I and II, and F. Rellich, *Perturbation Theory of Eigenvalue Problems*, Gordon and Breach, New York, 1969. Theorems XII.1 and XII.2 are discussed, and in particular Theorem XII.2 is proven, in K. Knopp *Theory of Functions*, Part II, Dover, New York, 1947.

Rellich's theorem was first proven in F. Rellich, "Störungstheorie der Spektralzerlegung, I," *Math. Ann.* **113** (1937), 600–619. To appreciate its depth, we note that it is false for analytic perturbations depending on two parameters, as is seen by the following example (a modification of one given by Rellich): Let $T(\beta, \lambda) = \beta\begin{bmatrix}1 & 0\\0 & -1\end{bmatrix} + \lambda\begin{bmatrix}0 & 1\\1 & 0\end{bmatrix}$. $T(\beta, \lambda)$ is analytic in β and λ and self-adjoint for β and λ real, but its eigenvalues $E(\beta, \lambda) = \pm\sqrt{\beta^2 + \lambda^2}$ are not analytic in the two variables β, λ. Rellich's theorem has been extended to various situations involving normal matrices by S. L. Jamison, "Perturbation of normal operators," *Proc. Amer. Math. Soc.* **5** (1954), 103–110, and F. Wolf, "Analytic perturbation of operators in Banach spaces," *Math. Ann.* **124** (1954), 317–333. In particular, Wolf has shown that if $A(\lambda)$ is an analytic family and $A(\lambda_n)$ is normal for a sequence $\lambda_n \to 0$, then the eigenvalues of $A(\lambda)$ are all analytic at $\lambda = 0$. Considerable light has been shed on this theorem by J. Butler, "Perturbation series for eigenvalues of analytic non-symmetric operators," *Arch. Math.* **10** (1959), 21–27; see Problem 21.

For 2×2 matrices, the connection between Jordan anomalous behavior and singularities of the eigenvalues is simple: If $T(\beta) = T_0 + T_1\beta + \cdots + T_n\beta^n + \cdots$ is an analytic 2×2 matrix-valued function with nonanalytic eigenvalues at $\beta = 0$, then for some n, $T_0, \ldots, T_{n-1}$ are multiples of $\mathbb{1}$ and T_n is nondiagonalizable. The general connection is much more complicated; it is implicit in Section II.2.3 of Kato's book; see also Problem 23.

The Rayleigh–Schrödinger series is named for fundamental work of Lord Rayleigh, *The Theory of Sound*, Vol. I, pp. 115–118, MacMillan, London, 1894, and of E. Schrödinger, "Quantisierung als Eigenwertproblem, IV. Störungstheorie mit Anwendung auf den Starkeffekt der Balmerlinien," *Ann. Physik* **80** (1926), 437–490. The projection technique that we use to discuss the Rayleigh–Schrödinger series is due to B. Sz-Nagy in "Perturbations des transformations autoadjointes dans l'espace de Hilbert," *Comment. Math. Helv.* **19** (1946/47), 347–366, and T. Kato in "On the convergence of the perturbation method, I, II" *Progr. Theoret. Phys.* **4** (1949), 514–523; **5** (1950), 95–101; 207–212.

Section XII.2 The original theory of regular perturbations appeared in F. Rellich, "Störungstheorie der Spektralzerlegung I–V," *Math. Ann.* **113** (1937), 600–619, 677–685; **116** (1939), 555–570; **117** (1940), 356–382; **118** (1942), 462–484, with simplifications due to Sz-Nagy and Kato (see the Notes to Section 1). Additional discussion can be found in the books of Rellich and Kato and in the book by K. O. Friedrichs, *Perturbation of Spectra in Hilbert Space*, Amer. Math. Soc., Providence, Rhode Island, 1965.

Our definition of analytic family is slightly different from that used by Kato in his book (p. 366) and by Rellich in the third paper of his series. In case $T(\beta)$ is closed and has a nonempty resolvent set, the definitions agree. The Kato–Rellich definition allows certain cases where $T(\beta)$ may have empty resolvent set; but since such cases rarely arise in practice, we have used a more restrictive but technically simpler definition.

For further discussion of the application of perturbation theory to atomic physics, see M. Mizushima, *Quantum Mechanics of Atomic Spectra and Atomic Structure*, Benjamin, New York, 1970. In particular, references for the hyperfine structure of hydrogen can be found therein.

Theorem XII.10 is proven in B. Simon and R. Höegh-Krohn, "Hypercontractive semigroups and two-dimensional self-coupled Bose fields," *J. Functional Analysis* **9** (1972), 121–180. Another technique for obtaining lower bounds on the radius of convergence of the Rayleigh–Schrödinger series in special cases of physical interest appears in D. Atkinson, "Bound state perturbation theory: A new approach," *Nuclear Phys.* **B20** (1970), 125–158.

Theorem XII.12 was first proven in T. Kato, "On the adiabatic theorem of quantum mechanics," *J. Phys. Soc. Japan* **5** (1950), 435–439. A weaker theorem that suffices for our application of Theorem XII.12 in Theorem XII.13 can be found in the paper of Sz-Nagy mentioned in the Notes to Section 1. Sz-Nagy gives an explicit formula for an invertible operator $W(\beta)$ defined for $|\beta|$ sufficiently small with $W(\beta)P(0)W(\beta)^{-1} = P(\beta)$. Namely (see Problem 19)

$$W(\beta) = [1 - (P(\beta) - P(0))^2]^{-1/2}[P(\beta)P(0) + (1 - P(\beta))(1 - P(0))]$$

The lemma to Theorem XII.12 can also be proven using fixed point theorems.

Section XII.3 The heuristic argument for the divergence of the perturbation series for the anharmonic oscillator is discussed in K. Gottfried, *Quantum Mechanics*, vol. I, pp. 361–362, Benjamin, New York, 1966. The model for all such heuristic arguments was given by F. Dyson, "Divergence of perturbation theory in quantum electrodynamics," *Phys. Rev.* **85** (1952), 631–632.

The detailed estimate necessary for proving divergence of the series ($|a_n| \geq AB^n\Gamma(n/2)$) is proven in C. Bender and T. T. Wu, "Anharmonic oscillator," *Phys. Rev.* **184** (1969), 1231–1260. Their proof is modeled on an analogous proof for the $(\varphi^4)_2$-field theory which is found in A. M. Jaffe, "Divergence of perturbation theory for bosons," *Comm. Math. Phys.* **1** (1965), 127–149.

The nature of the singularity of the anharmonic oscillator eigenvalues at $\beta = 0$ has been studied in detail. $E(\beta)$ has an analytic continuation to a three-sheeted surface on which $\beta = 0$ is *not* an isolated singularity. This was first suggested by approximate calculations of Bender and Wu (see above) and proven in B. Simon, "Coupling constant analyticity for the anharmonic oscillator," *Ann. Phys.* **58** (1970), 76–136.

A particularly clear discussion of the Gell'Mann–Low series and its *formal* derivation in quantum field theory may be found in J. Bjorken and S. Drell, *Relativistic Quantum Fields*, McGraw-Hill, New York, 1965. For a discussion of the general theory of asymptotic series, see W. Wasow, *Asymptotic Series for Ordinary Differential Equations*, Wiley (Interscience), New York, 1965. The table in Example 1 comes from the *Ann. Phys.* article by Simon.

That perturbation series are asymptotic was first proven by E. Titchmarsh, "Some theorems

XII: PERTURBATION OF POINT SPECTRA

on perturbation theory, I, II," *Proc. Roy. Soc.* **A200** (1949), 34–46; **A201** (1950), 473–479. Extensions of Titchmarsh's results are due to T. Kato, "On the convergence of the perturbation method," *J. Fac. Sci. Univ. Tokyo Sect. I*, **6** (1951), 145–226, and "Perturbation theory of semi-bounded operators," *Math. Ann.* **125** (1953), 435–447, and by V. Kramer "Asymptotic inverse series," *Proc. Amer. Math. Soc.* **7** (1956), 429–437, and "Asymptotic perturbation series," *Trans. Amer. Math. Soc.* **85** (1957), 88–105.

Some generalizations of the Titchmarsh results to nonsymmetric operators may be found in D. Huet, "Phénomènes de perturbation singulière," *C. R. Acad. Sci. Paris* **244** (1957), 1438–1440; **246** (1958), 2096–2098; **247** (1958), 2273–2276; **248** (1959), 58–60; and "Phénomènes de perturbation singulière dans les problèmes aux limites," *Ann. Inst. Fourier* **10** (1960), 1–96. An alternative method of proving perturbation series asymptotic which is applicable to positive self-adjoint perturbations of positive self-adjoint operators is discussed in Appendix II of Simon's *Ann. Phys.* article and in W. M. Greenlee, "Singular perturbation of eigenvalues," *Arch. Rational Mech. Anal.* **34** (1969), 143–164.

Asymptotic perturbation theory is discussed in Chapter VIII of Kato's book. Kato emphasizes that only stability of the eigenvalue and strong resolvent convergence is required; see our discussion of Theorem XII.16.5.

Theorems XII.15 and XII.16 are discussed and proven in B. Simon "Determination of eigenvalues by divergent perturbation series," *Advances in Math.* **7** (1971), 240–253.

The asymptotic nature of the perturbation series for spatially cutoff $(\varphi^{2n})_2$ field theories in sectors $\{\beta \mid |\arg \beta| \leq \theta\}$ for $\theta < \pi/2$ was first proven in the paper of B. Simon and R. Hoegh-Krohn, quoted in the Notes to Section 2. The extension to $\theta < \pi$ for $(\varphi^4)_2$ (using Theorem XII.15) first appeared in B. Simon, "Borel summability of the ground state energy in spatially cutoff $(\varphi^4)_2$," *Phys. Rev. Lett.* **25** (1970), 1583–1586. The extension to $\theta < \pi$ for general $(\varphi^{2n})_2$ theories is due to L. Rosen and B. Simon "The $(\varphi^{2n})_2$ Hamiltonian for complex coupling constant," *Trans. Amer. Math. Soc.* **165** (1972), 365–379. Various objects in the infinite volume $P(\varphi)_2$ field theory have been shown to have asymptotic series; see J. Dimock, "Asymptotic perturbation expansions in the $P(\varphi)_2$ quantum field theory," *Comm. Math. Phys.* **35** (1974), 347–356, for the earliest result of this sort.

There is a formal relation between vacuum energies and sums of connected Feynman diagrams with no external legs (see the book of Bjorken and Drell mentioned above). If one modifies the Feynman rules by placing a Fourier transform of g, the spatial cut-off, "at the vertex" instead of a δ function, one obtains the Rayleigh–Schrödinger series for the vacuum energy after explicitly doing the integrals over time.

The double well potential has evoked considerable interest. Previous work includes that of M. Kac and C. Thompson, "Phase transitions and eigenvalue degeneracy of a one dimensional anharmonic oscillator," *Stud. Appl. Math.* **48** (1969), 257–264, and D. Isacson, "The critical behavior of $(\phi^4)_1$," *Commun. Math. Phys.* (to appear). In particular Kac and Thompson prove that $E_i' - E_i$ goes to zero as $\exp(-b\beta^{-2})$. There has also been theoretical work on the question of Borel summability for the double well problem. E. Berezin, G. Parisi, and J. Zinn-Justin, in "Perturbation Theory at large orders for potentials with degenerate minima," *Phys. Rev.* **D16** (1977), 408 study the large n behavior of the Rayleigh–Schrödinger coefficients both numerically and using some not-yet-rigorous theory. Their results suggest that the series will not be Borel summable and this has been proven by A. Sokal, Princeton preprint, 1977.

There is an exactly soluble example similar to the double well potential discussed on pages 66–77 of E. Merzbacher, *Quantum Mechanics*, Wiley, New York, 1961, and pages 12–14 of U. P. Maslov, *Théorie des Perturbations et méthods Asymptotiques*, Dunod and Gauthier-Villars, Paris, 1972.

The special interest in the double well comes from the rather different behavior of the correspondng two-dimensional field theory. In the double well problem, even though $E_0' - E_0$ is small, it is nonzero and the ground state is symmetric about $x = -\frac{1}{2}\beta^{-1}$. In the corresponding field theory, if one chooses the limit theory which is symmetric, then the vacuum is degenerate and the (decomposed) theories with unique vacuum do not have the symmetry. This has been proven by J. Glimm, A. Jaffe, and T. Spencer, "Phase transitions for ϕ_2^4 quantum fields", *Commun. Math. Phys.* **45** (1975), 203–216. This distinction between one and two dimensions is analogous to the fact that the Ising model has a phase transition in two but not in one dimension; we return to these matters in a later volume.

Section XII.4 Carleman's theorem (Theorem XII.17) is proven in *Les Fonctions Quasi Analytiques*, Gauthier-Villars, Paris, 1926. The method of proof for the general case is very different from the special case given as Theorem XII.18. Our proof of Theorem XII.18 follows that given by G. H. Hardy, *Divergent Series*, Oxford Univ. Press, London and New York, 1949.

The idea of applying the notion of the strong asymptotic condition to divergent perturbation series, and in particular its application to the anharmonic oscillator, is due to S. Graffi, V. Grecchi, and B. Simon, "Borel summability: Application to the anharmonic oscillator," *Phys. Lett.* **32B** (1970), 631–634. Theorem XII.20 was first stated and proven in Simon's *Advances in Math.* paper quoted in the Notes to Section 3, and Example 2 was first discussed in Simon's *Phys. Rev. Lett.* article also mentioned in the above Notes. Basic references for two-dimensional Bose field theories are discussed in the Notes to Section X.7. Example 3 (x^{2n} oscillator) was first discussed in the above-mentioned paper by Graffi *et al.* It has been shown that certain asymptotic series for infinite volume $P(\varphi)_2$ theories are Borel summable. See J.-P. Eckmann, J. Magnen, and R. Sénéor, "Decay properties and Borel summability for the Schwinger functions in $P(\varphi)_2$ theories," *Comm. Math. Phys.* **39** (1975), 251–271. The perturbation series for the Zeeman effect (an atom in a constant magnetic field) has been proven to be Borel summable by J. Avron, I. Herbst, and B. Simon, "Schrödinger Operators with Magnetic Fields," 1977 preprint.

Watson's Theorem was first proven in G. Watson, "A theory of asymptotic series," *Philos. Trans. Roy. Soc. London Ser. A* **211** (1912), 279–313. Borel first proposed his method and showed that it worked in many examples in "Mémoire sur les séries divergentes," *Ann. Sci. École Norm. Sup.* **16** (1899), 9–136.

The Borel method is an example of a regular summability method. A summability method is a procedure for defining a sum α for certain formal series $\sum a_n$. The method is called regular if, whenever $\sum a_n$ is absolutely convergent to α_0, the procedure is applicable and yields α_0 as sum. The regularity of the Borel method, which follows from Watson's theorem, was first proven by G. Hardy, "On differentiation and integration of divergent series," *Trans. Cambridge Phil. Soc.* **19** (1904), 297–321.

One disadvantage of the Borel method, as it stands, is that it involves an analytic continuation that can be difficult to do computationally. This difficulty can be overcome by finding a conformal map that takes the region of analyticity given by Watson's theorem into a region including a unit circle in such a way that the image of the positive reals lies in this circle. One can then do the "continuation" by merely summing a power series in the image region. For additional details, see N.-E. Nörlund, *Leçons sur les séries d'interpolation*, Gauthier-Villars, Paris, 1926; G. Doetsch, *Handbuch der Laplace-Transformation*, Birkhäuser, Basel, 1955, Vol. 2, Chap. 11; B. Hirsbrunner and J. Loeffel, "Sur les séries asymptotiques sommables selon Borel," *Helv. Phys. Acta* **48** (1975), 546. The last reference contains an application to x^4 oscillators.

Another summability method, the method of Padé approximants, is known to correctly "sum" the perturbation series for x^4 and x^6 perturbations of $p^2 + x^2$. This method is discussed in G. A. Baker, "The theory and application of the Padé approximant method," *Advances Theoret. Phys.* **1** (1966), 1–58. The applicability of the method (which is not known to be regular) to x^4 and x^6 oscillators is proven in J. J. Loeffel, A. Martin, B. Simon, and A. S. Wightman, "Padé approximants and the anharmonic oscillator," *Phys. Lett.* **30B** (1969), 656–658. The advantage of the Padé method is its computational simplicity; one has explicit formulas for the approximants and explicit control on the errors. Its disadvantages are its limited applicability (numerical evidence suggests that it does not work for an x^8 oscillator) and the difficulty of proving convergence (it has not been proven that the method works for two-dimensional x^4 oscillators). The reader should compare the partial sums of the Rayleigh–Schrödinger series for the ground state of $p^2 + x^2 + 0.2x^4$ (exact value: $E_0 = 1.118292$) given in Section 3 with the $[N, N]$ Padé approximants:

$$E[1, 1] = 1.111111 \qquad E[5, 5] = 1.118288$$
$$E[2, 2] = 1.117541 \qquad E[6, 6] = 1.118292$$
$$E[3, 3] = 1.118183 \qquad E[7, 7] = 1.118292$$
$$E[4, 4] = 1.118272$$

The $[N, N]$ Padé approximant is computed using only the Rayleigh–Schrödinger coefficients $a_0, \ldots, a_{2N}$.

There is a family of perturbations of the discrete spectrum that leaves the spectrum discrete but which is more singular than anything we have discussed thus far. Typical of this class is $H_0 = -d^2/dx^2 + x^2$ on $L^2(\mathbb{R}, dx)$, where $V = x^{-\alpha}$. For $\alpha > 1$, this family is discontinuous at $\lambda = 0$ if $H_0 + \lambda V$ is defined as a form sum. Explicitly, $H_0 + \lambda V$ converges as $\lambda \downarrow 0$ in strong resolvent sense to an operator $\tilde{H}_0$ different from H_0. This phenomenon was discovered by J. Klauder, "Field structure through models studies," *Acta Phys. Austriaca. Supp.* **11** (1973), 341–387, and is further discussed by B. Simon, "Quadratic forms and Klauder's phenomenon: A remark on very singular perturbations," *J. Functional Analysis* **14** (1973), 295–298, and B. DeFacio and C. L. Hammer, "Remarks on the Klauder phenomenon," *J. Mathematical Phys.* **15** (1974), 1071–1077. If $H(\lambda)$ is set equal to $\tilde{H}_0$ when $\lambda = 0$, then the resulting family is analytic at $\lambda = 0$ so long as $1 \leq \alpha < 2$. For $2 \leq \alpha < 3$, the eigenvalues are given by asymptotic series to first order so long as $\lambda > 0$. But for $\alpha \geq 3$, the eigenvalues approach those for $\tilde{H}_0$ more slowly than linearly in λ. Explicitly, for $\alpha > 3$,

$$E(\lambda) - E(0) = c\lambda^{1/(\alpha-2)} + o(\lambda^{1/(\alpha-2)})$$

for $c \neq 0$, and for $\alpha = 3$,

$$E(\lambda) - E(0) = c\lambda \ln \lambda + O(\lambda)$$

These phenomena are further discussed in J. Klauder and L. Detwiler, "Supersingular quantum perturbations," *Phys. Rev.* **D11** (1975), 1436–1441 and in W. Greenlee, "Singular perturbation theory for semi-bounded operators," *Bull. Amer. Math. Soc.* **82** (1976), 341–344, and E. Harrell, II, "Singular Perturbation Potentials," *Ann. Phys.* **105** (1977), 379–406.

Section XII.5 That Schrödinger's theory of the Stark effect was not entirely satisfactory due to the occurrence of continuous spectrum in $(-\infty, \infty)$ was first pointed out by J. R. Oppenheimer in "Three notes on the quantum theory of aperiodic effects," *Phys. Rev.* **31** (1928), 66–81. One interpretation of this perturbation series, which we shall discuss below, is expounded in E. C. Titchmarsh "Some theorems on perturbation theory, III, IV, V," *Proc. Roy. Soc.* **A207** (1951), 321–328; **A210** (1951), 30–47; *J. Analyse Math.* **4** (1954/56), 187–208. In the last paper, "spectral concentration" was defined.

Progress toward the isolation of the notions and relations of spectral concentration and pseudo-eigenvectors was made due to work of K. Friedrichs, "Über die Spektralzerlegung eines Integraloperators," *Math. Ann.* **115** (1938), 249–272; and "On the perturbation of continuous spectra," *Comm. Pure Appl. Math.* **1** (1948), 361–406; T. Kato's *J. Fac. Sci. Univ. Tokyo* paper quoted in the notes to Section 3 and K. Friedrichs and P. Rejto, "On a perturbation through which a discrete spectrum can become continuous," *Comm. Pure Appl. Math.* **15** (1962), 219–235. The pth order analogue of Theorem XII.22 is due to R. C. Riddell, "Spectral concentration for self-adjoint operators," *Pacific J. Math.* **23** (1967), 377–401. Independently, one half of the extended theorem (that a pth order pseudo-eigenvector implies pth order spectral concentration) and all of Theorem XII.22 were proven by C. C. Conley and P. A. Rejto, "Spectral concentration II: General theory," in *Perturbation Theory and its Application in Quantum Mechanics* (C. H. Wilcox, ed.) Wiley, New York, 1966. Our proof of Theorem XII.23 follows that of Conley and Rejto, where there is also a detailed discussion of the Stark effect in order $p > 1$. For additional discussion of the theory of spectral concentration and some additional applications, see K. Veselić, "On spectral concentration for some classes of self-adjoint operators," *Glasnik Math. Ser. III* **4** (1969), 213–228 and "The nonrelativistic limit of the Dirac equation and spectral concentration," *Glasnik Math. Ser. III* **4** (1969), 231–240. In the last paper Veselić considers the Dirac Hamiltonian $H = H_0 + V(x) - mc^2$ where $V(x) \to \infty$ as $|x| \to \infty$. H has continuous spectrum in $(-\infty, \infty)$ for any c since the negative energy states see an effective potential going to $-\infty$ as $|x| \to \infty$. As $c \to \infty$, H has spectral concentration about the eigenvalues of $-\Delta + V$.

For a discussion of the spectral properties of the Stark effect Hamiltonian, see Section XIII.4, Example 8, and the Notes to that section.

For a class of models closely related to the Stark effect but having spherical symmetry so that ordinary differential equation methods are applicable (examples: $-d^2/dx^2 + |x| - \beta x^2$ and $-\Delta - 1/r - \beta |r|$), Titchmarsh found the following phenomena: H has continuous spectrum concentrated about pseudo-eigenvalues (for $\beta > 0$ in the two examples), but the pseudo-eigenvalues have a stronger meaning—the Green's function (kernel of the resolvent) has an analytic continuation below the real axis with a pole at $E(\beta)$ on the second sheet. Re $E(\beta)$ has the Rayleigh–Schrödinger series as asymptotic series, and Im $E(\beta)$ goes to zero faster than any β^n. This is to be compared with the phenomena discussed in Section 6. There is a second sheet pole in both cases, but in one case (Stark effect) the width is $o(\beta^n)$ for all n and in the other (Auger effect) it is $O(\beta^2)$. This is mirrored in the degree of spectral concentration: In one case (Stark effect) there is concentration to all orders and in the other case to second order and not to higher order (we discuss spectral concentration in this Auger case in the Notes to Section 6).

We note that in some physics literature, the imaginary part of this second sheet pole is called the "natural" width. There is also a contribution to the observed width due to the interaction with the radiation field (radiative width) and a contribution due to the thermal motion of the sources (thermal width or Doppler width).

Spectral concentration for the Stark effect in helium has been studied by P. Rejto in "Second order concentration near the binding energy of the helium Schrödinger operator," *Israel J. Math.* **6** (1968), 311–337, and "Spectral concentration for the helium Schrödinger operator," *Helv. Phys. Acta* **43** (1970), 652–667. Given Theorems X.38 and XIII.39, it is easy to verify the hypotheses of Theorem XII.22 in this case.

In discussing spectral concentration in electric fields, it is useful to know that the eigenfunctions of H_0 fall off faster than inverse polynomials in the sense that $\psi \in D(r^n)$ for all n. Results of this sort are proven in Section XIII.11.

There is a case where eigenvalues get "absorbed" by the continuous spectrum which is much less singular than the situations described in this section. This is the situation of an analytic family of type (A) where a discrete eigenvalue approaches the continuous spectrum as β approaches some critical coupling constant. Some information in this case may be found in B. Simon, "On the absorption of eigenvalues by continuous spectrum in regular perturbation problems," *J. Functional Analysis* (to appear).

Section XII.6   The association of resonances with second sheet poles in the scattering amplitude is an idea that was introduced in the early days of quantum mechanics by V. Weisskopf and E. P. Wigner "Berechnung der natürlichen Linienbreite auf Grund der Diracschen Lichttheorie," *Z. Phys.* **63** (1930), 54–73. The idea of looking instead at poles in the resolvent is a later refinement explicitly discussed by J. Schwinger, "Field theory of unstable particles," *Ann. Phys.* **9** (1960), 169–193; C. Lovelace, "Three particle systems and unstable particles," in *Strong Interactions and High Energy Physics: 1963 Scottish Universities' Summer School* (R. C. Moorhouse, ed.), Oliver and Boyd, 1964; and A. Grossman, "Nested Hilbert space in quantum mechanics, I," *J. Mathematical Phys.* **5** (1964), 1025–1037. There is, of course, a connection between this idea and the work of Titchmarsh described at the end of the Notes to Section 5.

Many of the earliest attempts at understanding resonances from a rigorous point of view involved ad hoc non-self-adjoint models, often with $H^* - H$ compact. Typical is M. Livsic, "The method of non-self-adjoint operators in dispersion theory," *Uspehi Mat. Nauk.* **12** (1957) [English translation: *Amer. Math. Soc. Transl. Ser. 2* **16** (1960), 427–433]. A summary of various attempts along these lines may be found in C. L. Dolph, "Recent developments in some non-self-adjoint problems of Mathematical Physics," *Bull. Amer. Math. Soc.* **67** (1961), 1–69.

Attempts at finding second sheet poles and related phenomena such as spectral concentration in the resolvents of self-adjoint operators obtained from perturbing operators with eigenvalues in the continuum were made by Friedrichs in his *Comm. Pure Appl. Math.* paper (see Notes to Section 5) and by J. S. Howland in a series of papers: "Perturbation of embedded eigenvalues by operators of finite rank," *J. Math. Anal. Appl.* **23** (1968), 575–584; "Embedded eigenvalues and virtual poles," *Pacific J. Math.* **29** (1969), 565–582; "Spectral concentration and virtual poles," *Amer. J. Math.* **91** (1969), 1106–1126; "On the Weinstein–Aronszajn formula," *Arch. Rational Mech. Anal.* **39** (1970), 323–339. All these papers use models where the perturbation is compact (and even finite rank in most cases). The perturbations of Section 6 are not even relatively compact. Recently (and approximately simultaneously to the work of Simon mentioned below), Howland has extended Friedrichs' method to treat models similar to the autoionizing model. This work is discussed in J. S. Howland "Perturbation of embedded eigenvalues" *Bull. Amer. Math. Soc.* **78** (1972), 380–383, and "Puiseux Series for Resonances at an Embedded Eigenvalue," *Pacific J. Math.* **55** (1974), 157–176. In the latter article, Howland shows that Rellich's theorem does not extend to resonance theory, i.e., one can have H_0 and V self-adjoint and an embedded degenerate eigenvalue E_0 of H_0 where $H_0 + \beta V$ has resonance at values $E_n(\beta)$ where the $E_n(\beta)$ are not analytic at $\beta = 0$ but have a Puiseux series with fractional terms. Howland has also explained how the notion of resonance associated with an embedded eigenvalue of H_0 is intrinsic to the pair $\{H_0, V\}$.

The basic technique sketched in this section is developed in Section XIII.10. Detailed references can be found in the notes to that section. Our discussion here is largely taken from B. Simon, "Resonances in N-body quantum systems with dilatation analytic potentials and the foundations of time-dependent perturbation theory," *Ann. Math.* **97** (1973), 247–274.

The reduction due to symmetry is described in Simon's paper. Basically, the idea is the following. H_0 and V both commute with rotations and the reflection P through the origin. Thus, they also commute with the generators J_x, J_y, J_z of rotations about the x, y, and z axes (angular momentum operators). The operators P, J_z, and $J^2 \equiv J_x^2 + J_y^2 + J_z^2$ commute with one another and all have discrete spectrum. Thus, $\mathscr{H}$ breaks up into a direct sum $\oplus \mathscr{H}_{p,j,m}$ with $p = \pm 1$, $j = 0, 1, 2, \ldots$, and $m = -j, -j+1, \ldots, j-1, j$, so that if $\psi \in \mathscr{H}_{p,j,m}$, then

$$P\psi = p\psi, \quad J^2\psi = j(j+1)\psi, \quad J_z\psi = m\psi$$

H_0 and V leave each $\mathscr{H}_{p,j,m}$ invariant.

For the case of the helium atom, we saw that $\sigma_{\text{ess}}(H_0) = [-1, \infty)$. However, it turns out that $\sigma_{\text{ess}}(H_0 \upharpoonright \mathscr{H}_{p,j,m}) = [-\frac{1}{4}, \infty)$ if $p = (-1)^{j+1}$ so that certain eigenvalues of H_0 lying in $[-1, -\frac{1}{4})$ that appear embedded are not embedded once the symmetry reduction is made. Of course, these eigenvalues do not disappear when the perturbation is turned on. It is expected that all other eigenvalues do disappear—the method of this section reduces the proof of this to explicitly evaluating Im a_2 to see if it is nonzero.

Spectral concentration has been studied in situations where a "virtual pole" (second sheet pole) of the resolvent of $H_0 + \beta V$ occurs at $E(\beta)$ with $E(0)$ a real eigenvalue of H_0 and $E(\beta)$ analytic. In a family of models, Howland studied the question in his *Pacific J. Math.* paper quoted above. In the Auger problem, Simon has studied the question (in the above reference). The result, related to the fact that Im $E = O(\beta^2)$, is that there is spectral concentration to first order: Explicitly, $P^{(H_0 + \beta V)}(E_0 + \beta E_1 - f(\beta), E_0 + \beta E_1 + f(\beta)) \to P_{\{E_0\}}$ strongly if and only if $f(\beta) \to 0$ and $f(\beta)/\beta^2 \to \infty$ (so, in some sense, there is concentration to order p for fractional $p < 2$).

Finally, let us summarize the "usual" physical (and *nonrigorous*) argument leading to the Fermi golden rule: This will enable us to understand how the formula

$$\Gamma = 2\pi \frac{d}{dE}(\Omega_0, V\tilde{P}(E)V\Omega_0)$$

is the Fermi formula, which is usually written in a more formal way. For further discussion of this derivation, see L. Landau and E. Lifshitz, *Quantum Mechanics: Non-Relativistic Theory*, pp. 140–153, Addison-Wesley, Reading, Massachusetts, 1958. One first needs to think of the bump as due to a "virtual process." That is, one pretends there actually are "bound states" near the energies E_n of H_0 which lie in the continuum. These states have characteristic lifetimes τ_n. Thus the scattering process $e + \text{He}^+ \to e + \text{He}^+$ is viewed as $e + \text{He}^+ \to \text{He} \to e + \text{He}^+$. A state of helium is formed and decays in time τ_n. This state has a definite energy E_n, but due to the uncertainty principle the initial energy E of $e + \text{He}^+$ need not be exactly E_n for the state to form. Only $\Delta E = E - E_n$ must be of order $1/\tau_n$ (we have taken $\hbar = 1$). Thus the bump due to formation of the excited state of helium has a characteristic width, $\Gamma_n \equiv \tau_n^{-1}$. Once the state of energy E_n is formed, it decays with a $|P_\ell(\cos\theta)|^2$ distribution if ℓ is the angular momentum of the resonance and if the final state of the He$^+$ ion is its ground state. As a result, the scattering cross section is enhanced by the occurrence of the resonance.

Now let φ_n be the bound state of H_0 at energy E_n. Suppose H_0 has "continuum eigenfunctions" $\varphi(E)$ with $H_0\varphi(E) = E\varphi(E)$ and $\int \varphi(E)(\varphi(E), \eta) \, dE = \eta$ for any $\eta \in \mathscr{H}$. Let us try to solve the Schrödinger equation $i\dot{\psi} = (H_0 + V)\psi$ with $\psi(0) = \varphi_n$. Write

$$\psi(t) = \int a(E; t)\varphi(E)e^{-iEt} \, dE + \sum a_k(t)e^{-iE_k t}\varphi_k$$

The probability of $\psi(t)$ not being in the state φ_n at time t (i.e., the probability of decay) is given by

$$P(t) = \int |a(E; t)|^2 \, dE + \sum_{k \neq n} |a_k(t)|^2$$

68   XII: PERTURBATION OF POINT SPECTRA

Formally, $a(E; t)$ and $a_k(t)$ obey the differential equations

$$i\dot{a}(E; t) = \int a(E'; t)(\varphi(E), V\varphi(E'))e^{-i(E'-E)t}\, dE' + \sum_k (\varphi(E), V\varphi_k)e^{-i(E_m-E)t}a_k(t)$$

$$i\dot{a}_m(t) = \int a(E', t)(\varphi_m, V\varphi(E'))e^{-i(E'-E_m)t}\, dE' + \sum_k (\varphi(E), V\varphi_k)e^{-i(E_m-E)t}a_k(t)$$

Formally, solve these equations in "powers of V." To lowest order ($V = 0$), we have $a(E; t) = 0$, $a_m(t) = 0$ if $m \neq n$, $a_n(t) = 1$. Thus to first order

$$i\dot{a}^{(1)}(E; t) = e^{-i(E_m-E)}(\varphi(E), V\varphi_m)$$

and similarly for a_k. Thus

$$a^{(1)}(E; t) = (\varphi(E), V\varphi_n)\left(\frac{e^{-i(E_n-E)t} - 1}{(E_n - E)}\right)$$

or

$$P^{(1)}(t) = t\int |(\varphi(E), V\varphi_n)|^2 \left(\frac{4\sin^2[(t/2)(E_n - E)]}{t(E_n - E)^2}\right) dE$$

$$+ \text{ a discrete sum.}$$

As $t \to \infty$,

$$\frac{4\sin^2((t/2)x)}{2\pi t x^2} \to \delta(x)$$

in the sense of distributions, so we can write

$$\lim_{t\to\infty} \frac{P^{(1)}(t)}{t} = 2\pi |(\varphi(E), V\varphi_n)|^2 \Big|_{E=E_n}$$

But to lowest order, $\Gamma = \lim_{t\to\infty} t^{-1}P^{(1)}(t)$. Thus

$$\Gamma = 2\pi |(\varphi(E), V\varphi_n)|^2 \Big|_{E=E_n}.$$

This is the Fermi golden rule in its more usual form. Sometimes, even more formally, one writes $P_{n\to E}(t) = |a(E; t)|^2$, $\Gamma_{n\to E} = \lim_{t\to\infty} t^{-1}|a(E; t)|^2$, $\Gamma = \int \Gamma_{n\to E}\, dE$ and

$$\Gamma_{n\to E} = 2\pi|(\varphi(E), V\varphi_n)|^2\,\delta(E - E_n)$$

To see that this is just a formal version of our result in Section 6, we note that formally speaking,

$$\tilde{P}(-\infty, E') = \int_{-\infty}^{E'} dE(\varphi(E), \cdot)\varphi(E)$$

Thus

$$\frac{d}{dE'}(V\varphi_n, \tilde{P}(-\infty, E')V\varphi_n)\Big|_{E'=E_n} = (V\varphi_n, \varphi(E'))(\varphi(E'), V\varphi_n)\Big|_{E'=E_n}$$

$$= |(V\varphi_n, \varphi(E'))|^2 \Big|_{E'=E_n}$$

PROBLEMS

†1. (a) Suppose that $f_0, f_2, f_3, \ldots, f_n$, and f_1^{-1} are all analytic in $\{\beta \mid |\beta - \beta_0| \leq R_0\}$. Let the α_k be determined by recursive substitution of $\lambda = \lambda_0 + \sum_{k=1}^{\infty} \alpha_k (\beta - \beta_0)^k$ in (1). Prove that $|\alpha_k| \leq AR^k$ for suitable A and R.
(b) Prove that any analytic function $\lambda(\beta)$ with $\lambda(\beta_0) = \lambda_0$ solving (1) near $\beta = \beta_0$ is equal to $\lambda_0 + \sum_{k=1}^{\infty} \alpha_k (\beta - \beta_0)^k$ near $\beta = \beta_0$.

†2. Let T be a matrix written in Jordan normal form with λ_0 an eigenvalue of T. Suppose that ε is so small that λ_0 is the only eigenvalue of T in $\{\lambda \mid |\lambda - \lambda_0| \leq \varepsilon\}$. Prove that

$$P = -(2\pi i)^{-1} \oint_{|\lambda - \lambda_0| = \varepsilon} (T - \lambda)^{-1} \, d\lambda$$

is the projection onto $\{v \mid (T - \lambda_0)^n v = 0 \text{ for some } n\}$ and $Pw = 0$ if $(T - \lambda_1)^n w = 0$ for $\lambda_1 \neq \lambda_0$. (Hint: Prove that $(a^{-1})_{ij} = b_{ij}$ with

$$a_{ij} = (\mu - \lambda)\delta_{ij} + \delta_{i+1, j}$$
$$b_{ij} = (\mu - \lambda)^{-1} - (\mu - \lambda)^{-2} \delta_{i+1, j} + (\mu - \lambda)^{-3} \delta_{i+2, j} - \cdots).$$

3. Let T be a finite matrix. Let 0 be an eigenvalue of T.
   (a) Prove that $(T - \lambda)^{-1}$ is analytic in $\{\lambda \mid 0 < |\lambda| < R\}$ for some R and that there are operators A_n so that $(T - \lambda)^{-1} = \sum_{n=-\infty}^{\infty} A_n \lambda^n$ with the sum converging if $0 < |\lambda| < R$.
   (b) Prove that $A_n = (2\pi i)^{-1} \oint_{|\lambda| = \varepsilon} \lambda^{-n-1} (T - \lambda)^{-1} d\lambda$ for $0 < \varepsilon < R$.
   *(c) Use (b) to prove that $A_n A_m = (\eta_n + \eta_m - 1) A_{n+m+1}$ where

$$\eta_n = \begin{cases} 1, & n \geq 0 \\ 0, & n < 0 \end{cases}$$

   (Hint: See the proof of Theorem XII.5.)
   (d) Let $P = -A_{-1}$; $N = -A_{-2}$; $S = A_0$. Prove that $A_n = S^{n+1}$ for $n \geq 0$, $A_{-n} = -N^{n-1}$ for $n \geq 2$ and that $P^2 = P$; $PN = NP = N$; $PS = SP = 0$.
   (e) Prove that $\sum_{m=0}^{\infty} \zeta^n N^n$ converges for all $|\zeta| < \infty$. (Hint: Use (a).)
   *(f) Prove that $N^k = 0$ for some k.
   (g) Prove that $TP = PT = N$.

4. Let T be an $n \times n$ matrix. Let $\lambda_1, \ldots, \lambda_\ell$ be its eigenvalues. Let

$$P_i = -(2\pi i)^{-1} \oint_{|\lambda - \lambda_i| = \varepsilon_i} (T - \lambda)^{-1} \, d\lambda$$

and

$$N_i = -(2\pi i)^{-1} \oint_{|\lambda - \lambda_i| = \varepsilon_i} (\lambda - \lambda_i)(T - \lambda)^{-1} \, d\lambda.$$

   (a) By mimicking Problem 3, prove that $P_i^2 = P_i$; $N_i P_i = P_i N_i = N_i$; $TP_i = P_i T = \lambda_i P_i + N_i$.
   *(b) Prove that $P_i P_j = 0$ if $i \neq j$ and $\sum_{i=1}^{\ell} P_i = 1$.
   (c) (abstract Jordan normal form). Prove that

$$T = \sum_{i=1}^{\ell} (\lambda_i P_i + N_i)$$

(d) Let N be nilpotent, i.e., $N^k = 0$ for some k. Show that there is a basis for the underlying vector space for which N has the matrix form

$$N = \begin{bmatrix} 0 & x & 0 & \cdots & 0 \\ 0 & 0 & x & \cdots & 0 \\ & & \ddots & \ddots & \\ 0 & 0 & 0 & \cdots & x \\ 0 & 0 & 0 & \cdots & 0 \end{bmatrix}$$

where each x is equal to 1 or 0.

(e) Show that in a suitable basis T has the Jordan normal matrix form discussed in Section 1.

Reference for Problems 3, 4: Kato, *Perturbation Theory for Linear Operators*, pp. 38–43, Springer-Verlag, Berlin and New York, 1966.

5. Let H_0 be a self-adjoint matrix with eigenvalues $E_0 < E_1 < E_2 < \cdots < E_k$. Suppose that E_0 is a nondegenerate eigenvalue. Let V be an arbitrary self-adjoint matrix and let $E_0 + \sum_{n=1}^{\infty} \alpha_n \beta^n$ be the Rayleigh–Schrödinger series for the eigenvalue of $H_0 + \beta V$ near E_0.
   (a) Prove that all the α_n are real.
   (b) Prove that $\alpha_2 \leq 0$.
   (c) Find an explicit example with $\alpha_4 > 0$.
   (d) If $V \geq 0$ in the sense of matrix inequality, prove that $\alpha_1 \geq 0$.
   (e) Find an explicit example with $V \geq 0$ but $\alpha_3 \leq 0$.

   (Hint for (c) and (e): Use an example of 2×2 matrices and solve the secular equation for 2×2 matrices explicitly.)

†6. (a) Let P be a bounded operator on a Banach space X with $P^2 = P$. Prove that $E = \text{Ran } P$ and $F = \text{Ker } P$ are closed subspaces in X with $E \cap F = \{0\}$ and $E + F = X$. Prove that any $x \in X$ can be uniquely written as $x = e + f$ with $e \in E, f \in F$.
   (b) Conversely, given closed subspaces E and F of a Banach space X with $E \cap F = \{0\}$ and $E + F = X$, prove there is a unique *bounded* operator P on X with $P^2 = P$ and $F = \text{Ker } P$; $E = \text{Ran } P$. (Hint: Use the closed graph theorem.)

7. Prove the following additional fact about the situation described in Theorem XII.6. If $\dim P = n$ and $v_1, \ldots, v_k$ are the eigenvalues of A in $\{v \mid |v - \lambda| < r\}$, then $\sum_{i=1}^{k} m_i = n$ if m_i is the algebraic multiplicity of v_i as an eigenvalue.

†8. Let $T(\beta)$ be a family of bounded operators defined on a region $R \subset \mathbb{C}$ so that $\sup_{\beta \in R} \|T(\beta)\| < \infty$. Prove that $T(\beta)$ is an analytic function in the sense of Section VI.3 if and only if $T(\beta)$ is an analytic family in the sense of Kato.

†9. A function F defined on an open set D of a Banach space X taking values in Y, another Banach space, is called analytic if and only if for all $x \in D$, there are functions $f_x^{(1)}, \ldots, f_x^{(n)}$, $\ldots$ with $f_x^{(j)}$ a j-linear function from $X \times \cdots \times X$ (j times) to Y so that

   (i) For some C and R, $\|f_x^{(j)}(y_1, \ldots, y_j)\| \leq CR^{-j}\|y_i\| \cdots \|y_j\|$.
   (ii) If $\|y\| < R$, $F(x + y) = \sum_{j=0}^{\infty} f_x^{(j)}(y, y, \ldots, y)$ where $f_x^{(0)} = F(x)$.

   (a) Let $\Omega \subset \mathbb{C}$. Let Y be a complex Banach space. Prove that $F : \Omega \to Y$ is analytic in the above sense if and only if it is analytic in the sense of Section VI.3.

(b) If F is an analytic function on $D \subset X$ taking values in Y and if g is an analytic function on a subset Ω of $\mathbb{C}$ taking values in D let $F \circ g: \Omega \to Y$. Prove $F \circ g$ is analytic in the sense of Section VI.3.

(c) Let $X = \mathscr{L}(Z)$ the bounded operators on some Banach space Z. Let $D = \{A \in X \,|\, A$ has a bounded inverse$\}$. Prove $F(A) = A^{-1}$ is an analytic function on D with values in X.

†10. Let $T(\beta)$ be an analytic family of type (A) in a neighborhood of $\beta = 0$.

(a) Prove that there exist operators T_n with $D(T_n) \supset D(T_0)$ and $T_0 \equiv T(0)$ so that (i) For some a, b, and c, and all $\psi \in D(T(0))$: $\|T_n \psi\| \leq c^{n-1}(a\|T(0)\psi\| + b\|\psi\|)$. (ii) For any $\psi \in D(T(0))$ and $|\beta|$ sufficiently small, $T(\beta)\psi = \sum_{n=0}^{\infty} \beta^n T_n \psi$ where the sum is norm convergent.

(b) Prove that a general analytic family of type (A) is an analytic family in the sense of Kato.

Reference for Problem 10: Kato's book (see Problem 4), pp. 375–381.

†11. Suppose $V < < H_0$, $W < < H_0$. Prove that $W < < H_0 + V$.

†12. In the context of Theorem XII.9, prove that for β small $(H_0 - \lambda)^{-1} \times [1 + \beta V(H_0 - \lambda)^{-1}]^{-1} = A$ is an inverse for $H_0 + \beta V - \lambda$. Explicitly Ran $A \subset D(H_0)$ and $(H_0 + \beta V - \lambda)A\psi = \psi$ for all $\psi \in \mathscr{H}$ and $A(H_0 + \beta V - \lambda)\psi = \psi$ for all $\psi \in D(H_0)$.

13. Let $A(\beta)$ be a compact operator-valued function on a connected open set $R \subset \mathbb{C}$. Let $f(\beta)$ be an analytic function on R. Suppose $f(\beta) \neq 0$ for all $\beta \in R$ and that $f(\beta_i)$ is an eigenvalue of $A(\beta_i)$ for $\beta_1, \ldots, \beta_n, \ldots \in R$ where $\beta_1, \ldots, \beta_n, \ldots$ have a limit point in R. Conclude that $f(\beta)$ is an eigenvalue of $A(\beta)$ for all $\beta \in R$. (Hint: Apply the analytic Fredholm theorem to $f(\beta)^{-1}A(\beta)$.)

†14. Prove Theorem XII.11. Reference: Kato's book, pp. 88–89, 379–381.

15. Let $H(\beta)$ be an analytic family in the sense of Kato in a simply connected region $R \subset \mathbb{C}$. Suppose that $E(\beta)$ is an isolated nondegenerate eigenvalue of $H(\beta)$ for each $\beta \in R$. Prove there is an analytic vector valued function ψ on R with $H(\beta)\psi(\beta) = E(\beta)\psi(\beta)$ and $\psi(\beta) \neq 0$ for all $\beta \in R$. (Hint: Use Theorem XII.12.)

16. Suppose that $P_i(\beta)$ is an analytic, projection-valued function of β for all β in R a connected, simply connected region of $\mathbb{C}$, for $i = 1, \ldots, k$. Suppose that $P_i(\beta)P_j(\beta) = 0$ if $i \neq j$ and $\beta \in R$ and $\sum_{i=1}^{k} P_i(\beta) = 1$. Suppose $0 \in R$. Find $U(\beta)$ analytic in R and invertible so that $U(\beta)P_i(0)U(\beta)^{-1} = P_i(\beta)$ for all $\beta \in R$ with $U(\beta)$ unitary for β real if the $P_i(\beta)$ are all orthogonal for β real. (Hint: Mimic the proof of Theorem XII.12 but take $Q(\beta) = \frac{1}{2}\sum_{i=1}^{k} [P_i'(\beta), P_i(\beta)]$.)

17. Let $T(\beta)$ be an analytic $n \times n$ matrix-valued function in a neighborhood of zero. Suppose that $T(\beta)$ is self-adjoint for all real β.

(a) Suppose that E_0 is an eigenvalue of multiplicity m for $T(0)$ and that $E_1(\beta), \ldots, E_k(\beta)$ are the distinct eigenvalues of $T(\beta)$ near E_0. Prove first that the projections $P_i(\beta)$ and $E_i(\beta)$ are meromorphic at $\beta = 0$. Then, using the self-adjointness, prove that the $P_i(\beta)$ are analytic at $\beta = 0$.

(b) Prove that there are analytic vector-valued functions $\psi_1(\beta), \ldots, \psi_n(\beta)$ in a neighborhood of zero so that the $\psi_i(\beta)$ are eigenvectors and $(\psi_i(\beta), \psi_j(\beta)) = \delta_{ij}$ for all real β. (Hint: Use Problem 16.)

†18. Complete the proof of the lemma to the proof of Theorem XII.12.

72   XII: PERTURBATION OF POINT SPECTRA

19. Let P and Q be (not necessarily orthogonal) projections on a Hilbert space with $\|P - Q\| < 1$. Let $A = (1 - P)(1 - Q) + PQ$; $B = (1 - Q)(1 - P) + QP$; $C = [1 - (P - Q)^2]$.
    (a) Prove that $AB = BA = C$.
    (b) Let $\sum_{n=0}^{\infty} \alpha_n x^n$ be the Taylor series for $(1 - x)^{-1/2}$ about $x = 0$ and let $D = \sum_{n=0}^{\infty} \alpha_n (P - Q)^n$ which converges since $\|P - Q\| < 1$. Prove that $D^2 C = CD^2 = DCD = 1$.
    (c) Prove that $P(P - Q)^2 = (P - Q)^2 P$ and $Q(P - Q)^2 = (P - Q)^2 Q$. Conclude that $DP = PD$; $QD = DQ$.
    (d) Let $W = DA$. Prove W^{-1} exists and $W^{-1} = BD$.
    (e) Prove that $WQ = PW$.
    (f) If P and Q are self-adjoint, prove that W is unitary.

20. Let $A(\beta)$ be a finite-matrix-valued analytic function defined near $\beta = 0$. Let E_0 be an eigenvalue of $\beta = 0$ of multiplicity m. Let $g_1(\beta), \ldots, g_k(\beta)$ be the multivalued analytic functions defined near $\beta = 0$ whose values are all the eigenvalues of $A(\beta)$ near E_0. Let $P_j(\beta)$ be the eigenprojection

$$(-2\pi i)^{-1} \int_{|\mu - g_j(\beta)| = \varepsilon} (A(\beta) - \mu)^{-1} \, d\mu$$

   (a) Prove that $P_j(\beta)$ is multivalued analytic near $\beta = 0$ and that $|\beta|^k \|P_j(\beta)\|$ is bounded as $\beta \to 0$ for some k.
   (b) Prove that if $A(\beta)$ is self-adjoint for β real, then $P_j(\beta)$ is single-valued analytic near and at $\beta = 0$.

21. The purpose of this problem is to prove Butler's theorem: If some eigenvalue $g_j(\beta)$ (see Problem 20) is not single valued, then $\|P_j(\beta)\| \to \infty$ as $\beta \to 0$.
    (a) Use Problem 20 to prove that if $\|P_j(\beta)\|$ does not diverge, then $P_j(\beta)$ has a Puiseux expansion

$$P_j(\beta) = A_0 + \beta^{1/m} A_1 + \cdots$$

   (b) Prove that $A_0^2 = A_0$. (Hint: $P_j(\beta)^2 = P_j(\beta)$.)
   (c) Prove that $A_0^2 = 0$. (Hint: $P_j(\beta e^{2\pi i}) P_j(\beta) = 0$.)
   (d) Conclude that $P_j(\beta) \to 0$ as $\beta \to 0$.
   (e) Prove that any nonzero projection Q obeys $\|Q\| \geq 1$.
   (f) Conclude that $\|P_j(\beta)\| \to \infty$.
   Reference: Butler's paper quoted in the Notes to Section 1.

22. Use Butler's theorem to prove Rellich's theorem (Theorem XII.3).

23. Let $A(\beta)$ be a finite-matrix-valued analytic function near $\beta = 0$. Suppose that E_0 is an eigenvalue of $A(0)$ with associated eigennilpotent $N \neq 0$ (see Problems 3, 4). Suppose that, for β near 0, the eigenvalues of $A(\beta)$ near E_0 have zero associated eigennilpotents. Prove that the projections $P_j(\beta)$ of Problem 20 have a singularity at $\beta = 0$.

24. Under the hypothesis of Theorem XII.14, prove there is a vector-valued function $\Omega(\beta)$, analytic in the region $\{\beta \,|\, |\arg \beta| < \theta,\ |\beta| < B\}$ for some $B > 0$ so that (i) $H(\beta)\Omega(\beta) = E(\beta)\Omega(\beta)$; (ii) $(\Omega_0, \Omega(\beta)) \equiv 1$; (iii) $\Omega(\beta)$ has an asymptotic series $\sum \varphi_n \beta^n$ as $|\beta| \downarrow 0$; $|\arg \beta| < \theta$, whose terms are given by expressions involving only V, $(H_0 - \lambda)^{-1}$, and Ω_0.

†25. Under the hypotheses of Theorem XII.14, for any vector $\Omega \in C^\infty(H_0)$ and any compact set $K \subset \rho(H_0)$, $[V(H_0 - E)^{-1}]^N \Omega$ is a norm continuous function of E in K. (Hint: Topologize $C^\infty(H_0)$ with the norms $\|\Omega\|_n = \|(|H_0| + 1)^n \Omega\|$. Prove that V is a continuous map from $C^\infty(H_0)$ to itself).

†26. A sequence $(a_0, a_1, \ldots)$ of complex numbers can be thought of as a **formal series** $\sum a_n z^n$. The set of all formal series F supports a sum $(a_0, \ldots) + (b_0, \ldots) = (c_0, \ldots)$ with $c_n = a_n + b_n$ and a multiplication $(a_0, \ldots) \cdot (b_0, \ldots) = (d_0, \ldots)$ with $d_n = \sum_{m=0}^n a_m b_{n-m}$.
 (a) Prove that F with these operations is an integral domain with $(1, 0, \ldots)$ as multiplicative identity.
 (b) Prove that $(b_0, b_1, \ldots)$ has a multiplicative inverse in F if and only if $b_0 \neq 0$.
 (c) Let A and B be formal series with $b_0 \neq 0$. Let C be the formal series $C = AB^{-1}$. Prove that C is the asymptotic series of f/g if A is an asymptotic series for f and B is an asymptotic series for g.
 (d) Prove the analogue of (c) with *strong asymptotic* replacing *asymptotic*.

†27. Let A and B be closed operators with $D(A) \cap D(B)$ dense. Prove that $C = A + B$ is closed on $D(A) \cap D(B)$ if and only if $A^*A + B^*B \leq \alpha(C^*C + 1)$ for some constant α.

†28. Using the creation and annihilation operators $A^\dagger$, A of Section V.3, prove the estimates (iii), (iv), (v) in Theorem XII.16 for Example 3 of Section 3.

†29. (a) Prove that any function $g(z)$ analytic in $R = \{z \mid 0 < |z| < B, |\arg z| < k\pi/2 + \varepsilon\}$ with $|g(z)| \leq A\sigma^N[k(N+1)]! |z|^N$ for all N and $z \in R$ is identically zero. (Hint: Let $h(w) = g(w^k)$ and apply Theorem XII.18 to h.)
 (b) Extend the Borel summability method and Watson's theorem to treat functions f analytic in the region R above obeying a strong asymptotic condition of order k.

30. Let V be given, a closed operator with $C^\infty(H_0) \subset D(V)$ for some self-adjoint operator H_0. Let K be a compact subset of $\rho(H_0)$ and let $H(\beta)$, closed operators, be defined for $0 < |\beta| < B$; $|\arg \beta| \leq \theta$ so that (i) $H(\beta) \upharpoonright C^\infty(H_0) = H_0 + \beta V$. (ii) for $|\beta| < B$, $K \subset \rho(H(\beta))$ for all β. Prove that

$$(H(\beta) - z)^{-1} \underset{\substack{|\beta| \to 0 \\ |\arg \beta| \leq \theta}}{\to} (H_0 - z)^{-1}$$

strongly for all $z \in K$ if and only if for each $z \in K$,

$$\limsup_{\substack{|\beta| \to 0 \\ |\arg \beta| \leq \theta}} \|(H(\beta) - z)^{-1}\| < \infty$$

†31. Let $P^{(n)}$ be a sequence of projection valued measures so that

$$P^{(n)}(a_0 + a_1\beta - f(\beta), a_0 + a_1\beta + f(\beta)) \overset{s}{\to} P_\infty$$

and

$$P^{(n)}(b_0 + b_1\beta - g(\beta), b_0 + b_1\beta + g(\beta)) \overset{s}{\to} P_\infty$$

where P_∞ is a nonzero projection and $f(\beta)/\beta \to 0$, $g(\beta)/\beta \to 0$. Prove that $a_0 = b_0$ and $a_1 = b_1$. (Hint: $P^{(n)}(\Omega)P^{(n)}(\Omega') = 0$ if $\Omega \cap \Omega' = \varnothing$.)

†32. Under hypotheses (i) and (ii) of Theorem XII.23, prove that
  (a) $(H_0 + i)[D(H_0) \cap D(V)]$ is dense in $\mathscr{H}$.
  (b) $(H_0 + i)^{-1} - (H_0 + \beta V + i)^{-1} \to 0$ strongly as $\beta \to 0$.

†33. Prove the formula
$$a_2 = (2\pi i)^{-1} \oint_{|E - E_0| = \varepsilon} \overline{(\Omega(\bar{\theta}; 0),} \; V[H_0(\theta) - E]^{-1} V\Omega(\theta; 0)) \frac{dE}{E - E_0}$$
for the situation described in Section 6. (Hint: Use the methods of Section 1 taking into account that $H_0(\theta)$ is not self-adjoint but that $H_0^*(\theta)\bar{f} = \overline{H_0(\bar{\theta})f}$.)

†34. Consider the situation described in Section 6 where $H(\theta)$ is an analytic family in θ with $H(\theta) = U(\theta) H U(\theta)^{-1}$ for θ real and
$$\sigma_{\text{ess}}(H(\theta)) = \{\lambda + xe^{-2\theta} \,|\, \lambda \in \Sigma; x \in \mathbb{R}_+\}$$
with $\Sigma = \{-1/n^2\}$. Suppose that $H(\theta)$ has a real eigenvalue $E_0 \notin \Sigma$ for all θ with Im $\theta > 0$. Write $H(0) \equiv H$.
  (a) Find vectors $\psi \in \mathscr{H}$ so that $(\psi, (H - z)^{-1}\psi)$ has an analytic continuation below the real axis near E_0 with a pole at E_0 with nonzero residue.
  (b) For $\psi \in \mathscr{H}$, prove that $\lim_{\varepsilon \downarrow 0} (z - E_0)(\psi, (H - z)^{-1}\psi)|_{z = E_0 + i\varepsilon} \ne 0$.
  (c) For any self-adjoint operator A, prove that
$$\text{s-lim}_{\varepsilon \downarrow 0} (z - E_0)(A - z)^{-1} \bigg|_{z = E_0 + i\varepsilon} = P_{\{E_0\}},$$
  the spectral projection onto the point E_0. (Hint: Use the functional calculus.)
  (d) Conclude that E_0 is an eigenvalue of H.
  (e) Prove that $(\psi, (H - z)^{-1}\psi)$ cannot have a double pole at E_0 for any $\psi \in \mathscr{H}$.
  (f) Let $\varphi = P(\theta)\varphi$ for some θ with Im $\theta > 0$, where
$$P(\theta) = -(2\pi i)^{-1} \oint_{|E - E_0| = \varepsilon(\theta)} (H(\theta) - E)^{-1} \, dE.$$
  Prove that $H(\theta)\varphi = E_0 \varphi$.
  Remark: In general, following Problem 4, $(H(\theta) - E_0)^n \varphi = 0$ for some n, but using (e), (f) can be proven.
  (g) Prove dim Ran$(P(\theta))$ = dim Ran$(P_{\{E_0\}})$.
  (h) If E_0' is an eigenvalue of H, prove it is an eigenvalue of $H(\theta)$ if Im $\theta > 0$.

†35. Under the conditions of Problem 34,
  *(a) Prove that $P(\theta) \to P_{\{E_0\}}$ strongly as $\theta \to 0$ with Im $\theta > 0$.
  (b) Prove that $P(\theta)$ extends to a function analytic in a strip $|\text{Im } \theta| \le b$. (Hint: Use the Schwarz reflection principle.)
  (c) For θ real, prove that $P(\theta) = U(\theta) P_{\{E_0\}} U(\theta)^{-1}$.
  (d) For $\Omega_0 \in \text{Ran } P_{\{E_0\}}$, prove that $U(\theta)\Omega_0$ has an analytic continuation to a strip $|\text{Im } \theta| \le b$ with $U(\theta)\Omega_0 \in D(H(\theta))$ for all θ.

Reference for Problems 34 and 35: The Balslev–Combes article mentioned in the Notes to Section XIII.10.

†36. Fill in the details of Example 6 of Section 3.

XIII: Spectral Analysis

If an atom can vibrate in more ways than one, it is certain that some connection must exist between the different periods, and this connection we may attempt to find out by trial. Or we may speculate on the causes which produce such vast differences in the chemical properties of some of the elements, while other elements have properties which resemble each other to an equally marked degree. We may be led on by such speculations to try whether we can trace any similarity in the periods of vibration of molecules which have similar chemical properties, or we may endeavour to classify the elements according to their spectra, and see whether such a classification would divide the elements into groups agreeing with those into which they have been divided by means of their chemical and physical behaviour.

A. Schuster, 1882

XIII.1 The min–max principle

In this chapter we describe methods for determining the spectral properties of a given self-adjoint operator H. We want information not only about the spectrum of H, $\sigma(H)$, but also about the subsets $\sigma_{ess}(H)$, $\sigma_{disc}(H)$, $\sigma_{ac}(H)$, $\sigma_{sing}(H)$, $\sigma_{pp}(H)$ defined in Sections VII.2 and VII.3. We are interested in quantitative information such as a precise determination of some or all of these subsets, and in qualitative information such as "$\sigma_{disc}(H)$ is finite" or "$\sigma_{pp} \subset (\sigma_{disc} \cup \partial \sigma_{ess})$."

In this section we shall develop a method for obtaining information about $\sigma_{disc}(H)$ and $\sigma_{ess}(H)$ from knowledge of the expectation values $(\psi, H\psi)$. This method is especially useful in the case that H has essential spectrum $[a, \infty)$ for some $a > -\infty$ and some eigenvalues in $(-\infty, a)$. This is the usual case encountered for Hamiltonians in quantum theories.

To understand the sort of result we would like, suppose that A and B are two 3×3 self-adjoint matrices with $A \leq B$ in the sense that $(\psi, A\psi) \leq (\psi, B\psi)$ for all $\psi \in \mathbb{C}^3$. Let $\lambda_1(A), \lambda_2(A), \lambda_3(A)$ be the three eigenvalues of A, listed so that $\lambda_1(A) \leq \lambda_2(A) \leq \lambda_3(A)$, and similarly for $\lambda_i(B), i = 1, 2, 3$. Since $A \leq B$, we expect $\lambda_i(A) \leq \lambda_i(B)$ for $i = 1, 2, 3$. Can we prove this? Let $\psi_1, \psi_2,$

76 XIII: SPECTRAL ANALYSIS

ψ_3 be the three eigenvectors of A chosen to be orthonormal. Writing $\psi = \alpha_1\psi_1 + \alpha_2\psi_2 + \alpha_3\psi_3$, we see that

$$\frac{(\psi, A\psi)}{(\psi, \psi)} = \frac{|\alpha_1|^2\lambda_1(A) + |\alpha_2|^2\lambda_2(A) + |\alpha_3|^2\lambda_3(A)}{|\alpha_1|^2 + |\alpha_2|^2 + |\alpha_3|^2}$$

so $\lambda_1(A) = \min_{\psi \neq 0}(\psi, A\psi)/(\psi, \psi)$ and $\lambda_3(A) = \max_{\psi \neq 0}(\psi, A\psi)/(\psi, \psi)$. These formulas and the similar formulas for $\lambda_1(B)$, $\lambda_3(B)$ imply $\lambda_1(A) \leq \lambda_1(B)$ and $\lambda_3(A) \leq \lambda_3(B)$ if $A \leq B$. What remains is to show that $\lambda_2(A) \leq \lambda_2(B)$. So we seek a formula for $\lambda_2(A)$ similar to our formula for $\lambda_1(A)$ in terms of $(\psi, A\psi)$. First, note that

$$\lambda_2 = \min_{\substack{\{\alpha_1, \alpha_2, \alpha_3\} \neq 0 \\ \alpha_1 = 0}} \frac{\alpha_1^2\lambda_1 + \alpha_2^2\lambda_2 + \alpha_3^2\lambda_3}{\alpha_1^2 + \alpha_2^2 + \alpha_3^2}$$

or equivalently,

$$\lambda_2(A) = U_A(\psi_1)$$

where

$$U_A(\psi) = \min_{\{\varphi | (\varphi, \psi) = 0, \varphi \neq 0\}} (\varphi, A\varphi)/(\varphi, \varphi)$$

This is not quite of the required form since $U_A(\psi_1)$ depends not only on $(\varphi, A\varphi)$ for all φ but on ψ_1. But, notice that for arbitrary ψ, we can find some $\varphi = \alpha_1\psi_1 + \alpha_2\psi_2$ with $\langle\alpha_1, \alpha_2\rangle \neq \langle 0, 0\rangle$ which is orthogonal to ψ. Thus there is a φ with $(\varphi, \psi) = 0$ and $(\varphi, A\varphi)/(\varphi, \varphi) \leq \lambda_2(A)$. So, for all ψ, $U_A(\psi) \leq \lambda_2(A)$. We conclude that

$$\lambda_2(A) = \max_{\psi} \min_{\substack{\varphi \in [\psi]^\perp \\ \varphi \neq 0}} (\varphi, A\varphi)/(\varphi, \varphi) \tag{1}$$

This expresses $\lambda_2(A)$ in terms of $(\varphi, A\varphi)$ only and can be used to prove $\lambda_2(A) \leq \lambda_2(B)$ (Problem 1).

We are now interested in extending (1) to the nth eigenvalue of a self-adjoint operator on an infinite-dimensional space. There are two problems: (1) the occurrence of nonpoint spectrum; (2) domain questions. These problems are overcome by a careful statement of the main result of this section:

Theorem XIII.1 (min–max principle, operator form) Let H be a self-adjoint operator that is bounded from below, i.e., $H \geq cI$ for some c. Define

$$\mu_n(H) = \sup_{\varphi_1, \ldots, \varphi_{n-1}} U_H(\varphi_1, \ldots, \varphi_{n-1})$$

where

$$U_H(\varphi_1, \ldots, \varphi_m) = \inf_{\substack{\psi \in D(H); \|\psi\|=1 \\ \psi \in [\varphi_1, \ldots, \varphi_m]^\perp}} (\psi, H\psi)$$

$[\varphi_1, \ldots, \varphi_m]^\perp$ is shorthand for $\{\psi \,|\, (\psi, \varphi_i) = 0, i = 1, \ldots, m\}$. Note that the φ_i are not necessarily independent.

Then, for each fixed n, *either*:

(a) there are n eigenvalues (counting degenerate eigenvalues a number of times equal to their multiplicity) below the bottom of the essential spectrum, and $\mu_n(H)$ is the nth eigenvalue counting multiplicity;

or

(b) μ_n is the bottom of the essential spectrum, i.e., $\mu_n = \inf\{\lambda \,|\, \lambda \in \sigma_{\text{ess}}(H)\}$ and in that case $\mu_n = \mu_{n+1} = \mu_{n+2} = \cdots$ and there are at most $n - 1$ eigenvalues (counting multiplicity) below μ_n.

Proof Let P_Ω be the projection-valued measure for H. We first prove that

$$\dim[\text{Ran}(P_{(-\infty, a)})] < n \quad \text{if} \quad a < \mu_n \tag{2a}$$

$$\dim[\text{Ran}(P_{(-\infty, a)})] \geq n \quad \text{if} \quad a > \mu_n \tag{2b}$$

For suppose (2a) is false. Then we can find an n-dimensional space $V \subset D(H)$ so that for any $\psi \in V$, $(\psi, H\psi) \leq a\|\psi\|^2$. That $V \subset D(H)$ is a consequence of the fact that H is bounded from below, which implies that $\text{Ran}(P_{(-\infty, a)}) \subset D(H)$ if $a < \infty$. But then given any $\varphi_1, \ldots, \varphi_{n-1}$, we can find $\psi \in V \cap [\varphi_1, \ldots, \varphi_{n-1}]^\perp$. Thus $U(\varphi_1, \ldots, \varphi_{n-1}) \leq a$ for any $\varphi_1, \ldots, \varphi_{n-1}$ so $\mu_n(H) \leq a$ contradicting the basic hypothesis. This proves (2a).

Next, suppose that (2b) is false. Then $\dim(\text{Ran}(P_{(-\infty, a)})) \leq n - 1$, so we can find $\varphi_1^{(0)}, \ldots, \varphi_{n-1}^{(0)}$ with $[\varphi_1^{(0)}, \ldots, \varphi_{n-1}^{(0)}] = \text{Ran}(P_{(-\infty, a)})$. Then any $\psi \in [\varphi_1^{(0)}, \ldots, \varphi_{n-1}^{(0)}]^\perp \cap D(H)$ is in $\text{Ran}\, P_{[a, \infty)}$, so $(\psi, H\psi) \geq a\|\psi\|^2$. Therefore, $U(\varphi_1^{(0)}, \ldots, \varphi_{n-1}^{(0)}) \geq a$ and $\mu_n \geq a$, contradicting the basic hypothesis. This proves (2b).

Notice that (2) and the fact that H is bounded from below imply that μ_n is finite. There are two distinct cases to consider:

Case 1 $\dim(\text{Ran}\, P_{(-\infty, \mu_n + \varepsilon)}) = \infty$ for all $\varepsilon > 0$. We claim we are then in the situation described by (b) in the theorem. For, by (2a), $\dim(\text{Ran}\, P_{(-\infty, \mu_n - \varepsilon]}) \leq n - 1$ and therefore $\dim P_{(\mu_n - \varepsilon, \mu_n + \varepsilon)} = \infty$ for all $\varepsilon > 0$. By the definition of σ_{ess}, $\mu_n \in \sigma_{\text{ess}}(H)$. On the other hand, again using (2a), if $a < \mu_n$ and $0 < \varepsilon < \mu_n - a$, then $\dim P_{(a-\varepsilon, a+\varepsilon)} < n < \infty$, so $a \notin \sigma_{\text{ess}}(H)$. Thus $\mu_n = \inf\{\lambda \,|\, \lambda \in \sigma_{\text{ess}}(H)\}$. Next notice that in general

$\mu_{n+1} \geq \mu_n$ since one can take $\varphi_{n+1} = \varphi_n$. Therefore $\mu_{n+1} = \mu_n$ for if $\mu_{n+1} > \mu_n$, then dim $P_{(-\infty, \frac{1}{2}\mu_n + \frac{1}{2}\mu_{n+1})} \leq n$ by (2a), contradicting the hypothesis that dim $P_{(-\infty, \mu_n + \varepsilon)} = \infty$. Finally, we note that if there were n eigenvalues strictly below μ_n, and a were the nth eigenvalue, then dim $P_{(-\infty, \frac{1}{2}a + \frac{1}{2}\mu_n)} \geq n$, contradicting (2a). Thus the situation in (b) holds.

Case 2 dim $P_{(-\infty, \mu_n + \varepsilon_0)} < \infty$ for some $\varepsilon_0 > 0$. We claim we are then in the situation described by (a) in the theorem. For by (2), dim $P_{(\mu_n - \varepsilon, \mu_n + \varepsilon)} \geq 1$ for all ε, so $\mu_n \in \sigma(H)$ by the proposition in Section VII.3. But dim $P_{(\mu_n - \varepsilon_0, \mu_n + \varepsilon_0)} < \infty$, so $\mu_n \in \sigma_{\text{disc}}(H)$ by definition. Therefore, μ_n is an eigenvalue and we can find δ with $(\mu_n - \delta, \mu_n + \delta) \cap \sigma(H) = \{\mu_n\}$. Then dim $P_{(-\infty, \mu_n]} = $ dim $P_{(-\infty, \mu_n + \delta)} \geq n$, so there are at least n eigenvalues $E_1 \leq \cdots \leq E_n \leq \mu_n$. If E_n were less than μ_n, then dim $P_{(-\infty, E_n)}$ would equal n, violating (2a). Thus $E_n = \mu_n$, i.e., μ_n is the nth eigenvalue. This shows we are in situation (a). ∎

There is another formulation of the min–max principle which differs on one technical point and which is also useful:

Theorem XIII.2 If H is self-adjoint and bounded from below, then

$$\mu_n = \sup_{\varphi_1, \ldots, \varphi_{n-1}} \inf_{\substack{\psi \in [\varphi_1, \ldots, \varphi_{n-1}]^\perp \\ \|\psi\| = 1; \psi \in Q(H)}} (\psi, H\psi) \tag{3}$$

where $Q(H)$ is the form domain of H.

Proof Let $\tilde{\mu}_n$ denote the right-hand side of (3). By mimicking the proof of Theorem XIII.1, we can prove that each $\tilde{\mu}_n$ obeys either condition (a) or condition (b) of Theorem XIII.1. These conditions determine the μ_n, so $\tilde{\mu}_n = \mu_n$. ∎

We shall turn to applications of the min–max principle in the next three sections. Let us however make a few general remarks:

(1) As we saw in our motivating example, the min–max principle is ideal for comparing eigenvalues of operators. This can have both quantitative and qualitative consequences.

(2) The method can be useful in locating where σ_{ess} begins. In particular, we shall see it tells us that σ_{ess} is empty in some cases (see Sections 4 and 14).

(3) In Sections 4 and 5, we shall prove that in certain cases $\sigma_{\text{ess}}(H) = [a, \infty)$ for explicit a. Suppose we know in addition that $\mu_n < a$. Then we can conclude that H has at least n eigenvalues! Thus the min–max principle can be used to prove the existence of discrete spectrum in some situations.

The following proposition shows how the min-max principle can be applied.

Proposition Let $A \geq 0$ and B be self-adjoint operators. Suppose that $Q(A) \cap Q(B)$ is dense and that B_-, the negative part of B, is relatively form bounded with respect to A with relative bound zero. For $\beta \geq 0$, let $A + \beta B$ denote the operator associated to the obvious form sum. Suppose that $\sigma_{\text{ess}}(A + \beta B) = [0, \infty)$ for all $\beta \geq 0$. Then $\mu_n(A + \beta B)$ is monotone nonincreasing on $[0, \infty)$.

Proof Since $\mu_n(A + \beta B) \leq 0$ for all n, by the hypothesis on $\sigma_{\text{ess}}(A + \beta B)$, we have

$$\mu_n(A + \beta B) = \max_{\varphi_1, \ldots, \varphi_{n-1}} \min_{\substack{\psi \in Q(A) \cap Q(B) \\ \|\psi\|=1, \psi \in [\varphi_1, \ldots, \varphi_{n-1}]^\perp}} [\min\{0, (\psi, (A + \beta B)\psi)\}]$$

Since $A \geq 0$, it is easy to see that $\min\{0, (\psi, (A + \beta B)\psi)\}$ is monotone nonincreasing in β. ∎

Example In Section 4, we shall prove that $\sigma_{\text{ess}}(-\Delta + \beta V) = [0, \infty)$ for a wide class of V's. The proposition shows that for such V's the negative eigenvalues and, in particular, $\inf \sigma(-\Delta + \beta V)$ are monotone decreasing in β and that the number of negative eigenvalues is monotone increasing.

XIII.2 Bound states of Schrödinger operators I: Quantitative methods

The min-max principle is useful for studying a wide variety of aspects of the point spectrum. In this section and the next we shall describe properties of the discrete spectrum of operators of the form $-\Delta + V$, the Hamiltonians of nonrelativistic quantum mechanics. These operators are often called **Schrödinger operators** and the eigenvectors associated to the discrete spectrum are called **bound states**. Points in the discrete spectrum are called **bound state energies** (or **energy levels**). In this section we develop the Rayleigh-Ritz technique for finding bound state energies to very high accuracy. In the next section we concentrate on qualitative aspects of the spectrum. Readers primarily interested in the min-max principle should read this section and the first part of the next.

The Hamiltonian of the elementary model for the helium atom is

$$H = -\Delta_1 - \Delta_2 - \frac{2}{|r_1|} - \frac{2}{|r_2|} + \frac{1}{|r_1 - r_2|}$$

on $L^2(\mathbb{R}^6)$. We have used atomic units, i.e., units with $\hbar = e^2 = 2\mu_e = 1$ where μ_e is the reduced mass of an electron–helium nucleus system. Unlike the hydrogen atom model, it is not possible to solve exactly the eigenvalue problem $H\psi = E\psi$. We wish to describe here methods for finding the lowest eigenvalue of H to high accuracy, but first we should like to explain the physical interest in this eigenvalue. The method also allows one to approximate higher eigenvalues.

The energy needed to ionize the helium atom can be measured. In the simplest model this ionization energy is just the difference between the lowest eigenvalue of H and the ground state energy of a helium ion. In the model the latter quantity is exactly calculable since it is related to the exactly soluble one-body Coulomb problem. In the early days of quantum theory, "crude" agreement (i.e., to about 0.01%) between this model computation and the measured ionization energy was an important experimental confirmation of quantum mechanics.

If one wants to compare experiment and theory to higher accuracy, it is necessary to deal with a more sophisticated model Hamiltonian than H. Since the electron spin becomes significant, it is necessary to use $L^2(\mathbb{R}^6) \otimes \mathbb{C}^4$ as the underlying Hilbert space where $\mathbb{C}^4 = \mathbb{C}^2 \otimes \mathbb{C}^2$ describes the spin of the two electrons. We denote $H \otimes I$ as H. The difference between H and the more sophisticated Hamiltonian is in several pieces, each called a correction. Below we shall briefly describe the physical basis for each of these corrections. Each of them has associated with it an operator A_α so that in the more complicated model the Hamiltonian is $H + \sum A_\alpha$. While we shall not be explicit about the form of the A_α, there are well-tested physical theories that lead to explicit operator corrections (see the Notes for references). The operator corrections are:

(1) *The **Hughes–Eckart** term described in Section XI.5.* The fractional change in the lowest eigenvalue due to this correction should be of order m_e/m_α where m_e is the mass of the electron and m_α is the mass of the helium nucleus. This number is about 1.3×10^{-4}.

(2) *The fine structure corrections* These are certain simple relativistic corrections: the **Sommerfeld correction** which is the difference between the relativistic kinetic energy $\sqrt{(m_e c^2)^2 + (pc)^2} - m_e c^2$ and the nonrelativistic kinetic energy $\frac{1}{2}p^2/m_e$; the **spin–orbit** correction which arises physically from the fact that the electron has a magnetic moment that interacts with the

magnetic field set up by the current the electron sees due to the motion of the nucleus relative to it; the **Darwin correction** which arises physically from the fact that a relativistic electron cannot be localized in regions smaller than $\hbar/mc$, so that the electron interacts not with $V(r)$ but an average of $V(r)$ over a sphere of radius approximately $\hbar/mc$; the **retardation term** which is due to the fact that electron one does not feel a force $\text{grad}_1 |r_1 - r_2|^{-1}$ with r_2 the instantaneous position of the second electron but rather due to the finiteness of the speed of light, it feels the force produced by an electron at the position electron two occupied slightly in the past; the **spin–spin** interaction caused by the interactions of the magnetic moments of the electrons. The spin–orbit and spin–spin corrections can be interpreted as relativistic corrections since the magnetic moment of the electron is a relativistic effect. These relativistic corrections are of order $(e^2/\hbar c)^2 \sim 5 \times 10^{-5}$.

Let us suppose that we knew the lowest eigenvalue E of H and the corresponding eigenfunction. Since the correction term $\sum A_\alpha$ is explicitly known, we could attempt to estimate the shift in the lowest eigenvalue by using low-order perturbation theory. In first order each A_α makes a separate contribution. In higher order there are cross terms, but these are smaller than the accuracy of the computations we shall discuss below. It is thus possible to estimate the shift in E due to $\sum A_\alpha$ as a sum of terms associated to each α. We emphasize that the usual form taken for the A_α is sufficiently singular so that the convergence theorems of Chapter XII are not applicable; in fact it is not known that $H + \sum A_\alpha$ is essentially self-adjoint! As a result, these computations are not on a rigorous footing. This problem could be partly eliminated by replacing a delta function nuclear charge distribution with a C^∞ charge distribution strongly peaked near the origin.

In addition to these operator corrections, there is believed to be a shift in the ionization energy due to interactions with the quantized electromagnetic field. This **Lamb shift** can be computed within the framework of the theory of quantum electrodynamics (QED). This theory does not lead to simple operator corrections to H but it does allow approximate calculations of the shift that is predicted to be about one-millionth of the simple model ionization energy.

Quantum electrodynamics is a theory whose internal structure is not completely understood and which is testable in a limited number of experiments. The physicist thus views the ionization energy of helium as a piece of raw data on which to test QED. Since the Lamb shift is only of order 10^{-6}, the experimental data must be obtained to high accuracy, and the other contributions to the ionization energy must be *computed* to great accuracy. Despite the questions of rigor in the perturbation theory calculation of the shifts due to the A_α, the physicist is willing to accept the predictions of Rayleigh–Schrödinger theory.

The main theoretical problem is then to compute the lowest eigenvalue of H to eight-place accuracy! One might try to start with the exactly soluble operator $-\Delta_1 - 2|r_1|^{-1} - \Delta_2 - 2|r_2|^{-1}$ and treat $|r_1 - r_2|^{-1}$ as a perturbation. In Section XII.2, we saw that the first-order term in the series disagreed with experiment by 15%. Since all the corrections contribute less than 1% to the shift, E presumably differs from first-order perturbation theory by over 10%. To hope to get a numerical answer to one part in 10^8 using perturbation theory is thus effectively impossible. However, one can compute E using a technique suggested by the min–max principle:

Theorem XIII.3 (the Rayleigh–Ritz technique)   Let H be a semibounded self-adjoint operator. Let V be an n-dimensional subspace, $V \subset D(H)$, and let P be the orthogonal projection onto V. Let $H_V = PHP$. Let $\hat{\mu}_1, \ldots, \hat{\mu}_n$ be the eigenvalues of $H_V \upharpoonright V$, ordered by $\hat{\mu}_1 \leq \hat{\mu}_2 \leq \cdots \leq \hat{\mu}_n$. Then

$$\mu_m(H) \leq \hat{\mu}_m, \quad m = 1, \ldots, n$$

In particular, if H has eigenvalues (counting multiplicity) $E_1, \ldots, E_k$ at the bottom of its spectrum with $E_1 \leq \cdots \leq E_k$, then

$$E_m \leq \hat{\mu}_m, \quad m = 1, \ldots, \min(k, n)$$

Proof   By the min–max principle, $H_V \upharpoonright V$ has eigenvalues given by

$$\hat{\mu}_m = \sup_{\varphi_1, \ldots, \varphi_{m-1} \in V} \inf_{\substack{\psi \in V; \|\psi\|=1 \\ \psi \in [\varphi_1, \ldots, \varphi_{m-1}]^\perp}} (\psi, H\psi)$$

$$= \sup_{\varphi_1, \ldots, \varphi_{m-1} \in \mathcal{H}} \inf_{\substack{\psi \in V; \|\psi\|=1 \\ \psi \in [P\varphi_1, \ldots, P\varphi_{m-1}]^\perp}} (\psi, H\psi)$$

$$= \sup_{\varphi_1, \ldots, \varphi_{m-1} \in \mathcal{H}} \inf_{\substack{\psi \in V; \|\psi\|=1 \\ \psi \in [\varphi_1, \ldots, \varphi_{m-1}]^\perp}} (\psi, H\psi)$$

$$\geq \sup_{\varphi_1, \ldots, \varphi_{m-1} \in \mathcal{H}} \inf_{\substack{\psi \in D(H); \|\psi\|=1 \\ \psi \in [\varphi_1, \ldots, \varphi_{m-1}]^\perp}} (\psi, H\psi)$$

$$= \mu_m(H)$$

In the third step we used the fact that for $\psi \in V$, $(\psi, P\varphi) = (\psi, \varphi)$.   ∎

Thus, to obtain upper bounds on the eigenvalues, and in particular on the lowest eigenvalue E_1, we pick an orthonormal set $\eta_1, \ldots, \eta_n \in D(H)$ and diagonalize the $n \times n$ matrix $(\eta_i, H\eta_j)$ on a computer. Before discussing the

XIII.2 Bound states: Quantitative methods

application of this method to the helium atom, let us consider two natural fundamental questions:

(i) We take an orthonormal basis $\{\eta_i\}_{i=1}^\infty$ and obtain an upper bound $\hat{\mu}_1^{(n)}$ on E_1 by diagonalizing $\{(\eta_i, H\eta_j)\}_{1 \le i, j \le n}$. Does $\hat{\mu}_1^{(n)}$ converge to E_1 as $n \to \infty$?
(ii) Is there any way of obtaining lower bounds on eigenvalues so that one can determine the accuracy of the upper bound?

To answer question (i), we note:

Theorem XIII.4 Let $\{\eta_i\}_{i=1}^\infty$ be an orthonormal basis for $\mathcal{H}$ with each $\eta_i \in D(H)$, the domain of a semibounded self-adjoint operator H. Suppose that $\mu_1(H)$ is an eigenvalue E_1 of H with a normalized eigenvector $\psi = \sum_{i=1}^\infty a_i \eta_i$. Suppose that

$$\lim_{N \to \infty} \left(\sum_{i=1}^N a_i \eta_i, H\left(\sum_{i=1}^N a_i \eta_i \right) \right) = \mu_1(H)$$

Then $\hat{\mu}_1^{(n)} \to E_1$, where $\hat{\mu}_1^{(n)}$ is the lowest eigenvalue of $\{(\eta_i, H\eta_j)\}_{1 \le i, j \le n}$.

Proof See Problem 3.

The importance of this theorem is that the convergence of $\lim_{N \to \infty} (\sum_{i=1}^N a_i \eta_i, H(\sum_{i=1}^N a_i \eta_i))$ to $\mu_1(H) = (\psi, H\psi)$ can sometimes be proven on abstract grounds. For example, it always holds if H is bounded since $\sum_{i=1}^N a_i \eta_i \to \psi$ in norm. The method needed to prove convergence in case H is not bounded is nicely illustrated by:

Example 1 Suppose that $H = H_0 + V$ with $V \in L^2(\mathbb{R}^3) + (L^\infty(\mathbb{R}^3))_\varepsilon$ and that $\mu_1(H) < 0$. Then one knows that $\mu_1(H)$ is an eigenvalue (see Section 4) which is nondegenerate (see Section 12) and that the corresponding eigenfunction is contained in $D(|x|^2)$ (see Section 11). Since $\psi \in D(H_0)$, $\psi \in D(H_0 + x^2) = D(H_0) \cap D(x^2)$. Let $\{\varphi_i\}$ be the eigenfunctions of the three-dimensional harmonic oscillator (see the Appendix to Section V.3). Then,

$$\lim_{N \to \infty} \left(\left(\sum_{i=1}^N a_i \varphi_i - \psi \right), (H_0 + x^2)\left(\sum_{i=1}^N a_i \varphi_i - \psi \right) \right) = 0$$

But since $V \in L^2 + L^\infty$, $|V| \le H_0 + b$ for suitable b so

$$H_0 + |V| \le 2H_0 + b \le 2(H_0 + x^2) + b$$

Therefore,
$$\lim_{N\to\infty}\left(\left(\sum_{i=1}^{N}a_i\varphi_i - \psi\right), (H_0 + V)\left(\sum_{i=1}^{N}a_i\varphi_i - \psi\right)\right) = 0$$
which implies
$$\lim_{N\to\infty}\left(\sum_{i=1}^{N}a_i\varphi_i, (H_0 + V)\sum_{i=1}^{N}a_i\varphi_i\right) = \mu_1$$
By Theorem XIII.4, $\hat{\mu}_1^{(n)} \to \mu_1$.

Example 2 Let $H = p^2 + x^2 + \beta x^4$ on $L^2(\mathbb{R})$ where $\beta > 0$. Let φ_i be the eigenfunctions of the harmonic oscillator and let ψ be the lowest eigenvector of H. Let $\psi = \sum_{i=1}^{\infty} a_i \varphi_i$. From the form estimates
$$0 \leq (p^2 + x^2 + \beta x^4) \leq C_1(p^2 + x^2)^2 \leq C_2(p^2 + x^2 + \beta x^4)^2$$
one can show that
$$\lim_{N\to\infty}\left(\sum_{i=1}^{N}a_i\varphi_i, H\sum_{i=1}^{N}a_i\varphi_i\right) = \mu_1(H)$$
so again $\hat{\mu}_1^{(n)} \to \mu_1$. The details are left to the reader as Problem 4.

Notice that it is not really necessary that $\{\eta_i\}_{i=1}^{\infty}$ be an orthonormal basis. What is important is that the ground state eigenvector ψ be in the space generated by the $\{\eta_i\}_{i=1}^{\infty}$ and that the convergence hypothesis in Theorem XIII.4 holds. The use of symmetries and this remark often allow one to simplify computations. For instance, in Example 2 above ψ is an even function of x and so $\psi = \sum_{i=1}^{\infty} a_{2i-1}\varphi_{2i-1}$ and we need only diagonalize $\{(\varphi_{2i-1}, H\varphi_{2j-1})\}_{1\leq i,j\leq N}$ to estimate $\mu_1(H)$ or any $\mu_{2k+1}(H)$.

The second question raised above is more interesting. There are several different lower bound techniques; we discuss the simplest one here and mention others briefly in the Notes.

Theorem XIII.5 (Temple's inequality) Let H be a self-adjoint operator that is bounded from below. Suppose that $\mu_1 < \mu_2$ and $(\psi, H\psi) < \check{\mu}_2$ where $\psi \in D(H)$, $\|\psi\| = 1$ and $\check{\mu}_2$ is some number less than μ_2. Then
$$\mu_1 \geq (\psi, H\psi) - \frac{(\psi, H^2\psi) - (\psi, H\psi)^2}{\check{\mu}_2 - (\psi, H\psi)}$$

Proof Since μ_1 is a discrete eigenvalue and $\sigma(H)\backslash\{\mu_1\} \subset [\check{\mu}_2, \infty)$, we have $(H - \mu_1)(H - \check{\mu}_2) \geq 0$. Thus,

$$(\psi, (H - \check{\mu}_2)H\psi) \geq \mu_1(\psi, (H - \check{\mu}_2)\psi)$$

By hypothesis, $(\psi, (H - \check{\mu}_2)\psi) < 0$, so

$$\mu_1 \geq \frac{\check{\mu}_2(\psi, H\psi) - (\psi, H^2\psi)}{\check{\mu}_2 - (\psi, H\psi)} = (\psi, H\psi) - \frac{(\psi, H^2\psi) - (\psi, H\psi)^2}{\check{\mu}_2 - (\psi, H\psi)} \quad \blacksquare$$

Notice that the lower bound for μ_1 given by Temple's inequality is close to the upper bound given by Rayleigh-Ritz if ψ is "almost" an eigenvector in the sense that $(\psi, (H - \langle H \rangle)^2 \psi)$ is small where $\langle H \rangle = (\psi, H\psi)$.

Temple's inequality requires as input a crude lower bound on the second eigenvalue. This can often be achieved as follows.

Definition Let A and B be self-adjoint operators that are bounded from below. Then we say $A \leq B$ if and only if $Q(B) \subset Q(A)$ and $(\varphi, A\varphi) \leq (\varphi, B\varphi)$ for all φ in $Q(B)$.

The reader is asked in Problem 1 to prove that if $A \leq B$, then $\mu_n(A) \leq \mu_n(B)$.

Example 3 Let H be the helium atom Hamiltonian and let

$$A = -\Delta_1 - \Delta_2 - \frac{2}{|r_1|} - \frac{2}{|r_2|}$$

Then $A \leq H$, so $\mu_2(H) \geq \mu_2(A) = -\frac{5}{4}$.

The application of the Rayleigh-Ritz method to the ground state of helium goes back to the earliest days of quantum theory when Hylleraas made a simple variational Rayleigh-Ritz calculation by hand. In those days the calculation was made as support for elementary quantum mechanics and a calculation with $n = 6$ was sufficient. With the advent of high-speed computers and the need for an accurate computation of $\mu_1(H)$ to test the Lamb shift, more elaborate calculations were needed. An early calculation involving the diagonalization of a 39×39 matrix was made by Kinoshita and one with 1078 parameters by Perkeris! Perkeris' results are summarized by the accompanying tabulation.

Contributions to the Ionization Energy (in units of cm^{-1})	
ΔE (purely Coulombic model)	198 317.374
Hughes–Eckart shift	-4.785
Relativistic corrections	-0.562
Lamb shift theory	-1.351 ± 0.02
Theoretical value	198 310.676 $\pm$ 0.02
Experimental value	198 310.82 $\pm$ 0.15

The error listed for the Lamb shift represents uncertainties in the calculations and mild disagreements between the calculations of various authors. We note that Temple's inequality gives an *upper* bound for $\Delta E = E_{\text{ion}} - E_{\text{atom}}$ of 198 317.866.

We thus see that to within experimental error the Lamb shift agrees with experiment to 1%. This verification of a QED prediction would have been impossible without the accurate calculation of the lowest eigenvalue made possible by the Rayleigh–Ritz method.

XIII.3 Bound states of Schrödinger operators II: Qualitative theory

In this section we discuss certain qualitative features of $N(V)$, the number of bound states of $-\Delta + V$ counting multiplicity. In the first part we investigate whether $N(V)$ is finite or infinite. In the latter parts we obtain bounds on $N(V)$. The min–max principle is used throughout, but to obtain bounds further analytic techniques are required.

A. Is $\sigma_{\text{disc}}(H)$ infinite or finite?

The min–max principle is useful for establishing the existence of eigenvalues, in particular in showing that $H_0 + V$ has an infinite discrete spectrum in certain cases. We shall first prove a result in the two-body case. Intuitively, if there are an infinite number of bound states, those with small binding energy should be spatially spread out and so should be more sensitive to the large x behavior of $V(x)$ than the small x behavior. We thus expect in the two-body case that the question of whether or not $H_0 + V$ has an infinity of bound states to depend on the large x behavior of V. We shall see that the borderline behavior is $|x|^{-2}$. The proof will depend on three things besides the min-max principle: (1) the "uncertainty principle lemma" of Section X.2, i.e., $-\Delta - \frac{1}{4}r^{-2} \geq 0$; (2) that $\sigma_{\text{ess}}(-\Delta + V) = [0, \infty)$ when $V \in R + (L^\infty)_\varepsilon$ where R is the Rollnik class: we shall prove this in the next section; (3) that $-\Delta + V$ has only finitely many bound states when $V \in R$; we shall prove this in Part C below.

Theorem XIII.6 Consider the operator $-\Delta + V$ on $L^2(\mathbb{R}^3)$.

(a) Suppose that $V \in R + (L^\infty)_\varepsilon$ and that V satisfies
$$V(x) \leq -ar^{-2+\varepsilon} \quad \text{if} \quad r \equiv |x| > R_0$$
for some R_0 and some $a > 0$, $\varepsilon > 0$. Then $\sigma_{\text{disc}}(-\Delta + V)$ is infinite.

(b) Suppose that $V \in R + (L^\infty)_\varepsilon$ and that V satisfies
$$V(x) \geq -\tfrac{1}{4}br^{-2} \quad \text{if} \quad r > R_0$$
for some R_0 and some $b < 1$. Then $\sigma_{\text{disc}}(-\Delta + V)$ is finite.

Proof (a) Since $\sigma_{\text{ess}}(-\Delta + V) = [0, \infty)$, it is sufficient to show $\mu_n < 0$ for each n. Pick a nonnegative C^∞ function ψ with support in $\{x \mid 1 < |x| < 2\}$ and $\|\psi\| = 1$. Let $\psi_R(x) = R^{-3/2}\psi(xR^{-1})$ so $\|\psi_R\| = 1$ and $\text{supp}\,\psi_R \subset \{x \mid R < r < 2R\}$. If $R > R_0$, we conclude that

$$(\psi_R, H\psi_R) = (\psi_R, -\Delta\psi_R) + (\psi_R, V\psi_R)$$
$$\leq (\psi_R, -\Delta\psi_R) - a(\psi_R, r^{-2+\varepsilon}\psi_R)$$
$$= R^{-2}(\psi, -\Delta\psi) - aR^{-2+\varepsilon}(\psi, r^{-2+\varepsilon}\psi)$$

Since $\varepsilon > 0$, this last expression is negative for large R. Thus we can find Q so that $(\psi_R, H\psi_R) < 0$ if $R > Q$. Now let $\varphi_n = \psi_{2^n Q}$, $n = 1, 2, \ldots$. The φ_n are orthonormal and $(\varphi_n, H\varphi_m) = 0$ if $n \neq m$ since φ_n and φ_m have disjoint supports if $n \neq m$. Thus, given N, and taking $V_N = \text{span}\{\varphi_1, \ldots, \varphi_N\}$, we see that $P_N H P_N \upharpoonright V_N$ has eigenvalues $\{(\varphi_n, H\varphi_n)\}_{n=1}^N$. By the Rayleigh-Ritz principle

$$\mu_N(H) \leq \sup_{1 \leq m \leq N} \{(\varphi_m, H\varphi_m)\} < 0$$

Since N is arbitrary and $\sigma_{\text{ess}} = [0, \infty)$, $-\Delta + V$ has an infinity of eigenvalues.

(b) Let $W = V + \tfrac{1}{4}br^{-2}$. Then as forms on $Q(H_0)$,

$$-\Delta + V = -(1-b)\Delta + W + b(-\Delta - \tfrac{1}{4}r^{-2})$$
$$\geq -(1-b)\Delta + W$$
$$\geq -(1-b)\Delta + \tilde{W}$$

where $\tilde{W} = \min(W, 0)$. Thus by the min–max principle

$$\mu_n(-\Delta + V) \geq \mu_n(-(1-b)\Delta + \tilde{W}) = (1-b)\mu_n(-\Delta + (1-b)^{-1}\tilde{W})$$

By hypothesis, $\tilde{W}$ has compact support and lies in $R + (L^\infty)_\varepsilon$. A simple exercise (Problem 70 of Chapter XI) shows that if a potential in $R + (L^\infty)_\varepsilon$ has compact support, it is in R. Thus $-\Delta + (1-b)^{-1}\tilde{W}$ has only finitely many bound states, so $\mu_n(-\Delta + (1-b)^{-1}\tilde{W}) = 0$ for $n \geq N_0$ for some N_0. Thus

$$0 \geq \mu_n(-\Delta + V) \geq (1-b)\mu_n(-\Delta + (1-b)^{-1}\tilde{W}) = 0$$

if $n \geq N_0$. Therefore, $-\Delta + V$ has at most N_0 eigenvalues in $(-\infty, 0)$. ∎

Looking back at the proof, we can see why r^{-2} is the critical borderline behavior. $-\Delta$ scales as a homogeneous function of degree -2 under the scaling $\psi \to \psi_R$. Thus, if V scales at large distances with degree $-2 + \varepsilon$, it wins out; and if it scales with degree $-2 - \varepsilon$, the kinetic energy wins out and spread-out states cannot have negative energy.

For N-body systems, the situation is more complicated and is not merely dependent on the long-range behavior of V. For example, there is a three-body system with two-body potentials, $V = V_{01}(r_1) + V_{02}(r_2) + V_{12}(r_1 - r_2)$, so that the number of bound states $N(\lambda)$ of $H_\lambda = -\Delta_1 - \Delta_2 + \lambda V$ has the following behavior: For $\lambda = 0$, $N(\lambda) = 0$; as λ is increased from 0, $N(\lambda)$ increases one by one until suddenly at $\lambda = \lambda_1$, $N(\lambda) = \infty$. As λ is increased further, a new critical value λ_2 is reached; $N(\lambda) = \infty$ if $\lambda_1 \leq \lambda \leq \lambda_2$. As λ is further increased, new critical values $\lambda_3, \lambda_4, \ldots$ enter so that $N(\lambda) = \infty$ if $\lambda \in [\lambda_1, \lambda_2] \cup [\lambda_3, \lambda_4] \cup [\lambda_5, \lambda_6] \cup \cdots$ and $N(\lambda) < \infty$ if $\lambda \in [0, \lambda_1) \cup (\lambda_2, \lambda_3) \cup (\lambda_4, \lambda_5) \cup \cdots$. A reference for this example appears in the Notes.

The complicating feature of N-body systems is that their essential spectrum is $[\Sigma, \infty)$ where $\Sigma < 0$ is possible (see Section 5). Thus, in the example mentioned above $\mu_n(H_{\lambda_2}) < \Sigma_{\lambda_2}$ for all n. As λ is increased past λ_2, $\mu_n(H_\lambda)$ decreases below $\mu_n(H_{\lambda_2})$; but Σ_λ decreases also and can overtake $\mu_n(H_\lambda)$ so that for $\lambda > \lambda_2$, $\mu_n(H_\lambda) = \Sigma_\lambda$ for n large.

Let us consider the Hamiltonian of the helium atom. Heuristically, states in the essential spectrum should not be bound and thus should involve clusterings of particles far from one another. For example, electron one might be bound while two goes to ∞. Such states will have energy beginning at -1, the ground state energy of $-\Delta_1 - 2|r_1|^{-1}$. Considering the other possibilities, one sees that $\sigma_{\text{ess}}(H) = [-1, \infty)$ is to be expected on physical grounds. In fact, we shall prove this in Section 5. Knowing this, we can now show how the min–max principle can be used in the many-body case:

Proposition (Kato)
$$H = -\Delta_1 - 2|r_1|^{-1} - \Delta_2 - 2|r_2|^{-1} + |r_1 - r_2|^{-1}$$
has an infinite discrete spectrum.

Proof We need only show that $\mu_n(H) < -1$ for all n since $\sigma_{\text{ess}}(H) = [-1, \infty)$. By the Rayleigh-Ritz method, we need only find, for any n, an n-dimensional space V_n with $\sup(\psi, H\psi) < -\|\psi\|^2$ for all $\psi \in V_n$. Intuitively, we should find V_n by putting electron one in its ground state and putting electron two in a suitable state. Let $\psi_1(r_1)$ be the normalized ground state of $-\Delta_1 - 2|r_1|^{-1}$, i.e., $(-\Delta_1 - 2|r_1|^{-1})\psi_1 = -\psi_1$, and let $\varphi_n(r_2)$ be the normalized nth eigenfunction of $-\Delta_2 - |r_2|^{-1}$, i.e. $(-\Delta_2 - |r_2|^{-1})\varphi_n = E_n \varphi_n$ where $E_1, E_2, \ldots = -\frac{1}{4}, -\frac{1}{16}, -\frac{1}{16}, \ldots$, where $-(4n^2)^{-1}$ is repeated n^2 times. Now let $\psi = \sum_{i=1}^{n} \alpha_i \psi_1(r_1)\varphi_i(r_2)$ with $\sum_{i=1}^{n} |\alpha_i|^2 = 1$. Then $(\psi, H\psi) = t_1 + t_2 + t_3$ where

$$t_1 = (\psi, (-\Delta_1 - 2|r_1|^{-1})\psi) = (\psi, -\psi) = -1$$

$$t_2 = (\psi, (-\Delta_2 - |r_2|^{-1})\psi) = \sum_{i=1}^{n} |\alpha_i|^2 E_i \leq E_n$$

$$t_3 = (\psi, |r_1 - r_2|^{-1} - |r_2|^{-1})\psi) \leq 0$$

The last inequality $t_3 \leq 0$ follows from the fact that $\psi_1(r_1)$ is spherically symmetric so

$$\int |r_1 - r_2|^{-1} |\psi_1(r_1)|^2 \, dr_1 = \int \min\{|r_1|^{-1}, |r_2|^{-1}\} |\psi_1(r_1)|^2 \, dr_1$$

$$\leq |r_2|^{-1}$$

Thus, letting $V_n = \text{span}\{\psi_1 \varphi_1, \ldots, \psi_1 \varphi_n\}$, we see that

$$(\psi, H\psi) \leq -1 + E_n$$

if $\psi \in V_n$. Thus $\mu_n(H) \leq -1 + E_n < -1$. This shows that $\sigma_{\text{disc}}(H)$ is an infinite set. ∎

The above arguments have been extended in various directions. Typical of the results is:

Theorem XIII.7 (Zhislin) Let H be an "atomic" Hamiltonian of the form

$$H = \sum_{i=1}^{n} \left(-\frac{\Delta_i}{2\mu_i} - \frac{n}{|r_i|} \right) + \sum_{i<j} \left(\frac{\nabla_i \cdot \nabla_j}{M} + \frac{1}{|r_i - r_j|} \right)$$

as an operator on $L^2(\mathbb{R}^{3n})$ where M, $\{\mu_i\}_{i=1}^n$ are arbitrary positive numbers. Then $\sigma_{\text{disc}}(H)$ is infinite.

H is the Hamiltonian of a system consisting of a nucleus of mass M and n electrons of masses $\mu_1, \ldots, \mu_n$ after the center of mass motion is removed.

B. Bounds on $N(V)$ in the central case

Recall that a **central potential** is a real-valued function that is a function of $r = |x|$ alone. In that case, if E is an eigenvalue of $H = -\Delta + V$, then $\{\psi \mid H\psi = E\psi\}$ is a rotationally invariant subspace of $L^2(\mathbb{R}^3)$ and so is spanned by vectors of the form $\psi(x) = r^{-1}f(r)Y_{\ell m}(\theta, \varphi)$ where $Y_{\ell m}$ are the spherical harmonics and $\int_0^\infty |f(r)|^2 \, dr < \infty$. ℓ is called the **angular momentum** of ψ. If $\psi \in D(H_0)$, it is bounded and continuous, so $f(r)$ is continuous and $f(0) = 0$. Moreover, f obeys the differential equation

$$\left(-\frac{d^2}{dr^2} + \frac{\ell(\ell+1)}{r^2} + V(r)\right)f(r) = Ef(r) \tag{4}$$

(4) holds in the sense of a differential operator on $L^2(0, \infty)$ with boundary condition $f(0) = 0$ rather than in the sense of classical differential equations. To avoid troubles with this fact, we shall suppose that V is C^∞ with compact support. The elliptic regularity theorem then implies that solutions of (4) are C^∞. We shall later see how to remove the C_0^∞ assumption when we pass to the bounds. Alternatively, we could use the Jost function methods of Section XI.8 and allow fairly general V's throughout.

For any integral ℓ and $E \leq 0$, let $u_\ell(r; E)$ be the solution of (4) obeying $u_\ell(0; E) = 0$ and $\lim_{r \downarrow 0} r^{-\ell-1} u_\ell(r; E) = 1$. The existence and uniqueness of such solutions when $V \in C_0^\infty$ is fairly easy (see Section XI.8). We are interested in the number of $E < 0$ for which $u_\ell(r; E)$ is square integrable. This is precisely $n_\ell(V)$, the "number" of discrete bound states of $-\Delta + V$ with angular momentum ℓ, in the sense that for fixed ℓ and m it is the number of discrete bound states of the form $r^{-1}f(r)Y_{\ell m}(\theta, \varphi)$. Since m can equal $-\ell$, $-\ell + 1, \ldots, \ell$ there are really $(2\ell + 1)n_\ell(V)$ discrete eigenfunctions of angular momentum ℓ, so $N(V) = \sum (2\ell + 1)n_\ell(V)$. There is a classical theorem about $n_\ell(V)$:

Theorem XIII.8 Suppose that $V \in C_0^\infty(0, \infty)$ and let $N_\ell(E; V)$ be the number of zeros of $u_\ell(r; E)$ different from $r = 0$. Then:

(a) If $E < 0$, then $N_\ell(E; V) < \infty$.

XIII.3 Bound states: Qualitative theory

(b) $N_\ell(E; V)$ is a monotone increasing function of E and a monotone decreasing function of ℓ.

(c) If $E_0 \leq 0$, $N_\ell(E_0; V)$ is identical to the number of square integrable solutions of (4) with $E < E_0$. In particular, $n_\ell(V) = N_\ell(0; V)$ and $\dim P_{(-\infty, E)} = \sum_{\ell=0}^{\infty} (2\ell + 1) N_\ell(E; V)$ for $E \leq 0$ where $\{P_\Omega\}$ are the spectral projections for $-\Delta + V(|x|)$.

We shall prove this theorem in a sequence of lemmas. While the classical proof uses notions from the theory of differential equations, we are able to avoid most of this machinery by appealing to the min–max principle several times. Before turning to the proof, let us explain why there should be a connection between N_ℓ and n_ℓ. Suppose E is so negative that $E - V$ is everywhere negative. By the differential equation, u and u'' have the same sign. So, if u and u' start out positive, they remain positive. Thus $N_\ell(E; V) = 0$ if E is very negative. We shall see that as E decreases from $E = 0$, the zeros of $u_\ell(r; E)$ move to larger r. They disappear by moving out to $r = \infty$. Since $N_\ell(E; V) = 0$ for E very negative, $N_\ell(0; V)$ is the number of zeros that have moved out to infinity. The point is that those energies at which zeros have just moved out to infinity are intuitively precisely those energies for which $\lim_{r \to \infty} u_\ell(r; E) = 0$. Since V has compact support, $u_\ell(r; E) \sim a \exp(-\sqrt{-E}\, r) + b \exp(+\sqrt{-E}\, r)$ for r large. If

$$\lim_{r \to \infty} u_\ell(r; E) = 0,$$

$u_\ell(r; E) \sim a \exp(-\sqrt{-E}\, r)$ for r large, so u_ℓ is square integrable. So the number of eigenvalues should be the number of zeros.

Lemma 1 Let $E_0 \leq 0$. If $N_\ell(E_0; V) \geq m_0$, then (4) has at least m_0 square integrable solutions with $E < E_0$. In particular, since $\sigma_{\text{ess}}(-\Delta + V) = [0, \infty)$, $N_\ell(E_0; V) < \infty$ if $E_0 < 0$.

Proof Let

$$H_\ell = -\frac{d^2}{dr^2} + \frac{\ell(\ell + 1)}{r^2} + V(r)$$

as an operator on $L^2(0, \infty)$. It is essentially self-adjoint with domain $\{u \mid u \in C_0^\infty[0, \infty); u(0) = 0\}$ if $\ell = 0$ and on $\{u \mid u \in C_0^\infty(0, \infty)\}$ if $\ell \neq 0$ (Problem 16). We shall first show if $N_\ell(E_0; V) \geq m_0$, then $\mu_{m_0}(H_\ell) \leq E_0$. For let $r_0 = 0 < r_1 < r_2 < \cdots < r_{m_0} < \infty$ be zeros of $u_\ell(r; E_0)$. Define ψ_i by

$$\psi_i(r) = \begin{cases} u_\ell(r; E_0), & r_{i-1} \leq r \leq r_i \\ 0, & \text{otherwise} \end{cases}$$

Then the ψ_i are continuous and piecewise C^1. While $\psi_i \notin D(-d^2/dr^2)$, it is in $Q(-d^2/dr^2)$ and so in $Q(H_\ell)$ (see Problem 17). Moreover

$$\left(\sum_{i=1}^{m_0} a_i\psi_i, H_\ell \sum_{i=1}^{m_0} a_i\psi_i\right) = \left(\frac{d}{dr}\left(\sum_{i=1}^{m_0} a_i\psi_i\right), \frac{d}{dr}\left(\sum_{i=1}^{m_0} a_i\psi_i\right)\right)$$
$$+ \left(\sum_{i=1}^{m_0} a_i\psi_i, \left(\frac{\ell(\ell+1)}{r^2} + V\right)\sum_{i=1}^{m_0} a_i\psi_i\right)$$
$$= \sum_{i=1}^{m_0} |a_i|^2 \int_{r_{i-1}}^{r_i} \left[|\psi_i'|^2 + \left(\frac{\ell(\ell+1)}{r^2} + V\right)|\psi_i|^2\right] dr$$
$$= \sum_{i=1}^{m_0} |a_i|^2 \int_{r_{i-1}}^{r_i} \bar{\psi}_i\left[-\psi_i'' + \left(\frac{\ell(\ell+1)}{r^2} + V\right)\psi_i\right] dr$$
$$= E_0\left(\sum_{i=1}^{m_0} a_i\psi_i, \sum_{i=1}^{m_0} a_i\psi_i\right)$$

Thus we have an m_0-dimensional subspace of $Q(H_\ell)$ on which the expectation value of H_ℓ is less than or equal to E_0, so $\mu_{m_0}(H_\ell) \leq E_0$, by Theorem XIII.2.

Next, we note that if $E_n \to E_\infty$, then $u_\ell(r; E_n)$ converges to $u_\ell(r; E_\infty)$ uniformly on compact subsets of $[0, \infty)$. This is because of the continuity of solutions of ordinary differential equations in their coefficients. Thus $u_\ell(r; E_0 - 1/n)$ converges uniformly on compacts to $u_\ell(r; E_0)$. It follows that if $u_\ell(r; E_0)$ has at least m_0 zeros, then for some n, $u_\ell(r; E_0 - 1/n)$ has at least m_0 zeros so $\mu_{m_0}(H_\ell) \leq E_0 - 1/n < E_0$. Since $\sigma_{\text{ess}}(H_\ell) \subset \sigma_{\text{ess}}(-\Delta + V) = [0, \infty)$, the lemma follows from the min-max principle. ∎

Lemma 2 (Sturm oscillation theorem) $N_\ell(E; V)$ is a monotone increasing function of E. If $E_0 < 0$ is an eigenvalue of H_ℓ, then $N_\ell(E; V) \geq N_\ell(E_0; V) + 1$ when $E > E_0$.

Proof Let $V_\ell = \ell(\ell+1)r^{-2} + V$ and let $r_0 = 0 < r_1 < \cdots < r_n < \infty$ be the zeros of $u_\ell(r; E_0)$. Let $E > E_0$. We shall show that $u_\ell(r, E)$ has at least one zero in each of the intervals $(0, r_1), (r_1, r_2), \ldots, (r_{n-1}, r_n)$ and when E_0 is an eigenvalue, also in the interval (r_n, ∞). For suppose $u_\ell(r; E)$ has no zero in (r_i, r_{i+1}). By replacing u_ℓ with $-u_\ell$ if necessary, we can suppose that both $u_\ell(r; E)$ and $u_\ell(r; E_0)$ are nonnegative in (r_i, r_{i+1}). In particular, $u_\ell'(r_i; E_0) \geq 0$ and $u_\ell'(r_{i+1}; E_0) \leq 0$. Let us compute

$$I = \int_{r_i}^{r_{i+1}} [u_\ell'(r; E_0)u_\ell(r; E) - u_\ell(r; E_0)u_\ell'(r; E)]' \, dr$$

On one hand

$$I = [u'_\ell(r; E_0)u_\ell(r; E) - u_\ell(r; E_0)u'_\ell(r; E)]\Big|_{r_i}^{r_{i+1}}$$
$$= u'_\ell(r_{i+1}; E_0)u_\ell(r_{i+1}; E) - u'_\ell(r_i; E_0)u_\ell(r_i; E)$$
$$\leq 0$$

On the other hand

$$I = \int_{r_i}^{r_{i+1}} [u''_\ell(r; E_0)u_\ell(r; E) - u_\ell(r; E_0)u''_\ell(r; E)]\,dr$$
$$= \int_{r_i}^{r_{i+1}} u_\ell(r; E_0)u_\ell(r; E)[(V_\ell - E_0) - (V_\ell - E)]\,dr$$
$$= (E - E_0)\int_{r_i}^{r_{i+1}} u_\ell(r; E_0)u_\ell(r; E)\,dr > 0$$

This contradiction shows that $u_\ell(r, E)$ must have a zero in (r_i, r_{i+1}). Now suppose E_0 is an eigenvalue and $\ell = 0$. Since V has compact support, $u_0(r, E_0) = a\exp(-\sqrt{-E_0}\,r)$ for r large and $|u_0(r; E)| + |u'_0(r; E)| \leq b\exp(+\sqrt{-E}\,r)$ for r large. As a result, the integrals I and the other integrals we wrote down above converge if we replace r_{i+1} with ∞ and r_i with r_n. Thus $u_0(r; E)$ has a zero in (r_n, ∞) by the same reasoning. A similar argument works for general ℓ if we replace the exponential by suitable Bessel functions. We conclude that the lemma is valid. ∎

Lemma 3 $N_\ell(\mu_n(H_\ell); V) = n - 1$ if $\mu_n(H_\ell) < 0$.

Proof If $N_\ell(\mu_n(H_\ell); V) > n - 1$, then, by Lemma 1, there are at least n eigenvalues in $(-\infty, \mu_n)$, so we see that $N_\ell(\mu_n(H_\ell)) \leq n - 1$ by the min-max principle. On the other hand, we shall prove that $N_\ell(\mu_n(H_\ell)) \geq n - 1$ by induction. Certainly $N_\ell(\mu_1(H_\ell)) \geq 0$. Suppose $\mu_n < 0$ and $N_\ell(\mu_{n-1}(H_\ell)) \geq n - 2$. Since $\mu_{n-1} \leq \mu_n < 0$, μ_{n-1} is an eigenvalue and μ_n is an eigenvalue, so $\mu_n > \mu_{n-1}$ since (4) can have at most one solution with $f(0) = 0$. Thus, by Lemma 2,

$$N_\ell(\mu_n(H_\ell)) \geq N_\ell(\mu_{n-1}(H_\ell)) + 1 = n - 1 \quad \blacksquare$$

We are now ready for

Proof of Theorem XIII.8 We have already proven (a) and that $N_\ell(E; V)$ is monotone in E. On one hand, by Lemma 1, the number of bound states with energy less than E is not smaller than $N_\ell(E; V)$. On the other hand, suppose it is larger than $N_\ell(E; V)$. This is impossible if $E = 0$ and $N_\ell = \infty$, so suppose $N_\ell(E; V) = n < \infty$. We are supposing that there are $n + 1$ eigenvalues below E so $\mu_{n+1}(H_\ell) < E$. But $N_\ell(\mu_{n+1}(H_\ell)) = n$ and μ_{n+1} is an eigenvalue; so by Lemma 2, $N_\ell(E; V) \geq n + 1$. This contradiction shows that $N_\ell(E; V)$ is the number of eigenvalues of H_ℓ in $(-\infty, E)$. Since $H_{\ell+1} \geq H_\ell$, we conclude $N_\ell(E; V) \geq N_{\ell+1}(E; V)$. ∎

We can use Theorem XIII.8 to prove a variety of bounds on $n_\ell(V)$ and on $\ell_{max}(V) = \max\{\ell \mid n_\ell(V) > 0\}$, the largest angular momentum that has bound states:

Theorem XIII.9 Let V be a central potential with $V \in R + L^\infty(\mathbb{R}^3)_\varepsilon$. Then:

(a) (Bargmann's bound)
$$n_\ell(V) \leq (2\ell + 1)^{-1} \int_0^\infty r|V(r)|\,dr$$

(b) (Calogero's bound) Suppose V is monotone increasing and negative. Then
$$n_\ell(V) \leq \frac{2}{\pi} \int_0^\infty |V(r)|^{1/2}\,dr$$

(c) (the GGMT bound) For any $p \geq 1$,
$$n_\ell(V) \leq c_p(2\ell + 1)^{-(2p-1)} \int_0^\infty r^{2p-1}|V(r)|^p\,dr$$
where $c_p = (p-1)^{p-1}\Gamma(2p)/p^p\Gamma(p)^2$ and Γ is the Euler gamma function.

(d) If $\int_0^\infty |V(r)|^{1/2}\,dr < \infty$ and if $V(r) < 0$ on some open set, then for each ℓ, there exist Λ, a, and b with $0 < a < b < \infty$ so that if $\lambda > \Lambda$,
$$a\lambda^{1/2} \leq n_\ell(\lambda V) \leq b\lambda^{1/2}$$

(e) If $\int_0^\infty r|V(r)|\,dr < \infty$ and $V(r) < 0$ on an open set, then there are c, d, and Λ with $0 < c < d < \infty$ so that if $\lambda > \Lambda$,
$$c\lambda^{1/2} \leq \ell_{max}(\lambda V) \leq d\lambda^{1/2}$$

Proof We shall prove only Bargmann's bound in case $\ell = 0$ (see Problem 19 and the references in the Notes for other parts of the proof). We first remark that it is enough to prove the bound when $V \leq 0$. For let V be

XIII.3 Bound states: Qualitative theory

arbitrary and let $V_- = \min\{V, 0\}$. Then by the min–max principle, $n_\ell(V_-) \geq n_\ell(V)$ since $-\Delta + V_- \leq -\Delta + V$. Thus, if we know $n_0(V_-) \leq \int_0^\infty r|V_-(r)|\,dr$, we can conclude

$$n_0(V) \leq n_0(V_-) \leq \int_0^\infty r|V_-(r)|\,dr \leq \int_0^\infty r|V(r)|\,dr \tag{5}$$

Next we remark that we need prove only the theorem for the case where $V \in C_0^\infty$ and is nonpositive, for an arbitrary $V \in R + (L^\infty)_\varepsilon$ with $V \leq 0$ and $\int_0^\infty r|V(r)|\,dr < \infty$ can be approximated by nonpositive $V_n \in C_0^\infty$ in such a way that $\int_0^\infty r|V_n(r)|\,dr \uparrow \int_0^\infty r|V(r)|\,dr$ and $V_n - V \to 0$ in $R + L^\infty$. Since $V_n - V \to 0$ in $R + L^\infty$, $-\Delta + V_n \to -\Delta + V$ in norm resolvent sense. Thus $P_{(-\infty, E)}^{(n)} \to P_{(-\infty, E)}$ for any $E < 0$ not an eigenvalue of $-\Delta + V$. Letting $P^{(\ell)}$ be the projection onto states of angular momentum ℓ, we conclude that

$$\lim_{E \uparrow 0} \mathrm{Tr}[P_{(-\infty, E)}^{(n)} P^{(\ell)}] = (2\ell + 1)n_\ell(V_n)$$

where the left-hand side converges monotonically and

$$\lim_{E \uparrow 0} \mathrm{Tr}[P_{(-\infty, E)} P^{(\ell)}] = (2\ell + 1)n_\ell(V_n)$$

If we know $n_0(V_n) \leq \int_0^\infty r|V_n(r)|\,dr$, we can conclude that

$$n_0(V) = \lim_{E \uparrow 0} \mathrm{Tr}[P_{(-\infty, E)} P^{(0)}] \leq \sup_{n,\, E<0} \mathrm{Tr}[P_{(-\infty, E)}^{(n)} P^{(0)}]$$

$$\leq \sup_n n_0(V_n) \leq \sup_n \int_0^\infty r|V_n(r)|\,dr$$

$$= \int_0^\infty r|V(r)|\,dr$$

It is the above argument that allows us to deal only with $V \in C_0^\infty$ in Theorem XIII.8. Henceforth we suppose that $V \in C_0^\infty$ and $V \leq 0$.

Let u be the solution of $-u'' + Vu = 0$ with $u(0) = 0$. Let us define

$$a(r) = \frac{u(r)}{u'(r)} - r$$

where $u'(r) \neq 0$. Then $u(r) = (a(r) + r)u'(r)$ so

$$u'(r) = (a'(r) + 1)u'(r) + (a(r) + r)u''(r)$$
$$= u'(r) + a'(r)u'(r) + V(r)(a(r) + r)u(r)$$
$$= u'(r) + a'(r)u'(r) + V(r)(a(r) + r)^2 u'(r)$$

Thus, when $u'(r) \neq 0$

$$a'(r) = -V(r)(a(r) + r)^2 \tag{6}$$

which is called the **Riccati equation**. The passage from the second-order linear equation $-u'' + Vu = 0$ to a first-order nonlinear differential equation like (6) is the first important element of the proof. The value of such equations is that they become first-order differential inequalities by dropping positive terms and can then be integrated to yield inequalities on the number of zeros of $u(r)$.

Since $V \leq 0$, a is monotone increasing by (6) and a goes to infinity at each zero of $u'(r)$. Since a is monotone, the number of poles of $a(r)$ is precisely the number of zeros of $a(r) + r$ (see Figure XIII.1). By Theorem XIII.8, the number of zeros of $u(r)$, which is the number of zeros of $a(r) + r$, is precisely $n_0(V)$. We have used the fact that u obeys a second-order equation so u and u' cannot both vanish simultaneously at a point $r \neq 0$.

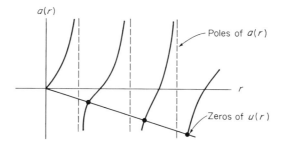

FIGURE XIII.1 The Ricati function $a(r)$.

The trick of proving both Bargmann's bound and Calogero's bound is to introduce an auxiliary function of a, write down its differential equation, and integrate a related differential inequality (see Problem 18 for Calogero's bound). Let $b(r) = a(r)/r$. Then (6) implies

$$b'(r) = -rV(r)(b(r) + 1)^2 - r^{-1}b(r)$$

Moreover, $b(0) = 0$ since $a(0) = \lim_{r \to 0} (u(r)/u'(r) - r) = 0$ and so $\lim_{r \to 0} b(r) = \lim_{r \to 0} a'(r) = 0$ by (6). Finally the poles of $b(r)$ are precisely the poles of $a(r)$, so the number of poles of b is $n_0(V)$. Suppose that b has zeros $z_1 = 0 < z_2 < \cdots < z_n$ and poles $p_1 < p_2 < \cdots < p_n$ with $z_i < p_i < z_{i+1}$. Then in each interval (z_i, p_i), $b(r) > 0$ so

$$b'(r) \leq -rV(r)(b(r) + 1)^2$$

or
$$-[(1 + b(r))^{-1}]' \leq -rV(r) = r|V(r)|$$
since $V \leq 0$. Integrating from z_i to p_i,
$$1 = \int_{z_i}^{p_i} -\frac{d}{dr}\left(\frac{1}{1 + b(r)}\right) dr \leq \int_{z_i}^{p_i} r|V(r)|\, dr$$
Summing over i,
$$n_0(V) \leq \sum_{i=1}^{n} \int_{z_i}^{p_i} r|V(r)|\, dr \leq \int_0^\infty r|V(r)|\, dr$$
which proves the bound. ∎

There are several remarks to be made about the bounds in Theorem XIII.9. First, by the argument leading to (5), the bounds
$$n_\ell(V) \leq (2\ell + 1)^{-1} \int_0^\infty r|V_-(r)|\, dr$$
and
$$n_\ell(V) \leq \frac{2}{\pi} \int_0^\infty |V_-(r)|^{1/2}\, dr$$

hold. Secondly, Bargmann's bound tells us that if $\int_0^\infty r|V(r)|\, dr < \infty$, then $n_\ell(V) = 0$ for ℓ large since $n_\ell(V) = 0$ if it is less than 1. Explicitly, Bargmann's bound implies
$$\ell_{\max}(V) \leq \tfrac{1}{2}\left[\int_0^\infty r|V(r)|\, dr - 1\right] \qquad (7)$$

Next we note that Bargmann's bound, Calogero's bound, and (7) are "best possible" in the sense that there exist potentials with $n_\ell(V)$ (or $\ell_{\max}(V)$) any preassigned integer and with the integral on the right-hand side arbitrarily close to $n_\ell(V)$ (or $\ell_{\max}(V)$). Thus, the constants $(2\ell + 1)^{-1}$, $2/\pi$, and $1/2$ cannot be replaced with smaller constants. On the other hand, Bargmann's bound and (7) are very poor for strong potentials in the sense that as $\lambda \to \infty$, the bounds grow like λ while $n_\ell(\lambda V)$ and $\ell_{\max}(\lambda V)$ grow like $\lambda^{1/2}$. Finally, we remark that (d) and (e) can be improved for restricted classes of potentials:
$$\lim_{\lambda \to \infty} \frac{n_\ell(\lambda V)}{\lambda^{1/2}} = \frac{1}{\pi} \int_0^\infty |V_-(r)|^{1/2}\, dr \qquad (8)$$

98   XIII: SPECTRAL ANALYSIS

for any ℓ,
$$\lim_{\lambda \to \infty} \frac{\ell_{\max}(\lambda V)}{\lambda^{1/2}} = -\min_r \{r^2 V(r)\}$$
and
$$\lim_{\lambda \to \infty} \frac{N(\lambda V)}{\lambda^{3/2}} = \frac{1}{6\pi^2} \int |V_-(r)|^{3/2} d^3r \qquad (9)$$

We shall prove (8) and (9) under very general circumstances in Section 15.

C. Bounds on $N(V)$ in the general two-body case

We now turn to proving bounds on $N(V)$, the total number of bound states, without supposing V is central. We first note two useful facts:

Lemma Let H_0 be self-adjoint and positive. Let V be a form-bounded perturbation of H_0 with relative bound zero. Let $H_0 + \lambda V$ be defined as the operator arising from the sum of forms ($\lambda \in \mathbb{R}$). Suppose that $[0, \infty) \subset \sigma(H_0 + \lambda V)$ for all $\lambda \in \mathbb{R}$. Then:

(a) $\mu_n(H_0 + \lambda V)$ is continuous in λ for $\lambda \in \mathbb{R}$;
(b) $\mu_n(H_0 + \lambda V)$ is monotone decreasing in λ for $\lambda \in [0, \infty)$ and strictly monotone once μ_n is negative.

This lemma follows from an application of the min–max principle (Problem 25a). In the case of interest where $H_0 = -\Delta$, $V \in R + (L^\infty)_\varepsilon$, one can also use the perturbation theory of Chapter XII (Problem 25b). R is the Rollnik class defined in Volume II.

Theorem XIII.10 (the Birman–Schwinger bound)   Let $V \in R$. Then
$$N(V) \leq \left(\frac{1}{4\pi}\right)^2 \int \frac{|V(x)||V(y)|}{|x-y|^2} d^3x \, d^3y$$
In particular, $N(V) < \infty$.

Proof As in the proof of Theorem XIII.9, it is sufficient to prove the bound when V is in C_0^∞ and $V \leq 0$. Let $E < 0$ and let $N_E(V) = \dim(\operatorname{Ran} P_{(-\infty, E)})$. For simplicity of notation, write $\mu_n(\lambda)$ for $\mu_n(-\Delta + \lambda V)$. Then
$$N_E(V) = \#\{n \mid \mu_n(1) < E\}$$

where $\#(A)$ is the cardinality of the set A. Since $\mu_n(\lambda)$ is monotone and continuous and $\mu_n(0) = 0$, $\mu_n(1) < E$ if and only if $\mu_n(\lambda) = E$ for some $0 < \lambda < 1$, and in that case $\mu_n(\lambda) = E$ for exactly one λ. Thus

$$N_E(V) = \#\{n \mid \mu_n(\lambda) = E \text{ for some } \lambda \in (0, 1)\}$$
$$\leq \sum_{\{\lambda \mid \mu_k(\lambda) = E;\, k = 1, \ldots, N_E(V)\}} \lambda^{-2} \leq \sum_{\{\lambda \mid \mu_k(\lambda) = E;\, k = 1, 2, \ldots\}} \lambda^{-2}$$

Next we note that $(H_0 + \lambda V - E)\psi = 0$ if and only if (Problem 26)

$$\lambda(|V|^{1/2}(H_0 - E)^{-1}|V|^{1/2})(|V|^{1/2}\psi) = (|V|^{1/2}\psi)$$

and this is true if and only if

$$\lambda \int \frac{|V(x)|^{1/2} e^{-\sqrt{-E}|x-y|} |V(y)|^{1/2}}{4\pi |x-y|} \varphi(y)\, dy = \varphi(x)$$

has a solution $\varphi \in L^2$ with $\varphi \neq 0$. Let K be the operator with integral kernel $|V(x)|^{1/2} \exp(-\sqrt{-E}|x-y|) |V(y)|^{1/2}/4\pi|x-y|$. Since $V \in R$, K is Hilbert–Schmidt and it is self-adjoint since the kernel is real and symmetric. As a result

$$\sum_{\{\mu \mid \mu \text{ is an eigenvalue of } K\}} \mu^2 = \text{Tr}(K^*K)$$
$$= \left(\frac{1}{4\pi}\right)^2 \int e^{-2\sqrt{-E}|x-y|} \frac{|V(x)||V(y)|}{|x-y|^2}\, dx\, dy$$

But a nonzero μ is an eigenvalue of K if and only if $\mu^{-1}K\varphi = \varphi$ has a solution, which by our above analysis is true if and only if $\mu_n(\lambda) = E$ with $\lambda = \mu^{-1}$. Thus

$$\sum_{\{\lambda \mid \mu_k(\lambda) = E\}} \lambda^{-2} = \sum_{\{\mu \mid \mu \text{ is an eigenvalue of } K\}} \mu^2$$

so

$$N_E(V) \leq \left(\frac{1}{4\pi}\right)^2 \int e^{-2\sqrt{-E}|x-y|} \frac{|V(x)||V(y)|}{|x-y|^2}\, dx\, dy$$

Since $N(V) = \lim_{E \uparrow 0} N_E(V)$, the theorem is proven. ∎

As in the case of the central bounds, if one defines $V_- = \min(V, 0)$,

$$N(V) \leq \left(\frac{1}{4\pi}\right)^2 \int \frac{V_-(x) V_-(y)}{|x-y|^2}\, dx\, dy$$

We also note that the Birman–Schwinger bound can be extended to include zero energy bound states, i.e., one can prove (Problem 28):

$$\dim(\operatorname{Ran} P_{(-\infty,\,0]}) \leq \left(\frac{1}{4\pi}\right)^2 \int \frac{|V(x)|\,|V(y)|}{|x-y|^2}\,dx\,dy$$

The method of proof used above can also be used to prove Bargmann's bound.

One consequence of the Birman–Schwinger bound is that, when $V \in C_0^\infty(\mathbb{R}^3)$, $N(\lambda V) = 0$ for sufficiently small λ. We shall see shortly that this is also true when $n > 3$. In fact, we shall prove in Section 7 that if $V \in L^{\frac{1}{2}n-\varepsilon}(\mathbb{R}^n) \cap L^{\frac{1}{2}n+\varepsilon}(\mathbb{R}^n)$, then $-\Delta + \lambda V$ and $-\Delta$ are unitarily equivalent for λ small. However, when $n = 1, 2$, the situation is different.

Theorem XIII.11 Let V be an everywhere nonpositive function in $C_0^\infty(\mathbb{R}^n)$, for $n = 1$ or 2, which is not identically zero. Then $-\Delta + \lambda V$ has a negative eigenvalue for all $\lambda > 0$.

Proof By the analysis in the proof of Theorem XIII.10, it suffices to prove that for any $\lambda > 0$, there is a κ for which $|V|^{1/2}(-\Delta + \kappa^2)^{-1}|V|^{1/2}$ has an eigenvalue larger than λ^{-1}. Since $|V|^{1/2}(-\Delta + \kappa^2)^{-1}|V|^{1/2}$ is a compact (in fact, Hilbert–Schmidt) operator that is positive and self-adjoint, it suffices to prove that

$$\lim_{\kappa^2 \to 0} \| |V|^{1/2}(-\Delta + \kappa^2)^{-1}|V|^{1/2}\| = \infty$$

To do this, we need only find $\eta \in L^2(\mathbb{R}^n)$ so that

$$\lim_{\kappa^2 \downarrow 0} (|V|^{1/2}\eta, (-\Delta + \kappa^2)^{-1}|V|^{1/2}\eta) = \infty$$

Choose any nonzero η in L^2 so that $\varphi(x) \equiv |V(x)|^{1/2}\eta(x) \geq 0$ and is not zero a.e. Then

$$(|V|^{1/2}\eta, (-\Delta + \kappa^2)^{-1}|V|^{1/2}\eta) = \int |\hat{\varphi}(p)|^2(p^2 + \kappa^2)^{-1}\,d^n p$$

Since $\hat{\varphi}(p) \neq 0$ for p near zero, and $n = 1$ or 2, this integral diverges as $\kappa^2 \to 0$. ∎

We shall discuss this phenomenon again in Section 17. There is more information in Problems 20–22. In particular, in the case just discussed it can be shown that $-\Delta + \lambda V$ has one negative eigenvalue for λ small.

We have already noted that the Bargmann bound has the wrong large coupling constant behavior in the sense that $n_\ell(\lambda V)$ grows like $c\lambda^{1/2}$ while the bound grows like $c\lambda$. Similarly, in the Birman–Schwinger bound $N(V)$ grows like $c\lambda^{3/2}$, while the bound grows like $c\lambda^2$. This defect is remedied by the following result, which is especially pretty because of the connection with the classical phase-space picture discussed in Section 15.

Theorem XIII.12 (the Cwikel–Lieb–Rosenbljum bound)   Let $n \geq 3$ and let $N(V)$ denote the number of bound states of $-\Delta + V$ on $L^2(\mathbb{R}^n)$. Then

$$N(V) \leq c_n \int |V_-(x)|^{n/2} d^n x \tag{10}$$

for suitable c_n.

Proof   We consider the case $n = 3$ first and then describe the modifications necessary for $n \geq 4$. The first remark is that, as in the proof of the Birman–Schwinger bound, we can suppose that $V \leq 0$ and that $V \in C_0^\infty$. We thus let $W = -V \geq 0$.

For $E < 0$, let $N_E(V)$ denote the number of eigenvalues of $-\Delta - W = -\Delta + V$ below E. Let $H_0 = -\Delta$ and $E = -\kappa^2$. We claim that

$$N_E(V) \leq 2 \operatorname{Tr}(W[(H_0 + \kappa^2)^{-1} - (H_0 + W + \kappa^2)^{-1}]) \tag{11}$$

For suppose that φ obeys $(H_0 - \lambda W)\varphi = E\varphi$. Then $\psi = W^{1/2}\varphi$ obeys

$$W^{1/2}(H_0 + \kappa^2)^{-1} W^{1/2}\psi = \lambda^{-1}\psi$$

$$W^{1/2}(H_0 + W + \kappa^2)^{-1} W^{1/2}\psi = (1 + \lambda)^{-1}\psi$$

and so letting

$$K = W^{1/2}[(H_0 + \kappa^2)^{-1} - (H_0 + W + \kappa^2)^{-1}]W^{1/2}$$

$$K\psi = [\lambda^{-1} - (1 + \lambda)^{-1}]\psi$$

As in the proof of the Birman–Schwinger bound, we see that $N_E(V)$, which is the number of λ's in $(0, 1]$ so that $(H_0 - \lambda W)\varphi = E\varphi$ has a solution (counting multiplicities), is bounded by the number of eigenvalues of K that are larger than $\frac{1}{2}$. Since K is a positive operator, this is bounded by $2 \operatorname{Tr}(K)$ which is given by (11) on account of cyclicity of the trace.

Using (X.98), which relates resolvents to semigroups, we see that

$$N_E(V) \leq 2 \int_0^\infty \operatorname{Tr}(W[e^{-tH_0} - e^{-t(H_0 + W)}])e^{Et}\, dt$$

where one can justify the interchange of the integral and Tr by using the bounds which we prove below. Since we shall also see that

$\mathrm{Tr}(W[e^{-tH_0} - e^{-t(H_0+W)}])$ is positive, and since $N(V) = \lim_{E \uparrow 0} N_E(V)$, we conclude that

$$N(V) \leq 2 \int_0^\infty \mathrm{Tr}(W[e^{-tH_0} - e^{-t(H_0+W)}]) \, dt \qquad (12)$$

by the monotone convergence theorem.

The key idea in the proof will be to realize $e^{-tH_0} - e^{-t(H_0+W)}$ as an integral operator using Wiener integrals. To do this, we need to expand slightly on our discussion of Wiener integrals in Section X.11. Let us do this for general n, rather than $n = 3$. Basic to the construction is the kernel

$$p(x, y; t) = (4\pi t)^{-n/2} e^{-(x-y)^2/4t}$$

of the integral operator e^{-tH_0}. In Section X.11, we constructed a measure μ_x on the continuous paths ω on $[0, \infty)$ with $\omega(0) = x$ so that for $0 < t_1 < \cdots < t_m$,

$$\mu_x\{\omega \,|\, \omega(t_1) \in A_1, \ldots, \omega(t_m) \in A_m\} = \int \prod_{i=1}^m p(x_{i-1}, x_i; t_i - t_{i-1}) \chi_{A_i}(x_i) \, d^n x_i$$

where χ_A is the characteristic function of A and $x_0 = x$, $t_0 = 0$. Because of the semigroup property, μ_x was a measure of total mass 1. By the same method as in X.11, one can construct a measure $\mu_{x,y;t}$ on the continuous paths on $[0, t]$ with $\omega(0) = x$, $\omega(t) = y$ so that for $0 < t_1 < \cdots < t_m < t$,

$$\mu_{x,y;t}\{\omega \,|\, \omega(t_1) \in A_1, \ldots, \omega(t_m) \in A_m\}$$

$$= \int \left(\prod_{i=1}^m p(x_{i-1}, x_i; t_i - t_{i-1}) \chi_{A_i}(x_i) \right) p(x_m, y; t - t_m) \, d^n x_i$$

$\mu_{x,y;t}$ is called **conditional Wiener measure**. It is related to μ_x: For if $f(\omega)$ is a function of the values the path takes on $[0, t]$, then

$$\int f(\omega) \, d\mu_x = \int dy \left[\int f(\omega) \, d\mu_{x,y;t} \right]$$

which is obvious for functions of $\omega(t_1), \ldots, \omega(t_n)$. $d\mu_{x,y;t}$ has total mass $p(x, y; t)$.

The point of conditional Wiener measure is that the Feynman–Kac formula now says that $e^{-t(H_0+W)}$ is an integral operator with kernel

$$e^{-t(H_0+W)}(x, y) = \int d\mu_{x,y;t} \, \exp\left(-\int_0^t W(\omega(s)) \, ds\right) \qquad (13)$$

XIII.3 Bound states: Qualitative theory

A priori (13) only holds a.e. in x, y; but when $W \in C_0^\infty$, it is not hard to show that the right-hand side of (13) is continuous in x and y (Problem 29). By the basic method used to prove the Feynman–Kac formula, one sees that

$$\left(e^{-s(H_0+W)} W e^{-(t-s)(H_0+W)}\right)(x, y) = \int d\mu_{x,y;t} W(\omega(s)) \exp\left(-\int_0^t W(\omega(s))\, ds\right)$$

Using the facts that the kernel on the right-hand side of this expression is continuous and the operator is trace class, one can show that (Problem 30):

$$\operatorname{Tr}\left(e^{-s(H_0+W)} W e^{-(t-s)(H_0+W)}\right)$$

$$= \int dx \int d\mu_{x,x;t} W(\omega(s)) \exp\left(-\int_0^t W(\omega(s))\, ds\right)$$

But $\operatorname{Tr}(e^{-s(H_0+W)} W e^{-(t-s)(H_0+W)})$ is independent of s by the cyclicity of the trace, so we have that

$$\operatorname{Tr}(W e^{-t(H_0+W)}) = \frac{1}{t} \int_0^t \operatorname{Tr}\left(e^{-s(H_0+W)} W e^{-(t-s)(H_0+W)}\right) ds$$

$$= \frac{1}{t} \int dx \int d\mu_{x,x;t} F\left(\int_0^t W(\omega(s))\, ds\right) \qquad (14)$$

where $F(u) = u e^{-u}$.

Armed with (14), we can rewrite the bound (12) as

$$N(V) \leq 2 \int_0^\infty dt \int dx \int d\mu_{0,0;t} t^{-1} G\left(\int_0^t W(x + \omega(s))\, ds\right) \qquad (15)$$

where $G(u) = u(1 - e^{-u})$ and we have used the covariance

$$\int d\mu_{x+a,\, y+a;\, t}\, f(\omega(s)) = \int d\mu_{x,y;t}\, f(\omega(s) + a)$$

Now the function G'' is easily seen to be positive on $(0, 2)$ and negative on $(2, \infty)$. Thus, φ given by

$$\varphi(u) = \begin{cases} u(1 - e^{-u}), & 0 < u \leq 2 \\ G(2) + (u - 2)G'(2), & 2 \leq u < \infty \end{cases}$$

is easily seen to obey

$$G(u) \leq \varphi(u) \qquad (16a)$$

$$\varphi(u) \sim \begin{cases} u^2 & \text{at} \quad u = 0 \\ u & \text{at} \quad u = \infty \end{cases} \qquad (16b)$$

$$\varphi(u) \text{ is convex} \qquad (16c)$$

XIII: SPECTRAL ANALYSIS

$\varphi(u)$ **convex** means that for $u, v \geq 0$, $0 \leq t \leq 1$, $\varphi(tu + (1-t)v) \leq t\varphi(u) + (1-t)\varphi(v)$. This follows from the fact that $\varphi'' \geq 0$. By induction

$$\varphi\left(\sum_{i=1}^{n} t_i u_i\right) \leq \sum_{i=1}^{n} t_i \varphi(u_i)$$

if $\sum t_i = 1$, and u_i, $t_i \geq 0$. By a simple limiting argument, we find that

$$\varphi\left(\int f(x)\, d\mu(x)\right) \leq \int \varphi(f(x))\, d\mu(x) \tag{17}$$

for any positive function f and measure $d\mu$ of total mass 1. On account of (16a) and (17),

$$G\left(\int_0^t W(x + \omega(s))\, ds\right) \leq t^{-1} \int_0^t ds\, \varphi(tW(x + \omega(s)))$$

Using Fubini's theorem, we can thus deduce from (15) that

$$N(V) \leq 2 \int_0^\infty dt\, t^{-2} \int_0^t ds \int d\mu_{0,0;t} \int dx\, \varphi(tW(x + \omega(s)))$$

$$= 2 \int_0^\infty t^{-2}\, dt \int_0^t ds \int d\mu_{0,0;t} \int dx\, \varphi(tW(x))$$

$$= 2 \int_0^\infty (4\pi t)^{-3/2} t^{-1}\, dt \int dx\, \varphi(tW(x))$$

$$= c \int W(x)^{3/2}\, dx$$

where $c = 2(4\pi)^{-3/2} \int_0^\infty u^{-5/2}\varphi(u)$ is finite by (16b). The first equality above follows from the invariance of Lebesgue measure dx; the second from the fact that the integrand is now independent of ω and s and $\int_0^t ds = t$, $\int d\mu_{0,0;t} = p(0,0,t) = (4\pi t)^{-3/2}$; the third by changing variables from (t, x) to (u, x) with $u = tW(x)$.

This proves the result for $n = 3$. For $n > 3$, this argument fails since $\int_0^\infty u^{-5/2}\varphi(u)$ is replaced by $\int_0^\infty u^{-n/2-1}\varphi(u)$, which diverges at $u = 0$. We introduce the following modifications to overcome this. For any m, we consider the function

$$H_m(y) = \sum_{j=0}^{m} (-1)^j \binom{m}{j}(1 + yj)^{-1}$$

It is useful to note that

$$H_m(y) = m!\, y^m/(1 + y)(1 + 2y) \cdots (1 + my)$$

which follows since the difference of the two functions is an entire function in y, which goes to zero at infinity. By mimicking the argument leading to (11) one obtains

$$N_E(V) \leq (m+1) \, \text{Tr}\left(W \sum_{j=0}^{m} (-1)^j \binom{m}{j} (H_0 + mW + \kappa^2)^{-1}\right) \quad (11')$$

In proving (11') the operator K is replaced by

$$K' = W^{1/2} \sum_{j=0}^{m} (-1)^j \binom{m}{j} (H_0 + mW + \kappa^2)^{-1} W^{1/2}$$

$$= W^{1/2}(H_0 + \kappa^2)^{-1/2} H_m(W^{1/2}(H_0+\kappa^2)^{-1}W^{1/2})(H_0+\kappa^2)^{-1/2}W^{1/2}$$

which is positive since H_m is. Moreover, $K\psi = (\lambda^{-1} - (1+\lambda)^{-1})\psi$ is replaced by

$$K'\psi = \left(\sum_{j=0}^{m}(-1)^j \binom{m}{j}(j+\lambda)^{-1}\right)\psi = \lambda^{-1} H_m(\lambda^{-1})\psi$$

Since

$$H_m(y) = \int_0^\infty e^{-t}(1-e^{-yt})^m \, dy$$

it is monotone in y. Thus, as λ goes from 0 to 1, $\lambda^{-1}H_m(\lambda^{-1})$ runs from ∞ to $H(1) = m!/(m+1)! = 1/(m+1)$, so (11') follows.

Rewriting (11') in the way we arrived at (15) leads to

$$N(V) \leq (m+1) \int_0^\infty dt \int dx \int d\mu_{0,0;t} \, t^{-1} G_m\left(\int_0^t W(x+\omega(s)) \, ds\right)$$

where $G_m(u) = u(1-e^{-u})^m$. G_m'' is seen to be positive on $(0, y_m)$ and negative on (y_m, ∞) for suitable y_m, so φ_m given by

$$\varphi_m(u) = \begin{cases} G_m(u), & 0 < u \leq y_m \\ G_m(y_m) + (u-y_m)G'(y_m), & y_m \leq u < \infty \end{cases}$$

obeys

$$G_m(u) \leq \varphi_m(u)$$

$$\varphi_m(u) \sim \begin{cases} u^{m+1} & \text{at} \quad u = 0 \\ u & \text{at} \quad u = \infty \end{cases}$$

$\varphi_m(u)$ is convex

We then find that for any n, m,

$$N(V) \leq C_{nm} \int W(x)^{m/2} \, d^n x$$

where $C_{nm} = 2(4\pi)^{-n/2} \int_0^\infty u^{-n/2-1} \varphi_m(u)$. So long as $m > (n/2) - 1$, $C_{nm} < \infty$, so choosing m suitably, the theorem is proven for arbitrary $n \geq 3$. ∎

This theorem can also be proven using Theorem XI.22.

XIII.4 Locating the essential spectrum I: Weyl's theorem

In this section and the next we shall discuss methods of determining $\sigma_{\text{ess}}(A)$ for a variety of operators A. In this section we prove a general perturbation theory result that tells us $\sigma_{\text{ess}}(A + C) = \sigma_{\text{ess}}(A)$ if C is compact, or more generally if C is "almost compact" in a sense to be made precise. This will enable us to find σ_{ess} for two-body Schrödinger operators with their center of mass removed. In the next section we shall describe a special theory to deal with the essential spectrum of N-body Schrödinger operators.

We are primarily interested in $\sigma_{\text{ess}}(A)$ when A is self-adjoint, but the natural setting for the results is for a general closed operator. $\sigma_{\text{disc}}(A)$ is defined in Section XII.2 and we define $\sigma_{\text{ess}}(A) = \sigma(A) \backslash \sigma_{\text{disc}}(A)$. The reader is cautioned that many other definitions of σ_{ess} appear in the literature.

The natural way to study $\sigma_{\text{ess}}(A)$ is to study $(A - z)^{-1}$ for some $z \notin \sigma(A)$. For a general self-adjoint operator, it is necessary to take z nonreal, so one is lead to the consideration of non-self-adjoint operators. If one were interested only in semibounded self-adjoint operators, then one could take z to be real and thus deal only with self-adjoint operators. In that case, the proofs below would be slightly simplified.

Our results in this section are based on use of the analytic Fredholm theorem (Theorem VI.14). The proof of our final result will have several complicating features, so let us first prove a very special case of it:

Proposition Let A and B be two bounded self-adjoint operators with empty discrete spectra and with $B - A$ compact. Then $\sigma_{\text{ess}}(A) = \sigma_{\text{ess}}(B)$.

Proof Write $A = B + C$ with C compact. Define $F(z) = C(A - z)^{-1}$ for $z \in \mathbb{C} \backslash \sigma(A)$. Then $F(z)$ is an analytic operator-valued function of z that is always compact since C is compact by hypothesis. Since A is bounded,

$$\|F(z)\| \leq \|C\| \, \|(A - z)^{-1}\| \to 0$$

in norm as $z \to \infty$. In particular $(1 - F(z))^{-1}$ exists if $|z|$ is large. By the analytic Fredholm theorem, $(1 - F(z))^{-1}$ exists for $z \in \mathbb{C} \backslash (\sigma(A) \cup D)$ where

D is a discrete subset of $\mathbb{C}\backslash\sigma(A)$. Further, $(1 - F(z))^{-1}$ is meromorphic in $\mathbb{C}\backslash\sigma(A)$ with finite rank residues at the points of D. If $z \notin \sigma(A)$, then $B - z = (1 - C(A - z)^{-1})(A - z)$ so if $(1 - F(z))$ is invertible, then $z \notin \sigma(B)$ and $(B - z)^{-1} = (A - z)^{-1}(1 - F(z))^{-1}$. Thus $\sigma(B) \subset D \cup \sigma(A)$. Moreover, $(B - z)^{-1}$ has finite rank residues at the points of D. Thus points of D are in $\sigma_{\text{disc}}(B)$, so $\sigma_{\text{ess}}(B) \subset \sigma(A)$. Similarly, $\sigma_{\text{ess}}(A) \subset \sigma(B)$. By hypothesis, $\sigma_{\text{ess}}(A) = \sigma(A)$ and $\sigma_{\text{ess}}(B) = \sigma(B)$, so we conclude that $\sigma_{\text{ess}}(B) = \sigma_{\text{ess}}(A)$. ∎

We want to extend this proposition in various directions. First we want to allow unbounded operators. Most crucially, we want to eliminate the hypothesis that $\sigma_{\text{disc}}(A) = \varnothing$. To eliminate this last hypothesis, we shall actually require a strengthening of the analytic Fredholm theorem which is of some interest in itself:

Theorem XIII.13 (meromorphic Fredholm theorem)   Let Ω be a connected open subset of $\mathbb{C}$. Let $A(z)$ be a meromorphic operator-valued function of z, i.e., A is analytic in $\Omega\backslash D$ where D is a discrete subset of Ω (a set with no limit points in Ω) and near any $z_0 \in D$,

$$A(z) = A_{-k}(z - z_0)^{-k} + \cdots + A_{-1}(z - z_0)^{-1} + \sum_{n=0}^{\infty} A_n(z - z_0)^n$$

Suppose in addition that:

(1) $A(z)$ is compact if $z \in \Omega\backslash D$.
(2) The coefficients $A_{-k}, \ldots, A_{-1}$ of the negative terms of the Laurent series of $A(z)$ at points $z_0 \in D$ are finite rank operators.

Then either:

(a) $1 - A(z)$ is invertible for no $z \in \Omega\backslash D$;

or

(b) There is a discrete set $D' \subset \Omega$ so that $(1 - A(z))^{-1}$ exists if $z \notin D \cup D'$ and extends to a function analytic in $\Omega\backslash D'$ and meromorphic in Ω such that the coefficients of the negative terms in the Laurent series at points $z_0 \in D'$ are finite rank operators.

Proof   Most of the ideas of the proof are those of Theorem VI.14, so we shall refer to that proof. As in that proof, it is enough to show that any $z_0 \in \Omega$ has a neighborhood N so that either (a) or (b) holds. If $z_0 \in \Omega\backslash D$, the proof that (a) or (b) holds in a neighborhood $N \subset \Omega\backslash D$ is identical to the one in Theorem VI.14. Thus suppose $z_0 \in D$. Near z_0 we know that

$$A(z) = A_{-k}(z - z_0)^{-k} + \cdots + A_{-1}(z - z_0)^{-1} + G(z)$$

where $G(z)$ is analytic at z_0 and compact for z near but not equal to z_0. Since G is norm continuous at z_0, $G(z_0)$ is compact also as a limit of compact operators. Now, let F be a finite rank operator and N be an open disk about z_0, so that $\|G(z_0) - F\| < \frac{1}{2}$ and $\|G(z) - G(z_0)\| < \frac{1}{2}$ if $z \in N$. Let

$$F(z) = A_{-k}(z - z_0)^{-k} + \cdots + A_{-1}(z - z_0)^{-1} + F$$

Then $\|A(z) - F(z)\| < 1$ in N so $C(z) = [1 - (A(z) - F(z))]^{-1}$ exists and is analytic in N. Thus $(1 - A(z))^{-1}$ exists if and only if $[1 - F(z)C(z)]^{-1}$ exists since

$$1 - A(z) = (1 - F(z)C(z))[1 - (A(z) - F(z))]$$

Since $F(z)$ has its range contained in Ran $A_{-k} + \cdots +$ Ran $A_{-1} +$ Ran $F \equiv R$, a fixed finite-dimensional space, $F(z)C(z)$ has its range in R. Thus, by Eq. (VI.5b) of Section VI.5, $1 - F(z)C(z)$ is invertible if and only if a certain determinant of matrix elements of $1 - C(z)F(z)$ is nonzero. This determinant is a meromorphic function in N, and so it has discrete zeros inside N if it is not identically zero. Moreover, in this case $(1 - C(z)F(z))^{-1}$ is given by $1 +$ finite cofactor matrix/determinant, so $(1 - C(z)F(z))^{-1}$ is meromorphic with finite rank residues in N. Thus we are in case (b) in N. If the determinant is identically zero, we are locally in case (a). ∎

Our condition that the operators $A_{-k}, \ldots, A_{-1}$ be finite rank is crucial. The theorem is false if we only suppose $A_{-k}, \ldots, A_{-1}$ are compact (Problem 34). Theorem XIII.13 provides the technique we need for eliminating the assumption $\sigma(A) = \sigma_{\text{ess}}(A)$ from the above proposition.

Lemma 1 Let A be a closed operator. Then $(A - z)^{-1}$ is a meromorphic function on $\mathbb{C} \setminus \sigma_{\text{ess}}(A)$ with singularities only at points $z \in \sigma_{\text{disc}}(A)$. The negative coefficients of the Laurent expansion at points $z_0 \in \sigma_{\text{disc}}(A)$ are finite rank operators.

Proof We already know that $(A - z)^{-1}$ is analytic in $\mathbb{C} \setminus \sigma(A)$. Let z_0 be an isolated point of $\sigma(A)$. Then $(A - z)^{-1}$ is analytic in a set $\{z \mid 0 < |z - z_0| < r\}$, and so in that set

$$(A - z)^{-1} = \sum_{n=-\infty}^{\infty} R_n (z - z_0)^n$$

where

$$R_n = (2\pi i)^{-1} \oint_{|z - z_0| = r/2} (z - z_0)^{-n-1} (A - z)^{-1} \, dz$$

XIII.4 Locating the essential spectrum I: Weyl's theorem

In Section XII.2, we used the resolvent identity to prove that $(-R_{-1})^2 = -R_{-1}$. In a similar way one proves (see Problem 3 of Chapter XII) that

$$R_{-n} = -(N)^{n-1} \quad (n \geq 2) \tag{18a}$$

where $N = -R_{-2}$, and

$$NP = PN = N \tag{18b}$$

where $P = -R_{-1}$ is the projection associated to z_0. Finally, since $\sum_{n=-\infty}^{0} R_n \mu^n$ is absolutely convergent for all $\mu \neq 0$, $\sum_{n=0}^{\infty} \lambda^n N^n$ converges for all λ, so $(1 - \lambda N)^{-1}$ exists for all λ. Thus

$$\sigma(N) = \{0\} \tag{18c}$$

Now suppose that $z_0 \in \sigma_{\text{disc}}(A)$ as well as being isolated. Then P is finite rank by definition. By (18b), Ran $N \subset$ Ran P, so each R_{-n} is finite rank. Moreover, by (18c) $N \restriction$ Ran P is a finite-dimensional operator with spectrum $\{0\}$. Thus $(NP)^k = 0$ for some k. Since $(NP)^k = N^k$ by (18b), $N^k = 0$ for some k. We thus conclude that $R_{-n} = 0$ if $n \geq k + 1$. Thus, $(A - z)^{-1}$ is meromorphic at z_0 with finite rank residues. ∎

We now have the major tools for extending the proposition at the beginning of the section to the case where $\sigma_{\text{disc}}(A) \cup \sigma_{\text{disc}}(B) \neq \emptyset$. The reader might stop at this point and try to use Theorem XIII.13 to extend the proposition in this way. Next, we introduce the tool for extending the proposition to unbounded operators:

Lemma 2 (strong spectral mapping theorem) Let A be a closed operator on a Hilbert space $\mathcal{H}$. Let $z_0 \notin \sigma(A)$ and $B = (A - z_0)^{-1}$. Then:

(a) If $z \neq 0$, then $z \in \sigma(B)$ if and only if $z_0 + z^{-1} \in \sigma(A)$; $z \in \sigma_{\text{ess}}(B)$ if and only if $z^{-1} + z_0 \in \sigma_{\text{ess}}(A)$ and $z \in \sigma_{\text{disc}}(B)$ if and only if $z^{-1} + z_0 \in \sigma_{\text{disc}}(A)$.
(b) $0 \in \sigma(B)$ if and only if A is not bounded. 0 is never in $\sigma_{\text{disc}}(B)$; so if $0 \in \sigma(B)$, then $0 \in \sigma_{\text{ess}}(B)$.

Proof Since we make no use of (b), we leave its proof to the problems (Problem 36). To prove (a), first note that if $z \neq 0$,

$$B - z = (A - z_0)^{-1} - z = [1 - z(A - z_0)](A - z_0)^{-1}$$
$$= -z[A - (z^{-1} + z_0)](A - z_0)^{-1}$$

If $z \notin \sigma(B)$, then $[A - (z^{-1} + z_0)]^{-1} = -z(A - z_0)^{-1}(B - z)^{-1}$ so $z^{-1} + z_0 \notin \sigma(A)$. If $z^{-1} + z_0 \notin \sigma(A)$, then $-z^{-1}(A - z_0)[A - (z^{-1} + z_0)]^{-1} = -z^{-1} - z^{-2}[A - (z^{-1} + z_0)]^{-1}$ is bounded and an inverse for $B - z$, so $z \notin \sigma(B)$. Thus $z \in \sigma(B)$ if and only if $z^{-1} + z_0 \in \sigma(A)$ and isolated points of the spectrum are mapped into one another by $z \mapsto z^{-1} + z_0$. Let $z_1 \in \sigma_{\text{disc}}(B)$. Then near $z = z_1$, $(B - z)^{-1}$ is meromorphic with finite rank residues. Thus for z near z_1, $[A - (z^{-1} + z_0)]^{-1} = -z(A - z_0)^{-1}(B - z)^{-1}$ is meromorphic with finite rank residues. Thus for λ near $z_1^{-1} + z_0$, $(A - \lambda)^{-1}$ is meromorphic with finite rank residues so $z_1^{-1} + z_0 \in \sigma_{\text{disc}}(A)$. Similarly if $\lambda \in \sigma_{\text{disc}}(A)$, one can show $z = (\lambda - z_0)^{-1} \in \sigma_{\text{disc}}(B)$. ∎

If we wanted to state our final theorem only for semibounded self-adjoint operators, we could apply the proposition at the start of this section to $(A + c)^{-1}$ for a suitable real number c. But we want to discuss self-adjoint operators with spectrum $\mathbb{R}$, and we shall need to look at $(A + i)^{-1}$, which is no longer self-adjoint. Thus we need to extend our first proposition to certain bounded non-self-adjoint operators.

The following example shows that it is not necessarily true that $\sigma_{\text{ess}}(A) = \sigma_{\text{ess}}(B)$ if $B - A$ is compact.

Example 1 Let $\mathscr{H} = \ell_2(-\infty, \infty)$. Let A be the left shift operator, that is, $(A\varphi)_n = \varphi_{n+1}$. Let $C: \mathscr{H} \to \mathscr{H}$ be defined by $(C\varphi)_n = \delta_{n,0} \varphi_1$. Thus C is a rank-one perturbation. Let $B = A - C$. We shall show that $\sigma(A) = \sigma_{\text{ess}}(A) = \{z \mid |z| = 1\}$ while $\sigma(B) = \sigma_{\text{ess}}(B) = \{z \mid |z| \leq 1\}$. To find $\sigma(A)$ map $\ell_2(-\infty, \infty)$ onto $L^2(0, 2\pi)$ by $U\{\varphi_n\} = \sum_{n=-\infty}^{\infty} (2\pi)^{-1/2} \varphi_n e^{inx}$. Then U is unitary and UAU^{-1} is multiplication by e^{-ix}, so $\sigma(A) = \sigma(UAU^{-1}) = \{z \mid |z| = 1\}$. Next we turn to B. If $z \notin \sigma(A)$, then $(B - z)^{-1}$ exists if and only if $(1 - C(A - z)^{-1})^{-1}$ exists. Since $C(A - z)^{-1}$ is compact, $(1 - C(A - z)^{-1})^{-1}$ will exist unless $C(A - z)^{-1}$ has eigenvalue 1 or equivalently if and only if z is an eigenvalue of B. Let $(B - z)\varphi = 0$. Then $\varphi_{n+1} - \delta_{n,0} \varphi_1 - z\varphi_n = 0$. Suppose $z \neq 0$. Letting $n = 0$, we see $\varphi_0 = 0$ and then letting $n = -1, -2, \ldots$ that $\varphi_n = 0$ for $n < 0$. Taking $n = 1, 2, \ldots$ we see $\varphi_n = z^{n-1}\varphi_1$ if $n \geq 1$. Thus if $|z| > 1$, no such eigenvalue exists and $B - z$ is invertible, i.e., $z \in \rho(B)$. If $|z| < 1$,

$$\langle 0, \varphi_1, z\varphi_1, \ldots, z^{n-1}\varphi_1, \ldots \rangle$$

is an eigenvector so $z \in \sigma(B)$.

It is interesting to notice where the proof that $\sigma_{\text{ess}}(B) = \sigma_{\text{ess}}(A)$ breaks down. One still knows $C(A - z)^{-1}$ is analytic in $\mathbb{C} \backslash \sigma(A)$. On the component of $\mathbb{C} \backslash \sigma(A)$ that goes out to infinity, $\|C(A - z)^{-1}\|$ is less than 1 somewhere so

XIII.4 Locating the essential spectrum I: Weyl's theorem

$(1 - C(A - z)^{-1})^{-1}$ exists on that component. On the other component, we do not know a priori that $(1 - C(A - z)^{-1})^{-1}$ exists at any point and so we are in the case (a) of the analytic Fredholm theorem where we obtain little information.

Because of Example 1 we need hypotheses on B that tell us that this situation of everywhere noninvertibility on a component of $\mathbb{C}\backslash\sigma(A)$ does not occur. If we turn Example 1 around, and look at A as a perturbation of B, we see the interior of $\sigma_{\text{ess}}(A)$ can disappear under perturbation, so we shall suppose $\sigma_{\text{ess}}(A)$ is nowhere dense:

Lemma 3 Let A and B be bounded operators with $A - B$ compact, so that:

(a) $\sigma(A)$ has empty interior as a subset of $\mathbb{C}$.
(b) Each component of $\mathbb{C}\backslash\sigma(A)$ contains a point in $\rho(B)$.

Then $\sigma_{\text{ess}}(A) = \sigma_{\text{ess}}(B)$.

Proof Let $C = A - B$. Since C is compact, $C(A - z)^{-1}$ is analytic and compact-valued in $\mathbb{C}\backslash\sigma(A)$ and, by Lemma 1, meromorphic in $\mathbb{C}\backslash\sigma_{\text{ess}}(A)$ with finite rank residues at points in $\sigma_{\text{disc}}(A)$. If $z \notin \sigma(A)$, then $(B - z)^{-1}$ exists if and only if $(1 - C(A - z)^{-1})^{-1}$ exists. Thus, by (b), in each component of $\mathbb{C}\backslash\sigma(A)$ we conclude that $(1 - C(A - z)^{-1})^{-1}$ is invertible somewhere. The components of $\mathbb{C}\backslash\sigma(A)$ and $\mathbb{C}\backslash\sigma_{\text{ess}}(A)$ are the same since $\sigma_{\text{disc}}(A)$ is discrete. By the meromorphic Fredholm theorem, we conclude that $(1 - C(A - z)^{-1})^{-1}$ exists on $\mathbb{C}\backslash\sigma_{\text{ess}}(A)$ except for a discrete set D' where it has finite rank residues. It therefore follows that B can only have discrete spectrum in $\mathbb{C}\backslash\sigma_{\text{ess}}(A)$, so $\sigma_{\text{ess}}(B) \subset \sigma_{\text{ess}}(A)$. Since $\sigma(A)$ has no interior, each component of $\mathbb{C}\backslash\sigma_{\text{ess}}(B)$ has a point not in $\sigma(A)$ nor in $\sigma(B)$. We can then turn the above argument around and conclude that $\sigma_{\text{ess}}(A) \subset \sigma_{\text{ess}}(B)$. ∎

We are now able to prove the main theorem. However, since the basic hypothesis will be that $(A - z)^{-1} - (B - z)^{-1}$ is compact, it is useful to note:

Lemma 4 Let A and B be closed operators. If $(A - z)^{-1} - (B - z)^{-1}$ is compact for some $z_0 \in \rho(A) \cap \rho(B)$, it is compact for all $z \in \rho(A) \cap \rho(B)$.

Proof By using the first resolvent formula, $(A - z)^{-1} - (B - z)^{-1}$ can be written as a finite sum of terms of the form $C[(A - z_0)^{-1} - (B - z_0)^{-1}]D$ with C and D bounded. ∎

We are now ready to prove the main theorem of this section.

Theorem XIII.14 (Weyl's essential spectrum theorem) Let A be a self-adjoint operator and let B be a closed operator so that:

(a) For some (and hence all) $z \in \rho(B) \cap \rho(A)$, $(A - z)^{-1} - (B - z)^{-1}$ is compact

and *either*

(b_1) $\sigma(A) \neq \mathbb{R}$ and $\rho(B) \neq \emptyset$

or

(b_2) There are points of $\rho(B)$ in both the upper and lower half-planes.

Then $\sigma_{\text{ess}}(B) = \sigma_{\text{ess}}(A)$.

Proof By hypothesis, some z_0 off the real axis is in $\rho(B)$. Let $D = (A - z_0)^{-1}$ and $E = (B - z_0)^{-1}$. By Lemma 2, we need only show $\sigma_{\text{ess}}(D) = \sigma_{\text{ess}}(E)$ to conclude that $\sigma_{\text{ess}}(A) = \sigma_{\text{ess}}(B)$. Since A is self-adjoint, $\sigma_{\text{ess}}(A) \subset \mathbb{R}$, so $\sigma_{\text{ess}}(D) \subset \{\lambda \,|\, \lambda^{-1} + z_0 \in \mathbb{R}\} \cup \{0\}$. As a consequence, $\sigma_{\text{ess}}(D)$ has empty interior and either $\sigma_{\text{ess}}(A) = \mathbb{R}$ in which case $\mathbb{C} \setminus \sigma_{\text{ess}}(D)$ has two components, or $\sigma_{\text{ess}}(A) \neq \mathbb{R}$ in which case $\mathbb{C} \setminus \sigma_{\text{ess}}(D)$ is connected. In either case, (b) and Lemma 2 imply that each component of $\mathbb{C} \setminus \sigma_{\text{ess}}(D)$ intersects $\rho(E)$. Since D and E are bounded with $D - E$ compact by Lemma 4, we can apply Lemma 3 and conclude that $\sigma_{\text{ess}}(D) = \sigma_{\text{ess}}(E)$. This proves the theorem. ∎

We note that Theorem XIII.14 can be extended to some cases where A is neither self-adjoint nor normal but that the hypotheses are a little complicated. In applications A is almost always self-adjoint, so we stated the theorem in the simple case. The following example shows that some condition of type (b) is needed.

Example 2 Let A and B be the operators of Example 1 so that $\sigma(A) = \{z \,|\, |z| = 1\}$ and $\sigma(B) = \{z \,|\, |z| \leq 1\}$. Notice that 1 is not an eigenvalue of A, A^*, B, or B^* so $A - 1$ and $B - 1$ are injective maps of ℓ_2 onto dense subsets of ℓ_2. Thus $(A - 1)^{-1}$ and $(B - 1)^{-1}$ are unbounded densely defined closed operators. Let $D = i(A + 1)(A - 1)^{-1} = i[1 + 2(A - 1)^{-1}]$ and
$$E = i(B + 1)(B - 1)^{-1} = i[1 + 2(B - 1)^{-1}].$$
Then D is self-adjoint since under the map U discussed in Example 1, D is multiplication by $i(e^{-i\theta} + 1)/(e^{-i\theta} - 1) = -\cot(\theta/2)$. Notice that
$$(D - i)^{-1} = \tfrac{1}{2}A - \tfrac{1}{2}$$

and $(E - i)^{-1} = \frac{1}{2}B - \frac{1}{2}$. Thus, by the strong spectral mapping principle (Lemma 2 above), $\sigma(D) = \mathbb{R}$ and $\sigma(E) = \{z \mid \text{Im } z \leq 0\}$. Moreover, $(D - i)^{-1} - (E - i)^{-1}$ is compact but $\sigma_{\text{ess}}(D) \neq \sigma_{\text{ess}}(E)$. This example is related to Example 1 by using an inverse Cayley transform (see the Notes to Section X.1).

In applying Theorem XIII.14 it is useful to replace (b) with simpler conditions. We thus state two corollaries:

Corollary 1 Let A and B be self-adjoint operators with $(A + i)^{-1} - (B + i)^{-1}$ compact, then
$$\sigma_{\text{ess}}(A) = \sigma_{\text{ess}}(B)$$
Proof Since B is self-adjoint, $\rho(B)$ contains points in both half-planes. ∎

To state Corollary 2, we need a new notion:

Definition Let A be self-adjoint. An operator C with $D(A) \subset D(C)$ is called **relatively compact** with respect to A if and only if $C(A + i)^{-1}$ is compact.

We note that if C is relatively compact, then $C(A - z)^{-1}$ is compact for all $z \in \rho(A)$ and that if $C(A - z)^{-1}$ is compact for some $z \in \rho(A)$, then C is relatively compact (Problem 38a). Also relative compactness can be phrased in terms of the Hilbert space $D(A)$ with the norm $\|\varphi\|_A^2 = \|\varphi\|^2 + \|A\varphi\|^2$: C is relatively compact if and only if C is compact as a map from $\langle D(A), \|\cdot\|_A \rangle$ to $\langle \mathcal{H}, \|\ \| \rangle$.

Corollary 2 Let A be a self-adjoint operator and let C be a relatively compact perturbation of A. Then:

(a) $B = A + C$ defined with $D(B) = D(A)$ is a closed operator.
(b) If C is symmetric, B is self-adjoint.
(c) $\sigma_{\text{ess}}(A) = \sigma_{\text{ess}}(B)$.

Proof $C(A + i\lambda)^{-1} = [C(A + i)^{-1}][(A + i)(A + i\lambda)^{-1}]$ for λ real and positive. If we pass to a spectral representation for A, $(A \pm i)(A \pm i\lambda)^{-1}$ is multiplication by $(x \pm i)(x \pm i\lambda)^{-1}$ so s-$\lim_{\lambda \to \infty} (A \pm i)(A \pm i\lambda)^{-1} = 0$. Since $C(A + i)^{-1}$ is compact, and $[(A + i)(A + i\lambda)^{-1}]^*$ goes strongly to zero, $C(A + i)^{-1}[(A + i)(A + i\lambda)^{-1}]$ goes to zero in norm. Thus
$$\lim_{\lambda \to \infty} \|C(A + i\lambda)^{-1}\| = 0$$

As a result, C is A-bounded with relative bound zero, so (a) and (b) follow immediately (see Section X.2). Moreover, for λ large, both $[1 + C(A + i\lambda)^{-1}]^{-1}$ and $[1 + C(A - i\lambda)^{-1}]^{-1}$ exist, so $(B + i\lambda)^{-1}$ and $(B - i\lambda)^{-1}$ exist for λ large. Thus $\rho(B)$ contains points in each half-plane. Finally, for λ large,

$$(B + i\lambda)^{-1} - (A + i\lambda)^{-1} = (A + i\lambda)^{-1}[(1 + C(A + i\lambda)^{-1})^{-1} - 1]$$
$$= -(B + i\lambda)^{-1} C(A + i\lambda)^{-1}$$

Since $C(A + i\lambda)^{-1}$ is compact, $(B + i\lambda)^{-1} - (A + i\lambda)^{-1}$ is compact, so Theorem XIII.14 is applicable. Thus $\sigma_{\text{ess}}(B) = \sigma_{\text{ess}}(A)$. ∎

There is also a quadratic form version of relative compactness and an analog of Corollary 2 (Problem 39).

By doing some additional work, we can strengthen Corollary 2:

Corollary 3 Let A be a self-adjoint operator and let C be a symmetric operator so that C is a relatively compact perturbation of A^n for some positive integer n. Suppose further that $B = A + C$ is self-adjoint on $D(A)$. Then $\sigma_{\text{ess}}(A) = \sigma_{\text{ess}}(B)$.

Proof Note that since $D(A) \subset D(C)$, C is A-bounded by the closed graph theorem. Suppose first that $n = 2$. For φ, $\eta \in C^\infty(A)$, $(\varphi, (A^2 + 1)^{-z} C(A^2 + 1)^{-1+z} \eta)$ is analytic in the strip $0 < \operatorname{Re} z < 1$, continuous on the closure, and $C(A^2 + 1)^{-1+iy}$ is compact, so

$$(A^2 + 1)^{-1/2} C(A^2 + 1)^{-1/2}$$

is compact by the lemma below. Thus $(A + i)^{-1} C(A + i)^{-1}$ is compact. Since $D(A) = D(B)$, $(A - i)(B - i)^{-1}$ is a bounded operator, so $(A + C + i)^{-1} C(A + i)^{-1}$ is compact. Because

$$(A + i)^{-1} - (B + i)^{-1} = (B + i)^{-1} C(A + i)^{-1}$$

we can apply Theorem XIII.14 to see that $\sigma_{\text{ess}}(A) = \sigma_{\text{ess}}(B)$.

We now prove the case of general n by proving that if the hypotheses hold for some $n > 2$, then they hold for $n = 2$. For let $\varphi, \eta \in C^\infty(A)$. Then

$$(\varphi, C(A^2 + 1)^{-g(z)} \eta)$$

where $g(z) = \frac{1}{2}z + \frac{1}{2}n(1 - z)$, is analytic in the strip $0 < \operatorname{Re} z < 1$ and continuous on the closure. Since $C(A^2 + 1)^{-g(iy)}$ is compact for y real and $C(A^2 + 1)^{-g(1+iy)}$ is bounded, $C(A^2 + 1)^{-g(t)}$ is compact for $t = (n - 2)/(n - 1)$ by the lemma below. Thus C is A^2-compact. ∎

XIII.4 Locating the essential spectrum I: Weyl's theorem

We note that Corollary 3 is false if we require only essential self-adjointness on $D(A)$. For example, suppose that A has compact resolvent and that $C = -A$. Then $\sigma_{\text{ess}}(A) = \emptyset$ and $\sigma_{\text{ess}}(B) = \{0\}$. In the proof we used:

Lemma Let D be dense in some Hilbert space $\mathscr{H}$ and suppose that for each $z \in S = \{z \mid 0 \leq \text{Re } z \leq 1\}$ we have a quadratic form $a(z)$ on $D \times D$ so that:

(1) $(\eta, a(z)\varphi)$ is analytic in S^{int} and continuous on S for all $\eta, \varphi \in D$.
(2) For $z = iy$ (respectively, $1 + iy$), $a(z)$ is the quadratic form of a compact (respectively, bounded) operator $A(z)$.
(3) $\sup_y \{\|A(iy)\|, \|A(1+iy)\|\} \equiv M < \infty$.

Then, for any $z \in S^{int}$, $a(z)$ is the quadratic form of a compact operator.

Proof By the Hadamard three lines theorem (see the Appendix to Section IX.4) $|(\eta, a(z)\varphi)| \leq M\|\eta\|\|\varphi\|$ for any z, so $a(z)$ is the quadratic form of a bounded operator $A(z)$. The map $z \to A(z)$ is analytic in S^{int} and weakly continuous on S. We must show only that $A(z)$ is compact in S^{int}. We shall give the proof for $z = \frac{1}{2}$; the proof for other z is the same. Let $A = A(\frac{1}{2})$. To show that A is compact, it is sufficient to show that $\||A|^{1/2}\varphi_n\| \to 0$ as $n \to \infty$ for any orthonormal set $\{\varphi_n\}$. Since $\||A|^{1/2}\varphi_n\|^2 = (U^*\varphi_n, A\varphi_n)$ for a suitable partial isometry U, we need only show that for any orthonormal set $\{\varphi_n\}$ and any sequence $\{\eta_n\}$ with $\|\eta_n\| \leq 1$, we have $(\eta_n, A\varphi_n) \to 0$ as $n \to \infty$.

Let C be the contour $C_1 \cup C_2 \cup C_3 \cup C_4$ depicted in Figure XIII.2. Then, for any entire function $f(z)$ with $f(\frac{1}{2}) = 1$ we have

$$|(\eta_n, A\varphi_n)| \leq (2\pi)^{-1} \int_C |z - \tfrac{1}{2}|^{-1} |f(z)| |(\eta_n, A(z)\varphi_n)| |dz|$$

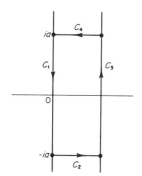

FIGURE XIII.2

Fix $\varepsilon > 0$ and let $f(z) = \exp[z^2 - \frac{1}{4} - B(z - \frac{1}{2})]$. Then $f(\frac{1}{2}) = 1$ and by choosing B and a appropriately we can arrange that

$$(2\pi)^{-1} M \int_{C_2 \cup C_3 \cup C_4} |z - \tfrac{1}{2}|^{-1} |f(z)| \, d|z| \leq \varepsilon$$

But, for each iy on C_1, $\lim_{n \to \infty} |(\eta_n, A(iy)\varphi_n)| = 0$ by the compactness of $A(iy)$. So, since the integrand is uniformly bounded,

$$\lim_{n \to \infty} (2\pi)^{-1} \int_{C_1} |z - \tfrac{1}{2}|^{-1} |f(z)| |(\eta_n, A(z)\varphi_n)| \, d|z| = 0$$

by the dominated convergence theorem. Thus, for any $\varepsilon > 0$,

$$\overline{\lim_n} \, |(\eta_n, A\varphi_n)| \leq \varepsilon$$

so $(\eta_n, A\varphi_n) \to 0$ as $n \to \infty$. ∎

There is a quadratic form version of Corollary 3, which is often useful:

Corollary 4 Let A be a positive self-adjoint operator and let C be a self-adjoint operator with $Q(C) \supset Q(A)$. Suppose that the sum of quadratic forms $A + C$ is bounded from below and closed on $Q(A)$ and let B be the corresponding self-adjoint operator. Suppose further that *either*:

(i) C is a relatively compact perturbation of A^n for some positive integer n;

or

(ii) $Q(|C|) \supset Q(A)$ and $|C|$ is a relatively form-compact perturbation of A^n for some positive integer n, i.e., $(A+1)^{-n/2}|C|(A+1)^{-n/2}$ is compact.

Then $\sigma_{\text{ess}}(A) = \sigma_{\text{ess}}(B)$.

Proof As in the case of Corollary 3, it suffices to prove that $(A + \lambda)^{-1} C (B + \lambda)^{-1}$ is compact where $\lambda > 0$ is chosen with $-\lambda < \inf \sigma(B)$. Writing

$$(A + \lambda)^{-1} C (B + \lambda)^{-1} = [(A + \lambda)^{-1} C (A + \lambda)^{-1/2}][(A + \lambda)^{+1/2}(B + \lambda)^{-1/2}]$$
$$\times (B + \lambda)^{-1/2}$$

and noting that, since $Q(A) = Q(B)$, $(A + \lambda)^{1/2}(B + \lambda)^{-1/2}$ is bounded, we conclude that it suffices to prove that $(A + 1)^{-1} C (A + 1)^{-1/2}$ is compact. Under hypothesis (i), this follows by interpolating between the fact that $(A + 1)^{-n} C (A + 1)^{-1/2}$ is compact and that $(A + 1)^{-1/2} C (A + 1)^{-1/2}$ is bounded. Under hypothesis (ii), we note that $|C|^{1/2}(A + 1)^{-1/2}$ is bounded

XIII.4 Locating the essential spectrum I: Weyl's theorem

and $|C|^{1/2}(A-1)^{-n/2}$ is compact, so by interpolation, $|C|^{1/2}(A+1)^{-1}$ is compact. Thus

$$(A+1)^{-1}C(A+1)^{-1/2} = [|C|^{1/2}(A+1)^{-1}]^*(\operatorname{sgn} C)[|C|^{1/2}(A+1)^{-1/2}]$$

is compact. ∎

We are now ready to consider various examples where Weyl's theorem is applicable:

Example 3 (classical Weyl theorem)   If A is self-adjoint and C is compact, then $\sigma_{\mathrm{ess}}(A) = \sigma_{\mathrm{ess}}(A+C)$. This holds because C is automatically relatively compact.

Example 4   Let $V \in C_0^\infty(0, \infty)$ and let H be the operator $-d^2/dx^2 + V$ on $C_0^\infty(0, \infty)$. H is not self-adjoint but it has deficiency indices $\langle 1, 1 \rangle$ (see the Appendix to Section X.1). Let A and B be two different self-adjoint extensions of H. If $D = D(\bar{H})$, then $\operatorname{Ran}(\bar{H} + i) \equiv R$ is a closed subspace codimension 1 since H has deficiency indices $\langle 1, 1 \rangle$. If $\psi \in R$, then $\psi = (\bar{H} + i)\varphi$ for some φ and thus

$$[(A+i)^{-1} - (B+i)^{-1}]\psi = (A+i)^{-1}(\bar{H}+i)\varphi - (B+i)^{-1}(\bar{H}+i)\varphi = 0$$

since $\bar{H} \subset A$ and $\bar{H} \subset B$. Thus $(A+i)^{-1} - (B+i)^{-1}$ vanishes on R. Since R has codimension 1, it follows that $(A+i)^{-1} - (B+i)^{-1}$ is rank 1 and so compact. Thus $\sigma_{\mathrm{ess}}(A) = \sigma_{\mathrm{ess}}(B)$.

Example 5   In general, let H be an operator with deficiency indices $\langle d, d \rangle$ with $d < \infty$. If A and B are two self-adjoint extensions of H, then $(A+i)^{-1} - (B+i)^{-1}$ is a rank d operator as in Example 4 so $\sigma_{\mathrm{ess}}(A) = \sigma_{\mathrm{ess}}(B)$. If $d = \infty$, the conclusion need not hold (Problem 40).

Example 6   Let $H_0 = -\Delta$ on $L^2(\mathbb{R}^3)$. By using the Fourier transform one can easily see that $\sigma_{\mathrm{ess}}(-\Delta) = [0, \infty)$. Let V be in $L^2 + (L^\infty)_\varepsilon$; then $V(H_0 + 1)^{-1}$ is compact. For, we can find $V_n \in L^2$ with $V - V_n \in L^\infty$ and $\lim_{n \to \infty} \|V_n - V\|_\infty = 0$. Thus $V_n(H_0 + 1)^{-1}$ converges in norm to $V(H_0 + 1)^{-1}$, so we need only show that $V_n(H_0 + 1)^{-1}$ is compact for each n. But $V_n(H_0 + 1)^{-1}$ is an integral operator with kernel $V_n(x)e^{-|x-y|}/4\pi|x-y|$, which is in $L^2(\mathbb{R}^6)$. Thus $V_n(H_0 + 1)^{-1}$ is Hilbert–Schmidt and so compact. Since $V(H_0 + 1)^{-1}$ is compact, V is relatively compact and so $\sigma_{\mathrm{ess}}(-\Delta + V) = \sigma_{\mathrm{ess}}(-\Delta) = [0, \infty)$.

For an extension of this result to n-dimensions ($n \neq 3$), see Problem 41.

Example 7 More generally, let $V \in R + (L^\infty)_\varepsilon$. In general V is not relatively compact since $D(V)$ may not contain $D(H_0)$. However, it is a form-compact perturbation and so the theory of Problem 39 can be used. Instead, we note that if $W \in R$, then

$$\lim_{E \to +\infty} \| |W|^{1/2}(H_0 + E)^{-1}|W|^{1/2}\| = 0$$

since $|W|^{1/2}(H_0 + E)^{-1}|W|^{1/2}$ is a Hilbert–Schmidt operator with kernel

$$|W(x)|^{1/2} \exp(-\sqrt{E}|x - y|)|W(y)|^{1/2}/4\pi|x - y|$$

and this kernel has L^2-norm going to 0 by the dominated convergence theorem. Thus, for E large and positive

$$(H + E)^{-1} - (H_0 + E)^{-1}$$

$$= \sum_{n=0}^{\infty} ((H_0 + E)^{-1}|W|^{1/2})[-W^{1/2}(H_0 + E)^{-1}|W|^{1/2}]^n (W^{1/2}(H_0 + E)^{-1})$$

where $H = H_0 + W$ and $W^{1/2} \equiv W/|W|^{1/2}$. Each term in this norm convergent sum is compact. For, letting $A = (H_0 + E)^{-1/2}|W|^{1/2}$, A^*A is compact, and so A is compact. Thus if $W \in R$, then $(H_0 + W + E)^{-1} - (H_0 + E)^{-1}$ is compact. The extension from R to $R + (L^\infty)_\varepsilon$ is as above.

Example 8 Consider the operator $H \equiv -\Delta + \mathbf{a} \cdot \mathbf{x} + V(x)$ associated to the Stark effect problem. By Theorem X.29 and Theorem X.38, we know that H is essentially self-adjoint on $C_0^\infty(\mathbb{R}^n)$ if $V \in L^p(\mathbb{R}^n) + L^\infty(\mathbb{R}^n)$ where $p = 2$ if $n \leq 3$, $p > 2$ if $n = 4$ and $p = n/2$ if $n \geq 5$. Because the $\mathbf{a} \cdot \mathbf{x}$ term tends to $-\infty$ in a way that cannot be compensated for by the $-\Delta$, one can show that H is not bounded below and one expects that $\sigma(H) = (-\infty, \infty)$. Under the stronger hypothesis that $V = V_1 + V_2$ with $V_1 \in L^p$ having compact support and $V_2 \in L^\infty$ going to zero at infinity, we shall show that V is a relatively compact perturbation of $H_0 = -\Delta + \mathbf{a} \cdot \mathbf{x}$. Let p_a be the momentum operator in the direction parallel to $\mathbf{a}$ and $p_a^\perp$ the momentum orthogonal. Then on $\mathscr{S}(\mathbb{R}^n)$, an elementary calculation (see Section XI.4) shows that

$$\exp(-i\alpha p_a^3) H_0 \exp(+i\alpha p_a^3) = (p_a^\perp)^2 + \mathbf{a} \cdot \mathbf{x} \equiv \tilde{H}_0$$

where $\alpha = (3|a|)^{-1}$. $\tilde{H}_0$ can be "diagonalized" by Fourier transforming in the directions orthogonal to $\mathbf{a}$. It follows that H_0 is essentially self-adjoint on $\mathscr{S}(\mathbb{R}^n)$ and $\sigma(H_0) = (-\infty, \infty)$ so that if V is relatively compact, we have a new proof that $H_0 + V$ is essentially self-adjoint on $\mathscr{S}(\mathbb{R}^n)$ and moreover that $\sigma_{\text{ess}}(H) = \sigma_{\text{ess}}(H_0) = (-\infty, \infty)$, so that $\sigma(H) = (-\infty, \infty)$. As above, the V_2 part of V is easy to add if we show that $V_1(H_0 + i)^{-1}$ is compact. Let $T = -\Delta$. Then on $\mathscr{S}$ which is a core for both H_0 and T,

$$\begin{aligned}(H_0 + i)^{-1} &= (T + i)^{-1} - (T + i)^{-1}(\mathbf{a} \cdot \mathbf{x})(H_0 + i)^{-1} \\ &= (T + i)^{-1} - (\mathbf{a} \cdot \mathbf{x})(T + i)^{-1}(H_0 + i)^{-1} \\ &\quad + 2i|a|(T + i)^{-1}p_a(T + i)^{-1}(H_0 + i)^{-1}\end{aligned}$$

because $[\mathbf{a} \cdot \mathbf{x}, T] = 2i|a|p_a$. Since V_1 and $(\mathbf{a} \cdot \mathbf{x})V_1$ are relatively compact perturbations of T (Problem 41) and $p_a(T + i)^{-1}$ is bounded, $V_1(H_0 + i)^{-1}$ is compact.

We summarize the above results together with Problem 41:

Theorem XIII.15 Let $H = -\Delta + \mathbf{a} \cdot \mathbf{x} + V$ on $L^2(\mathbb{R}^n)$. Then:
(a) If $a = 0$, $n = 3$, $V \in R + L^\infty$, then $\sigma_{\text{ess}}(H) = [0, \infty)$.
(b) If $a = 0$, $V \in L^p + L_\varepsilon^\infty$ where $p \geq \max\{n/2, 2\}$ if $n \neq 4$, $p > 2$ if $n = 4$, then $\sigma_{\text{ess}}(H) = [0, \infty)$
(c) If $a \neq 0$, $V = V_1 + V_2$ where $V_1 \in L^p$ (p as above) has compact support and $V_2 \in L^\infty$ goes to zero at infinity, then $\sigma(H) = \sigma_{\text{ess}}(H) = (-\infty, \infty)$.

Using Corollary 4, certain other potentials can be handled.

Example 9 Let $V = |r|^{-2}$ on $L^2(\mathbb{R}^5)$ and let $H_0 = -\Delta$. Then V cannot be H_0-compact since it does not have relative bound zero (see Example 4 in Section X.2). However, by the following general criterion, V is H_0^2-compact.
Suppose that $V \in L^2(\mathbb{R}^5) + (L^\infty(\mathbb{R}^5))_\varepsilon$. Then for each $\varepsilon > 0$ we may write $V = V_{1,\varepsilon} + V_{2,\varepsilon}$ where $V_{1,\varepsilon} \in L^2$ and $\sup_x |V_{2,\varepsilon}(x)| \leq \varepsilon$. Thus, $V_{1,\varepsilon}(H_0^2 + 1)^{-1}$ converges in norm to $V(H_0^2 + 1)^{-1}$, so we need only show that $V_{1,\varepsilon}(H_0^2 + 1)^{-1}$ is compact for each $\varepsilon > 0$. But this is immediate since $V_{1,\varepsilon} \in L^2$ and $(p^4 + 1)^{-1} \in L^2$ imply that $V_{1,\varepsilon}(H_0^2 + 1)^{-1}$ is Hilbert–Schmidt.
Returning to the case $V = |r|^{-2}$, we see from Example 4 of Section X.2 that if $\lambda \geq -2.25$, then $Q(-\Delta + \lambda V) = Q(-\Delta)$ and $-\Delta + \lambda V$ is bounded from below. For such λ, Corollary 4 is applicable, so $\sigma_{\text{ess}}(-\Delta + \lambda V) = [0, \infty)$. Similar considerations apply to the case $V = |r|^{-1}$ when $H_0 = \sqrt{-\Delta + m^2}$. When H_0 is a Dirac operator, H_0 is not bounded from below, so more subtle methods are needed.

Theorem XIII.15 shows that if V is a bounded function of compact support on $\mathbb{R}^3$, then $-\Delta + V$ has only finitely many eigenvalues in $(-\infty, -1]$. And, it is not hard to extend this to $\mathbb{R}^n$ (Problem 41). It can be proven (indeed when $n \geq 3$ we have already proven in Theorem XIII.12) that there are only finitely many eigenvalues in $(-\infty, 0)$, but the simpler $(-\infty, -1]$ result suffices for the following application:

Theorem XIII.16 Let V be a locally bounded positive function with $V(x) \to \infty$ as $|x| \to \infty$. Define $-\Delta + V$ as a sum of quadratic forms. Then $-\Delta + V$ has purely discrete spectrum.

Proof Let $H = -\Delta + V$. By the min-max principle, it suffices to prove that $\mu_n(H) \to \infty$ as $n \to \infty$. Given c, find a ball S with $V(x) \geq c$ if $x \notin S$. Such a ball exists since $V(x) \to \infty$ as $x \to \infty$. Let W be the potential that is $-c$ if $x \in S$ and 0 if $x \notin S$. Then $V \geq c + W$ so that

$$\mu_n(H) \geq c + \mu_n(-\Delta + W)$$

Since W is a bounded potential of compact support, there is an N with $\mu_n(-\Delta + W) \geq -1$ if $n \geq N$. Thus $\mu_n(H) \geq c - 1$ if $n \geq N$. Since c was arbitrary, $\mu_n(H) \to \infty$ as $n \to \infty$. ∎

We shall return to the general question of operators with purely discrete spectra in Sections 14 and 15.

XIII.5 Locating the essential spectrum II: The HVZ theorem

The Weyl theorem suffices to prove that $\sigma_{\text{ess}}(-\Delta + V) = [0, \infty)$ for a large class of two body Schrödinger operators. In this section we analyze the spectrum of N-body Schrödinger operators of the type discussed in Section XI.5. We use without comment the notation introduced there.

In the two-body case a key role was played by the condition that $V(\mathbf{x})$ go to zero at infinity at least in the average sense implied by $V \in R + (L^\infty)_\varepsilon$. In the N-body case this is no longer true. Even if V_{ij} is in C_0^∞, $V = \sum_{i<j} V_{ij}(\mathbf{r}_i - \mathbf{r}_j)$ fails to go to zero at infinity in tubes where $\sum_{i=1}^{n-1} |\mathbf{r}_i - \mathbf{r}_n|^2 \to \infty$ while some $|\mathbf{r}_i - \mathbf{r}_j|$ remains finite. This is reflected in the technical fact that $(H_0 + I)^{-1}V$ is not compact. The important new idea will be to replace

$$(H - E)^{-1} = (H_0 - E)^{-1} - (H_0 - E)^{-1} V (H - E)^{-1}$$

with an equation

$$(H - E)^{-1} = D(E) + I(E)(H - E)^{-1}$$

where $I(E)$ is compact.

We will give two distinct proofs of the basic result of this section. The first, based on Lemmas 1–6, uses resolvent equations of the type just discussed. The second, using Lemmas 1, 2, and 7–11 is based on geometric ideas of the type indicated above when we discussed the special tubes in which V fails to go to zero. Each proof has its advantages: the first is especially appropriate in cases where H is not self-adjoint such as the situation arising in Sections 10 and 11. As we shall see below, the second proof is particularly convenient when H is restricted to an invariant subspace.

For each cluster decomposition $D = \{C_\ell\}_{\ell=1}^k$ of $\{1, \ldots, n\}$, we defined in Section XI.5 a Hamiltonian $H_D = H - I_D$ where I_D is the sum of all potentials V_{ij} with i and j in different clusters. For potentials $V_{ij} \in L^2 + L^p$ $(p < 3)$, we proved that $\sigma(H_D) \subset \sigma(H)$ since the wave operators $\Omega_D^\pm$ provide a unitary equivalence of H_D and $H \restriction \operatorname{Ran} \Omega_D^\pm$. To find $\sigma(H_D)$, we write $\mathscr{H} = \mathscr{H}(C_1) \otimes \cdots \otimes \mathscr{H}(C_k) \otimes \mathscr{H}_D$ corresponding to the breakup of $\mathscr{H}$ resulting from choosing internal coordinates $\zeta_1^{(C_1)}, \ldots, \zeta_{n_1-1}^{(C_1)}$ for $C_1, \ldots, \zeta_1^{(C_k)}, \ldots, \zeta_{n_k-1}^{(C_k)}$ for C_k and $\zeta_1, \ldots, \zeta_{k-1}$ for the relative coordinates of the clusters. Then

$$H_D = H(C_1) + \cdots + H(C_k) + T_D$$

where

$$H(C_\ell) = I \otimes \cdots \otimes h_{C_\ell} \otimes \cdots \otimes I$$

is the Hamiltonian of cluster C_ℓ with its center of mass removed and T_D is the kinetic energy of the clusters, i.e., the sum of the kinetic energies of the centers of mass of the C_ℓ minus the kinetic energy of the center of mass of the entire system. Thus by Theorem VIII.33,

$$\sigma(H_D) = \left\{ \sum_{\ell=1}^k \lambda_\ell + \tau \,\middle|\, \lambda_\ell \in \sigma(H(C_\ell)),\ \tau \in \sigma(T_D) \right\}$$

Since $\sigma(T_D) = [0, \infty)$, we have $\sigma(H_D) = [\Sigma_D, \infty)$, where

$$\Sigma_D = \sum_{\ell=1}^k \inf \sigma(H(C_\ell)) \tag{19}$$

Since we expect that $\mathscr{H}$ should consist only of bound states and scattering states, we expect that $\sigma_{\mathrm{ess}}(H) = [\Sigma, \infty)$ where

$$\Sigma = \inf_{D,\ \#(D) \geq 2} \Sigma_D \tag{20a}$$

In fact, this is true under much more general conditions than $V \in L^2 + L^p$.

Theorem XIII.17 (the HVZ theorem) Let H be the operator on $L^2(\mathbb{R}^{3N-3})$ obtained by removing the center of mass motion from

$$-\sum_{i=1}^N (2\mu_i)^{-1} \Delta_i + \sum_{i<j} V_{ij}(\mathbf{r}_{ij}), \quad \text{where} \quad V_{ij} \in R + (L^\infty)_\varepsilon, \quad \text{and} \quad \mathbf{r}_{ij} = \mathbf{r}_i - \mathbf{r}_j$$

For each cluster decomposition $D = \{C_1, \ldots, C_k\}$, define Σ by (19) and (20a). Then

$$\sigma_{\text{ess}}(H) = [\Sigma, \infty)$$

The remainder of this section is concerned with the proof of the HVZ theorem. The techniques that we develop are useful in the study of many aspects of N-body systems with more than one channel, i.e., systems for which some H_D possesses a bound state. In particular, these techniques play an important role in Section 10. Before proving this theorem, we note that it follows inductively that $\inf\{\Sigma_D \mid \#(D) = m+1\} \geq \inf\{\Sigma_D \mid \#(D) = m\}$, so (20a) can be replaced by (Problem 42):

$$\Sigma = \inf\{\Sigma_D \mid \#(D) = 2\} \tag{20b}$$

We shall show that $[\Sigma, \infty) \subset \sigma(H)$ by first assuming that the V_{ij} are nice and then using an approximation argument.

Lemma 1 If $V_{ij} \in C_0^\infty(\mathbb{R}^3)$, then $[\Sigma, \infty) \subset \sigma(H)$.

Proof Since the wave operators $\Omega_D^\pm$ exist (Theorem XI.34), $\sigma(H_D) \subset \sigma(H)$. Since $\sigma(H_D) = [\Sigma_D, \infty)$ if $\#(D) \geq 2$, the result holds. ∎

For an alternative proof of Lemma 1, avoiding the use of wave operators, see Problem 45.

Lemma 2 Under the hypotheses of Theorem XIII.17, $[\Sigma, \infty) \subset \sigma(H)$.

Proof For each i, j, we can find $V_{ij}^{(n)} \in C_0^\infty$ so that, as forms, $|V_{ij} - V_{ij}^{(n)}| \leq n^{-1}(-\Delta + 1)$ (Problem 44). Letting $H^{(n)} = H_0 + \sum_{i<j} V_{ij}^{(n)}$ we see that $H^{(n)} \to H$ in norm resolvent sense and that $H_D^{(n)} \to H_D$ similarly. Thus, by Theorem VIII.20, $\exp(-H_D^{(n)}) \to \exp(-H_D)$ in norm. Since $\Sigma_D^{(n)} = -\log\|e^{-H_D^{(n)}}\|$, we conclude that $\Sigma_D^{(n)} \to \Sigma_D$ and thus $\Sigma^{(n)} \to \Sigma$. Let $\lambda > \Sigma$. For n sufficiently large, we conclude that $\lambda > \Sigma^{(n)}$, so $\lambda \in \sigma(H^{(n)})$ by Lemma 1. Since $H^{(n)} \to H$ in norm resolvent sense, $\lambda \in \sigma(H)$ by Theorem VIII.23. Thus $(\Sigma, \infty) \subset \sigma(H)$. Since $\sigma(H)$ is closed, $\Sigma \in \sigma(H)$. ∎

The proof of the other direction of Theorem XIII.17, namely that if $\lambda < \Sigma$, then either $\lambda \in \sigma_{\text{disc}}(H)$ or $\lambda \in \rho(H)$, is considerably harder than Lemma 2 and requires the development of some machinery. To illustrate the ideas, let us first discuss the case where $N = 3$ and $V_{ij} \in L^2$. In the two-body case, we used the key equation $(H - E)^{-1} = (H_0 - E)^{-1} - (H_0 - E)^{-1} V (H - E)^{-1}$

and the critical fact that $(H_0 - E)^{-1}V$ is compact. In the three-body case we still have

$$(H - E)^{-1} = (H_0 - E)^{-1} - (H_0 - E)^{-1}(V_{12} + V_{13} + V_{23})(H - E)^{-1}$$

but it is no longer true that $(H_0 - E)^{-1}(V_{12} + V_{13} + V_{23})$ is compact. To see that $(H_0 - E)^{-1}V_{12}$ is not compact, notice that it commutes with the one-parameter group of translations $\langle r_1, r_2, r_3 \rangle \mapsto \langle r_1 - a, r_2 - a, r_3 \rangle$. Such an operator cannot be compact (see Problem 43).

Thus we seek a decomposition of $(H - E)^{-1}$ into the form $A(E) + B(E) \times (H - E)^{-1}$ with $B(E)$ compact, which must be more subtle than the decomposition given by the second resolvent equation. To get an idea for the proper decomposition take E so negative that the perturbation series obtained by iterating the second resolvent equation converges. Then

$$\begin{aligned}(H - E)^{-1} &= (H_0 - E)^{-1} - (H_0 - E)^{-1}V_{12}(H_0 - E)^{-1} \\ &\quad - (H_0 - E)^{-1}V_{13}(H_0 - E)^{-1} \\ &\quad - (H_0 - E)^{-1}V_{23}(H_0 - E)^{-1} \\ &\quad + (H_0 - E)^{-1}V_{12}(H_0 - E)^{-1}V_{12}(H_0 - E)^{-1} + \cdots \quad (21)\end{aligned}$$

We introduce a graphical notation for individual terms in (21). $\equiv$ represents a factor of $(H_0 - E)^{-1}$ and a vertical line joining lines i and j represents a factor of $-V_{ij}$. Thus, the diagram

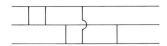

represents the operator

$$(-1)^5 (H_0 - E)^{-1} V_{12}(H_0 - E)^{-1} V_{12}(H_0 - E)^{-1} V_{23}(H_0 - E)^{-1}$$
$$\times V_{13}(H_0 - E)^{-1} V_{23}(H_0 - E)^{-1}$$

In terms of diagrams, it is easy to describe which terms in (21) are compact:

Lemma 3A A diagram represents a compact operator if and only if it is connected. In fact, if a diagram G is connected and if the last factor of $(H_0 - E)^{-1}$ is removed, then the operator remains compact.

Proof We remark that we are interpreting the symbol $(H_0 - E)^{-1}V_{ij}$ as the (bounded) closure of the operator defined on $D(H_0)$ or alternatively as the adjoint of the everywhere defined operator $V_{ij}(H_0 - \bar{E})^{-1}$. If a diagram is disconnected, by relabeling the particles we can suppose that line 3 is not linked to lines 1 and 2. Then the associated operator commutes with the unitary operator induced by the map $\langle r_1, r_2, r_3 \rangle \mapsto \langle r_1 + a, r_2 + a, r_3 \rangle$, so the operator is not compact (Problem 43).

On the other hand, if G is connected, it must contain two consecutive links that are different. By relabeling, we may suppose they are V_{12} and V_{23}. The operator associated with G after the last $(H_0 - E)^{-1}$ removed is then of the form $A[(H_0 - E)^{-1} V_{12} (H_0 - E)^{-1} V_{23}] B$ where A and B are products of terms of the form $(H_0 - E)^{-1} V_{ij}$. Since A and B are bounded, we need only prove that $C = (H_0 - E)^{-1} V_{12} (H_0 - E)^{-1} V_{23}$ is compact. Let us work in momentum space with momentum variables p, q representing the relative momenta of 1 to 2 and 2 to 3, i.e., we consider the operator $D = \mathscr{F} C \mathscr{F}^{-1}$ where $\mathscr{F}$ is the Fourier transform, in terms of the variables $\langle p, q \rangle$ dual to $\langle r_{12}, r_{23} \rangle$. Let $h_0(p, q)$ be the kinetic energy written in terms of p and q. Explicitly,

$$h_0(p, q) = (2m_{12})^{-1} p^2 + (2m_{23})^{-1} q^2 - \mu_2^{-1} p \cdot q$$

where $m_{ij}^{-1} = \mu_i^{-1} + \mu_j^{-1}$. Then

$$(D\psi)(p, q) = (2\pi)^{-3} \int (h_0(p, q) - E)^{-1} \hat{V}_{12}(p - p')$$
$$\times (h_0(p', q) - E)^{-1} \hat{V}_{23}(q - q') \psi(p', q') \, dp' \, dq' \quad (22)$$

Since E is negative, $|h_0(p, q) - E| > c(p^2 + q^2 + 1)$ for some constant c. Thus the kernel in (22) is bounded by

$$(2\pi)^{-3} c^{-2} (p^2 + 1)^{-1} \hat{V}_{12}(p - p')(q^2 + 1)^{-1} \hat{V}_{23}(q - q')$$

Since $\hat{V}_{12}, \hat{V}_{23} \in L^2$, this kernel is in $L^2(\mathbb{R}^{12})$ and so the integral operator in (22) is Hilbert–Schmidt. Therefore D and C are compact. ∎

This suggests that in attempting to find an expansion $(H - E)^{-1} = A(E) + B(E)(H - E)^{-1}$ with $B(E)$ compact, we should choose $A(E)$ to be the sum of all disconnected diagrams in the formal expansion (21) when E is negative and then determine $A(E)$ by analytic continuation to other E. To make this continuation, we need a closed expression for this sum of disconnected diagrams. The disconnected diagrams fall naturally into four classes according to whether no horizontal lines are connected to each other or whether only one of the three pairs is connected to each other. Only $\equiv = (H_0 - E)^{-1}$ is in the first class. The class where 1 and 2 are connected together has a sum

$$\overline{\underline{\quad\rule{0pt}{6pt}\quad}} + \overline{\underline{\quad\rule{0pt}{6pt}\quad}} + \overline{\underline{\quad\rule{0pt}{6pt}\quad}} + \cdots \quad (23)$$

If we add ≡ to this sum, we obtain the sum of all graphs for the case with $V_{23} = V_{13} = 0$. Thus, the sum (23) is $(H_0 + V_{12} - E)^{-1} - (H_0 - E)^{-1}$. Introducing the notation

$$G_0(E) = (H_0 - E)^{-1}$$
$$G_{ij}(E) = (H_0 + V_{ij} - E)^{-1}$$
$$G(E) = (H_0 + V - E)^{-1}$$

We can rewrite (23) as $-G_{12}(E)V_{12}G_0(E)$. This suggests:

Definition Let $E \notin [\Sigma, \infty)$ be a complex number. The **disconnected part of the resolvent** or, for short, the **disconnected part** is defined by

$$D(E) = G_0(E) - \sum_{1 \leq i < j \leq 3} G_{ij}(E)V_{ij}G_0(E) \tag{24a}$$

Similarly, we can analyze the connected diagrams. A diagram is called **barely connected** if it becomes disconnected when its last leg is removed. The sum of the barely connected diagrams falls into six parts, one of which is

and the other classes are obtained by permuting the labels 1, 2, 3. Each connected diagram can be broken at a unique point in such a way that a barely connected diagram occurs to the left of the break. We can obtain all connected diagrams by placing an arbitrary barely connected diagram to the left of the break and an arbitrary diagram to the right. Thus, formally

$$\sum \text{(all connected diagrams)} = \left(\sum \text{(all barely connected diagrams with the final } G_0(E) \text{ removed))}\right)$$
$$\times \left(\sum \text{(all diagrams)}\right)$$

if we do not distinguish a diagram and its associated operator. We have used the symbol $\sum (\ )$ to denote the sum of all diagrams with property $(\ )$ in the formal expansion (21). Since $\sum$ (all diagrams) is $(H - E)^{-1}$, we seek a closed expression for the first factor. The sum associated with the class of barely connected diagrams drawn above is clearly

$$-[G_{12}(E) - G_0(E)]V_{23} = G_{12}(E)V_{12}G_0(E)V_{23}$$

Definition Let $E \notin [\Sigma, \infty)$ be a complex number. The **Weinberg connected interaction** or **Weinberg kernel** is defined by

$$I(E) = \sum_{\substack{j \neq k \neq i \\ i < j}} G_{ij}(E)V_{ij}G_0(E)(V_{jk} + V_{ik}) \tag{24b}$$

We are now prepared for:

Lemma 4A The functions $D(E)$ and $I(E)$ are analytic operator-valued functions in the region $\mathbb{C}\backslash[\Sigma, \infty)$. If $E \notin \sigma(H)$, then

$$(H - E)^{-1} = D(E) + I(E)(H - E)^{-1} \qquad (24c)$$

Moreover, $I(E)$ is compact for all $E \in \mathbb{C}\backslash[\Sigma, \infty)$, and $\lim_{E \to -\infty} \|I(E)\| = 0$. (24) is called the **Weinberg–Van Winter equation**.

Proof Since $\mathbb{C}\backslash[\Sigma, \infty) \subset \bigcup_{i<j} \rho(H_0 + V_{ij})$, $D(E)$ and $I(E)$ are analytic if we interpret $G_0(E)V_{ij}$ as the adjoint of the bounded operator $V_{ij} G_0(E)^*$ and similarly for $G_{ij}(E)V_{kl}$. Choose E so negative that the perturbation series (21) converges absolutely in operator norm. Then we can rearrange sums of diagrams at will and see that (24c) holds. By analytic continuation, (24c) extends to any $E \notin \sigma(H) \cup [\Sigma, \infty)$ which is equal to $\sigma(H)$ by Lemma 2. Since $\|G_0(E)V_{ij}\| \to 0$ as $E \to -\infty$, $\lim_{E \to -\infty} \|I(E)\| = 0$. Finally, we notice that if E is so negative that the perturbation series for $I(E)$ converges, $I(E)$ is compact by Lemma 3A and the basic fact that the compact operators are norm closed (Theorem VI.12a). That $I(E)$ is compact for all $E \in \mathbb{C}\backslash[\Sigma, \infty)$ follows from Lemma 5 below. ∎

Lemma 5 Let $f(z)$ be an operator-valued function on a connected region $R \subset \mathbb{C}$. If $f(z)$ is compact for z in an open set $S \subset R$, then $f(z)$ is compact for all $z \in R$.

Proof Suppose that $f(z_0)$ is not compact. Since the compact operators are closed, we can find, by the Hahn–Banach theorem, a linear functional $L \in \mathcal{L}(\mathcal{H})^*$ so that $L(f(z_0)) = 1$ and $L(A) = 0$ if A is compact. Then $L(f(z))$ is analytic, $L(f(z)) = 0$ if $z \in S$ and $L(f(z_0)) = 1$. Since S is open, this is impossible. We conclude that $f(z_0)$ is compact for all z_0. ∎

This last lemma can also be proven using the idea of analytic continuation by power series.

Proof of Theorem XIII.17 in case $N = 3$, $V_{ij} \in L^2$ Since $I(E)$ is a compact operator-valued function on $\mathbb{C}\backslash[\Sigma, \infty)$ and $1 - I(E)$ is invertible if E is real and very negative, the analytic Fredholm theorem (Theorem VI.14) implies that there is a discrete set $S \subset \mathbb{C}\backslash[\Sigma, \infty)$ so that $(1 - I(E))^{-1}$ exists and is analytic in $\mathbb{C}\backslash([\Sigma, \infty) \cup S)$ and is meromorphic in $\mathbb{C}\backslash[\Sigma, \infty)$ with finite rank residues. Thus $[1 - I(E)]^{-1} D(E) \equiv f(E)$ is analytic in $\mathbb{C}\backslash([\Sigma, \infty) \cup S)$ with finite rank residues at points of S. Let $E \notin S$, Im $E \neq 0$. Then, by (24),

$f(E) = (H - E)^{-1}$. In particular, $f(E)(H - E)\psi = \psi$ for any $\psi \in D(H)$. By analytic continuation, this holds for any $E \notin S \cup [\Sigma, \infty)$. We conclude that, for any such E, $H - E$ has a bounded inverse. Thus any point of $\sigma(H)\backslash[\Sigma, \infty)$ must be in S, so $\sigma(H)\backslash[\Sigma, \infty)$ consists of isolated points with only Σ as a possible limit point. Finally, since $(H - E)^{-1}$ has finite rank residues at any point $\lambda \in S$, we conclude that the spectral projection

$$P_\lambda = (-2\pi i)^{-1} \oint_{|E - \lambda| = \varepsilon} (H - E)^{-1} \, dE$$

is finite dimensional, i.e., any $\lambda \in \sigma(H)\backslash[\Sigma, \infty)$ is in $\sigma_{\mathrm{disc}}(H)$. ∎

The proof in the general case involves no additional analytic ideas but we need to overcome two complications: (1) V_{ij} need only be in $R + (L^\infty)_\varepsilon$; (2) the diagrammatic analysis is not so simple if $N > 3$ and it is necessary to develop some combinatorial machinery. Since V_{ij} is, in general, only form bounded, we must deal with expressions like $(H_0 - E)^{-1/2} V_{ij} (H_0 - E)^{-1/2}$ rather than $V_{ij}(H_0 - E)^{-1}$. We therefore revise our graphical rules: Given a diagram with successive pairwise links of n horizontal lines, associate a factor of $-G_0^{1/2}(E)V_{ij}G_0^{1/2}(E)$ with each link between i and j and no additional factors corresponding to the horizontal lines. This changes the operator that we associated to a diagram before by removing a factor of $G_0^{1/2}$ from each end. By $G_0^{1/2}$ we mean the operator that acts in momentum space as multiplication by $(p^2 - E)^{-1/2}$ where we take the square root branch that is positive if $E < 0$. We shall consider the case of general $N \geq 3$, and (21) will henceforth denote the analogous general expansion for such N.

Lemma 3B Let $V_{ij} \in R + (L^\infty)_\varepsilon$. Let G be a diagram in the formal expansion (21). Then the operator associated with G according to the graphical rules *just discussed* is compact if and only if G is connected.

Proof As before, if the diagram is disconnected, it is noncompact, so suppose that G is a fixed connected diagram. Let $V_{ij} \in \mathscr{S}(\mathbb{R}^3)$ for each ij. We shall first show that under such circumstances (actually $V_{ij} \in L^2$ suffices) the associated operator G is Hilbert–Schmidt and thus compact.

The proof will be by induction on N, the number of lines in the graph. For $N = 1$, the Hilbert space is $\mathbb{C}$ so the operator (which is 1) is certainly Hilbert–Schmidt. Thus, we suppose that all connected graphs with fewer than N lines are known to be Hilbert–Schmidt.

We first note that we may as well suppose that G is barely connected since any G can be written as $G = G_1 G_2$ with G_2 bounded and G_1 barely connected. By relabeling lines we can suppose that the last link in G connects lines i and $i+1$ and that if this last line is removed, then the graph remaining has two connected pieces, one containing lines $1, \ldots, i$; the other $i+1$, $\ldots, N$.

Pick new coordinates $R_1, \ldots, R_{N-1}$ by $R_j = r_{j+1} - r_j$ and let $P_1, \ldots, P_{N-1}$ be the corresponding momentum operators, i.e., P_j is $-i \nabla_{R_j}$, the partial derivative being taken with $R_1, \ldots, R_{j-1}, R_{j+1}, \ldots, R_{N-1}$ fixed. Let W_{jk} be the function given by $\hat{W}_{jk} = |\hat{V}_{jk}|$ and define the operator $A(jk)$ as

$$A(jk) = \begin{cases} \left(\sum_{\ell \leq i-1} P_\ell^2 + 1\right)^{-1/2} W_{jk} \left(\sum_{\ell \leq i-1} P_\ell^2 + 1\right)^{-1/2} & \text{if } j, k \leq i \\ (P_i^2 + 1)^{-1/2} W_{i,i+1} (P_i^2 + 1)^{-1/2} & \text{if } j = i, k = i+1 \\ \left(\sum_{\ell \geq i+1} P_\ell^2 + 1\right)^{-1/2} W_{jk} \left(\sum_{\ell \geq i+1} P_\ell^2 + 1\right)^{-1/2} & \text{if } j, k \geq i+1 \end{cases}$$

Let $\tilde{G}$ be the operator obtained by replacing each factor $G_0^{1/2}(E) V_{jk} G_0^{1/2}(E)$ in G by $A(jk)$. We first claim that $\tilde{G}$ is Hilbert–Schmidt: For under the decomposition

$$L^2(\mathbb{R}^{3N-3}) = L^2(\mathbb{R}^{3i-3}) \otimes L^2(\mathbb{R}^3) \otimes L^2(\mathbb{R}^{3N-3i-3})$$

corresponding to $\langle R_1, \ldots, R_{i-1} \rangle$, R_i, $\langle R_{i+1}, \ldots, R_{N-1} \rangle$, $\tilde{G}$ is a tensor product $A \otimes B \otimes C$ where A and C are operators associated to graphs with i and $N-i$ lines respectively. B is explicitly seen to be Hilbert–Schmidt on $L^2(\mathbb{R}^3)$, for example by writing its kernel in p space (i.e., looking at $\mathscr{F}B\mathscr{F}^{-1}$) and A and C are, by the induction hypothesis, Hilbert–Schmidt on $L^2(\mathbb{R}^{3i-3})$ and $L^2(\mathbb{R}^{3N-3i-3})$, respectively. Thus $\tilde{G}$ is Hilbert–Schmidt.

Now let $g = \mathscr{F}G\mathscr{F}^{-1}$ and $\tilde{g} = \mathscr{F}\tilde{G}\mathscr{F}^{-1}$. Since $|\hat{V}_{ij}| \leq \hat{W}_{ij}$ and $|H_0 - E| \geq c_0(\sum_{\ell=1}^{N+1} P_\ell^2 + 1)$ for some constant c_0, the kernel of g is majorized by a multiple of the kernel of $\tilde{g}$. Since $\tilde{g}$ is Hilbert–Schmidt, so is g and thus so is G.

Now suppose that $V_{ij} \in R + (L^\infty)_\varepsilon$. We can find $V_{ij}^{(n)} \in \mathscr{S}(\mathbb{R}^3)$ so that $\|(H_0 - E)^{-1/2}(V_{ij} - V_{ij}^{(n)})(H_0 - E)^{-1/2}\| \to 0$ as $n \to \infty$ (Problem 44). Given any connected diagram G, let $g^{(n)}$ be the associated operator if potentials $V_{ij}^{(n)}$ are used and let g be the associated operator if V_{ij} is used. Then $g^{(n)} \to g$ in norm. Since each $g^{(n)}$ is compact, g is compact. ∎

We now turn to the combinatorics necessary to find closed expressions for $D_R(E)$, the sum of all disconnected diagrams, and $I_R(E)$, the sum of all barely connected diagrams.

XIII.5 The HVZ theorem

Definition Let G be a diagram with N labeled horizontal lines and some vertical links of pairs. G is called k-**connected** if it has k connected components. The **cluster decomposition** $D(G)$ associated with G is the family of sets of lines linked to one another.

Thus, the diagram in Figure XIII.3 is 2-connected with $D(G) = \{\{1, 2, 4\}, \{3, 5\}\}$.

Definition The **associated string** of a k-connected diagram is a set of cluster decompositions $D_N, \ldots, D_k$ defined as follows. Suppose G has ℓ links. Let $G_0, \ldots, G_\ell$ be the diagrams obtained by keeping the first $0, 1, \ldots, \ell$ links (counting from the left). Consider the cluster decompositions $D(G_0), \ldots, D(G_\ell)$. D_m is the $(N + 1 - m)$th *distinct* decomposition in the family $D(G_0), \ldots, D(G_\ell)$. It has exactly m clusters. We write $S(G)$ for the associated string of G.

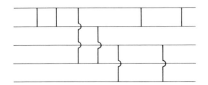

FIGURE XIII.3 A 2-connected diagram.

Thus, the diagram in Figure XIII.3 has

$$D_5 = \{\{1\}, \{2\}, \{3\}, \{4\}, \{5\}\}$$
$$D_4 = \{\{1, 2\}, \{3\}, \{4\}, \{5\}\}$$
$$D_3 = \{\{1, 2, 4\}, \{3\}, \{5\}\}$$
$$D_2 = \{\{1, 2, 4\}, \{3, 5\}\} \equiv D(G)$$

Notice that $D_5 = D(G_0)$; $D_4 = D(G_1) = D(G_2)$; $D_3 = D(G_3) = D(G_4)$; $D_2 = D(G_5) = D(G_6) = D(G_7) = D(G_8)$.

Finally, we define:

Definition A string S is a family $D_N, D_{N-1}, \ldots, D_k$ of cluster decompositions of $\{1, \ldots, N\}$ so that $D_{j+1} \rhd D_j$ and so that D_j has j clusters. The string is called connected if $k = 1$ and **disconnected** if $k > 1$. k will be called the **index** of S and we write $i(S) = k$.

We recall from Section XI.5 that $D \triangleright D'$ means that each cluster in D' is a union of clusters in D. Thus, if S is a string, D_j is obtained from D_{j+1} by lumping one pair of clusters together. Also, iDm means that i and m are indices in the same cluster of D and $\sim iDm$ means that they are in different clusters.

To sum up all disconnected diagrams, we shall first sum all diagrams with the same associated string and then sum over the (finitely many) strings. In the following, the symbols

$$\sum_D \quad \text{and} \quad \sum_{D \sim D'}$$

stand respectively for the sum over those pairs ij with $i < j$ and iDj (respectively iDj and not $iD'j$).

Definition Let $S_0 = \langle D_N, D_{N-1}, \ldots, D_k \rangle$ be a fixed string and let E be so negative that

$$\sum_{1 \leq i < j \leq N} \|(H_0 - E)^{-1/2} V_{ij} (H_0 - E)^{-1/2}\| < 1 \tag{25}$$

Given any cluster decomposition D, we define the **reduced resolvent** $R_D(E)$ by

$$R_D(E) = \left[1 + (H_0 - E)^{-1/2} \left(\sum_D V_{ij} \right)(H_0 - E)^{-1/2} \right]^{-1} \tag{26}$$

Define $I_{D_m D_{m-1}}$ by

$$I_{D_m D_{m-1}} = (H_0 - E)^{-1/2} \left(- \sum_{D_{m-1} \sim D_m} V_{ij} \right)(H_0 - E)^{-1/2} \tag{27}$$

Finally, let $R_{S_0}(E)$ be the convergent sum of all diagrams G with $S(G) = S_0$. The sum (27) can be described as follows. D_{m-1} differs from D_m by having two clusters C_ℓ and C_p in D_m being replaced in D_{m-1} by the single cluster $C_q = C_p \cup C_\ell$. The sum in (27) is over those $i < j$ with $i \in C_\ell, j \in C_p$ or $i \in C_p, j \in C_\ell$.

Lemma 6

$$R_{S_0}(E) = R_{D_N}(E) I_{D_N D_{N-1}} R_{D_{N-1}}(E) I_{D_{N-1} D_{N-2}} \cdots I_{D_{k+1} D_k} R_{D_k}(E) \tag{28}$$

Proof Any arbitrary diagram G with $S(G) = S_0$ can be described as follows. The first link must change the cluster decomposition from D_N to D_{N-1} and so must be of the form $-(H_0 - E)^{-1/2} V_{ij} (H_0 - E)^{-1/2}$ with $\sim iD_N j$ and $iD_{N-1} j$. Then there are an arbitrary number of links that do not change the associated clustering. These must be i, j links with $iD_{N-1} j$. Next follows a link that changes the clustering from D_{N-1} to D_{N-2}, etc. If we write $W_{ij} = -(H_0 - E)^{-1/2} V_{ij} (H_0 - E)^{-1/2}$, we see that

$$R_{S_0}(E) = \left(\sum_{D_{N-1} \sim D_N} W_{ij} \right) \left(\sum_{\ell=0}^{\infty} \left(\sum_{D_{N-1}} W_{ij} \right)^{\ell} \right) \left(\sum_{D_{N-2} \sim D_{N-1}} W_{ij} \right) \cdots$$

Since $\sum_{i<j} \|W_{ij}\| < 1$, all the infinite sums converge and (28) results. ∎

Let $\mathscr{H}_{+1}$, $\mathscr{H}_{-1}$ be the scale of spaces associated with H_0 (see Section VIII.6). Let D be a cluster decomposition with $\#(D) \geq 2$ and suppose that $E \in \mathbb{C} \setminus [\Sigma, \infty)$. Then $E \notin \sigma(H_D)$. By the construction of sums of quadratic forms, $(H_D - E)^{-1}$ maps $\mathscr{H}_{-1}$ boundedly onto $\mathscr{H}_{+1}$, so

$$(H_0 - E)^{+1/2} (H_D - E)^{-1} (H_0 - E)^{+1/2}$$
$$= \left[1 + (H_0 - E)^{-1/2} \sum_D V_{ij} (H_0 - E)^{-1/2} \right]^{-1}$$

defines a bounded operator, which we denote by $R_D(E)$ consistently with the definition (26). For $E \notin [\Sigma, \infty)$, we *define* $R_{S_0}(E)$ by (28) whenever S_0 is a string with $i(S_0) \geq 2$. Lemma 6 shows that when E is very negative this definition is consistent with the definition as the sum of all graphs with string S_0. We are now ready to define the objects required for a general Weinberg–Van Winter equation. The symbol $\sum_{S, i(S)=2}$ means the sum over all strings of index 2.

Definition The **reduced disconnected part** is defined for $E \notin [\Sigma, \infty)$ by

$$D_R(E) = \sum_{S, i(S) \geq 2} R_S(E) \qquad (29a)$$

The **symmetrized Weinberg kernel** is defined by

$$I_R(E) = - \sum_{S, i(S)=2} R_S(E) \left[\sum_{\sim D_2} (H_0 - E)^{-1/2} V_{ij} (H_0 - E)^{-1/2} \right] \qquad (29b)$$

In (29b), D_2 refers to the last decomposition in the string S involved in the particular summand. Note that in (29) the sum over strings is a *finite* sum.

Lemma 4B $D_R(E)$ and $I_R(E)$ are analytic in $\mathbb{C} \setminus [\Sigma, \infty)$ and $I_R(E)$ is compact operator valued with $\lim_{E \to -\infty} \|I_R(E)\| = 0$. For $E \notin \sigma(H)$, let

$$R(E) \equiv R_{\{\{1, \ldots, N\}\}}(E) = \left[1 + (H_0 - E)^{-1/2} \left(\sum_{i<j} V_{ij} \right) (H_0 - E)^{-1/2} \right]^{-1}$$

XIII: SPECTRAL ANALYSIS

where $\{\{1, \ldots, N\}\}$ refers to the cluster decomposition D with one cluster. Then for all $E \notin \sigma(H)$,

$$R(E) = D_R(E) + I_R(E)R(E) \tag{29c}$$

Proof Any D with $\#(D) \geq 2$ has $\sigma(H_D) \subset [\Sigma, \infty)$ so $I_R(E)$ and $D_R(E)$ are analytic in $\mathbb{C}\backslash[\Sigma, \infty)$. For all E with sufficiently negative real part, $I_R(E)$ can be expanded into the norm convergent sum of all barely connected diagrams. By Lemma 3B, $I_R(E)$ is compact for all such E, so, by Lemma 5, $I_R(E)$ is compact for all $E \in \mathbb{C}\backslash[\Sigma, \infty)$. From the perturbation series, it also follows that $\lim_{E \to -\infty} \|I_R(E)\| = 0$.

Next notice that there is a one-one correspondence between strings S with $i(S) = 2$ and strings $\tilde{S}$ with $i(\tilde{S}) = 1$ for the only decomposition with one element is $D_1 \equiv \{\{1, \ldots, N\}\}$ and that $R_{D_1} = R$. Thus if $S = \langle D_N, \ldots, D_2 \rangle$ and $\tilde{S} = \langle D_N, \ldots, D_1 \rangle$, Lemma 6 implies that

$$R_{\tilde{S}}(E) = -R_S(E)\left[\sum_{\sim D_2}(H_0 - E)^{-1/2}V_{ij}(H_0 - E)^{-1/2}\right]R(E)$$

Summing over all $\tilde{S}$ (equivalently all S), we see that

$$\sum_{\tilde{S}, i(\tilde{S})=1} R_{\tilde{S}}(E) = I_R(E)R(E)$$

For all sufficiently negative E, the perturbation series (21) converges so $R(E)$ is the sum over all diagrams; and thus summing first over diagrams in each fixed string and then over strings, we conclude that

$$R(E) = \sum_{S, i(S) \geq 2} R_S(E) + \sum_{\tilde{S}, i(\tilde{S})=1} R_{\tilde{S}}(E)$$

This proves (29c) when E is very negative. By analytic continuation it holds for $E \notin \sigma(H) \cup [\Sigma, \infty) = \sigma(H)$ (by Lemma 2). ∎

(29) is called the **Weinberg–Van Winter equation**.

Proof of Theorem XIII.17, general case The proof is essentially the same as in the special case $N = 3$; $V_{ij} \in L^2$. By Lemma 4B and the analytic Fredholm theorem, $(I - I_R(E))^{-1}$ exists and is analytic in $\mathbb{C}\backslash([\Sigma, \infty) \cup S)$ where S is a discrete set in $\mathbb{C}\backslash[\Sigma, \infty)$. Thus

$$f(E) = (H_0 - E)^{-1/2}[1 - I_R(E)]^{-1}D_R(E)(H_0 - E)^{-1/2}$$

exists in $\mathbb{C}\backslash([\Sigma, \infty) \cup S)$ with finite rank residues at points of S. When $\operatorname{Im} E \neq 0$, $f(E) = (H - E)^{-1}$ so for all $\psi \in D(H)$, $f(E)(H - E)\psi = \psi$. By analytic continuation $\sigma(H) \subset [\Sigma, \infty) \cup S$ and the points of S have finite spectral projections. Thus $\sigma_{\text{ess}}(H) \subset [\Sigma, \infty)$. By Lemma 2, $[\Sigma, \infty) \subset \sigma(H)$ so $\sigma_{\text{ess}}(H) = [\Sigma, \infty)$. ∎

As an immediate corollary of the HVZ theorem, we have

Theorem XIII.18 If H is an N-body Schrödinger operator with center of mass removed and if $V_{ij} \in L^2 + L^r$ ($r < 3$) for all i, j, then $\sigma_{\mathrm{ac}}(H) = [\Sigma, \infty)$ with Σ given by (20b).

Proof From Theorem XI.34, and the relations

$$e^{-iHt}\Omega_D^\pm = \Omega_D^\pm e^{-iH_D t} \quad \text{and} \quad (\Omega_D^+)^*\Omega_D^+ = 1$$

it follows that $\sigma_{\mathrm{ac}}(H) \supset \sigma_{\mathrm{ac}}(H_D)$ for all D. If $\#(D) \geq 2$, then $\sigma_{\mathrm{ac}}(H_D) = \sigma(H_D) = [\Sigma_D, \infty)$ so $\sigma_{\mathrm{ac}}(H) \supset [\Sigma, \infty)$. On the other hand, $\sigma_{\mathrm{ac}}(H) \subset \sigma_{\mathrm{ess}}(H) = [\Sigma, \infty)$ by the HVZ theorem. ∎

We want to conclude this section by giving a second proof of the "hard" part of Theorem XIII.17, i.e., that $\sigma_{\mathrm{ess}}(H) \subset [\Sigma, \infty)$. This proof uses very different ideas but is also based on a compactness condition, namely:

Lemma 7 Let H be a self-adjoint operator so that $f(H)$ is compact for every continuous function f of compact support with $\operatorname{supp} f \subset (-\infty, \Sigma)$. Then $\sigma_{\mathrm{ess}}(H) \subset [\Sigma, \infty)$.

Proof This is a part of Theorem XIII.77 which is proven in Section 14. ∎

Our second proof that $\sigma_{\mathrm{ess}}(H) \subset [\Sigma, \infty)$ is based on two ideas. First, bounded regions cannot contribute to $\sigma_{\mathrm{ess}}(H)$ and, secondly, near infinity, H must look like some H_D. To express the first idea, we pick a function $\varphi \in C_0^\infty(\mathbb{R}^{3N-3})$ with $0 \leq \varphi \leq 1$, so that $\varphi(x) = 1$ if $|x| \leq 1$ and $\varphi(x) = 0$ if $|x| \geq 2$. Define $j_{\leq n}(x) = \varphi(xn^{-1})$ and $j_{\geq n} = 1 - j_{\leq n}$. The first idea is expressed by:

Lemma 8 Let H obey the hypotheses of Theorem XIII.17. Then for each continuous function f of compact support and for each n, $j_{\leq n} f(H)$ is compact.

Proof Pick $C < \inf \sigma(H)$ and a bounded continuous function g so that $g(x) = (x - c)^{1/2} f(x)$ for $x \in \sigma(H)$. Then

$$j_{\leq n} f(H) = j_{\leq n}(H - c)^{-1/2} g(H) =$$
$$j_{\leq n}(H_0 + 1)^{-1/2}[(H_0 + 1)^{1/2}(H - c)^{-1/2}]g(H)$$

Thus it suffices to prove that $j_{\leq n}(H_0 + 1)^{-1/2}$ is compact. This follows from Theorem XI.20 or Problem 41. ∎

XIII: SPECTRAL ANALYSIS

To express the notion that H looks like some H_D near infinity, at least in suitable tubes, the following is useful:

Lemma 9 Let A_n be a family of bounded operators with $\sup_n \|A_n\| < \infty$ and let H_1, H_2 be two self-adjoint operators with $\sigma(H_1)$ and $\sigma(H_2)$ contained in $[c, \infty)$ so that

$$\lim_{n\to\infty} \|A_n[(H_1 - z)^{-1} - (H_2 - z)^{-1}]\| = 0$$

for all z in some open subset of $\mathbb{C}\backslash[c, \infty)$. Then

$$\lim_{n\to\infty} \|A_n[f(H_1) - f(H_2)]\| = 0$$

for any continuous function f of compact support.

Proof The proof is similar to that of Theorem VIII.20a. By the Vitali convergence theorem, the first limiting statement holds for all z in $\mathbb{C}\backslash[c, \infty)$ and uniformly on compact subsets of $\mathbb{C}\backslash[c, \infty)$. Using the Cauchy integral formula, we conclude that

$$\lim_{n\to\infty} \|A_n[(H_1 - c + 1)^{-m} - (H_2 - c + 1)^{-m}]\| = 0$$

The lemma now follows from the Stone–Weierstrass theorem which implies that $f(x)$ can be uniformly approximated on $[c, \infty)$ by polynomials in $(x - c + 1)^{-1}$. ∎

Next we single out the regions in which H and H_D look alike. Only D's with $\#(D) = 2$ will be needed, so let D_a, $a = 1, 2, \ldots, 2^{N-1} - 1$ label all the distinct ways of decomposing $\{1, \ldots, N\}$ into two nonempty sets $C_a^{(1)}$ and $C_a^{(2)}$. Define, $H_a = H_{D_a}$, $I_a = I_{D_a}$, and

$$|x|_0 = \left(\sum_{i \leq j} |x_i - x_j|^2\right)^{1/2}$$

$$|x|_a = \min\{|x_i - x_j| \mid i \in C_a^{(1)}, j \in C_a^{(2)}\}$$

Lemma 10 Let $d_N = (N-1)^{-3/2} N^{-1/2} \sqrt{2}$. Then for any $x \in \mathbb{R}^{3N-3}$, there is an a so that $|x|_a \geq d_N |x|_0$. In particular, there exist functions j_a, $a = 1, \ldots, 2^{N-1} - 1$ so that
(a) $0 \leq j_a \leq 1$, $\sum_a j_a(x) \equiv 1$.
(b) $j_a(\lambda x) = j_a(x)$, all $\lambda \in (0, \infty)$.
(c) j_a is C^∞ on $\mathbb{R}^{3N-3}\backslash\{0\}$.
(d) $\text{Supp } j_a \subset \{x \mid |x|_a \geq \tfrac{1}{2} d_N |x_0|\}$.

Proof This is a geometric proof identical to one given in our discussion of the Haag–Ruelle scattering theory (Section XI.16). Given x, let i and j be two points with $|x_i - x_j| \geq \sqrt{2N^{-\frac{1}{2}}(N-1)^{-\frac{1}{2}}}|x|_0$. Let $\pi_1, \ldots, \pi_N$ be the N planes through $x_1, \ldots, x_N$ perpendicular to $x_i - x_j$. Clearly they divide the region between π_i and π_j into at most $(N-1)$ slabs one of which must have thickness at least $(N-1)^{-1}|x_i - x_j|$. Choosing $C_a^{(1)}$ and $C_a^{(2)}$ to be the particles on each side of this slab, we have $|x|_a \geq (N-1)^{-1}|x_i - x_j|$.

To construct the j_a's, we first consider the open subsets of the sphere $S \equiv \{x \mid |x|_0 = 1\}$ of the form $R_a = \{x \mid |x|_a \geq \frac{1}{2}d_N, |x|_0 = 1\}$ which by the above cover S. In the usual way, we find C^∞ functions $\tilde{j}_a$ on S with $\operatorname{supp} \tilde{j}_a \subset R_a$, $0 \leq \tilde{j}_a$ and $\sum \tilde{j}_a \equiv 1$. We take $j_a(x) = \tilde{j}_a(|x|_0^{-1}x)$. ∎

We can now give a precise meaning to the fact that "near infinity" H looks like some H_a.

Lemma 11 Fix $a = 1, \ldots, 2^{N-1} - 1$. For any $z \notin \sigma(H) \cup \sigma(H_a)$,
$$\lim_{n \to \infty} \|j_{\geq n} j_a [(H-z)^{-1} - (H_a - z)^{-1}]\| = 0$$

Proof Writing
$$j_{\geq n} j_a [(H-z)^{-1} - (H_a - z)^{-1}] \equiv -(H-z)^{-1} I_a j_a j_{\geq n}(H_a - z)^{-1}$$
$$+ [(H-z)^{-1}, j_{\geq n} j_a] I_a (H_a - z)^{-1}$$

and using $[(H-z)^{-1}, f] = -(H-z)^{-1}[H_0, f](H-z)^{-1}$, we see that it suffices to prove that
$$\|(H_0 + 1)^{-1/2} I_a j_a j_{\geq n}(H_0 + 1)^{-1/2}\| \to 0 \tag{29.1}$$

and
$$\|(H_0 + 1)^{-1/2} [H_0, j_{\geq n} j_a](H_0 + 1)^{-1/2}\| \to 0 \tag{29.2}$$

Since $j_a j_{\geq n}$ is supported in the region where $|x|_a \geq \frac{1}{2} d_N n$, (29.1) follows from the fact that V_{ij} is in $R + L_\varepsilon^\infty$. (29.2) follows from the fact that
$$\|\nabla(j_{\geq n} j_a)\|_\infty \leq \|j_{\geq n} \nabla j_a\|_\infty + \|\nabla j_{\geq n}\|_\infty = C n^{-1} \to 0 \quad \blacksquare$$

Second proof of Theorem XIII.17 We give a second proof that $\sigma_{\text{ess}}(H) \subset [\Sigma, \infty)$. Since $\sum j_a = j_{\leq n} + j_{\geq n} = 1$, we have
$$f(H) = f(H) j_{\leq n} + \sum_a [f(H) - f(H_a)] j_a j_{\geq n} + \sum_a f(H_a) j_a j_{\geq n} \tag{29.3}$$

Let f be a continuous function of compact support with $\operatorname{supp} f \subset (-\infty, \Sigma]$. Since $\sigma(H_a) \subset [\Sigma, \infty)$, we have that $f(H_a) = 0$, so the last term in (29.3) is zero. By Lemmas 9 and 11, the second term in (29.3) goes to zero in norm as $n \to \infty$. By Lemma 8, the first term is compact. Thus, for such f's, $f(H)$ is compact. By Lemma 7, we conclude that $\sigma_{\text{ess}}(H) \subset [\Sigma, \infty)$. ∎

Finally, we note that (29.3) is useful for considering the case where we restrict to suitable symmetry subspaces. The abstract theorem is

Theorem XIII.17′ Let H be an operator obeying the hypotheses of Theorem XIII.17. Let P be the projection onto an invariant subspace for H and P_a, $a = 1, \ldots, 2^{N-1} - 1$ the projections onto some invariant subspace for H_a containing $P\mathscr{H}$. Let $\Sigma_P = \min_a \inf \sigma(H_a P_a)$. Then $\sigma_{\text{ess}}(HP) \subset [\Sigma_P, \infty)$.

Proof Clearly $\Sigma_P \leq 0$ since $0 \in \sigma(H_a P_a)$. Thus, for any f supported in $(-\infty, \Sigma_P)$, $Pf(H) = f(PH)$ and $P_a f(H_a) = f(P_a H_a) = 0$. Since $P = PP_a$, by hypothesis, we have that

$$f(PH) = Pf(H)j_{\leq n} + P \sum_a [f(H) - f(H_a)]j_a j_{\geq n}$$

on account of (29.3). As in the above proof, we conclude that $f(PH)$ is compact and thus that $\sigma_{\text{ess}}(HP) \subset [\Sigma_P, \infty)$. ∎

Example Consider H, the Hamiltonian of an atom, that is, the operator obtained by removing the center of mass from

$$\tilde{H} = -\sum_{i=1}^{N-1} (2\mu)^{-1} \Delta_i - (2M)^{-1} \Delta_N - Z \sum_{i=1}^{N-1} |r_i - r_N|^{-1} + \sum_{i<j\leq N-1} |r_i - r_j|^{-1}$$

and let P be the projection onto all functions antisymmetric under interchange of the r_i's, $i = 1, \ldots, N - 1$. For each a, let P_a be the projection onto all functions antisymmetric under interchange of two electrons in the same $C_j^{(a)}$. Then, Theorem XIII.17′ implies that $\sigma_{\text{ess}}(HP) \subset [\Sigma_P, \infty)$ which is generally strictly smaller than $[\Sigma, \infty)$. One can show easily (Problem 46) that $[\Sigma_P, \infty) \subset \sigma_{\text{ess}}(HP)$ in this case.

XIII.6 The absence of singular continuous spectrum I: General theory

Spectral analysis of an operator A concentrates on identifying the five sets $\sigma_{\text{ess}}(A)$, $\sigma_{\text{disc}}(A)$, $\sigma_{\text{ac}}(A)$, $\sigma_{\text{sing}}(A)$, $\sigma_{\text{pp}}(A)$. For large classes of Schrödinger operators H, we have succeeded in identifying $\sigma_{\text{ess}}(H)$ and $\sigma_{\text{ac}}(H)$. A precise determination of $\sigma_{\text{disc}}(H)$ is a detailed question, but we have seen how to use the min–max principle to obtain a lot of information about it. That leaves $\sigma_{\text{sing}}(H)$ and $\sigma_{\text{pp}}(H)$. We shall discuss the question of proving $\sigma_{\text{pp}} = \sigma_{\text{disc}}$ in Section 13. The next five sections involve the difficult study of $\sigma_{\text{sing}}(H)$. Our discussion of asymptotic completeness in Section XI.3 suggests that $\sigma_{\text{sing}}(H) = \emptyset$ and our main goal will be the proof of this fact for various classes of Schrödinger operators.

As we have already emphasized, there are close connections between proving the absence of singular continuous spectrum and scattering theory. In fact, the development of eigenfunction expansions in Section XI.6 already tells us something about the singular spectrum. In general when $V \in L^1 \cap R$ we know that $\sigma_{\text{sing}} \subset \mathscr{E}$, the exceptional set, so that σ_{sing} has Lebesgue measure 0. And in two cases, first when $V \in L^1 \cap R$ and $\|V\|_R < 4\pi$, and secondly, if $Ve^{a|x|} \in R$ for some $a > 0$, we know that $\mathscr{E}$ is discrete so that $\sigma_{\text{sing}} = \emptyset$. In this section we shall develop a fundamental criterion for the absence of singular spectrum and show how it allows us to recover these two results from first principles without very much effort. In Sections 7 and 8, we shall use the connection between scattering theory and spectral analysis in reverse: The techniques developed to prove that there is no singular continuous spectrum will yield very strong results on asymptotic completeness.

The fundamental criterion for the absence of singular spectrum is very simple. In part it will depend on Stone's formula (Theorem VII.13):

$$\tfrac{1}{2}(\varphi, (E_{[a,b]} + E_{(a,b)})\varphi) = \lim_{\varepsilon \downarrow 0} \pi^{-1} \int_a^b \operatorname{Im}(\varphi, R(x + i\varepsilon)\varphi)\, dx$$

where $R(\lambda)$ is the resolvent $(H - \lambda)^{-1}$ of some self-adjoint H and $\{E_\Omega\}$ is its family of spectral projections. In particular, one has that

$$\sup_{0 < \varepsilon < 1} \int_a^b |\operatorname{Im}(\varphi, R(x + i\varepsilon)\varphi)|^p\, dx < \infty \qquad (30)$$

in the case $p = 1$ for any $\varphi \in \mathscr{H}$.

Theorem XIII.19 Let H be a self-adjoint operator with resolvent $R(\lambda) = (H - \lambda)^{-1}$. Let (a, b) be a bounded interval and $\varphi \in \mathscr{H}$. Suppose that there is a $p > 1$ for which (30) holds. Then $E_{(a,b)}\varphi \in \mathscr{H}_{ac}$.

Proof By Stone's formula and the fact that $E_{(c,d)} \leq E_{[c,d]}$

$$(\varphi, E_{(c,d)}\varphi) \leq \frac{1}{\pi} \lim_{\varepsilon \downarrow 0} \int_c^d \operatorname{Im}(\varphi, R(x + i\varepsilon)\varphi)\, dx$$

for each open interval (c, d). Let S be an open set in (a, b) so that $S = \bigcup_{i=1}^{N} (a_i, b_i)$ is a union of disjoint open intervals. Suppose first that $N < \infty$. Then

$$(\varphi, E_S \varphi) \leq \frac{1}{\pi} \sum_{i=1}^{N} \lim_{\varepsilon \downarrow 0} \int_{a_i}^{b_i} \operatorname{Im}(\varphi, R(x + i\varepsilon)\varphi) \, dx$$

$$\leq \frac{1}{\pi} \lim_{\varepsilon \downarrow 0} \sum_{i=1}^{N} \int_{a_i}^{b_i} \operatorname{Im}(\varphi, R(x + i\varepsilon)\varphi) \, dx$$

$$\leq \frac{1}{\pi} \lim_{\varepsilon \downarrow 0} \left[\int_{a}^{b} |\operatorname{Im}(\varphi, R(x + i\varepsilon)\varphi)|^p \, dx \right]^{1/p} |S|^{1/q}$$

$$\leq C |S|^{1/q}$$

where $|S|$ is the Lebesgue measure of S and q is the conjugate index to p. If N is infinite, let $S_m = \bigcup_{i=1}^{m} (a_i, b_i)$. Then

$$(\varphi, E_S \varphi) = \lim_{m \to \infty} (\varphi, E_{S_m} \varphi) \leq \lim_{m \to \infty} C |S_m|^{1/q} = C |S|^{1/q}$$

Let I be an arbitrary set of Lebesgue measure 0 inside (a, b). Since Lebesgue measure is outer-regular, we can find an open set $S^{(k)}$ with $I \subset S^{(k)}$ and $|S^{(k)}| < 1/k$. Thus

$$(\varphi, E_I \varphi) \leq \inf_k (\varphi, E_{S^{(k)}} \varphi) \leq C \inf_k |S^{(k)}|^{1/q} = 0$$

Thus the measure $\Omega \mapsto (\varphi, E_\Omega \varphi)$ is absolutely continuous on (a, b) so $E_{(a,b)} \varphi \in \mathscr{H}_{\text{ac}}$. ∎

Theorem XIII.20 Let H be a self-adjoint operator with resolvent $R(\lambda) = (H - \lambda)^{-1}$. Let (a, b) be a bounded interval. Suppose that there is a dense set D in $\mathscr{H}$ so that, for each $\varphi \in D$, (30) holds for some $p > 1$. Then H has purely absolutely continuous spectrum on (a, b), i.e., $\operatorname{Ran} E_{(a,b)} \subset \mathscr{H}_{\text{ac}}$. Conversely, if $\operatorname{Ran} E_{(a,b)} \subset \mathscr{H}_{\text{ac}}$, then there is a set D, dense in $\operatorname{Ran} E_{(a,b)}$, so that (30) holds for any $p \geq 1$ including $p = \infty$ when $\varphi \in D$.

Proof The first half of the theorem follows from Theorem XIII.19. For the second half, let D be the set of vectors φ for which the spectral measure $d\mu_\varphi$ for H is of the form $f(x) \, dx$ where $f \in L^\infty$ with compact support in (a, b). By hypothesis, D is dense in $\operatorname{Ran} E_{(a,b)}$. Moreover,

$$\operatorname{Im}(\varphi, R(x + i\varepsilon)\varphi) = \int_{-\infty}^{\infty} g_\varepsilon(x - y) f(y) \, dy$$

where $g_\varepsilon(y) = \varepsilon(y^2 + \varepsilon^2)^{-1}$. Since $\|g_\varepsilon\|_1 = \pi$ independently of ε, (30) holds with $p = \infty$ by Young's inequality and thus for all p since (a, b) is finite. ∎

In most applications of Theorem XIII.20, one actually proves that $(\varphi, R(\lambda)\varphi)$ is bounded on $M = \{x + i\varepsilon | \varepsilon \in (0, 1), x \in (a, b)\}$ or more strongly that $(\varphi, R(\lambda)\varphi)$ has a continuous extension to $\bar{M}$. A typical application is:

Theorem XIII.21 Let $V \in R$, the Rollnik class, and let $H = -\Delta + V$ on $L^2(\mathbb{R}^3)$. Suppose that *either*:

(a) $\|V\|_R < 4\pi$;

or

(b) $Ve^{a|x|} \in R$ for some $a > 0$.

Then $\sigma_{\text{sing}}(H) = \emptyset$.

Proof (a) Since $\|V\|_R < 4\pi$, $|V|^{1/2}(H_0 - \lambda)^{-1}V^{1/2}$ has a Hilbert–Schmidt norm less than one uniformly for $\lambda \in \mathbb{C}\backslash[0, \infty)$ (see Section XI.6). Let $f \in C_0^\infty$ and note that $|f|^{1/2} \in R$. Then

$$|f|^{1/2}(H - \lambda)^{-1}|f|^{1/2} = |f|^{1/2}(H_0 - \lambda)^{-1}|f|^{1/2}$$
$$+ \sum_{n=0}^{\infty} (-1)^{n+1}(|f|^{1/2}(H_0 - \lambda)^{-1}V^{1/2})$$
$$\times (|V|^{1/2}(H_0 - \lambda)^{-1}V^{1/2})^n$$
$$\times (|V|^{1/2}(H_0 - \lambda)^{-1}|f|^{1/2})$$

converges uniformly in Hilbert–Schmidt norm to an operator with norm less than $C_1 + C_2(1 - (4\pi)^{-1}\|V\|_R)^{-1}$. Thus

$$|(f, (H - \lambda)^{-1}f)| \leq \| |f|^{1/2} \|_2 \| |f|^{1/2}(H - \lambda)^{-1}|f|^{1/2} \|$$

is bounded on $\mathbb{C}\backslash[0, \infty)$. The fundamental criterion, Theorem XIII.19, thus implies that Ran $E_{(a,b)} \subset \mathcal{H}_{\text{ac}}$ for *all* (a, b) so σ_{pp} and σ_{sing} are empty.

(b) As in our discussion in Section XI.6, $(1 + |V|^{1/2}(H_0 - \lambda)^{-1}V^{1/2})^{-1}$ exists for all $\lambda \in \mathbb{C}\backslash[0, \infty)$ and has continuous boundary values as $\lambda \to x + i0$ as long as x avoids a finite set $\mathcal{E}$ of real numbers. Let $[a, b]$ be disjoint from $\mathcal{E}$. Then

$$|f|^{1/2}(H - \lambda)^{-1}|f|^{1/2} = |f|^{1/2}(H_0 - \lambda)^{-1}|f|^{1/2}$$
$$- (|f|^{1/2}(H_0 - \lambda)^{-1}V^{1/2})$$
$$\times (1 + |V|^{1/2}(H_0 - \lambda)^{-1}V^{1/2})^{-1}|V|^{1/2}$$
$$\times (H_0 - \lambda)^{-1}|f|^{1/2}$$

is a uniformly bounded operator on $\{x + i\varepsilon | 0 < \varepsilon < 1; x \in [a, b]\}$ whenever $f \in C_0^\infty$. As in (a), this implies that Ran $E_{(a,b)} \subset \mathcal{H}_{\text{ac}}$ so $\sigma_{\text{sing}} \subset \mathcal{E}$, a finite set. This implies that $\sigma_{\text{sing}} = \emptyset$. ∎

140   XIII: SPECTRAL ANALYSIS

Given the simple criterion, Theorem XIII.20, for $\sigma_{\text{sing}}(H)$ to be empty, one can reasonably ask why the problem of controlling the singular continuous spectrum is so much more difficult than that of identifying $\sigma_{\text{ess}}(H)$ (Sections 4 and 5) or $\sigma_{\text{ac}}(H)$ (Section XI.3). The reason is that σ_{sing} is much less stable under perturbations. To illustrate this, we describe the Aronszajn–Donoghue theory of the behavior of σ_{sing} under rank one perturbations. This theory should be compared with the Weyl theorem on invariance of σ_{ess} and the Kato–Birman theory of the invariance of σ_{ac}. Basic to the Aronszajn–Donoghue theory is the following combination of classical results of Fatou and de la Vallée Poussin:

Proposition   Let v be a finite Borel measure on $\mathbb{R}$ and let

$$F(z) = \int (x - z)^{-1} \, dv(x)$$

for $\operatorname{Im} z > 0$. Let $A_v = \{x \mid \lim_{\varepsilon \downarrow 0} F(x + i\varepsilon) = \infty\}$ and let

$$B_v = \{x \mid \lim_{\varepsilon \downarrow 0} F(x + i\varepsilon) = \Phi(x), \text{ a finite number with } \operatorname{Im} \Phi(x) \neq 0\}$$

Then $v(\mathbb{R} \setminus (A_v \cup B_v)) = 0$, $v \upharpoonright A_v$ is singular relative to Lebesgue measure, and $v \upharpoonright B_v$ is absolutely continuous.

Proofs of this result may be tracked down by consulting the reference in the notes. Now let H_0 be some self-adjoint operator on $\mathcal{H}$ and let $\varphi \in \mathcal{H}$ be a unit vector. Let P denote the projection onto φ and let

$$H_\alpha = H_0 + \alpha P$$

Let $\mathcal{H}'$ be the cyclic subspace for φ generated by H_0. Clearly all the H_α equal H_0 on $(\mathcal{H}')^\perp$ so we may as well suppose that $\mathcal{H}' = \mathcal{H}$, that is, we henceforth suppose that φ is cyclic for H_0. It follows that φ is also cyclic for H_α since

$$(H_\alpha - z)^{-1}\varphi = (H_0 - z)^{-1}\varphi - \alpha(\varphi, (H_0 - z)^{-1}\varphi)(H_\alpha - z)^{-1}\varphi$$

so that the span of $\{(H_\alpha - z)^{-1}\varphi\}$ is identical to the span of $\{(H_0 - z)^{-1}\varphi\}$. By this cyclicity result, H_α is unitarily equivalent to multiplication by x on $L^2(\mathbb{R}, dv_\alpha)$ for a suitable measure dv_α.

Clearly,

$$F_\alpha(z) \equiv \int (x - z)^{-1} \, dv_\alpha(x) = (\varphi, (H_\alpha - z)^{-1}\varphi)$$

obeys

$$F_\alpha(z) = F_\beta(z) + (\beta - \alpha)F_\alpha(z)F_\beta(z)$$

on account of the resolvent equation
$$(H_\alpha - z)^{-1} = (H_\beta - z)^{-1} + (\beta - \alpha)(H_\alpha - z)^{-1} P(H_\beta - z)^{-1}$$
Thus, we have the basic equation:
$$F_\alpha(z) = F_\beta(z)(1 + (\alpha - \beta)F_\beta(z))^{-1} \tag{30.5}$$
Note that if $\lim_{\varepsilon \downarrow 0} F_\beta(z) = \infty$, then $\lim_{\varepsilon \downarrow 0} F_\alpha(z) = (\alpha - \beta)^{-1} \neq \infty$ for $\alpha \neq \beta$. Applying the proposition, we have proven the following striking *non-invariance* of σ_{sing}:

Theorem XIII.21.5 Let H_0 be a self-adjoint operator and let φ be a cyclic vector for H_0. Let
$$H_\alpha = H_0 + \alpha(\varphi, \cdot)\varphi$$
Then for $\alpha \neq \beta$ the singular (i.e., the union of pure point and singular continuous) parts of the spectral measures for H_α and H_β are mutually singular.

Example Let $d\nu_0 = dx \upharpoonright [0, 1] + d\mu_C$, where $d\mu_C$ is the Cantor measure. Then $\sigma_{\text{ess}}(H_\alpha) = [0, 1]$ for all α on account of Weyl's theorem. Moreover, $F_0 = F_L + F_C$ in the obvious way, and
$$\lim_{\varepsilon \downarrow 0} \operatorname{Im} F_L(x + i\varepsilon) = \begin{cases} 1, & x \in (0, 1) \\ \frac{1}{2}, & x = 0, 1 \\ 0, & x \notin [0, 1] \end{cases}$$
Since $\operatorname{Im} F_C(z) \geq 0$ for all z with $\operatorname{Im} z > 0$, we see that $\lim_{\varepsilon \downarrow 0} F_0(x + i\varepsilon)$ is never real for any $x \in [0, 1]$. It follows from the basic equation (30.5) that $\lim_{\varepsilon \downarrow 0} F_\beta(x + i\varepsilon)$ is never infinite for $\beta \neq 0$, $x \in [0, 1]$. Thus, each H_β for $\beta \neq 0$ has no singular continuous spectrum while $\sigma_{\text{sing}}(H_0) \neq \emptyset$. Thus, we have two bounded operators H_0 and H_1 such that $H_0 - H_1$ has rank one, with $\sigma_{\text{sing}}(H_1) = \emptyset \neq \sigma_{\text{sing}}(H_0)$!

XIII.7 The absence of singular continuous spectrum II: Smooth perturbations

Theorem XIII.20 suggests that one study L^p bounds on expectation values of the resolvent and our experience suggests that $p = 2$ will be particularly easy to study.

XIII: SPECTRAL ANALYSIS

Definition Let H be a self-adjoint operator with resolvent $R(\mu) = (H - \mu)^{-1}$. Let A be a closed operator. A is called **H-smooth** if and only if for each $\varphi \in \mathcal{H}$ and each $\varepsilon \neq 0$, $R(\lambda + i\varepsilon)\varphi \in D(A)$ for almost all $\lambda \in \mathbb{R}$ and moreover

$$\|A\|_H^2 = \sup_{\substack{\|\varphi\|=1 \\ \varepsilon > 0}} \frac{1}{4\pi^2} \int_{-\infty}^{\infty} (\|AR(\lambda + i\varepsilon)\varphi\|^2 + \|AR(\lambda - i\varepsilon)\varphi\|^2) \, d\lambda < \infty \quad (31)$$

Since A is closed, it is enough that (31) hold for a dense set of φ (Problem 47a). More interestingly, the uniform boundedness principle shows that if for each φ, $\int_{-\infty}^{\infty} \|AR(\lambda \pm i\varepsilon)\varphi\|^2 \, d\lambda \leq M_\varphi^2$ for a constant M_φ (independent of $\varepsilon > 0$ and of $\pm$, but dependent on φ), then A is H-smooth (Problem 47b,c).

The H-smooth operators are an especially nice class of operators. For example, we shall see that they are H-bounded with relative bound 0. They present in microcosm the close connection between scattering theory and spectral analysis. On one hand, we shall prove a basic result on the existence and completeness of wave operators $\Omega^\pm(H_1, H_0)$ when $H_1 - H_0$ is the product of an H_1-smooth operator and an H_0-smooth operator (Theorem XIII.24; see also Theorem XIII.31). On the other hand, we will see that if A is H-smooth, then $\overline{\text{Ran}(A^*)} \subset \mathcal{H}_{\text{ac}}(H)$ (Theorem XIII.23), which will lead to several theorems on the absence of singular continuous spectrum (see Theorems XIII.26 and XIII.28).

This section is divided into two parts. First, we shall develop the abstract theory of smooth perturbations, and in the second part we apply the theory to a variety of Schrödinger operators. The first two applications involve situations where the wave operators are unitary, i.e., where H and H_0 are unitarily equivalent. These are the cases of Schrödinger operators $H_0 + \lambda V$ with λ small or V repulsive. In the latter case, the theory of smooth perturbations will lead only to a proof that H has purely absolutely continuous spectrum. In our last application we shall establish the existence of wave operators for the case of repulsive potentials with some fall-off hypotheses by developing a generalization of the notion of H-smoothness. Part of this generalized theory will play a role in the next section.

One of our first main goals will be the reformulation of H-smoothness in a number of equivalent forms. We shall need the following vector-valued version of the Plancherel theorem.

Lemma 1 Let $\varphi(\cdot)$ be a (weakly measurable) function from $\mathbb{R}$ to a separable Hilbert space $\mathcal{H}$. Suppose that $\int \|\varphi(x)\| \, dx < \infty$. Define $\hat{\varphi}: \mathbb{R} \to \mathcal{H}$ by

$$\hat{\varphi}(p) = (2\pi)^{-1/2} \int_{\mathbb{R}} e^{-ipx} \varphi(x) \, dx$$

XIII.7 Smooth perturbations

(where, by our standard convention, the integral is a weak integral). Let A be a closed operator on $\mathcal{H}$. Then

$$\int \|A\hat{\varphi}(p)\|^2 \, dp = \int \|A\varphi(x)\|^2 \, dx \tag{32}$$

where the integrals in (32) are set equal to ∞ if $\hat{\varphi}(p)$ (respectively, $\varphi(x)$) is not in $D(A)$ almost everywhere.

Proof First suppose that A is bounded. Then for any $\psi \in \mathcal{H}$, $(\psi, A\hat{\varphi}(p)) = (A^*\psi, \hat{\varphi}(p))$ is the (ordinary) Fourier transform of $(A^*\psi, \varphi(x)) = (\psi, A\varphi(x))$, so by the Plancherel theorem

$$\int |(\psi, A\hat{\varphi}(p))|^2 \, dp = \int |(\psi, A\varphi(x))|^2 \, dx$$

(32) follows by summing over ψ in an orthonormal basis. Next let A be self-adjoint and let $\{E_\Omega\}$ be its family of spectral projections. Then $AE_{(-a, a)}$ is bounded so

$$\int \|AE_{(-a, a)}\hat{\varphi}(p)\|^2 \, dp = \int \|AE_{(-a, a)}\varphi(x)\|^2 \, dx$$

Suppose that one of the integrals in (32) is finite—without loss suppose the right-hand side is finite. Then $\varphi(x) \in D(A)$ a.e. in x, so $\|AE_{(-a, a)} \varphi(x)\|^2$ converges monotonically to $\|A\varphi(x)\|^2$ as $a \to \infty$. Thus,

$$\int \lim_{a \to \infty} \|AE_{(-a, a)}\hat{\varphi}(p)\|^2 \, dp = \int \|A\varphi(x)\|^2 \, dx < \infty$$

In particular, $\lim_{a \to \infty} \|AE_{(-a, a)} \hat{\varphi}(p)\|^2 < \infty$ a.e. in p. It follows that $\hat{\varphi}(p) \in D(A)$ a.e. and (32) holds. Finally, let A be an arbitrary closed operator. By Theorem VIII.32, there is a self-adjoint operator $|A|$ with $D(|A|) = D(A)$ and $\||A|\psi\| = \|A\psi\|$. Thus (32) follows from the case where A is self-adjoint. ∎

Example 1 Let $H = -i \, d/dx$ on $L^2(\mathbb{R})$ so that $(e^{-iHt}\varphi)(x) = \varphi(x - t)$. Let g be in $L^2(\mathbb{R})$ and let A be multiplication by g. By a change of variables, $\int |g(x)\varphi(x - t)|^2 \, dx \, dt = \|g\|_2^2 \|\varphi\|_2^2$ so, by Fubini's theorem, for any $\varphi \in L^2$, and almost all t, $e^{-iHt}\varphi \in D(A)$ and $\int_{-\infty}^{\infty} \|Ae^{-iHt}\varphi\|^2 \, dt = \|g\|_2^2 \|\varphi\|_2^2$. Fix $\varepsilon > 0$; then

$$\int_0^\infty e^{-\varepsilon t}e^{i\lambda t}e^{-iHt}\varphi \, dt = -iR(\lambda + i\varepsilon)\varphi$$

Thus, by the lemma

$$\int_{-\infty}^{\infty} \|AR(\lambda + i\varepsilon)\varphi\|^2 \, d\lambda = 2\pi \int_{0}^{\infty} e^{-2\varepsilon t} \|Ae^{-itH}\varphi\|^2 \, dt$$

Using a similar computation for $\lambda - i\varepsilon$, we see that

$$\int_{-\infty}^{\infty} (\|AR(\lambda + i\varepsilon)\varphi\|^2 + \|AR(\lambda - i\varepsilon)\varphi\|^2) \, d\lambda = 2\pi \int_{-\infty}^{\infty} e^{-2\varepsilon|t|} \|Ae^{-itH}\varphi\|^2 \, dt \quad (33)$$

(33) and the bound on $\int_{-\infty}^{\infty} \|Ae^{-iHt}\varphi\|^2 \, dt$ imply that g is H-smooth and $\|g\|_H = (2\pi)^{-1/2} \|g\|_2$.

Example 2 Let H be any self-adjoint operator and let $A = I$, the identity operator. Again, we use Lemma 1 to conclude that

$$\int_{-\infty}^{\infty} \|R(\lambda + i\varepsilon)\varphi\|^2 \, d\lambda = 2\pi \int_{0}^{\infty} e^{-2\varepsilon t} \|e^{-itH}\varphi\|^2 \, dt = \pi \varepsilon^{-1} \|\varphi\|^2 \quad (34)$$

so that $\sup_{\varepsilon > 0} \int_{-\infty}^{\infty} \|R(\lambda + i\varepsilon)\varphi\|^2 \, d\lambda = \infty$ for any $\varphi \neq 0$. Thus I is never an H-smooth operator.

One reformulation of H-smoothness is in terms of the unitary group e^{itH}. As an immediate consequence of (33) we have:

Lemma 2 A is H-smooth if and only if for all $\varphi \in \mathcal{H}$, $e^{itH}\varphi \in D(A)$ for almost every $t \in \mathbb{R}$ and for some constant C,

$$\int_{-\infty}^{\infty} \|Ae^{-iHt}\varphi\|^2 \, dt \leq C\|\varphi\|^2$$

C can be chosen equal to $(2\pi)\|A\|_H^2$ and no smaller.

This lemma has several important consequences:

Theorem XIII.22 If A is H-smooth, then A is H-bounded with relative bound zero.

Proof Let $\psi \in D(A^*)$. By the resolvent formula

$$-i(A^*\psi, R(\lambda + i\varepsilon)\varphi) = \int_{0}^{\infty} (A^*\psi, e^{-iHt}\varphi) e^{i\lambda t} e^{-\varepsilon t} \, dt$$

Thus, by the Schwarz inequality

$$|(A^*\psi, R(\lambda + i\varepsilon)\varphi)|^2 \leq \frac{1}{2\varepsilon} \int_0^\infty |(A^*\psi, e^{-iHt}\varphi)|^2 \, dt$$

$$\leq \frac{1}{2\varepsilon} \|\psi\|^2 \int_0^\infty \|Ae^{-iHt}\varphi\|^2 \, dt$$

$$\leq \frac{\pi}{\varepsilon} \|A\|_H^2 \|\psi\|^2 \|\varphi\|^2$$

by Lemma 2. Thus, $R(\lambda + i\varepsilon)\varphi \in D(A^{**}) = D(A)$ and $\|AR(\lambda + i\varepsilon)\|^2 \leq (\pi/\varepsilon)\|A\|_H^2$. We conclude that $D(H) \subset D(A)$ and for $\psi \in D(H)$

$$\|A\psi\| \leq \pi^{1/2}\|A\|_H(\varepsilon^{-1/2}\|H\psi\| + \varepsilon^{1/2}\|\psi\|)$$

Since ε is arbitrary, the theorem is proven. ∎

For a strengthening of Theorem XIII.22, see Problem 49.

Theorem XIII.23 If A is H-smooth, then $\text{Ran}(A^*) \subset \mathscr{H}_{\text{ac}}(H)$.

Proof Since $\mathscr{H}_{\text{ac}}(H)$ is closed, we need only show $\overline{\text{Ran}(A^*)} \subset \mathscr{H}_{\text{ac}}(H)$. Let $\varphi \in D(A^*)$, $\psi = A^*\varphi$, and let $d\mu_\psi$ be the spectral measure for H associated with ψ. Define

$$F(t) = (2\pi)^{-1/2} \int e^{-itx} \, d\mu_\psi(x) = (2\pi)^{-1/2}(A^*\varphi, e^{-itH}\psi)$$

Then $|F(t)| \leq (2\pi)^{-1/2}\|\varphi\| \|Ae^{-itH}\psi\|$, so by Lemma 2, $F \in L^2(\mathbb{R})$. By the Plancherel theorem, $\check{F} \in L^2$ so $d\mu_\psi = \check{F} \, dx$ is absolutely continuous with respect to Lebesgue measure. ∎

It is an instructive exercise to prove Theorem XIII.23 using Theorem XIII.20 and the direct definition of H-smoothness. See also condition (5) of Theorem XIII.25 below. Lemma 2 has important consequences for scattering theory.

Theorem XIII.24 Let H and H_0 be self-adjoint operators. Suppose that $H = H_0 + \sum_{i=1}^n A_i^* B_i$ in the following sense:

(1) $D(H) \subset D(A_i)$; $D(H_0) \subset D(B_i)$; $i = 1, \ldots, n$.
(2) If $\varphi \in D(H)$ and $\psi \in D(H_0)$, then

$$(H\varphi, \psi) = (\varphi, H_0\psi) + \sum_{i=1}^n (A_i\varphi, B_i\psi) \tag{35}$$

If each A_i is H-smooth and each B_i is H_0-smooth, then the wave operators $s\text{-}\lim_{t \to \mp\infty} e^{iHt}e^{-iH_0 t}$ exist and are unitary.

Proof Suppose first that $n = 1$. Let $\varphi \in D(H_0)$, $w(t) = e^{iHt}e^{-iH_0 t}\varphi$, and $\psi \in D(H)$. Then $(\psi, w(t))$ is differentiable and

$$\frac{d}{dt}(\psi, w(t)) = i(Ae^{-iHt}\psi, Be^{-iH_0 t}\varphi)$$

Therefore, if $t > s$,

$$|(\psi, w(t) - w(s))| = \int_s^t |(Ae^{-i\tau H}\psi, Be^{-iH_0 \tau}\varphi)| \, d\tau$$

$$\leq \left(\int_{-\infty}^{\infty} \|Ae^{-i\tau H}\psi\|^2 \, d\tau\right)^{1/2} \left(\int_s^t \|Be^{-iH_0 \tau}\varphi\|^2 \, d\tau\right)^{1/2}$$

Since ψ is an arbitrary element of $D(H)$,

$$\|w(t) - w(s)\| \leq \sqrt{2\pi}\|A\|_H \left(\int_s^t \|Be^{-iH_0 \tau}\varphi\|^2 \, d\tau\right)^{1/2}$$

Since the integrand on the right-hand side of the last equation is in $L^1(\mathbb{R})$, this last estimate shows that $w(s)$ is Cauchy as $s \to +\infty$ or as $s \to -\infty$. As a result, the limits exist on a dense set and so on all of $\mathcal{H}$. Since we also have $H_0 = H - B^*A$ by symmetry, the s-$\lim e^{iH_0 t}e^{-iHt}$ exist and so are unitary. The proof is similar for $n > 1$. ∎

Notice that the limit in Theorem XIII.24 does *not* involve $P_{\text{ac}}(H_0)$ (see Section XI.3 for comparison). Thus, if $H_0\psi = E\psi$, then $H\psi = E\psi$. This follows directly from the hypothesis, for $\text{Ran}(B^*) \subset \mathcal{H}_{\text{ac}}(H_0)$ implies that $\mathcal{H}_{\text{pp}}(H_0) \subset \text{Ker}(B)$, which implies that $H\psi = H_0\psi$.

As a final result in the general theory, we prove that a variety of forms of smoothness are equivaent.

Theorem XIII.25 Let A be a closed operator and let H be a self-adjoint operator. Then the following are equivalent:

(0) A is H-smooth.
(1) For all $\varphi \in \mathcal{H}$, $e^{-itH}\varphi \in D(A)$ for almost all t and

$$c_1 = \sup_{\|\varphi\|=1} \frac{1}{2\pi} \int_{-\infty}^{\infty} \|Ae^{-iHt}\varphi\|^2 \, dt < \infty$$

XIII.7 Smooth perturbations

(2) $D(H) \subset D(A)$ and

$$c_2 = \sup_{\substack{\|\varphi\|=1 \\ -\infty < a < b < \infty}} \frac{\|AE_{(a,b]}\varphi\|^2}{|b-a|} < \infty$$

where E_Ω is the spectral family for H.

(3) $$c_3 = \sup_{\substack{\|\varphi\|=1;\, \varphi \in D(A^*) \\ -\infty < a < b < \infty}} \frac{\|E_{(a,b]}A^*\varphi\|^2}{|b-a|} < \infty$$

(4) $$c_4 = \frac{1}{2\pi} \sup_{\substack{\mu \notin \mathbb{R};\, \varphi \in D(A^*) \\ \|\varphi\|=1}} |(A^*\varphi, [R(\mu) - R(\bar\mu)]A^*\varphi)| < \infty$$

(5) $$c_5 = \frac{1}{\pi} \sup_{\substack{\mu \notin \mathbb{R};\, \varphi \in D(A^*) \\ \|\varphi\|=1}} \|R(\mu)A^*\varphi\|^2 |\operatorname{Im}\mu| < \infty$$

(6) $D(H) \subset D(A)$ and

$$c_6 = \frac{1}{\pi} \sup_{\mu \notin \mathbb{R},\, \|\varphi\|=1} \|AR(\mu)\varphi\|^2 |\operatorname{Im}\mu| < \infty$$

(7) $D(H) \subset D(A)$ and

$$c_7 = \frac{1}{\pi} \sup_{\mu \notin \mathbb{R},\, \|\varphi\|=1} (\|AR(\mu)\varphi\|^2 + \|AR(\bar\mu)\varphi\|^2) |\operatorname{Im}\mu| < \infty$$

Moreover, if any (and thus all) of $c_1 - c_7$ are finite, then all c_i equal $\|A\|_H^2$.

Before proving the theorem, we note that in conditions (5) and (6) one may replace the sup over $\mathbb{C}\setminus\mathbb{R}$ by a sup over $0 < \operatorname{Im}\mu < \alpha$ for any $\alpha > 0$. In the original definition one may replace $\varepsilon > 0$ by $0 < \varepsilon < \alpha$. This follows from the details of the proof if one notes (by (33)) that the expression in (31) is monotone increasing as ε decreases.

The following condition, which is stronger than smoothness, comes from condition (4).

148 XIII: SPECTRAL ANALYSIS

Corollary If $AR(\mu)A^*$ is bounded for each $\mu \notin \mathbb{R}$ in the sense that
$$\sup_{\substack{\psi, \varphi \in D(A^*) \\ \|\psi\| = \|\varphi\| = 1}} |(A^*\varphi, R(\mu)A^*\psi)| < \infty \quad \text{and} \quad \Gamma \equiv \sup_{\mu \notin \mathbb{R}} \|AR(\mu)A^*\| < \infty$$
then A is H-smooth and $\|A\|_H \leq \Gamma/\pi$.

Proof of Theorem XIII.25 We first note that for every closed operator A, every bounded operator B, every $\varphi \in D(A)$, and $\psi \in \mathcal{H}$, we have
$$(\psi, BA\varphi) = (B^*\psi, A\varphi)$$
It follows that $\|BA\varphi\| \leq c\|\varphi\|$ for all $\varphi \in D(A)$ if and only if Ran $B^* \subset D(A^*)$ and $\|A^*B^*\| \leq c$. As a consequence $c_2 = c_3$ and $c_5 = c_6$. Moreover, since $(A^*\varphi, [R(\mu) - R(\bar{\mu})]A^*\varphi) = 2\,\mathrm{Im}\,\mu \|R(\mu)A^*\varphi\|^2$, $c_5 = c_4$. Further, $c_0 = c_1$ by Lemma 2. Thus we need only prove that $c_0 = c_3 = c_4 = c_7$ where $c_0 \equiv \|A\|_H^2$. We shall show that
$$c_0 \leq c_4 \leq c_3 \leq c_0 \quad \text{and} \quad c_3 \leq c_7 \leq c_1.$$

$c_0 \leq c_4$: $(2\pi i)^{-1}[R(\mu) - R(\bar{\mu})] = \pi^{-1}(\mathrm{Im}\,\mu)R(\bar{\mu})R(\mu) \geq 0$ if $\mathrm{Im}\,\mu > 0$. Let $K(\mu)$ be its positive square root. By the definition of c_4, $\|K(\mu)A^*\varphi\|^2 \leq c_4\|\varphi\|^2$ if $\varphi \in D(A^*)$. Thus, if $c_4 < \infty$, the remark above implies that Ran $K(\mu) \subset D(A)$ and $\|AK(\mu)\|^2 \leq c_4$. Therefore
$$\int_{-\infty}^{\infty} \|A[R(\lambda + i\varepsilon) - R(\lambda - i\varepsilon)]\varphi\|^2 \, d\lambda$$
$$= 4\pi^2 \int_{-\infty}^{\infty} \|AK(\lambda + i\varepsilon)K(\lambda + i\varepsilon)\varphi\|^2 \, d\lambda$$
$$\leq 4\pi^2 c_4 \int_{-\infty}^{\infty} \|K(\lambda + i\varepsilon)\varphi\|^2 \, d\lambda$$
$$= 4\pi^2 c_4 \frac{1}{2\pi i} \int_{-\infty}^{\infty} (\varphi, [R(\lambda + i\varepsilon) - R(\lambda - i\varepsilon)]\varphi) \, d\lambda$$
$$= 4\pi^2 c_4 \|\varphi\|^2$$
since in terms of the spectral measure $d\mu_\varphi$ for H,
$$(2\pi i)^{-1} \int_{-\infty}^{\infty} (\varphi, [R(\lambda + i\varepsilon) - R(\lambda - i\varepsilon)]\varphi) \, d\lambda$$
$$= \frac{\varepsilon}{\pi} \int_{-\infty}^{\infty} \int_{-\infty}^{\infty} \frac{1}{(x-\lambda)^2 + \varepsilon^2} \, d\mu_\varphi(x) \, d\lambda$$
$$= \int_{-\infty}^{\infty} d\mu_\varphi(x) = \|\varphi\|^2$$

Further,
$$[R(\lambda + i\varepsilon) - R(\lambda - i\varepsilon)]\varphi = i \int_{-\infty}^{\infty} e^{-\varepsilon|t|} e^{i\lambda t} e^{-iHt} \varphi \, dt$$
so Lemma 1 implies that
$$\int_{-\infty}^{\infty} \|A[R(\lambda + i\varepsilon) - R(\lambda - i\varepsilon)]\varphi\|^2 \, d\lambda = 2\pi \int_{-\infty}^{\infty} e^{-2\varepsilon|t|} \|Ae^{-iHt}\varphi\|^2 \, dt \quad (36)$$
so $c_0 = c_1 \le c_4$.

$c_4 \le c_3$: Let $\varphi \in D(A^*)$ and let $d\mu_{A^*\varphi}$ be the spectral measure for H with respect to $A^*\varphi$. Then, by (3)
$$d\mu_{A^*\varphi}(a, b] \le c_3 |b - a| \, \|\varphi\|^2$$
If I is an arbitrary Borel set, let $|I|$ be its Lebesgue measure. It follows that $d\mu_{A^*\varphi}(I) \le c_3 |I| \, \|\varphi\|^2$ for arbitrary I, for it is true for open sets and thus by outer regularity for arbitrary sets. Therefore $d\mu_{A^*\varphi}$ is absolutely continuous w.r.t. dx and the Radon–Nikodym derivative $g(x)$ obeys $\|g\|_\infty \le c_3 \|\varphi\|^2$. Thus if $\mu = \lambda + i\varepsilon$, then
$$|(A^*\varphi, [R(\mu) - R(\bar\mu)]A^*\varphi)| = \int_{-\infty}^{\infty} \frac{2|\varepsilon|}{(x - \lambda)^2 + \varepsilon^2} g(x) \, dx$$
$$\le \|g\|_\infty \int_{-\infty}^{\infty} \frac{2|\varepsilon|}{(x - \lambda)^2 + \varepsilon^2} \, dx \le c_3(2\pi) \|\varphi\|^2$$
We conclude that $c_4 \le c_3$.

$c_3 \le c_0$: Let $\varphi \in D(A^*)$. Suppose that a and b are not eigenvalues of H. By Stone's formula,
$$|(A^*\varphi, E_{(a,b]}\psi)|^2$$
$$= \frac{1}{4\pi^2} \lim_{\varepsilon \downarrow 0} \left| \int_a^b (A^*\varphi, [R(\lambda + i\varepsilon) - R(\lambda - i\varepsilon)]\psi) \, d\lambda \right|^2$$
$$\le \frac{1}{4\pi^2} \|\varphi\|^2 \, |b - a| \lim_{\varepsilon \downarrow 0} \int_{-\infty}^{\infty} \|A[R(\lambda + i\varepsilon) - R(\lambda - i\varepsilon)]\psi\|^2 \, d\lambda$$
$$\le |b - a| \, \|A\|_H^2 \, \|\varphi\|^2 \, \|\psi\|^2$$

In the last step we have used (36). If a and/or b are eigenvalues, $a + \delta$ and $b + \delta$ are not eigenvalues for almost all small δ and $E_{(a,b]} = s\text{-}\lim_{\delta \to 0} E_{(a+\delta, b+\delta]}$ so we see that $c_3 \le c_0$.

$c_3 \le c_7$: We already know that $c_3 = c_6$, and $c_6 \le c_7$ is obvious.

$c_7 \le c_1$: This is essentially the argument used in the proof of Theorem XIII.22. For by that argument,
$$|(A^*\psi, R(\lambda + i\varepsilon)\varphi)|^2 \le \frac{1}{2\varepsilon} \|\psi\|^2 \int_0^\infty \|Ae^{-iHt}\varphi\|^2 \, dt$$

so

$$\|AR(\lambda + i\varepsilon)\varphi\|^2 \leq \frac{1}{2\varepsilon} \int_0^\infty \|Ae^{-iHt}\varphi\|^2 \, dt$$

A similar argument for $R(\lambda - i\varepsilon)$ proves that

$$\|AR(\lambda + i\varepsilon)\varphi\|^2 + \|AR(\lambda - i\varepsilon)\varphi\|^2 \leq \frac{1}{2\varepsilon} \int_{-\infty}^\infty \|Ae^{-iHt}\varphi\|^2 \, dt$$

Thus $c_7 \leq c_1$. ∎

Notice that criterion (3) provides another proof that $\operatorname{Ran}(A^*) \subset \mathscr{H}_{ac}(H)$ if A is H-smooth, and that criterion (6) implies Theorem XIII.22. The equality of c_6 and c_7 at first sight seems very mysterious. Some of this mystery can be removed by looking at the proof of a closely related equality:

$$\sup_{\substack{\varepsilon > 0 \\ \|\varphi\| = 1}} \int_{-\infty}^\infty \|AR(\lambda + i\varepsilon)\varphi\|^2 \, d\lambda = \sup_{\substack{\varepsilon > 0 \\ \|\varphi\| = 1}} \left\{ \int_{-\infty}^\infty (\|AR(\lambda + i\varepsilon)\varphi\|^2 + \|AR(\lambda - i\varepsilon)\varphi\|^2) \, d\lambda \right\}$$

By (33), this equality is equivalent to

$$\sup_{\|\varphi\|=1} \int_0^\infty \|Ae^{-itH}\varphi\|^2 \, dt = \sup_{\|\varphi\|=1} \int_{-\infty}^\infty \|Ae^{-itH}\varphi\|^2 \, dt$$

which follows by noting that if $\varphi_s = e^{-isH}\varphi$, then

$$\int_0^\infty \|Ae^{-itH}\varphi_s\|^2 \, dt = \int_s^\infty \|Ae^{-itH}\varphi\|^2 \, dt$$

converges to $\int_{-\infty}^\infty \|Ae^{-itH}\varphi\|^2 \, dt$ as $s \to -\infty$.

Before turning to applications, we give two more examples of smooth operators:

Example 3 Let $H_0 = -\Delta$ on $L^2(\mathbb{R}^3)$ and let A be multiplication by $V \in R$, the Rollnik class. By the estimate,

$$\| |V|^{1/2}(H_0 - \lambda)^{-1}|V|^{1/2}\|_{\text{oper}} \leq \| |V|^{1/2}(H_0 - \lambda)^{-1}|V|^{1/2}\|_2$$
$$\leq (4\pi)^{-1}\|V\|_R$$

and the corollary to Theorem XIII.25, $|A|^{1/2}$ is H-smooth.

Example 4 Let H be multiplication by x on $L^2([\alpha, \beta], dx)$ with $\alpha, \beta \in \mathbb{R}$. Let A be a bounded operator on $\mathcal{H}$ so that A^*A is an integral operator of the form

$$(A^*A\psi)(x) = \int_\alpha^\beta K(x, y)\psi(y)\, dy \tag{37}$$

with $\|K\|_\infty \equiv \operatorname{ess\,sup}_{[\alpha, \beta] \times [\alpha, \beta]} |K(x, y)| < \infty$. Then A is H-smooth with $\|A\|_H \leq \|K\|_\infty^{1/2}$. For, let $(a, b]$ be an interval and $\|\varphi\| = 1$. Then

$$\|AE(a, b]\varphi\|^2 = (\varphi, E(a, b]A^*AE(a, b]\varphi)$$

$$= \int_a^b \int_a^b K(x, y)\overline{\varphi(x)}\varphi(y)\, dx\, dy$$

$$\leq \left(\int_a^b \int_a^b |K(x, y)|^2\, dx\, dy\right)^{1/2} \left(\int_a^b \int_a^b |\bar{\varphi}(x)\varphi(y)|^2\, dx\, dy\right)^{1/2}$$

$$\leq |b - a|\,\|K\|_\infty \|\varphi\|^2$$

By criterion (2), $\|A\|_H \leq \|K\|_\infty^{1/2}$. By further analysis (Problems 50, 51) one can show that for every H-smooth operator A, A^*A is of the form (37) and that $\|A\|_H = \|K\|_\infty^{1/2}$. In particular, we note that the analysis above shows that any integral operator

$$(A\psi)(x) = \int_\alpha^\beta A(x, y)\psi(y)\, dy$$

with $\|A\|_\infty < \infty$ is H-smooth. We have thus found many H-smooth operators and, in particular, H-smooth operators A with $\operatorname{Ker}(A) = \{0\}$. Example 1 is related to this example although the operator H in Example 1 is not bounded. For if $f \in L^2$, then $A = $ multiplication by f has $A^*A = $ multiplication by $g \equiv |f|^2 \in L^1$. Passing to the representation in which $-i\, d/dx$ is diagonal (by using Fourier transform), A^*A is an integral operator with kernel equal to $(2\pi)^{-1/2}\hat{g}(x - y) \in L^\infty$.

We turn now to various applications:

A. Weakly coupled quantum systems

Suppose that H_0 is a positive self-adjoint operator and that C is self-adjoint. If $|C|^{1/2}(H_0 + I)^{-1}|C|^{1/2}$ is a bounded operator with bound a, then for any $\varphi \in D(|C|^{1/2})$, $\|(H_0 + I)^{-1/2}|C|^{1/2}\varphi\|^2 \leq a\|\varphi\|^2$ so

$(H_0 + I)^{-1/2}|C|^{1/2}$ is bounded. Taking adjoints, $Q(H_0) \subset Q(C)$ and $\||C|^{1/2}(H_0 + I)^{-1/2}\|^2 \leq a$. It follows that $(\varphi, |C|\varphi) \leq a(\varphi, (H_0 + I)\varphi)$ for all $\varphi \in Q(H_0)$. Thus C is H_0-form bounded and if $a < 1$, the form sum $H = H_0 + C$ is self-adjoint. Such form sums are the subject of the following theorem.

Theorem XIII.26 (Kato's smoothness theorem) Let H_0 be a positive self-adjoint operator on a Hilbert space $\mathcal{H}$ and let $C_1, \ldots, C_n$ be self-adjoint operators with

$$\alpha_{ij} \equiv \sup_{\mu \notin \mathbb{R}} \| |C_i|^{1/2}(H_0 - \mu)^{-1}|C_j|^{1/2} \| < \infty$$

Suppose that $\{\alpha_{ij}\}_{1 \leq i, j \leq n}$ is the matrix of an operator of norm less than 1 on $\mathbb{C}^n$ (with the natural Hilbert space norm). Then:

(a) The form sum $H = H_0 + \sum_{i=1}^n C_i$ is a closed form on $Q(H_0)$.
(b) The wave operators s-$\lim_{t \to \mp\infty} e^{iHt}e^{-iH_0t}$ exist and are unitary.

In particular, H and H_0 are unitarily equivalent operators.

Proof Introduce the Hilbert space $\mathcal{H}' = \oplus_{i=1}^n \mathcal{H}_i$ where each $\mathcal{H}_i$ is a copy of $\mathcal{H}$. By the discussion preceding the theorem and the hypothesis $\alpha_{ii} < \infty$, we have that each $|C_i|$ is H_0-form bounded, so that $|C_i|^{1/2}$ is defined from $Q(H_0)$ to $\mathcal{H}$. As a result, we can define $B: Q(H_0) \to \mathcal{H}'$ by $(B\psi)_i = |C_i|^{1/2}\psi$. The hypothesis of the theorem implies that

$$\alpha \equiv \sup_{\mu \notin \mathbb{R}} \| B(H_0 - \mu)^{-1}B^* \| < 1 \tag{38}$$

Let $B = U|B|$ be the polar decomposition of B. Then, since $|B| = U^*B$ and U is a partial isometry, (38) implies that $|B|(H_0 + I)^{-1}|B|$ is a bounded operator of norm smaller than α. Again, it follows from the argument preceding the theorem that

$$(\varphi, |B|^2\varphi) \leq \alpha(\varphi, (H_0 + I)\varphi)$$

or that

$$\left(\varphi, \sum_{i=1}^n |C_i|\varphi\right) \leq \alpha(\varphi, (H_0 + I)\varphi) \tag{39}$$

From (39), we conclude that $\sum_{i=1}^n C_i$ is H_0-form bounded with relative bound $\alpha < 1$ proving (a).

Next, we note that for $\mu \notin [0, \infty)$,

$$(H - \mu)^{-1} = (H_0 - \mu)^{-1}$$
$$- \sum_{n=0}^{\infty} (-1)^n (H_0 - \mu)^{-1} B^* W (B(H_0 - \mu)^{-1} B^* W)^n B(H_0 - \mu)^{-1} \quad (40)$$

where $W: \mathcal{H}' \to \mathcal{H}'$ by $(W\varphi)_i = (\text{sgn } C_i)\varphi_i$. (40) holds since the sum is convergent by (38) and one checks that, in the language of quadratic forms (Section VIII.6), the right-hand side maps $\mathcal{H}_{-1}$ to $\mathcal{H}_{+1}$ and is an inverse for $(H - \mu): \mathcal{H}_{+1} \to \mathcal{H}_{-1}$. From (40) one easily sees that

$$\sup_{\mu \notin \mathbb{R}} \|B(H - \mu)^{-1} B^*\| \leq \alpha(1 - \alpha)^{-1} \quad (41)$$

so each $|C_i|^{1/2}$ and $C_i^{1/2} \equiv |C_i|^{1/2} \text{sgn } C_i$ is H-smooth by the corollary to Theorem XIII.25. Similarly, by the basic hypotheses and that corollary, each $|C_i|^{1/2}$ is H_0-smooth. Thus, by Theorem XIII.24, both s-$\lim_{t \to \mp\infty} e^{iHt} e^{-iH_0 t}$ and s-$\lim_{t \to \mp\infty} e^{iH_0 t} e^{-iHt}$ exist. These maps are all isometries and inverses of one another. This proves (b). ∎

As an application of this theorem, the reader should prove the strong version of Theorem XIII.21 found in Problem 56. If one does not want to worry about detailed estimates, it is often useful to state Kato's smoothness theorem in the form:

Corollary Let H_0 be self-adjoint and positive. Let $C_1, \ldots, C_n$ be self-adjoint and let $C = \sum_{i=1}^{n} C_i$. Suppose that for each i and j

$$\sup_{\mu \notin \mathbb{R}} \| |C_i|^{1/2} (H_0 - \mu)^{-1} |C_j|^{1/2} \| < \infty$$

Then there exists $\Lambda > 0$ so that for all $\lambda \in (-\Lambda, \Lambda)$:
(a) $H(\lambda) = H_0 + \lambda C$ is a closed form on $Q(H_0)$.
(b) The wave operators $\Omega_\lambda^\pm = $ s-$\lim_{t \to \mp\infty} e^{itH(\lambda)} e^{-itH_0}$ exist and are unitary.

It is actually possible to prove that $\Omega_\lambda^\pm$ are analytic in λ (see the Notes and Problems 53, 54). Now we can easily handle "weakly coupled" N-body Schrödinger operators.

Theorem XIII.27 (the Iorio–O'Carroll theorem)   Let $m \geq 3$, $N \geq 2$. Let $\tilde{H}_0$ be the operator

$$\tilde{H}_0 = \sum_{i=1}^{N} (-2\mu_i)^{-1} \Delta_i$$

on $L^2(\mathbb{R}^{Nm})$ where each $\mu_i > 0$ and $r \in \mathbb{R}^{Nm}$ is written $r = \langle r_1, \ldots, r_N \rangle$ with $r_i \in \mathbb{R}^m$. Let H_0 be $\tilde{H}_0$ with the center of mass motion removed (see Section XI.5). Let $V = \sum_{i<j} V_{ij}(r_i - r_j)$ where each $V_{ij} \in L^{m/2+\varepsilon}(\mathbb{R}^m) \cap L^{m/2-\varepsilon}(\mathbb{R}^m)$ for some fixed $\varepsilon > 0$. Then for all $\lambda \in \mathbb{R}$ with $|\lambda|$ sufficiently small, $H_0 + \lambda V$ (defined as a form sum) is unitarily equivalent to H_0. The wave operators provide unitary equivalence. In particular, $H_0 + \lambda V$ has no bound states, no singular spectrum and has complete scattering.

Before proving the Iorio–O'Carroll theorem, we make several remarks. First, it is merely for contact with the notions of N-body quantum theory that we remove the center of mass motion. Secondly, we note that only when $m = 3$ is the form sum necessary. If $m \geq 4$, Theorem X.20 tells us that the operator sum $H_0 + \lambda V$ is self-adjoint on $C_0^\infty(\mathbb{R}m(N-1))$ if λ is real. Thirdly, we note that, by Theorem XIII.11, there can be no weak coupling theorem when $m = 1$ or 2, but there is a result on $L^2([0, \infty))$ (Problem 58). Finally, the conditions on V_{ij} cannot be weakened much; for if V_{ij} has non-$L^{m/2}$ local singularities at finite points (e.g., $r^{-2-\varepsilon}$ behavior at $r = 0$), then self-adjointness is lost and if V_{ij} has non-$L^{m/2}$ behavior at infinity (e.g., only $r^{-2+\varepsilon}$ falloff at ∞), then it can happen that $H_0 + \lambda V$ has bound states no matter how small λ is (see Theorem XIII.6).

Proof of Theorem XIII.27   Consider first the case $N = 2$. Then $H_0 = (-2\nu)^{-1} \Delta$ on $L^2(\mathbb{R}^m)$, so by Theorem IX.30,

$$\|e^{-iH_0 t}\varphi\|_r \leq (ct)^{-m[\frac{1}{2} - r^{-1}]} \|\varphi\|_{r'} \tag{42}$$

where $2 \leq r \leq \infty$, $r^{-1} + r'^{-1} = 1$ and c is chosen suitably. Thus, if $f \in L^p$ and $p > 2$, then by Hölder's inequality and (42)

$$\|fe^{-iH_0 t}f\varphi\|_2 \leq (ct)^{-mp^{-1}} \|f\|_p^2 \|\varphi\|_2 \tag{43}$$

Thus, if $f \in L^{m-\varepsilon} \cap L^{m+\varepsilon}$ (and $m > 2 + \varepsilon$),

$$\int_{-\infty}^{\infty} \|fe^{-iH_0 t}f\varphi\|_2 \, dt \leq d(\|f\|_{m+\varepsilon} + \|f\|_{m-\varepsilon})^2 \|\varphi\|_2$$

for a suitable constant d. Since

$$f(H_0 - z)^{-1}f\varphi = i \int_0^\infty e^{izt}(fe^{-iH_0 t}f)\varphi \, dt$$

XIII.7 Smooth perturbations   155

if Im $z > 0$, we conclude that

$$\|f(H_0 - z)^{-1}f\varphi\|_2 \leq d(\|f\|_{m+\varepsilon} + \|f\|_{m-\varepsilon})^2 \|\varphi\|_2$$

if Im $z > 0$. A similar argument holds if Im $z < 0$. Letting $f = |V|^{1/2}$, we conclude that

$$\sup_{z \notin \mathbb{R}} \| |V|^{1/2}(H_0 - z)^{-1}|V|^{1/2}\| < \infty$$

The theorem (in case $N = 2$) then follows from the corollary to Theorem XIII.26.

To prove the theorem in the general case, we need only prove that

$$\sup_{z \notin \mathbb{R}} \| |V_{ij}|^{1/2}(H_0 - z)^{-1}|V_{k\ell}|^{1/2}\| < \infty$$

for all i, j, k, ℓ and apply the corollary to Theorem XIII.26. It is necessary to consider three distinct cases:

Case 1 $(ij) = (k\ell)$ Without loss of generality suppose that $i = k = 1$ and $j = \ell = 2$. Let $H_0^{(12)} = (-2\mu_{12})^{-1}\Delta_{12}$ where $\mu_{12}^{-1} = \mu_1^{-1} + \mu_2^{-1}$ and Δ_{12} is the Laplacian with respect to r_{12} in a Jacobi coordinate system (see Section XI.5). Thus, $H_0 - H_0^{(12)}$ depends only on the remaining Jacobi coordinates $\zeta_2, \ldots, \zeta_{N-1}$ and so commutes with any function of $\zeta_1 = r_2 - r_1$. If $f \in L^p(\mathbb{R}^m)$ and f_1 is multiplication by $f(\zeta_1)$, then

$$\|f_1 e^{-itH_0}f_1\varphi\|_2 = \|f_1 e^{-itH_0^{(12)}}f_1\varphi\|_2 \qquad (44)$$

Let $\varphi \in \mathscr{S}(\mathbb{R}^{m(N-1)})$. By the basic two-body estimate (43),

$$\int |f_1(e^{-iH_0^{(12)}t}f_1\varphi(\zeta_1, \zeta_2, \ldots, \zeta_{N-1})|^2 \, d\zeta_1 \leq (ct)^{-2mp^{-1}}\|f\|_p^4 \int |\varphi(\zeta)|^2 \, d\zeta_1$$

Integrating over $\zeta_2, \ldots, \zeta_{N-1}$ and using (44) we see that (43) still holds. Thus, by mimicking the two-body proof,

$$\sup_{z \notin \mathbb{R}} \| |V_{12}|^{1/2}(H_0 - z)^{-1}|V_{12}|^{1/2}\| < \infty$$

Case 2 $j = k$; i, j, ℓ distinct Without loss of generality, suppose that $i = 1$; $j = k = 2$; $\ell = 3$. Again use Jacobi coordinates with $\zeta_1 = r_2 - r_1$ and with $r_{23} = \alpha\zeta_1 + \beta\zeta_2$ (with $\alpha, \beta \neq 0$). Since $H_0 - H_0^{(12)}$ commutes with functions of ζ_1:

$$\| |V_{12}|^{1/2}e^{-itH_0}|V_{23}|^{1/2}\varphi\| = \| |V_{12}|^{1/2}e^{-itH_0^{(12)}}|V_{23}|^{1/2}\varphi\|$$

Fix $\zeta' \equiv \langle \zeta_2, \ldots, \zeta_{N-1} \rangle$ and let $\varphi \in \mathscr{S}(\mathbb{R}^{m(N-1)})$. Then by the basic two-body estimate (43),

$$\int |V_{12}(\zeta_1)| \, |e^{-itH_0(12)} V_{23}^{1/2}(\alpha\zeta_1 - \beta\zeta_2)\varphi(\zeta_1, \zeta')|^2 \, d\zeta_1$$

$$\leq (ct)^{-2mp^{-1}} \|V_{12}\|_{p/2} \|V_{23}\|_{p/2} \alpha^{-2m/p} \int |\varphi(\zeta)|^2 \, d\zeta_1$$

where we have used the fact that

$$\int |V_{23}(\alpha\zeta_1 - \beta\zeta_2)|^{p/2} \, d\zeta_1 = \alpha^{-1} \int |V_{23}(x)|^{p/2} \, dx$$

independently of ζ_2. Integrating over ζ', we find that

$$\| |V_{12}|^{1/2} e^{-itH_0(12)} |V_{23}|^{1/2} \varphi \|_2^2 \leq \alpha^{-2/p}(ct)^{-2mp^{-1}} \|V_{12}\|_{p/2} \|V_{23}\|_{p/2} \|\varphi\|_2^2$$

From this estimate, it follows that

$$\sup_{z \notin \mathbb{R}} \| |V_{12}|^{1/2}(H_0 - z)^{-1} |V_{23}|^{1/2} \| < \infty$$

as in the two-body case.

Case 3 i, j, k, ℓ all distinct Without loss of generality suppose that $i = 1, j = 2, k = 3, \ell = 4$. Then,

$$|(\varphi, |V_{12}|^{1/2} e^{-itH_0} |V_{34}|^{1/2} \psi)|$$
$$= |(e^{+itH_0(34)}\varphi, |V_{12}|^{1/2} e^{-it(H_0 - H_0(34))} |V_{34}|^{1/2} \psi)|$$
$$= |(|V_{34}|^{1/2} e^{itH_0(34)}\varphi, |V_{12}|^{1/2} e^{-it(H_0 - H_0(34))} \psi)|$$
$$\leq \| |V_{34}|^{1/2} e^{itH_0(34)}\varphi \| \, \| |V_{12}|^{1/2} e^{-itH_0(12)} \psi \|$$

In the first step, we have used the fact that $|V_{12}|^{1/2}$ and $H_0^{(34)}$ commute; in the second step that $|V_{34}|^{1/2}$ commutes with $H_0 - H_0^{(34)}$ and $|V_{12}|^{1/2}$; and in the final step that $|V_{12}|^{1/2}$ and $H_0 - H_0^{(34)} - H_0^{(12)}$ commute. By step (1) and the corollary to Theorem XIII.25, $|V_{ij}|^{1/2}$ is $H_0^{(ij)}$-smooth. Thus,

$$\int_{-\infty}^{\infty} \| |V_{34}|^{1/2} e^{+itH_0(34)}\varphi \| \, \| |V_{12}|^{1/2} e^{-itH_0(12)} \psi \| \, dt$$

$$\leq \left(\int_{-\infty}^{\infty} \| |V_{34}|^{1/2} e^{itH_0(34)}\varphi \|^2 \, dt \right)^{1/2}$$

$$\times \left(\int_{-\infty}^{\infty} \| |V_{12}|^{1/2} e^{-itH_0(12)} \psi \|^2 \, dt \right)^{1/2}$$

$$\leq \| |V_{34}|^{1/2} \|_{H_0(34)} \, \| |V_{12}|^{1/2} \|_{H_0(12)} \, \|\varphi\| \, \|\psi\|$$

so that
$$\int_{-\infty}^{\infty} |(\varphi, |V_{12}|^{1/2} e^{-itH_0} |V_{34}|^{1/2} \psi)| \, dt \leq c \|\varphi\| \|\psi\|$$

Since
$$(\varphi, |V_{12}|^{1/2}(H_0 - z)^{-1} |V_{34}|^{1/2} \psi)$$
$$= -i \int_0^{\infty} e^{izt} (\varphi, |V_{12}|^{1/2} e^{-iH_0 t} |V_{23}|^{1/2} \psi) \, dt$$

if Im $z > 0$, we conclude that
$$\sup_{z \notin \mathbb{R}} \| |V_{12}|^{1/2} (H_0 - z)^{-1} |V_{34}|^{1/2} \| < \infty$$

This completes the proof of case 3. ∎

If one of the particles has infinite mass, Case 2 cannot be handled as above since $\alpha = 0$ if $\mu_2 = \infty$. In that event, the method of case 3 will work and the theorem remains true.

B. Positive commutators and repulsive potentials

As a second application of smoothness techniques, we develop a method that will allow us to prove that Hamiltonians with repulsive potentials have purely absolutely continuous spectrum.

Theorem XIII.28 (the Putnam–Kato theorem) Let H and A be bounded self-adjoint operators. Suppose that $C = i[H, A]$ is positive. Then $C^{1/2}$ is H-smooth. In particular, if Ker $C = \{0\}$, then H has purely absolutely continuous spectrum.

Proof The second statement follows from the first by Theorem XIII.23 and the fact that $\text{Ker}(C^{1/2}) = [\text{Ran}(C^{1/2})]^\perp = \{0\}$. We compute:
$$\frac{d}{dt} e^{itH} A e^{-itH} \varphi = i e^{itH}[H, A] e^{-itH} \varphi = e^{itH} C e^{-itH} \varphi$$

Thus
$$\int_s^t (\varphi, e^{i\tau H} C e^{-i\tau H} \varphi) \, d\tau = (\varphi, e^{itH} A e^{-itH} \varphi) - (\varphi, e^{isH} A e^{-isH} \varphi)$$

so

$$\int_s^t \|C^{1/2}e^{-i\tau H}\varphi\|^2 \, d\tau \leq 2\|A\| \, \|\varphi\|^2$$

Since t and s are arbitrary, $C^{1/2}$ is H-smooth and

$$\|C^{1/2}\|_H^2 \leq \|A\|/\pi \quad \blacksquare$$

For an alternative method of proof, see Problem 59.

Example 5 Since examples of positive commutators are not easy to construct directly, we present examples to show that the hypotheses of Theorem XIII.28 are sometimes obeyed. In fact, there is a sort of converse to the fact that $C^{1/2}$ is H-smooth: If H is bounded and B is any bounded H-smooth operator, there exists a bounded operator A with $i[H, A] = B^*B$. Since there are H-smooth operators with zero kernel when H has purely absolutely continuous spectrum (see Example 4), one can construct many pairs H, A for which the hypotheses of Theorem XIII.28 are obeyed. If B is H-smooth, then by following the argument in Theorem XIII.24, one can show that $\Gamma_H^+(B^*B) \equiv \text{s-lim}_{s \to \infty} i \int_0^s e^{itH} B^* B e^{-itH} \, dt$ exists. Now,

$$[H, \Gamma_H^+(B^*B)] = \underset{s \to \infty}{\text{s-lim}} \, i \int_0^s e^{itH}[H, B^*B]e^{-itH} \, dt$$

$$= \underset{s \to \infty}{\text{s-lim}} \, (e^{isH} B^* B e^{-isH} - B^*B)$$

$$= -B^*B$$

where we have used the fact that $e^{itH} B^* B e^{-itH} \varphi \in L^2$ with a uniformly bounded derivative to conclude that $\text{s-lim}_{s \to \infty} (e^{isH} B^* B e^{-isH}) = 0$ (Problem 62). Letting $A = i\Gamma_H^+(B^*B)$, we see that $i[H, A] = B^*B$.

We want to apply the idea of positive commutators to prove that N-body Schrödinger operators with repulsive potentials have purely absolutely continuous spectrum. A repulsive potential is a function on $\mathbb{R}^m$ so that $V(r\hat{e}) \leq V(r'\hat{e})$ for each unit vector $\hat{e}$ and all $r > r'$. Thus repulsive potentials tend to push particles apart, so we expect no bound states and thus only absolutely continuous spectra. There is another way of describing the fact that V is repulsive that makes the connection with positive commutators clearer. Let $U(\theta)$ be the family of dilations, $(U(\theta)\psi)(r) = e^{+m\theta/2}\psi(e^\theta r)$. Then $[U(\theta)VU(\theta)^{-1}](r) = V(e^\theta r)$, so repulsive potentials obey the condition that $V_\theta \equiv U(\theta)VU(\theta)^{-1}$ is monotone decreasing as θ increases. Since $U(\theta)H_0 U(\theta)^{-1} = e^{-2\theta}H_0$, the kinetic energy also decreases monotonically under dilations. Thus, formally,

$$\frac{d}{d\theta} U(\theta)(H_0 + V)U(\theta)^{-1} \leq 0$$

Letting

$$D = \frac{i}{2} \sum_{i=1}^{m} \left(x_i \frac{\partial}{\partial x_i} + \frac{\partial}{\partial x_i} x_i \right)$$

be the generator of $U(\theta)$ (i.e., $U(\theta) = \exp(-i\theta D)$), we see that $i[D, H] \geq 0$ formally.

Physically, there is another way of looking at the fact that $i[D, H] \geq 0$. For $D = -(i/2)[H, x^2]$. Therefore, if one defines the Heisenberg picture moment of inertia

$$I(t) = e^{+iHt}x^2 e^{-iHt}$$

then on a formal level $i[D, H] \geq 0$ is equivalent to $\ddot{I}(t) \geq 0$. It is easy to see that classically the repulsive potential obeys this condition and indeed we have used similar ideas in Theorem XI.3.

Theorem XIII.28 is not directly applicable for three reasons: (1) The computations were formal; as usual this requires care with domains and cores, but it turns out that no extra technical assumption is needed. (2) H is not bounded. It turns out that since H is positive, this in itself would not be very serious (see Problem 61). (3) D is not bounded. This is more serious. The d/dx piece of D is not hard to control since it is H-bounded but the factor of x is hard to control directly. In the proof of the theorem below, we will use a cutoff x which will have the effect of making the computations more complicated. Since V being repulsive is related to $dV/dr \leq 0$, it is the derivative in D rather than the x that is important—thus cutting off x will not destroy $i[A, H] \geq 0$.

Theorem XIII.29 (Lavine's theorem)   Let $H_0 = -\Delta$ on $L^2(\mathbb{R}^m)$ ($m \geq 3$). Let V be a function on $\mathbb{R}^m$ so that

(i) Multiplication by V is H_0-bounded with relative bound zero.
(ii) The distributional derivative $\sum_{i=1}^{n} x_i \, \partial V/\partial x_i$ is negative.

Then $H = H_0 + V$ has purely absolutely continuous spectrum.

XIII: SPECTRAL ANALYSIS

Proof Choose α obeying $\frac{1}{2} < \alpha < \frac{3}{2}$ and define $g(r) = \int_0^r (1 + \rho^2)^{-\alpha} \, d\rho$. Let A be the operator

$$Af = i \sum_{k=1}^m \left[x_k \frac{g(r)}{r} \frac{\partial f}{\partial x_k} + \frac{\partial}{\partial x_k}\left(x_k \frac{g(r)f}{r}\right) \right] \quad (45)$$

Since $g(r)r^{-1}$ is C^∞, A maps C_0^∞ into itself. If g were replaced by r in (45), the operator resulting would be the generator of dilations, so (45) is a partially cutoff version of this generator. In fact, A has a direct geometrical interpretation. Let $T_\gamma \colon \mathbb{R}^n \to \mathbb{R}^n$ be defined by the conditions: $T_0(x) = x$, and $y(\gamma) = T_\gamma x$ solves $\dot{y} = |y|^{-1}g(|y|)y$. Then,

$$(e^{-i\gamma A}f)(x) = N_\gamma(x)f(T_\gamma x)$$

where $N_\gamma(x)$ is a normalizing factor (the square root of a Jacobian determinant).

Our first goal is to prove that for some constant $c > 0$,

$$i[A, H] \geq c(1 + r^2)^{-\alpha - 1}$$

in the sense that for all $\varphi \in C_0^\infty(\mathbb{R}^n)$,

$$i(A\varphi, H\varphi) - i(H\varphi, A\varphi) \geq c(\varphi, (1 + r^2)^{-\alpha - 1}\varphi) \quad (46)$$

First, we compute

$$i(A\varphi, V\varphi) - i(V\varphi, A\varphi)$$

$$= 2 \operatorname{Re} \sum_{i=1}^m \int V(x) \left[\overline{\varphi(x)} \left\{ \frac{x_i}{r} g(r) \frac{\partial}{\partial x_i} + \frac{\partial}{\partial x_i}\left(\frac{x_i}{r} g(r)\right) \right\} \varphi(x) \right] dx$$

$$= 2 \int V(x) \sum_{i=1}^m \frac{\partial}{\partial x_i} x_i \left(\frac{g(r)}{r} |\varphi(x)|^2\right) dx \geq 0$$

by hypothesis (ii). Thus we need only verify (46) when $V = 0$, a tedious but not too difficult task. As a preliminary, we note that for all r,

$$rg'(r) - g(r) \leq 0 \quad (47a)$$

$$g''(r) = -2\alpha r(1 + r^2)^{-\alpha - 1} \quad (47b)$$

$$g'''(r) + (2\alpha + 1)r^{-1}g''(r) \leq 0 \quad (47c)$$

To prove (47a), we remark that $(1 + r^2)^{-\alpha}$ is monotone decreasing, so its average value $g(r)r^{-1} = r^{-1}\int_0^r (1 + \rho^2)^{-\alpha} d\rho$ is also monotone decreasing. Since $rg' - g = r^2(r^{-1}g)'$, (47a) is proven. (47b) and (47c) follow from the explicit calculations:

$$g''(r) = -2\alpha r(1 + r^2)^{-\alpha - 1}$$

$$g'''(r) + (2\alpha + 1)r^{-1}g''(r) = -4\alpha(\alpha + 1)(1 + r^2)^{-\alpha - 2}$$

Next we prove that

$$-\Delta\left(\sum_{j=1}^{m} \frac{\partial g_j}{\partial x_j}\right) \geq c(1+r^2)^{-\alpha-1} \tag{48}$$

where $g_j(x) = x_j r^{-1} g(r)$. For

$$\frac{\partial g_j}{\partial x_k} = r^{-1}g(\delta_{jk} - r^{-2}x_j x_k) + r^{-2} x_j x_k g' \tag{49}$$

Thus

$$\sum_{j=1}^{m} \frac{\partial g_j}{\partial x_j} = (n-1)r^{-1}g + g'$$

Using the fact that for a spherically symmetric function h,

$$\Delta h = r^{-1}(rh)'' + (n-3)r^{-1}h'$$

we find that

$$-\Delta\left(\sum_{j=1}^{n} \frac{\partial g_j}{\partial x_j}\right) = -(g''' + (2\alpha+1)r^{-1}g'') - (2n - 2\alpha - 3)r^{-1}g''$$

$$- (n-1)(n-3)r^{-3}(rg' - g)$$

Since $n \geq 3$ and $\alpha < \tfrac{3}{2}$, (48) follows from (47). Finally, we compute on C_0^∞ functions that

$$-\left[g_k \frac{\partial}{\partial x_k}, H_0\right] = \sum_{i=1}^{n}\left[g_k \frac{\partial}{\partial x_k}, \frac{\partial^2}{\partial x_i^2}\right]$$

$$= \sum_{i=1}^{n} 2p_i \frac{\partial g_k}{\partial x_i} p_k + i\frac{\partial^2 g_k}{\partial x_i^2} p_k$$

where $p_k = i^{-1} \partial/\partial x_k$. Using

$$-\left[g_k \frac{\partial}{\partial x_k} + \frac{\partial}{\partial x_k} g_k, H_0\right] = -\left[g_k \frac{\partial}{\partial x_k}, H_0\right] - \left[g_k \frac{\partial}{\partial x_k}, H_0\right]^*$$

we see that

$$i[A, H_0] = \sum_{i,k} 2p_k\left(\frac{\partial g_i}{\partial x_k} + \frac{\partial g_k}{\partial x_i}\right)p_i - \Delta\left(\sum_j \frac{\partial g_j}{\partial x_j}\right)$$

Thus by (48) the proof of (46) has been reduced to proving

$$\sum_{i,k} 2p_k\left(\frac{\partial g_i}{\partial x_k} + \frac{\partial g_k}{\partial x_i}\right)p_i \geq 0 \tag{50}$$

By (49) the left-hand side of (50) is

$$4 \sum_{i,k} \{p_k[(r^{-1}g)(\delta_{ik} - r^{-2}x_i x_k)]p_i + p_k(r^{-2}g')x_i x_k p_i\}$$

Fix $x \in \mathbb{R}^n$. Then, by the Schwarz inequality, the matrix $\{\delta_{ik} - r^{-2}x_i x_k\}$ is positive definite. Further, $\{x_i x_k r^{-2}\}$ is clearly positive definite, so (50) holds and this completes the proof of (46).

Next, we need to know that $A < < H$, i.e., that A is H-bounded with relative bound zero. Since $V < < H_0$ by hypothesis, we need only prove that $A < < H_0$. But $\partial/\partial x_i < < H_0$, $x_i g r^{-1}$ is bounded, and

$$A = 2i \sum_{i=1}^{n} x_i g r^{-1} \partial/\partial x_i + i(n-1)r^{-1}g + ig'$$

so $A < < H_0$ is proven. Since C_0^∞ is a core for H and $A < < H$, (46) holds for all $\varphi \in D(H)$.

Now let $\varphi \in D(H)$ and let B be multiplication by $c^{1/2}(1 + r^2)^{-(\alpha+1)/2}$. Then we claim that

$$\int_{-\infty}^{\infty} \|Be^{-itH}\varphi\|^2 \, dt \le d\|(H + I)\varphi\|^2 \tag{51}$$

for a suitable constant d. For, by (46),

$$\int_{-\infty}^{\infty} \|Be^{-itH}\varphi\|^2 \, dt$$

$$= \int_{-\infty}^{\infty} (e^{-itH}\varphi, B^*Be^{-itH}\varphi) \, dt$$

$$\le \int_{-\infty}^{\infty} i\{(Ae^{-itH}\varphi, He^{-itH}\varphi) - (He^{-itH}\varphi, Ae^{-itH}\varphi)\} \, dt$$

$$\le 2 \sup_{-\infty < s < \infty} |(e^{-isH}\varphi, Ae^{-isH}\varphi)|$$

$$\le 2\|\varphi\| \, \|A(H + I)^{-1}\| \, \|(H + I)\varphi\|$$

which proves (51).

To conclude the proof we notice that (51) implies that $B(H + I)^{-1}$ is H-smooth. Since $\operatorname{Ran}(B^*)$ and $\operatorname{Ran}(H + I)^{-1}$ are dense, the closure of $\operatorname{Ran}(H + I)^{-1}B^*$ is $\mathcal{H}$. Theorem XIII.23 thus implies that $\mathcal{H} \subset \mathcal{H}_{ac}$. Therefore H has purely absolutely continuous spectrum. ∎

Corollary Let $m(N-1) \geq 3$. Let $\tilde{H}_0 = \sum_{i=1}^{N}(-2\mu_i)^{-1}\Delta_i$ on $L^2(\mathbb{R}^{Nm})$ and let H_0 be $\tilde{H}_0$ with its center of mass motion removed. For each i, j suppose that $V_{ij}: \mathbb{R}^m \to \mathbb{R}$ and

(i) $V_{ij} \in L^p(\mathbb{R}^m) + L^\infty(\mathbb{R}^m)$ ($p = 2$ if $m \leq 3$, $p > 2$ if $m = 4$; $p = m/2$ if $m \geq 5$).
(ii) $\sum_{k=1}^{m} x_k \, \partial V_{ij}/\partial x_k \leq 0$ in the distributional sense.

Then the operator $H = H_0 + \sum_{i<j} V_{ij}(r_i - r_j)$ on $L^2(\mathbb{R}^{m(N-1)})$ has purely absolutely continuous spectrum.

Proof Let ζ be a Jacobi coordinate system so $H_0 = \sum_{i=1}^{N-1}(-2M_i)^{-1}\Delta_{\zeta_i}$ and let $q_i = (2M_i)^{1/2}\zeta_i$ so $H_0 = -\sum_{i=1}^{N-1}\Delta_{q_i}$. If we can show that $\sum_{i=1}^{N-1} q_i \cdot \nabla_{q_i} V_{jk} \leq 0$ for all j, k, then the result follows from Theorem XIII.29. Suppose first that each V_{jk} is smooth. Then $V_{jk} = V_{jk}(\sum_{i=1}^{N-1} \alpha_i q_i)$ for suitable α_i. So

$$\sum_i q_i \cdot \nabla_{q_i} V_{jk}\left(\sum_\ell \alpha_\ell q_\ell\right) = \left(\sum_i \alpha_i q_i\right) \cdot (\nabla_r V_{jk})(r)\Big|_{r=\sum \alpha_\ell q_\ell} = (r \cdot \nabla_r V_{jk})(r)\Big|_{r=\sum \alpha_\ell q_\ell}$$

This is nonpositive by hypothesis. For arbitrary V_{jk}, we average with test functions and mimic the above argument. ∎

C. Local smoothness and wave operators for repulsive potentials

As a final topic in the theory of smooth operators, we shall discuss an extension of Theorem XIII.24. The "trouble" with that theorem is that its conclusion is too strong—H and H_0 are unitarily equivalent. In particular, the theorem cannot be applied to quantum Hamiltonians that have any pure point spectrum. We thus introduce a weaker notion than smoothness.

Definition Let H be a self-adjoint operator with spectral projections E_Ω. We say that A is *H-smooth on Ω*, a Borel set, if and only if AE_Ω is H-smooth.

Theorem XIII.30 Let H be self-adjoint with resolvent R and spectral projections, $\{E_\Omega\}$. Let $\Omega \subset \mathbb{R}$. Suppose that $D(A) \supset D(H)$ and that *either*

(a) $\sup_{0<|\varepsilon|<1, \lambda \in \Omega} |\varepsilon| \, \|AR(\lambda + i\varepsilon)\|^2 < \infty$

or

(b) $\sup_{0<\varepsilon<1, \lambda \in \Omega} \|AR(\lambda + i\varepsilon)A^*\| < \infty$

Then A is H-smooth on $\bar{\Omega}$, the closure of Ω.

Proof (a) For each $\varepsilon \neq 0$, $\|AR(\lambda + i\varepsilon)\|$ is continuous in λ, so the bound
$$C = \sup\{|\varepsilon| \|AR(\lambda + i\varepsilon)\|^2 \, | \, \lambda \in \Omega, \, 0 < |\varepsilon| < 1\}$$
extends to all $\lambda \in \bar{\Omega}$. Suppose $\lambda \in \mathbb{R}\backslash\bar{\Omega}$. Choose $\lambda_0 \in \bar{\Omega}$ with $|\lambda - \lambda_0| = \text{dist}(\lambda, \bar{\Omega})$. Then

$$|\varepsilon| \|AR(\lambda + i\varepsilon)E_\Omega \varphi\|^2$$
$$= |\varepsilon| \|AR(\lambda_0 + i\varepsilon)[I - (\lambda_0 - \lambda)R(\lambda + i\varepsilon)]E_{\bar{\Omega}} \varphi\|^2$$
$$\leq |\varepsilon| \|AR(\lambda_0 + i\varepsilon)\|^2 \|(I - (\lambda_0 - \lambda)R(\lambda + i\varepsilon))E_{\bar{\Omega}}\|^2 \|\varphi\|^2$$
$$\leq 4|\varepsilon| \|AR(\lambda_0 + i\varepsilon)\|^2 \|\varphi\|^2$$

since $|\lambda_0 - \lambda| |x - \lambda + i\varepsilon|^{-1} < 1$ for any $x \in \bar{\Omega}$. Thus
$$\sup_{\substack{\lambda \in \mathbb{R} \\ \varepsilon \neq 0}} |\varepsilon| \|AR(\lambda + i\varepsilon)E_{\bar{\Omega}}\| < \infty$$
so $AE_{\bar{\Omega}}$ is H-smooth by Theorem XIII.25 and the remark following its statement.

(b) Since $(AR(\lambda + i\varepsilon)A^*)^* = AR(\lambda - i\varepsilon)A^*$,
$$c = \sup_{\substack{\lambda \in \Omega \\ 0 < |\varepsilon| < 1}} \|AR(\lambda + i\varepsilon)A^*\| < \infty$$

Thus
$$\sup_{\substack{\lambda \in \Omega \\ 0 < |\varepsilon| < 1}} |\varepsilon| \|R(\lambda + i\varepsilon)A^*\|^2$$
$$= \sup_{\substack{\lambda \in \Omega \\ 0 < |\varepsilon| < 1}} |\varepsilon| \|AR(\lambda + i\varepsilon)R(\lambda - i\varepsilon)A^*\|$$
$$= \sup_{\substack{\lambda \in \Omega \\ 0 < |\varepsilon| < 1}} \tfrac{1}{2}\|A[R(\lambda + i\varepsilon) - R(\lambda - i\varepsilon)]A^*\| \leq c < \infty$$

Since $\|R(\lambda + i\varepsilon)A^*\| = \|AR(\lambda - i\varepsilon)\|$, the conditions of (a) hold. ∎

We can now state the extension of Theorem XIII.24.

Theorem XIII.31 Let H and H_0 be self-adjoint operators with spectral projections E_Ω and $E_\Omega^{(0)}$. Suppose that
$$H - H_0 = A^*B$$

XIII.7 Smooth perturbations

in the sense of (35). Suppose that A is H-bounded and H-smooth on some bounded open interval $\Omega \subset \mathbb{R}$ and that B is H_0-bounded and H_0-smooth on Ω. Then the limits

$$W_{\pm} = \operatorname*{s-lim}_{t \to \mp \infty} e^{iHt} e^{-iH_0 t} E_\Omega^{(0)}$$

and

$$\tilde{W}_{\pm} = \operatorname*{s-lim}_{t \to \mp \infty} e^{iH_0 t} e^{-iHt} E_\Omega$$

exist. Moreover,

$$W_{\pm}^* = \tilde{W}_{\pm} \tag{52}$$

$$\tilde{W}_{\pm} W_{\pm} = E_\Omega^{(0)}; \qquad W_{\pm} \tilde{W}_{\pm} = E_\Omega \tag{53}$$

Proof Suppose that we prove that $W_{\pm}$ exist and that $\operatorname{Ran} W_{\pm} \subset E_\Omega$. Then, by symmetry, and the fact that $H_0 - H = -B^*A$, $\tilde{W}_{\pm}$ exist and $\operatorname{Ran} \tilde{W}_{\pm} \subset E_\Omega^{(0)}$. (52) is obvious, and (53) follows from the fact that $W_{\pm}$ (respectively $\tilde{W}_{\pm}$) is a partial isometry with initial space $E_\Omega^{(0)}$ (respectively E_Ω). We shall show that $W_{\pm}$ exist and $\operatorname{Ran} W_{\pm} \subset E_\Omega$ by proving that $\operatorname*{s-lim}_{t \to \mp \infty} E_\Omega e^{iHt} e^{-iH_0 t} E_\Omega^{(0)}$ exists and

$$\operatorname*{s-lim}_{t \to \mp \infty} E_{\mathbb{R} \setminus \Omega} e^{iHt} e^{-iH_0 t} E_\Omega^{(0)} = 0 \tag{54}$$

The proof that the first limit exists is identical to the proof of Theorem XIII.24. It is obviously sufficient to prove (54) with $E_\Omega^{(0)}$ replaced by $E_I^{(0)}$ for an arbitrary compact subinterval $I \subset \Omega$. Given such an I, let C be the contour in Figure XIII.4. Then

$$E_{\mathbb{R} \setminus \Omega} e^{iHt} e^{-iH_0 t} E_I^{(0)} \varphi$$

$$= \frac{1}{2\pi i} \oint_C E_{\mathbb{R} \setminus \Omega} (H - z)^{-1} e^{iHt} e^{-iH_0 t} E_I^{(0)} \varphi \, dz$$

$$- \frac{1}{2\pi i} \oint_C E_{\mathbb{R} \setminus \Omega} e^{iHt} e^{-iH_0 t} (H_0 - z)^{-1} E_I^{(0)} \varphi \, dz$$

$$= \frac{1}{2\pi i} \oint_C E_{\mathbb{R} \setminus \Omega} e^{iHt} [(H - z)^{-1} - (H_0 - z)^{-1}] e^{-iH_0 t} E_I^{(0)} \varphi \, dz$$

Since this last integrand is uniformly bounded on C, it is sufficient, by the dominated convergence theorem, to prove that

$$\operatorname*{s-lim}_{t \to \mp \infty} [(H - z)^{-1} - (H_0 - z)^{-1}] e^{-iH_0 t} E_I^{(0)} = 0 \tag{55}$$

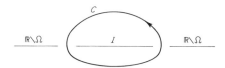

FIGURE XIII.4 A contour of integration.

for all nonreal z. But

$$|(\psi, [(H-z)^{-1} - (H_0-z)^{-1}]e^{-iH_0t}E_I^{(0)}\varphi)|$$
$$= |(A(H-z)^{-1}\psi, B(H_0-z)^{-1}e^{-iH_0t}E_I^{(0)}\varphi)|$$
$$\leq \|A(H-z)^{-1}\| \|B(H_0-z)^{-1}e^{-iH_0t}E_I^{(0)}\varphi\| \|\psi\|$$

Thus, to prove (55) we need only show

$$\lim_{t \to \pm\infty} \|B(H_0-z)^{-1}e^{-iH_0t}E_I^{(0)}\varphi\| = 0 \tag{56}$$

But, by the smoothness hypothesis, the function of t in (56) is square-integrable. Moreover, it has a uniformly bounded derivative, so (56) holds (Problem 62). ∎

Corollary Suppose that H and H_0 are self-adjoint with spectral projections E_Ω and $E_\Omega^{(0)}$ and that

$$H - H_0 = A^*B$$

where B is H_0-bounded and A is H-bounded. Let $S \subset \mathbb{R}$ with $S = \bigcup_{i=1}^\infty \Omega_i$ and each Ω_i a bounded open interval. Suppose that:

(i) A is H-smooth on each Ω_i and B is H_0-smooth on each Ω_i.
(ii) Both $\sigma(H)\backslash S$ and $\sigma(H_0)\backslash S$ have Lebesgue measure zero.

Then the generalized wave operators s-$\lim_{t \to \mp\infty} e^{iHt}e^{-iH_0t}E_{\rm ac}^{(0)}$ exist and are complete.

Proof Since $S = \bigcup_{i=1}^\infty \Omega_i$ and $E_{\rm ac}^{(0)} = E_S^{(0)}$, existence follows if we can prove that $\lim_{t \to \mp\infty} e^{iHt}e^{-iH_0t}\varphi$ exists for $\varphi \in \bigcup_{i=1}^\infty \operatorname{Ran} E_{\Omega_i}^{(0)}$. This is a direct consequence of the theorem. Since the inverse wave operators exist by symmetry, the wave operators are complete. ∎

One application of this corollary will be given in Section 8. Here we apply it to prove a result in the theory of repulsive potentials. This result is not the best possible (see the Notes).

XIII.7 Smooth perturbations

Theorem XIII.32 Let H be an operator of the form given in the corollary of Theorem XIII.29. Suppose that each V_{ij} is a function of $|x|$ obeying $|V_{ij}(x)| \leq C(1 + |x|)^{-\frac{3}{2}-\varepsilon}$ for each i and j and some $\varepsilon > 0$. Then the wave operators $\Omega^\pm = \text{s-lim}_{t \to \mp\infty} e^{iHt} e^{-iH_0 t}$ exist and are unitary.

Proof Since we already know that H has purely absolutely continuous spectrum, it is sufficient to prove that the wave operators exist and are complete. In the notation used in the proof of Theorem XIII.29, let

$$A_{jk} = +i\left[g(r_j - r_k)\frac{d}{dr_{jk}} + \frac{d}{dr_{jk}}g(r_j - r_k)\right], \qquad A = \sum_{j=k} A_{jk}$$

where $r_{jk} = |r_j - r_k|$ and d/dr_{jk} is defined by

$$r_{jk}\frac{d}{dr_{jk}} = (r_j - r_k) \cdot (\nabla_j - \nabla_k)$$

Now, $i[A_{ij}, V_{ij}] \geq 0$ as in the proof of Theorem XIII.29 and $i[A_{k\ell}, V_{ij}] = 0$ for i, j, k, ℓ distinct. Finally, for i, j, k distinct,

$$i[A_{ik} + A_{kj}, V_{ij}(r_{ij})] = \mathbf{a}(i, j, k) \cdot (\nabla V_{ij})(r_{ij})$$

where $\mathbf{a}(i, j, k) = g(r_{ik})\mathbf{e}_{ik} + g(r_{kj})\mathbf{e}_{kj}$ and $\mathbf{e}_{ik} = r_{ik}^{-1}(\mathbf{r}_i - \mathbf{r}_k)$. Since V_{ij} is only a function of $|x|$ and $(\mathbf{x} \cdot \nabla)V_{ij} \leq 0$, we can conclude that $i[A_{ik} + A_{kj}, V_{ij}] \geq 0$ if $(\mathbf{r}_i - \mathbf{r}_j) \cdot (\mathbf{a}(i, j, k)) \geq 0$. But

$$(\mathbf{r}_i - \mathbf{r}_j) \cdot \mathbf{a}(i, j, k) = r_{ik}g(r_{ik}) + r_{kj}g(r_{kj})$$
$$+ (\mathbf{e}_{ik} \cdot \mathbf{e}_{kj})(r_{ik}g(r_{kj}) + r_{kj}g(r_{ik}))$$
$$\geq (r_{ik} - r_{kj})(g(r_{ik}) - g(r_{jk}))$$
$$\geq 0$$

since g is monotone. It follows that $i[A, V] \geq 0$ so that computations identical to those used in Theorem XIII.29 show that (Problem 64):

$$i[A, H] \geq c(1 + r_{jk})^{-3-\varepsilon}$$

for a suitable constant c. Again, following the proof of Theorem XIII.29, we conclude that $(1 + r_{jk})^{-\frac{3}{2}-\varepsilon}(H + I)^{-1}$ is H-smooth. Since

$$\| |V_{ij}|^{3/5}(H + I)^{-1}e^{-iHt}\varphi\| \leq c^{1/2}\|(1 + r_{jk})^{-\frac{3}{2}-\varepsilon}(H + I)^{-1}e^{-iHt}\varphi\|$$

$|V_{ij}|^{3/5}(H + I)^{-1}$ is H-smooth and thus $|V_{ij}|^{3/5}$ is H-smooth on any compact set. By hypothesis, $|V_{ij}|^{2/5} \in L^{m+\varepsilon}(\mathbb{R}^m) \cap L^{m-\varepsilon}(\mathbb{R}^m)$, so by the proof of Theorem XIII.27, $|V_{ij}|^{2/5}$ is H_0-smooth. The result now follows from the previous corollary. ∎

XIII.8 The absence of singular continuous spectrum III: Weighted L^2 spaces

We have seen that a sufficient condition for $\sigma_{\text{sing}}(H)$ to be empty is that there be a dense set of vectors $X \subset \mathscr{H}$ so that $(\varphi, (H-z)^{-1}\varphi)$ remains bounded as $z \to \lambda \in \mathbb{R}$ for any $\varphi \in X$. There is a natural way of approaching this condition. Suppose that $X \subset \mathscr{H}$ is dense and that there is a norm $|||\cdot|||_+$ on X so that X is a Banach space and so that $|||\varphi|||_+ \geq \|\varphi\|$ for any $\varphi \in X$. Then the inner product on $\mathscr{H}$ allows one to realize $\mathscr{H}$ in a natural way as a subset of X^*. If $|||\cdot|||_-$ denotes the norm on X^*, then $\|\varphi\| \geq |||\varphi|||_-$ for any $\varphi \in \mathscr{H}$. We have already discussed such a situation and related ideas in Section VIII.6 and the appendix to Section XI.6. Let $z \in \mathbb{C}$ with Im $z \neq 0$. Then $(H-z)^{-1}$ takes $\mathscr{H}$ into $\mathscr{H}$ and so X into X^*. Of course the norm of $(H-z)^{-1}$ as a map from $\mathscr{H}$ to $\mathscr{H}$ diverges as z approaches $\sigma(H)$, but suppose that it remains bounded as a map of X into X^*. Then

$$|(\varphi, (H-z)^{-1}\varphi)| \leq |||\varphi|||_+^2 \, \|(H-z)^{-1}\|_{+,-}$$

where

$$\|A\|_{+,-} = \sup_{\psi \neq 0, \, \psi \in X} \|A\psi\|_- / \|\psi\|_+$$

so we can conclude that $\sigma_{\text{sing}}(H) = \emptyset$, by Theorem XIII.20.

It is natural to try to implement this idea with perturbation techniques. That is, we consider $H = H_0 + V$ and begin by proving bounds on $(H_0 - z)^{-1}$ as a map from X to X^*. We shall consider Schrödinger operators and take $H_0 = -\Delta$. To motivate our choice of X, let $f \in \mathscr{S}(\mathbb{R}^n)$ and consider

$$\lim_{y \downarrow 0} (f, [(H_0 - x + iy)^{-1} - (H_0 - x - iy)^{-1}]f)$$
$$= \lim_{y \downarrow 0} \int |\hat{f}(p)|^2 2i \, \text{Im}[(p^2 - x + iy)^{-1}] \, d^n p$$
$$= -2i\pi \int |\hat{f}(p)|^2 \, \delta(p^2 - x) \, d^n p$$

by Eq. (V.4). Thus, for $(f, (H_0 - x + iy)^{-1}f)$ to have a limit as $y \downarrow 0$, $\hat{f}$ must have a "natural" restriction to the sphere of radius $x^{1/2}$. Because of our discussion in Section IX.9, it is natural to choose X as an L^2_δ space (with $\delta > \tfrac{1}{2}$) so that X^* is $L^2_{-\delta}$:

$$L^2_\delta(\mathbb{R}^n) = \left\{ f(x) \, \middle| \, \|f\|_\delta^2 \equiv \int (1+x^2)^\delta |f(x)|^2 \, dx < \infty \right\}$$

We shall lean heavily on the estimates proven in Section IX.9, especially Theorems IX.39 and IX.41.

XIII.8 Weighted L^2 spaces

Definition A multiplication operator is called an **Agmon potential** if $V(x) = (1 + x^2)^{-\frac{1}{2} - \varepsilon} W(x)$ for some $\varepsilon > 0$ and some W that is a relatively compact perturbation of $-\Delta$.

The Agmon potentials form a vector space of $-\Delta$-bounded perturbations of relative bound zero (see Problem 20 of Chapter X).

Example 1 Let $p > n/2$, $p < \infty$, and $p \geq 2$. Then any $W \in L^p(\mathbb{R}^n)$ is relatively compact (Problem 41), so any V with $(1 + x^2)^{\frac{1}{2} + \varepsilon} V \in L^p$ is an Agmon potential.

Example 2 Let $W \in L^\infty(\mathbb{R}^n)$ and define $V(x) = (1 + x^2)^{-\frac{1}{2} - \varepsilon} W(x)$. To show that V is an Agmon potential we need only prove that $U(x) = (1 + x^2)^{-\varepsilon/2} W(x)$ is relatively compact. But this is true since $U(x)$ is in $L^p + (L^\infty)_\varepsilon$ for all p (Problem 41).

The main theorem of this section is:

Theorem XIII.33 (the Agmon–Kato–Kuroda theorem) Let V be an Agmon potential and let $H = H_0 + V$ where $H_0 = -\Delta$. Then:

(a) The set $\mathscr{E}_+$ of positive eigenvalues of H is a discrete subset of $(0, \infty)$, and each eigenvalue has finite multiplicity.
(b) If C is any compact subinterval of $(0, \infty) \backslash \mathscr{E}_+$ and if $\delta > \frac{1}{2}$, then
$$\sup_{\lambda \in C,\, 0 < y < 1} \sup_{\psi,\, \varphi \in L_\delta^2;\, \|\varphi\|_\delta \leq 1,\, \|\psi\|_\delta \leq 1} |(\psi, (H - \lambda - iy)^{-1} \varphi)| < \infty$$
(c) $\sigma_{\text{sing}}(H) = \varnothing$.
(d) The wave operators $\Omega^\pm(H, H_0)$ exist and are complete.

The proof of Theorem XIII.33, while elegant, is rather long so we break it up into a series of lemmas. After proving (a), we develop a few technical estimates that will allow us to prove (c) and (d) from (b); then we prove (b) in case $V = 0$; finally, we use the estimates proven for H_0, Theorem IX.39, Theorem IX.41, and a bootstrap argument to complete the proof. Throughout we let $\rho(x) \equiv (1 + x^2)^{1/2}$. The use of the weighted L^2 spaces and Theorem IX.39 is illustrated in:

Proof of (a) of Theorem XIII.33 Suppose that $\varphi \in D(H)$ and $H\varphi = \lambda \varphi$ with $\lambda > 0$. First, we shall show that $\|\varphi\|_\delta \leq c \|\varphi\|$ for some $\delta > 0$ where c depends only on λ and remains bounded as λ varies through compact subsets of

(0, ∞). Since W is H_0-compact, it is H-bounded so that $\|W\varphi\| \leq a\|H\varphi\| + b\|\varphi\| = (a\lambda + b)\|\varphi\|$. As a result $\psi \equiv V\varphi = \rho^{-1-\varepsilon}W\varphi$ is in $L^2_{1+\varepsilon}$. In particular, by Theorem IX.40, $\hat{\psi}$ has restrictions to each sphere S_E, and the restriction is Hölder continuous in E. Since $(H_0 - \lambda)\varphi = -\psi$, $\hat{\varphi} = -(k^2 - \lambda)^{-1}\hat{\psi}$. If $\hat{\psi} \upharpoonright S_{\sqrt{\lambda}}$ were not identically zero, then $\hat{\varphi}$ could not be in L^2, so we conclude that $\hat{\psi} \upharpoonright S_{\sqrt{\lambda}} = 0$. As a result, Theorem IX.41 is applicable, so

$$\|\varphi\|_{\varepsilon/2} \leq c_\lambda \|\psi\|_{1+\varepsilon} = c_\lambda \|W\varphi\| \leq d_\lambda \|\varphi\|$$

Let $\eta \equiv \rho^{\varepsilon/2}(-\Delta + 1)\varphi$. Then $H\varphi = \lambda\varphi$ implies that

$$\|\eta\| \leq \|\rho^{\varepsilon/2}(\lambda + 1)\varphi\| + \|\rho^{\varepsilon/2}V\varphi\|$$
$$\leq (\lambda + 1)\|\varphi\|_{\varepsilon/2} + \|W\varphi\| \leq c'_\lambda \|\varphi\|$$

Since $\varphi = (-\Delta + 1)^{-1}\rho^{-\varepsilon}\eta$, we conclude that for any compact subset K in $(0, \infty)$, any solution of $H\varphi = \lambda\varphi$ with $\lambda \in K$ and $\|\varphi\| = 1$ is of the form $\varphi = A\eta$ where: (i) $A = (-\Delta + 1)^{-1}\rho^{-\varepsilon}$ and (ii) $\|\eta\| \leq c$, a constant only depending on K. By Problem 41, A is a compact operator, so the set $M = \{\varphi = A\eta \mid \|\eta\| \leq c\}$ is compact. If any eigenvalue $\lambda \in K$ were of infinite multiplicity or if there were infinitely many eigenvalues in K, M would contain an infinite orthonormal set. Since M is compact, K contains only finitely many eigenvalues and each is of finite multiplicity. ∎

We note that we shall show that $\mathscr{E}_+ = \varnothing$ under some additional regularity hypotheses on V in Section 13.

Lemma 1 Let F and G be any two real-valued multiplication operators that are H_0-bounded with relative bound zero. Then for any $\mu \in \mathbb{C}\setminus\mathbb{R}$, $(H_0 - \mu)^{-1}$, $(H_0 + G - \mu)^{-1}$, $F(H_0 - \mu)^{-1}$, and $F(H_0 + G - \mu)^{-1}$ are bounded on each L^2_δ. Moreover, if F is also H_0-relatively compact, then $F(H_0 - \mu)^{-1}$ and $F(H_0 + G - \mu)^{-1}$ are compact on each L^2_δ.

Proof We shall prove the lemma for $(H_0 - \mu)^{-1}$ in the case $|\delta| \leq 1$. The other cases are similar and are left to the problems (Problem 66). Introduce the symbol ∂_j for the operator $\partial/\partial x_j$ and consider the formal computation

$$[(H_0 - \mu)^{-1}, \rho^\delta] = -(H_0 - \mu)^{-1}[H_0, \rho^\delta](H_0 - \mu)^{-1}$$
$$= \sum_{i=1}^n \{(H_0 - \mu)^{-1}\,\partial_i\}(\partial_i\rho^\delta)(H_0 - \mu)^{-1}$$
$$+ (H_0 - \mu)^{-1}(\partial_i\rho^\delta)\{\partial_i(H_0 - \mu)^{-1}\}$$

Applied to vectors in $\mathscr{S}$, all computations are legitimate. Moreover, if $\delta \leq 1$, $\partial_i(\rho^\delta)$ is bounded. Since $(H_0 - \mu)^{-1}$ and $\partial_i(H_0 - \mu)^{-1}$ are bounded on L^2 we conclude that

$$\|[(H_0 - \mu)^{-1}, \rho^\delta]\psi\| \leq \text{const}\|\psi\|$$

if ψ is in $\mathscr{S}$ and so for arbitrary ψ in L^2. Suppose $1 \geq \delta > 0$. Then

$$\|(H_0 - \mu)^{-1}\psi\|_\delta = \|\rho^\delta(H_0 - \mu)^{-1}\psi\|$$
$$\leq \|(H_0 - \mu)^{-1}\| \|\psi\|_\delta + \|[(H_0 - \mu)^{-1}, \rho^\delta]\psi\|$$
$$\leq d\|\psi\|_\delta$$

so $(H_0 - \mu)^{-1}$ is bounded from L^2_δ to L^2_δ. By duality, $(H_0 - \bar\mu)^{-1}$ is bounded from $L^2_{-\delta}$ to $L^2_{-\delta}$. ∎

Lemma 2 Let $H = H_0 + V$ as in Theorem XIII.33. Suppose that F is an Agmon potential and that for some compact interval $\Omega \subset \mathbb{R}$, and each $\delta > \frac{1}{2}$,

$$\sup_{x \in \Omega, \, 0 < y < 1} \sup_{\psi, \varphi \in L_\delta^2; \, \|\varphi\|_\delta \leq 1, \, \|\psi\|_\delta \leq 1} |(\psi, (H - \lambda - iy)^{-1}\varphi)| < \infty \quad (57)$$

Then $|F|^{1/2}$ is H-smooth on Ω.

Proof By Theorem XIII.30, we need only prove that

$$\sup_{\substack{\lambda \in \Omega \\ 0 < y < 1}} \| |F|^{1/2}(H - \lambda - iy)^{-1}|F|^{1/2}\| < \infty$$

(57) implies that $(H - \lambda - iy)^{-1}$ is uniformly bounded from L^2_δ to $L^2_{-\delta}$ for $\lambda + iy \in \Omega \times (0, 1)$. Write $F = \rho^{-1-\varepsilon}G$ where G is H_0-compact and let $\delta = \frac{1}{2} + (\varepsilon/2)$. Then, by Lemma 1, $(H - i)^{-1}|G|^{1/2}$ is bounded from L^2_δ to L^2_δ so $|G|^{1/2}(H - i)^{-1}(H - \lambda - iy)^{-1}(H - i)^{-1}|G|^{1/2}$ is uniformly bounded from L^2_δ to $L^2_{-\delta}$ for $\lambda + iy \in \Omega \times (0, 1)$. Thus

$$|F|^{1/2}(H - i)^{-1}(H - \lambda - iy)^{-1}(H - i)^{-1}|F|^{1/2}$$

is uniformly bounded from L^2 to L^2 for $\lambda + iy \in \Omega \times (0, 1)$. Writing

$$(H - z)^{-1} = (H - i)^{-1} + (z - i)(H - i)^{-2}$$
$$+ (z - i)^2(H - i)^{-1}(H - z)^{-1}(H - i)^{-1}$$

and using the fact that $\| |F|^{1/2}(H - i)^{-1}|F|^{1/2}\| < \infty$, we see that

$$\sup_{\lambda + iy \in \Omega \times (0, 1)} \| |F|^{1/2}(H - \lambda - iy)^{-1}|F|^{1/2}\| < \infty. \quad ∎$$

Reduction of Theorem XIII.33 to the proof of (b)   We want to show that once we prove (b), then (c) and (d) follow. Suppose that (b) holds, that $\varphi \in L_\delta^2$ for $\delta > \frac{1}{2}$, and that K is a compact subinterval of $(0, \infty)\backslash\mathscr{E}_+$. Then $E_K \varphi \in \mathscr{H}_{ac}$ by Theorem XIII.20. Thus $K \cap \sigma_{sing} = \varnothing$. By the second corollary to Theorem XIII.14, $\sigma_{ess} = [0, \infty)$, so we conclude that $\sigma_{sing} \cap (-\infty, 0) = \varnothing$. Therefore, $\sigma_{sing} \subset (\mathscr{E}_+ \cup \{0\})$. $\mathscr{E}_+$ is countable by (a), so $\sigma_{sing} = \varnothing$. Applying Lemma 2 to the case $V = 0$, we see that $|V|^{1/2}$ is H_0-smooth on Ω for any interval $[a, b]$ with $a > 0$. In addition, $|V|^{1/2}$ is H-smooth on any such Ω with $\Omega \cap \mathscr{E}_+ = \varnothing$. (d) now follows from the corollary of Theorem XIII.31. ∎

The remainder of this section is devoted to proving (b). We first prove the case $V = 0$, which is an estimate quite similar to Theorem IX.41. The proof is also quite similar.

Lemma 3   Let $\varepsilon > 0$. Then there exists a constant c so that for all $\lambda \in \mathbb{C}$ and all $\varphi \in \mathscr{S}(\mathbb{R})$

$$\|\varphi\|_{-\frac{1}{2}-\varepsilon} \leq c \left\|\left(\frac{d}{dx} - \lambda\right)\varphi\right\|_{\frac{1}{2}+\varepsilon} \tag{58}$$

Proof   Suppose that Re $\lambda \leq 0$. Let $\psi = (d/dx - \lambda)\varphi$. Then

$$\varphi(x) = \int_{-\infty}^{x} e^{\lambda(x-y)} \psi(y) \, dy$$

Thus

$$\|\varphi\|_{L^\infty} \leq \|\psi\|_{L^1} = ((1+x^2)^{-\frac{1}{4}-\frac{1}{2}\varepsilon}, (1+x^2)^{\frac{1}{4}+\frac{1}{2}\varepsilon}\psi)$$
$$\leq c_1 \|\psi\|_{\frac{1}{2}+\varepsilon}$$

Since

$$\|\varphi\|_{-\frac{1}{2}-\varepsilon} \leq \|\varphi\|_{L^\infty} \|(1+x^2)^{-\frac{1}{4}-\frac{1}{2}\varepsilon}\|_{L^2}$$

(58) follows in the case Re $\lambda \leq 0$. A similar argument works for Re $\lambda \geq 0$. ∎

Lemma 4   Let n be fixed. Then for all $\varepsilon > 0$, there is a constant d so that for all $\lambda \in \mathbb{C}$, each $j = 1, \ldots, n$, and all $\varphi \in \mathscr{S}(\mathbb{R}^n)$,

$$\|\partial_j \varphi\|_{-\frac{1}{2}-\varepsilon} \leq d \|(-\Delta - \lambda)\varphi\|_{\frac{1}{2}+\varepsilon} \tag{59}$$

Proof Consider first the case $n = 1$. Let $\lambda = -\mu^2$. Then by Lemma 3,

$$\left\| \frac{d}{dx} \varphi \right\|_{-\frac{1}{2}-\varepsilon} \leq \frac{1}{2} \left\| \left(\frac{d}{dx} - \mu \right) \varphi \right\|_{-\frac{1}{2}-\varepsilon} + \frac{1}{2} \left\| \left(\frac{d}{dx} + \mu \right) \varphi \right\|_{-\frac{1}{2}-\varepsilon}$$

$$\leq c \left\| \left(-\frac{d^2}{dx^2} - \lambda \right) \varphi \right\|_{\frac{1}{2}+\varepsilon}$$

Now let $\varphi \in \mathscr{S}(\mathbb{R}^n)$ and let $\psi(x_1, k_2, \ldots, k_n)$ be the partial Fourier transform of φ with respect to $x_2, \ldots, x_n$,

$$\psi(x_1, k_2, \ldots, k_n) = (2\pi)^{-(n-1)/2} \int e^{-i \Sigma_2^n k_j x_j} \varphi(x_1, \ldots, x_n) \, dx_2 \cdots dx_n$$

Using the one-dimensional result, for each fixed $k_2, \ldots, k_n$,

$$\int (1 + x_1^2)^{-\frac{1}{2}-\varepsilon} |\partial_1 \psi(x_1, k_2, \ldots, k_n)|^2 \, dx_1$$

$$\leq c \int (1 + x_1^2)^{\frac{1}{2}+\varepsilon} |(-\partial_1^2 + k_2^2 + \cdots + k_n^2 - \lambda)\psi|^2 \, dx_1$$

Integrating with respect to $k_2, \ldots, k_n$ and using the Plancherel theorem, we see that

$$\int (1 + x_1^2)^{-\frac{1}{2}-\varepsilon} |\partial_1 \varphi(x)|^2 \, d^n x \leq c \int (1 + x_1^2)^{\frac{1}{2}+\varepsilon} |(-\Delta - \lambda)\varphi|^2 \, d^n x$$

Since $(1 + x^2)^{-\frac{1}{2}-\varepsilon} \leq (1 + x_1^2)^{-\frac{1}{2}-\varepsilon}$ and $(1 + x_1^2)^{\frac{1}{2}+\varepsilon} \leq (1 + x^2)^{\frac{1}{2}+\varepsilon}$, (59) follows. ∎

Lemma 5 Let n be fixed. For any compact set $K \subset \mathbb{C}\setminus\{0\}$, there is a constant c so that for all $\lambda \in K$ and $\varphi \in \mathscr{S}(\mathbb{R}^n)$,

$$\|\varphi\|_{-\frac{1}{2}-\varepsilon} \leq c \|(-\Delta - \lambda)\varphi\|_{\frac{1}{2}+\varepsilon} \tag{60}$$

Proof We can find $c_1 > 0$ satisfying

$$\inf_{x \in \mathbb{R}, \lambda \in K} [|x^2 - \lambda|^2 + |x|^2] \geq c_1^{-2}$$

Therefore, given $\psi \in \mathscr{S}(\mathbb{R}^n)$

$$|\hat{\psi}(k_1, \ldots, k_n)|^2 \leq c_1^2 \left[|(\sum k_i^2 - \lambda)\hat{\psi}(k_1, \ldots, k_n)|^2 + \sum_{i=1}^n |k_i \hat{\psi}(k_1, \ldots, k_n)|^2 \right]$$

Integrating and using the Plancherel theorem

$$\|\psi\|^2 \leq c_1^2 \left[\|(-\Delta - \lambda)\psi\|^2 + \sum_{j=1}^n \|\partial_j \psi\|^2 \right] \tag{61a}$$

for all $\lambda \in K$. Let α be a positive real number which we shall adjust below. Set $\delta = \frac{1}{2} + \varepsilon$ and $\rho_\alpha = (1 + \alpha x^2)^{1/2}$. Finally, let $\varphi = (\rho_\alpha)^\delta \psi$. Then, by (61a)

$$\|\rho_\alpha^{-\delta}\varphi\| \leq c_1 \|(-\Delta - \lambda)\rho_\alpha^{-\delta}\varphi\| + c_1 \sum_{j=1}^n \|\partial_j \rho_\alpha^{-\delta}\varphi\| \tag{61b}$$

Now $\partial_j \rho_\alpha^{-\delta}\varphi = \rho_\alpha^{-\delta} \partial_j \psi - \delta \alpha x_j \rho_\alpha^{-\delta-1}\varphi$ so

$$\|\partial_j \rho_\alpha^{-\delta}\varphi\| \leq \|\rho_\alpha^{-\delta} \partial_j \varphi\| + \alpha^{1/2} \delta \|\rho_\alpha^{-\delta}\varphi\|$$

since $|\alpha^{1/2} x_j \rho^{-1}| \leq 1$ for all x. Similarly (Problem 67),

$$\|(-\Delta - \lambda)\rho_\alpha^{-\delta}\varphi\| \leq \|\rho_\alpha^{-\delta}(-\Delta - \lambda)\varphi\| + 2\delta\alpha^{1/2} \sum_{j=1}^n \|\rho_\alpha^{-\delta} \partial_j \varphi\|$$

$$+ d_{n,\varepsilon} \alpha \|\rho_\alpha^{-\delta}\varphi\| \tag{62}$$

where $d_{n,\varepsilon}$ is only dependent on ε and n. Pick α so small that $c_1(n \delta \alpha^{1/2} + d_{n,\varepsilon} \alpha) < \frac{1}{2}$ and $\alpha < 1$. Then, by (61), for all $\varphi \in \mathscr{S}$ and $\lambda \in K$,

$$\tfrac{1}{2}\|\rho_\alpha^{-\delta}\varphi\| \leq c_2 \left[\|\rho_\alpha^{-\delta}(-\Delta - \lambda)\varphi\| + \sum_{j=1}^n \|\rho_\alpha^{-\delta} \partial_j \varphi\| \right]$$

Since $\rho^{-\delta} \leq \rho_\alpha^{-\delta} \leq \alpha^{-\delta/2}\rho^{-\delta}$, we have that

$$\|\varphi\|_{-\delta} \leq c_3 [\|(-\Delta - \lambda)\varphi\|_{-\delta} + \sum_j \|\partial_j \varphi\|_{-\delta}]$$

where c_3 is independent of $\lambda \in K$ and $\varphi \in \mathscr{S}$. Since $\|\cdot\|_{-\delta} \leq \|\cdot\|_\delta$ and $\|\partial_j \varphi\|_{-\delta} \leq c\|(-\Delta - \lambda)\varphi\|_\delta$ by Lemma 4, (60) follows. ∎

If now Im $\lambda \neq 0$ and $\delta > \frac{1}{2}$, then by Lemma 1, $(H_0 - \lambda)^{-1}$ is a bounded map from L_δ^2 to L_δ^2. Lemma 5 assures us that, for $\lambda \in K = [a, b] \times (0, 1]$ ($a > 0$), we have the basic estimate

$$\|(H_0 - \lambda)^{-1}\varphi\|_{-\delta} \leq c\|\varphi\|_\delta \tag{63}$$

Given (63), it is natural to consider the boundary values $\lim_{y \downarrow 0} (H_0 - x - iy)^{-1}$ as maps of L_δ^2 to $L_{-\delta}^2$. Such boundary values are not strictly necessary for the proof, but they help to make it more conceptual, so we introduce them. As preparation, we need

Lemma 6 Let $\delta > \frac{3}{2}$ and let $0 < a < b$. Then there is a constant c so that for all $\varphi \in L_\delta^2$ and $\lambda = x + iy$ with $x \in [a, b]$ and $y \in (0, 1]$,

$$\|(H_0 - \lambda)^{-2}\varphi\|_{-\delta} \leq c\|\varphi\|_\delta$$

Proof Let A be the operator $\sum_{j=1}^{n} x_j \partial_j$. Then $[A, (H_0 - \lambda)] = -2H_0$ so that

$$[A, (H_0 - \lambda)^{-1}] = -(H_0 - \lambda)^{-1}[A, (H_0 - \lambda)](H_0 - \lambda)^{-1}$$
$$= 2H_0(H_0 - \lambda)^{-2}$$
$$= 2(H_0 - \lambda)^{-1} + 2\lambda(H_0 - \lambda)^{-2}$$

where all the computations are legitimate when applied to vectors in $\mathscr{S}(\mathbb{R}^n)$. Since $(H_0 - \lambda)^{-1}$ is uniformly bounded from L_δ^2 to $L_{-\delta}^2$, we need only prove that $[A, (H_0 - \lambda)^{-1}]$ is bounded from L_δ^2 to $L_{-\delta}^2$, uniformly for λ satisfying $a \leq \operatorname{Re} \lambda \leq b$, $0 < \operatorname{Im} \lambda \leq 1$. It is thus sufficient to prove that $\rho^{-\delta}(x_j \partial_j) \times (H_0 - \lambda)^{-1}\rho^{-\delta}$ is bounded on L^2 uniformly in λ. Write

$$(H_0 - \lambda)^{-1} = (H_0 + 1)^{-1} + (\lambda + 1)(H_0 + 1)^{-1}(H_0 - \lambda)^{-1}$$

Certainly $(\rho^{-\delta}x_j)[\partial_j(H_0 + 1)^{-1}]\rho^{-\delta}$ is bounded on L^2. Moreover

$$\rho^{-1}x_j[\rho^{-\delta+1}(\partial_j(H_0 + 1)^{-1})\rho^{\delta-1}][\rho^{-\delta+1}(H_0 - \lambda)^{-1}\rho^{-\delta+1}]\rho^{-1}$$

is bounded for the first and last factor are trivially bounded, the third is bounded by Lemma 5, and the second is bounded by mimicking the proof of Lemma 1 (Problem 66c). ∎

Lemma 7 Let $\delta > \frac{1}{2}$ and let $x > 0$. Then $(H_0 - x - i0)^{-1} \equiv \lim_{y \downarrow 0} (H_0 - x - iy)^{-1}$ exists in norm as a map from L_δ^2 to $L_{-\delta}^2$. Moreover:

(a) $V(H_0 - x - i0)^{-1}$ is compact as a map of L_δ^2 to L_δ^2 if V is an Agmon potential such that $\rho^{2\delta}V = W$ is relatively H_0-compact.
(b) $\operatorname{Im}(\varphi, (H_0 - x - i0)^{-1}\varphi) = (\pi/2)x^{\frac{1}{2}n-1} \int_{S^{n-1}} |\hat{\varphi}(x^{1/2}\Omega)|^2 \, d\Omega$ where $\hat{\varphi} \upharpoonright S_{x^{1/2}}$ is defined by Theorem IX.39, and $d\Omega$ is the usual surface measure on the sphere.

Proof Let $\delta' = \delta + 1$. Then as operators from $L_{\delta'}^2$ to $L_{-\delta'}^2$,

$$\|(H_0 - \lambda_1)^{-1} - (H_0 - \lambda_2)^{-1}\|$$
$$\leq |\lambda_2 - \lambda_1| \sup_{0 \leq t \leq 1} \|[H_0 - t\lambda_1 - (1-t)\lambda_2]^{-2}\|$$

By Lemma 6, we see that $(H_0 - x - iy)^{-1}$ is norm Cauchy as $y \downarrow 0$. Let $\delta'' = \frac{1}{2}(\delta + \frac{1}{2})$. Then by Lemma 5, $(H_0 - x - iy)^{-1}$ is norm bounded as $y \downarrow 0$ as a map from $L_{\delta''}^2$ to $L_{-\delta''}^2$. Since $\delta'' < \delta < \delta'$, we can interpolate between the

δ' and δ'' results (see Example 3 of the appendix to Section IX.4) and conclude that as maps from L_δ^2 to $L_{-\delta}^2$, $(H_0 - x - iy)^{-1}$ is norm Cauchy as $y \downarrow 0$. To prove (a) we write

$$W(H_0 - x - i0)^{-1} = W(H_0 + 1)^{-1}$$
$$+ (x + 1)W(H_0 + 1)^{-1}(H_0 - x - i0)^{-1}$$

and so conclude by Lemma 1 that $W(H_0 - x - i0)^{-1}$ is compact as a map from L_δ^2 to $L_{-\delta}^2$. Since $\rho^{-2\delta}$ is an isometry from $L_{-\delta}^2$ to L_δ^2, (a) follows. Finally, (b) holds for $\varphi \in \mathscr{S}$ because of (V.4) (see Problem 22 of Chapter V). By Theorem IX.39, it extends to all $\varphi \in L_\delta^2$. ∎

Lemma 8 Let $\delta > \tfrac{1}{2}$ and let $\varphi \in L_\delta^2$ satisfy $\varphi = -V(H_0 - x - i0)^{-1}\varphi$ where $x > 0$, and V is an Agmon potential so that $\rho^{2\delta}V = W$ is relatively H_0-compact, and where $V(H_0 - x - i0)^{-1}$ is interpreted as the composition of maps $W(H_0 - x - i0)^{-1}$ from L_δ^2 to $L_{-\delta}^2$ and $\rho^{-2\delta}$ from $L_{-\delta}^2$ to L_δ^2. Then:

(a) $\psi \equiv (H_0 - x - i0)^{-1}\varphi$ is in L^2.
(b) If $\varphi \neq 0$, then x is an eigenvalue of $H = H_0 + V$ as an operator on L^2.

Proof The argument is very similar to the proof of (a) of Theorem XIII.33. By the formula

$$(H_0 - x - i0)^{-1} = (H_0 + 1)^{-1} + (x + 1)(H_0 + 1)^{-1}(H_0 - x - i0)^{-1}$$

we see that $\psi \in (H_0 + 1)^{-1}[L_{-\delta}^2]$ so that $W\psi \in L_{-\delta}^2$ by Lemma 1. As a result, the integral $\int V(\xi)|\psi(\xi)|^2 \, d\xi = (\rho^{-\delta}\psi, \rho^{-\delta}W\psi)$ is absolutely convergent and obviously real. But $V\psi = -\varphi$ so we conclude that $(\varphi, (H_0 - x - i0)^{-1}\varphi)$ is real. By Lemma 7, part (b), $\hat{\varphi} \restriction S_{x^{1/2}} \equiv 0$. As a result, Theorem IX.41 is applicable and we have the following "bootstrap" argument: Let $\delta = \tfrac{1}{2} + \varepsilon$. Since $\varphi \in L_\delta^2$, $\psi \in L_{\delta-1-\varepsilon}^2$ by Theorem IX.41. Using Lemma 1 and

$$W\psi = W(H_0 + 1)^{-1}\varphi + (x + 1)W(H_0 + 1)^{-1}\psi$$

we see that $W\psi \in L_{\delta-1-\varepsilon}^2$ also. Thus $\varphi = -V\psi = -\rho^{2\delta}W\psi$ is in $L_{\delta-1-\varepsilon+2\delta}^2 = L_{\delta+\varepsilon}^2$. By this argument we have improved the estimate $\varphi \in L_\delta^2$ to $\varphi \in L_{\delta+\varepsilon}^2$. There is nothing to stop us from repeating it! Thus, $\varphi \in L_{\delta+n\varepsilon}^2$ for all n so $\psi \in L_{\delta-1+(n-1)\varepsilon}^2$ for all n. In particular $\psi \in L^2$. For $\eta \in \mathscr{S}$,

$$(H_0\eta, \psi) = \lim_{y \downarrow 0} (\eta, H_0(H_0 - x - iy)^{-1}\varphi) = \lim_{y \downarrow 0} (\eta, (x + iy)\psi + \varphi)$$
$$= (\eta, x\psi + \varphi)$$

Therefore, $\psi \in D(H_0)$ and $H_0\psi = x\psi + \varphi = x\psi - V\psi$ so x is an eigenvalue of H as an operator on L^2. ∎

Completion of the proof of Theorem XIII.33 Choose $\delta > \frac{1}{2}$ so that $\rho^{2\delta} V = W$ is relatively H_0-compact. Given a compact subinterval $K \subset (0, \infty) \backslash \mathscr{E}_+$, consider the operator-valued function $A(\mu) = V(H_0 - \mu)^{-1}$ on $K \times [0, 1]$ where $(H_0 - \mu)^{-1}$ is interpreted as $(H_0 - x - i0)^{-1}$ if Im $\mu = 0$. Then $A(\mu)$ is a function with values in the compact operators on L_δ^2, continuous on $K \times [0, 1]$ and analytic in its interior. Moreover, $A(\mu)\varphi = -\varphi$ has no nonzero solutions for $\mu \in K \times [0, 1]$. When Im $\mu = 0$, this follows from the hypothesis $K \cap \mathscr{E}_+ = \varnothing$ and Lemma 8. For Im $\mu \neq 0$, this follows from the facts that $(1 + V(H_0 - \mu)^{-1})(H_0 - \mu) = H - \mu$ and that both $H_0 - \mu$ and $H - \mu$ are invertible as maps of L_δ^2 to L_δ^2 by Lemma 1. It follows by a simple extension of the analytic Fredholm theorem (Theorem VI.14) that $(1 + A(\mu))^{-1}$ is a continuous function on $K \times [0, 1]$; in particular, it is uniformly bounded. But for Im $\mu \neq 0$, $(H - \mu)^{-1} = (H_0 - \mu)^{-1} \times (1 + A(\mu))^{-1}$. Since $(H_0 - \mu)^{-1}$ is uniformly bounded from L_δ^2 to $L_{-\delta}^2$ by Lemma 5 and $(1 + A(\mu))^{-1}$ is uniformly bounded from L_δ^2 to L_δ^2, $(H - \mu)^{-1}$ is uniformly bounded from L_δ^2 to $L_{-\delta}^2$ for $\mu \in K \times [0, 1]$. This is just a rephrasing of (b). ∎

XIII.9 The spectrum of tensor products

In Section VIII.10 we proved that if A and B are self-adjoint operators on Hilbert spaces $\mathscr{H}_1$ and $\mathscr{H}_2$ with domains D_1 and D_2, then $A \otimes I + I \otimes B$ is essentially self-adjoint on $D_1 \otimes D_2$ and $\sigma(A \otimes I + I \otimes B) = \sigma(A) + \sigma(B)$. The purpose of this section is to extend this result to the case where A and B are m-sectorial (see Section VIII.6). We shall exploit the connection between m-sectorial operators and bounded holomorphic semigroups (see Section X.8). In the self-adjoint case the proof was "easy" in the sense that it was a relatively straightforward consequence of the spectral theorem. In the m-sectorial case the use of the spectral theorem is replaced by the theory of commutative Banach algebras and the Laplace transform formulas relating bounded holomorphic semigroups to the resolvents of their generators. The reader is referred to the Notes for references about commutative Banach algebras. This section is divided into two parts. In the first part we prove that $\sigma(A \otimes B) = \sigma(A) \otimes \sigma(B)$ if A and B are bounded operators on $\mathscr{H}_1$ and $\mathscr{H}_2$. In the second part we apply this result to $e^{-tA} \otimes e^{-tB}$ where A and B are the generators of bounded holomorphic semigroups in order to obtain the desired result for m-sectorial operators.

Let A be a bounded operator on a Hilbert space $\mathscr{H}$. We shall denote by $\mathscr{R}(A)$ the Banach algebra of operators generated by the identity and all the resolvents of A, i.e., $\mathscr{R}(A)$ is just the closure in the operator norm of the family of polynomials in finitely many variables in the resolvents of A at different points. $\mathscr{R}(A)$ is a commutative Banach algebra that contains A since $\lambda^2(\lambda - A)^{-1} - \lambda I \to A$ in norm as $\lambda \to \infty$. Since all the resolvents of A are in $\mathscr{R}(A)$ we have $\sigma(A) = \sigma_{\mathscr{R}(A)}(A)$ where $\sigma_{\mathscr{R}(A)}(A)$ denotes the Gel'fand spectrum of A with respect to $\mathscr{R}(A)$.

Let $\mathscr{R}(A)$ and $\mathscr{R}(B)$ be the resolvent algebras of A and B on $\mathscr{H}_1$ and $\mathscr{H}_2$, respectively. Then if $C \in \mathscr{R}(A)$ and $D \in \mathscr{R}(B)$, $C \otimes D$ is a well-defined bounded operator on $\mathscr{H}_1 \otimes \mathscr{H}_2$ and $\|C \otimes D\| = \|C\| \|D\|$ (see the second proposition in Section VIII.10). We denote by $\mathscr{A}$ the norm closure in $\mathscr{L}(\mathscr{H}_1 \otimes \mathscr{H}_2)$ of the finite linear combinations of such operators $C \otimes D$. $\mathscr{A}$ is a commutative Banach algebra and the maps $A \mapsto A \otimes I$, $B \mapsto I \otimes B$ are isometric isomorphisms imbedding $\mathscr{R}(A)$ and $\mathscr{R}(B)$ into $\mathscr{A}$. If $\lambda \in \sigma(A \otimes B)$, then a fortiori λ is in the spectrum of $A \otimes B$ with respect to the algebra $\mathscr{A}$, so by the Gel'fand theory, there is a multiplicative linear functional ℓ on $\mathscr{A}$ so that $\lambda = \ell(A \otimes B) = \ell(A \otimes I)\ell(I \otimes B)$. Since the restrictions of ℓ to $\mathscr{R}(A) \otimes I$ and $I \otimes \mathscr{R}(B)$ are multiplicative, $\ell(A \otimes I) \in \sigma(A)$ and $\ell(I \otimes B) \in \sigma(B)$. This shows that

$$\sigma(A \otimes B) \subset \sigma(A)\sigma(B) \equiv \{\lambda_1 \lambda_2 \,|\, \lambda_1 \in \sigma(A), \lambda_2 \in \sigma(B)\} \quad (64)$$

In order to prove the reverse inclusion we shall have to work a little harder. First, we introduce a new subset of the spectrum.

Definition Let A be a closed linear operator on a Hilbert space $\mathscr{H}$. Let S denote the set of $\lambda \in \mathbb{C}$ so that for some $c_\lambda > 0$, $\|(A - \lambda)\varphi\| \geq c_\lambda \|\varphi\|$ for all $\varphi \in D(A)$. We define the **approximate point spectrum** by $\sigma_{\mathrm{ap}}(A) \equiv \mathbb{C}\setminus S$. We also define $\boldsymbol{\sigma_r}(A) \equiv \sigma(A)\setminus\sigma_{\mathrm{ap}}(A)$.

$\sigma_{\mathrm{ap}}(A)$ is called the approximate point spectrum because if $\lambda \in \sigma_{\mathrm{ap}}(A)$, there is a sequence $\varphi_n \in D(A)$ with $\|\varphi_n\| = 1$ so that $(A - \lambda)\varphi_n \to 0$. The reader can check that σ_r is a subset of the residual spectrum defined in Section VI.3.

The following lemma summarizes two important properties of $\sigma_r(A)$.

Lemma 1

(a) If $\lambda \in \sigma_r(A)$, then $\bar\lambda \in \sigma_{\mathrm{ap}}(A^*)$.
(b) $\sigma_r(A)$ is open.

Proof (a) is easy. For let $\lambda \in \sigma_r(A)$. Since $A - \lambda$ is not invertible but $\|(A - \lambda)\varphi\| \geq c_\lambda \|\varphi\|$ for some $c_\lambda > 0$, $\mathrm{Ran}(\lambda - A)$ is not dense. Thus $\bar\lambda$ is in the point spectrum of A^*.

XIII.9 The spectrum of tensor products 179

To prove (b), suppose that $\lambda \in \sigma_r(A)$. Then since $\|(A - \lambda)\varphi\| \geq c_\lambda \|\varphi\|$, $\mathrm{Ran}(A - \lambda)$ is a closed subspace of $\mathscr{H}$, but since $\lambda \in \sigma_r(A)$, $\mathrm{Ran}(A - \lambda) \neq \mathscr{H}$. Further, if $|z| \leq c_\lambda/2$, then $\|(A - (\lambda + z))\varphi\| \geq \frac{1}{2}c_\lambda\|\varphi\|$ so to prove that $\lambda + z \in \sigma_r(A)$ for z small enough we need only show that $\mathrm{Ran}(A - (\lambda + z)) \neq \mathscr{H}$. Suppose $\mathrm{Ran}(A - (\lambda + z)) = \mathscr{H}$ and let $\psi \in [\mathrm{Ran}(A - \lambda)]^\perp$ with $\|\psi\| = 1$. Then there is a $\varphi_z \in \mathscr{H}$ so that $(A - (\lambda + z))\varphi_z = \psi$ and $\|(A - (\lambda + z))\varphi_z\| \geq \frac{1}{2}c_\lambda\|\varphi_z\|$ if $|z| \leq c_\lambda/2$. Thus $\|\varphi_z\| \leq 2/c_\lambda$. In addition,

$$1 = \|\psi\|^2 = ((A - (\lambda + z))\varphi_z, \psi) = -z(\varphi_z, \psi) \leq |z|\|\varphi_z\| \leq |z|\frac{2}{c_\lambda}$$

For small z this is a contradiction, so $\mathrm{Ran}(A - (\lambda + z)) \neq \mathscr{H}$ if z is small enough. ∎

Actually, as we have already noted in the proof of Theorem X.1, this proof of (b) shows that the codimension of $\mathrm{Ran}(A - \lambda)$ is constant on each component of $\sigma_r(A)$.

Lemma 2 Let A and B be bounded operators. Let $\langle \lambda_1, \lambda_2 \rangle$ be an interior point of $\sigma(A) \times \sigma(B)$ with $\lambda_1 \neq 0$. Then there is a point $\langle \lambda_1', \lambda_2' \rangle$ on the topological boundary of $\sigma(A) \times \sigma(B)$ so that $\lambda_1 \lambda_2 = \lambda_1' \lambda_2'$.

Proof Let $s_1 = \sup\{t \in \mathbb{R} \mid t\lambda_1 \in \sigma(A)\}$, $s_2 = \sup\{t \in \mathbb{R} \mid \lambda_2/t \in \sigma(B)\}$, and $s = \min\{s_1, s_2\}$. Since A is bounded, $s_1 < \infty$ so $s < \infty$. Since either $s\lambda_1$ is on $\partial\sigma(A)$, the topological boundary of $\sigma(A)$, or λ_2/s is on $\partial\sigma(B)$, $\langle s\lambda_1, \lambda_2/s \rangle$ is on the topological boundary of $\sigma(A) \times \sigma(B)$. ∎

Theorem XIII.34 Let A and B be bounded operators on Hilbert spaces $\mathscr{H}_1$ and $\mathscr{H}_2$, respectively. Then $\sigma(A \otimes B) = \sigma(A)\sigma(B)$.

Proof We have already shown (64) so we need only show that $\sigma(A)\sigma(B) \subset \sigma(A \otimes B)$. Let $\lambda_1 \in \sigma(A)$ and $\lambda_2 \in \sigma(B)$. If $\lambda_1 = 0$, then it is easy to see that $0 \in \sigma(A \otimes B)$ so assume $\lambda_1 \neq 0 \neq \lambda_2$. By Lemma 2, we may assume that $\langle \lambda_1, \lambda_2 \rangle$ is on the topological boundary of $\sigma(A) \times \sigma(B)$. Without loss we assume $\lambda_1 \in \partial\sigma(A)$. Since $\sigma(A)$ is closed, $\lambda_1 \in \sigma(A)$. On the other hand, $\sigma_r(A)$ is open so $\lambda_1 \in \sigma_{\mathrm{ap}}(A)$ and similarly, $\bar{\lambda}_1 \in \sigma_{\mathrm{ap}}(A^*)$. There are now two cases to handle. Suppose $\lambda_2 \in \sigma_{\mathrm{ap}}(B)$. Then there exist sequences $\varphi_n \in \mathscr{H}_1$, with $\|\varphi_n\| = 1$, $(A - \lambda_1)\varphi_n \to 0$, and $\psi_n \in \mathscr{H}_2$, with $\|\psi_n\| = 1$, $(B - \lambda_2)\psi_n \to 0$. Let $\eta_n = \varphi_n \otimes \psi_n$. Then

$$(A \otimes B - \lambda_1 \lambda_2 I)\eta_n = (A - \lambda_1)\varphi_n \otimes B\psi_n + \lambda_1 \varphi_n \otimes (B - \lambda_2)\psi_n$$
$$\to 0 \quad \text{as} \quad n \to \infty$$

Thus, $\lambda_1 \lambda_2 \in \sigma_{\mathrm{ap}}(A \otimes B)$. On the other hand, suppose that $\lambda_2 \in \sigma_r(B)$. Then $\bar{\lambda}_2 \in \sigma_{\mathrm{ap}}(B^*)$ and $\bar{\lambda}_1 \in \sigma_{\mathrm{ap}}(A^*)$. The same argument as above now shows that $\bar{\lambda}_1 \bar{\lambda}_2 \in \sigma_{\mathrm{ap}}((A \otimes B)^*) \subset \sigma(A^* \otimes B^*)$. Thus $\lambda_1 \lambda_2 \in \sigma(A \otimes B)$ by Theorem VI.7. ∎

We shall use Theorem XIII.34 to prove that $\sigma(A \otimes I + I \otimes B) = \sigma(A) + \sigma(B)$ by first deriving spectral mapping formulas that relate the spectrum of the generator of a bounded holomorphic semigroup to the spectrum of its resolvent and the semigroup itself. Let C generate a bounded holomorphic semigroup on a Banach space X and define $\mathscr{C} \equiv \{e^{-tC} | t \geq 0\}''$ where $\{\cdot\}'$ denotes the commutant of a family of operators and $\{\cdot\}''$ the double commutant. Explicitly,

$$\mathscr{A}' = \{B | AB = BA \text{ for all } A \in \mathscr{A}\}$$
$$\mathscr{A}'' = \{\mathscr{A}'\}'$$

Lemma 3 If C is the generator of a bounded holomorphic semigroup, then
$$\mathscr{C} \equiv \{e^{-tC} | t \geq 0\}'' = \{R_\lambda(C) | \lambda \in \rho(C)\}''$$
Moreover, $\mathscr{C}$ is an abelian Banach algebra.

Proof It is sufficient to prove that
$$\{e^{-tC} | t \geq 0\}' = \{R_\lambda(C) | \lambda \in \rho(C)\}'$$
This follows easily from (X.98)
$$(\lambda + C)^{-1} \varphi = \int_0^\infty e^{-\lambda t} e^{-tC} \varphi \, dt$$
and (X.102)
$$e^{-zC} = -\frac{1}{2\pi i} \int_\Gamma e^{-\lambda z} (\lambda - C)^{-1} \, d\lambda$$
where Γ is a suitable path.

The last statement in the lemma is a general fact about operator algebras. Since $\{e^{-tC} | t \geq 0\}$ is abelian, $\{e^{-tC} | t \geq 0\} \subset \{e^{-tC} | t \geq 0\}'$. Thus, $\{e^{-tC} | t \geq 0\}'' \subset \{e^{-tC} | t \geq 0\}'$. Therefore if A and B are in $\{e^{-tC} | t \geq 0\}''$, A is also in $\{e^{-tC} | t \geq 0\}'$ so A and B commute. ∎

Since $\mathscr{C}$ is a commutant, it has the following very important property: The Gel'fand spectrum with respect to the algebra of each $D \in \mathscr{C}$ is the same as its spectrum as an operator on X. This is because if D commutes with

$\{e^{-tC} | t > 0\}'$, then all the resolvents of D also commute with $\{e^{-tC} | t > 0\}'$, so they are all in $\mathscr{C}$. Therefore, for an element D in $\mathscr{C}$ we shall just write $\sigma(D)$ for its spectrum with respect to $\mathscr{C}$.

The following is essentially a restatement of Lemma 2 from Section 4.

Lemma 4 Let C be a closed operator with nonempty resolvent set on a Banach space X. Suppose that $\lambda \in \rho(C)$ and regard $h: z \mapsto (\lambda - z)^{-1}$ as a map of the extended complex plane onto itself. Then:

(a) If C is bounded, h is a homeomorphism of $\sigma(C)$ onto $\sigma((\lambda - C)^{-1})$.
(b) If C is unbounded, h is a homeomorphism of $\sigma(C) \cup \{\infty\}$ onto $\sigma((\lambda - C)^{-1})$.

We can now prove a spectral mapping theorem for the semigroup generated by C.

Lemma 5 Let C be the generator of a bounded holomorphic semigroup on a Banach space X. Then:

(a) $\sigma(e^{-tC}) = e^{-t\sigma(C)} \equiv \{e^{-tz} | z \in \sigma(C)\}$ if C is bounded.
(b) $\sigma(e^{-tC}) = e^{-t\sigma(C)} \cup \{0\}$ if C is unbounded.

Proof If $\ell \in \sigma(\mathscr{C})$, the set of multiplicative linear functionals on $\mathscr{C}$, then the first resolvent equation shows that either $\ell((\lambda - C)^{-1}) = 0$ for all $\lambda \in \rho(C)$ or for no $\lambda \in \rho(C)$. Let $\mathscr{M}_\infty = \{\ell \in \sigma(\mathscr{C}) | \ell((\lambda - C)^{-1}) = 0\}$ and $\mathscr{M}_0 = \sigma(\mathscr{C}) \setminus \mathscr{M}_\infty$. For $\ell \in \mathscr{M}_0$, we define

$$\hat{C}(\ell) \equiv \lambda - [\ell((\lambda - C)^{-1})]^{-1}$$

It follows from the first resolvent equation that this definition is independent of which $\lambda \in \rho(C)$ is chosen. Since $\ell((\lambda - C)^{-1}) = 0$ if $\ell \in \mathscr{M}_\infty$ and

$$\ell((\lambda - C)^{-1}) = (\lambda - \hat{C}(\ell))^{-1}, \qquad \ell \in \mathscr{M}_0$$

it follows that $\sigma((\lambda - C)^{-1}) = \{(\lambda - \hat{C}(\ell))^{-1} | \ell \in \mathscr{M}_0\} \cup \{0\}$ in the case where C is unbounded. Thus by Lemma 4, $\sigma(C) = \operatorname{Ran} \hat{C} \upharpoonright \mathscr{M}_0$. The same holds if C is bounded. Now, suppose that $\ell \in \mathscr{M}_0$, then by (X.102)

$$\ell(e^{-tC}) = \frac{-1}{2\pi i} \int_\Gamma e^{-\lambda t} \ell((\lambda - C)^{-1}) \, d\lambda$$
$$= e^{-t\hat{C}(\ell)}$$

by Cauchy's theorem. If $\ell \in \mathscr{M}_\infty$, the same formula shows that $\ell(e^{-tC}) = 0$ for all $t > 0$. Since $\sigma(e^{-tC}) = \{\ell(e^{-tC}) | \ell \in \sigma(\mathscr{C})\}$, we conclude that $\sigma(e^{-tC}) = e^{-t\sigma(C)} \cup \{0\}$ if C is unbounded and $\sigma(e^{-tC}) = e^{-t\sigma(C)}$ if C is bounded. ∎

182 XIII: SPECTRAL ANALYSIS

The following theorem and its proof extend to the Banach space case. We state and prove it only in the Hilbert space case since we have not discussed the tensor product of Banach spaces and the Hilbert space case is all we need in Section 10.

Theorem XIII.35 Let A and B be the generators of bounded holomorphic semigroups on Hilbert spaces $\mathcal{H}_1$ and $\mathcal{H}_2$. Let C be the closure of the operator $A \otimes I + I \otimes B$ defined on $D(A) \otimes D(B)$. Then C generates a bounded holomorphic semigroup and

$$\sigma(C) = \sigma(A) + \sigma(B)$$

Proof Suppose that e^{-zA} and e^{-zB} are bounded holomorphic semigroups of angle θ_1 and θ_2, respectively. Then $W(z) = e^{-zA} \otimes e^{-zB}$ is a bounded holomorphic semigroup of angle $\theta = \min\{\theta_1, \theta_2\}$. Let G be the generator of $W(z)$. $W(t)$ is strongly differentiable on $D(A) \otimes D(B)$ and $G \upharpoonright D(A) \otimes D(B) = A \otimes I + I \otimes B$. Since $W(z): D(A) \otimes D(B) \to D(A) \otimes D(B)$, Theorem X.49 implies that $D(A) \otimes D(B)$ is a core for G. Thus $G = C$.

By Theorem XIII.34, $\sigma(e^{-tC}) = \sigma(e^{-tA})\sigma(e^{-tB})$. Therefore, by Lemma 5,

$$e^{-t\sigma(C)} = \tilde{\sigma}(e^{-tC}) = \tilde{\sigma}(e^{-tA})\tilde{\sigma}(e^{-tB}) = e^{-t(\sigma(A)+\sigma(B))}$$

where $\tilde{\sigma}(\cdot)$ denotes $\sigma(\cdot)\backslash\{0\}$. Thus, if $\mu \in \sigma(A)$ and $\lambda \in \sigma(B)$, then for all t there is a $\gamma_t \in \sigma(C)$ so that $e^{-t\gamma_t} = e^{-t(\mu+\lambda)}$, i.e., $\gamma_t = \mu + \lambda + t^{-1}2\pi i n_t$, where n_t is an integer. Since $\gamma_t \in \overline{S}_{\frac{1}{2}\pi-\theta} = \{z \mid |\arg z| \leq \frac{1}{2}\pi - \theta\}$, n_t must be zero if t is small enough. Thus, $\gamma = \mu + \lambda \in \sigma(C)$. Conversely, suppose $\gamma \in \sigma(C)$. Then, for each t, there is a $\mu_t \in \sigma(A) \subset \overline{S}_{\frac{1}{2}\pi-\theta}$ and a $\lambda_t \in \sigma(B) \subset \overline{S}_{\frac{1}{2}\pi-\theta}$ so that $\gamma = \mu_t + \lambda_t + t^{-1}2\pi i n_t'$ where n_t' is an integer. Since $\gamma, \mu_t, \lambda_t \in \overline{S}_{\frac{1}{2}\pi-\theta}$ we again see that if t is small $\gamma = \mu_t + \lambda_t$. Thus $\sigma(C) = \sigma(A) + \sigma(B)$. ∎

Corollary 1 If A and B are bounded operators on Hilbert spaces $\mathcal{H}_1$ and $\mathcal{H}_2$, then

$$\sigma(A \otimes I + I \otimes B) = \sigma(A) + \sigma(B)$$

Proof $A + 2\|A\|$ and $B + 2\|B\|$ generate bounded holomorphic semigroups so by the above theorem,

$$\sigma(A \otimes I + I \otimes B) + 2\|A\| + 2\|B\|$$
$$= \sigma((A + 2\|A\|) \otimes I + I \otimes (B + 2\|B\|))$$
$$= \sigma(A + 2\|A\|) + \sigma(B + 2\|B\|)$$
$$= \sigma(A) + \sigma(B) + 2\|A\| + 2\|B\| \quad \blacksquare$$

Corollary 2 (Ichinose's lemma)   Let $\bar{S}_{\omega,\varphi,\theta}$ denote the sector $\{z \,|\, \varphi - \theta \leq \arg(z - \omega) \leq \varphi + \theta;\ \theta < \pi/2\}$. Let A and B be strictly m-sectorial operators on Hilbert space $\mathcal{H}_1$ and $\mathcal{H}_2$ with sectors $\bar{S}_{\omega_1,\varphi,\theta_1}$ and $\bar{S}_{\omega_2,\varphi,\theta_2}$ (same φ!). Let C denote the closure of $A \otimes I + I \otimes B$ on $D(A) \otimes D(B)$. Then C is a strictly m-sectorial operator with sector $S_{\omega_1+\omega_2,\varphi,\min\{\theta_1,\theta_2\}}$ and $\sigma(C) = \sigma(A) + \sigma(B)$.

Proof   The proof follows immediately by translating and rotating A and B so that they become strictly m-accretive and then applying Corollary 1 of Theorem X.52 and the above theorem. ∎

XIII.10 The absence of singular continuous spectrum IV: Dilation analytic potentials

Thus far, we have proven the absence of singular continuous spectrum for three types of Schrödinger operators: a wide range of two-body operators, n-body systems with repulsive potentials, and n-body systems with weak potentials. All these systems have the property of possessing only one scattering channel; equivalently, their subsystems have no bound states. In this section we discuss a method for proving the absence of singular continuous spectrum in multichannel n-body systems with $n \geq 3$. The class of pair interactions that we can treat is very restrictive but it does include Coulomb and generalized Yukawa potentials.

Definition   The group of unitary operators $u(\theta)$ on $L^2(\mathbb{R}^3)$ given by

$$(u(\theta)\psi)(r) = e^{3\theta/2}\psi(e^\theta r)$$

is called the **group of dilation operators on** $\mathbb{R}^3$.

The factor $e^{3\theta/2}$ is included to make u unitary. Throughout this section we use $u(\theta)$ to stand for this family of operators.

The idea behind using dilations for spectral analysis is the following. The kinetic energy $H_0 = -\Delta$ transforms very simply under dilations; in fact,

$$u(\theta)H_0 u(\theta)^{-1} = e^{-2\theta}H_0 \equiv H_0(\theta)$$

This last expression implies that $u(\theta)H_0 u(\theta)^{-1}$, defined a priori when θ is real, has an analytic continuation to complex θ. We shall restrict our potentials so that the same is true when $H = -\Delta + V$ replaces H_0. We shall see below that $H(\theta)$ has a discrete spectrum that is "locally" independent of θ.

However as is clear from the case of H_0 where $\sigma(H_0(\theta)) = \{z \mid \arg z = -2 \operatorname{Im} \theta\}$, the continuous spectrum changes considerably as θ varies. This allows one to separate the continuous spectrum from the real axis:

Definition Let $\alpha > 0$. A quadratic form V on $L^2(\mathbb{R}^3)$ is said to be in class $\mathscr{F}_\alpha$ if and only if:

(1) V is a symmetric form with $Q(V) \supset Q(H_0)$ where $H_0 = -\Delta$.
(2) $(H_0 + I)^{-1/2} V (H_0 + I)^{-1/2}$ is compact.
(3) The family of operators
$$F(\theta) = (H_0 + I)^{-1/2}(u(\theta)Vu(\theta)^{-1})(H_0 + I)^{-1/2}$$
defined for $\theta \in \mathbb{R}$ has an extention to an analytic bounded operator-valued function into the strip B_α where
$$B_\alpha \equiv \{\theta \mid |\operatorname{Im} \theta| < \alpha\}$$

The set $\bigcup_{\alpha > 0} \mathscr{F}_\alpha$ is called the family of **dilation analytic potentials**.

If the function $F(\theta)$ in (3) has an extension to $\{\theta \mid |\operatorname{Im} \theta| \leq \alpha\}$ analytic in the interior and norm continuous up to the boundary, we say that V is in class $\overline{\mathscr{F}}_\alpha$.

In the above definition $u(\theta)Vu(\theta)^{-1} \equiv V(\theta)$ denotes the quadratic form $V(\theta)(\psi, \varphi) = V(u(-\theta)\psi, u(-\theta)\varphi)$. The definition of dilation analytic potentials can be conveniently rephrased in terms of the scale of spaces $\mathscr{H}_{+1} \subset \mathscr{H} \subset \mathscr{H}_{-1}$ defined by the quadratic form of H_0 (see Section VIII.6). For (1) and (2) say that V defines a compact operator from $\mathscr{H}_{+1}$ to $\mathscr{H}_{-1}$ that is self-dual. To state (3), we first note that
$$(H_0 + 1)^{1/2} u(\theta)(H_0 + 1)^{-1/2} = (H_0 + 1)^{1/2}(e^{-2\theta}H_0 + 1)^{-1/2} u(\theta)$$
so that $u(\theta)$ defines a bounded map from $\mathscr{H}_{+1}$ to $\mathscr{H}_{+1}$ and, since $u(-\theta)^* = u(\theta)$, from $\mathscr{H}_{-1}$ to $\mathscr{H}_{-1}$. Thus, since V is bounded from $\mathscr{H}_{+1}$ to $\mathscr{H}_{-1}$, for any real θ, $u(\theta)Vu(\theta)^{-1}$ is defined as a compact operator from $\mathscr{H}_{+1}$ to $\mathscr{H}_{-1}$. Statement (3) then says that this function has an $\mathscr{L}(\mathscr{H}_{+1}, \mathscr{H}_{-1})$-valued analytic continuation.

When V is dilation analytic and $\operatorname{Im} \theta < \alpha$, we shall let $V(\theta)$ denote the quadratic form associated with the operator in $\mathscr{L}(\mathscr{H}_{+1}, \mathscr{H}_{-1})$ obtained by the above continuation. Since the above argument shows that $V(\theta)$ is a compact operator from $\mathscr{H}_{+1}$ to $\mathscr{H}_{-1}$ when θ is real and since the analytic continuation of an operator-valued function with compact values on the real axis is compact (Lemma 5 in Section 5), we conclude that $V(\theta)$ is a relatively form compact perturbation of H_0 for all θ and in particular, it is a form-bounded perturbation with relative bound zero (see Problem 38). As a result, we have the following:

Proposition 1 Let $V \in \mathscr{F}_\alpha$ and let $H_0 = -\Delta$. Let $H_0(\theta) = e^{-2\theta}H_0$ for $\theta \in B_\alpha$. Define the quadratic form $H(\theta) = H_0(\theta) + V(\theta)$ on $Q(H_0)$ as the sum of the forms $H_0(\theta)$ and $V(\theta)$. Then:

(a) For any $\theta \in B_\alpha$, $H(\theta)$ is a strictly m-sectorial form; explicitly, for any ε, there is a z_0 so that

$$S_{z_0,\,\text{Im}\,\theta,\,\varepsilon} = \{\omega \,|\, 2\,\text{Im}\,\theta - \varepsilon < \arg(\omega - z_0) < 2\,\text{Im}\,\theta + \varepsilon\}$$

is a sector for $H(\theta)$.
(b) $H(\theta)$ is an analytic family of type (B) in the region B_α.
(c) For any real φ, $u(\varphi)H(\theta)u(\varphi)^{-1} = H(\theta + \varphi)$.

Proof Let $A(\theta) \equiv e^{2\theta}H(\theta) = H_0 + e^{2\theta}V(\theta)$. Since H_0 is self-adjoint and $e^{2\theta}V(\theta)$ is a form-bounded perturbation of H_0 with relative bound 0, for any δ, we can find b so that

$$|(\varphi, e^{2\theta}V(\theta)\varphi)| < \delta(\varphi, (H_0 + b)\varphi)$$

Letting $\varepsilon = \sin^{-1}\delta$, we have $\arg[(\varphi, (H_0 + e^{2\theta}V(\theta) + b)\varphi)] < \varepsilon$. This proves (a). $A(\theta)$ is clearly analytic of type (B) so the same is true of $H(\theta)$. To prove (c), fix $\varphi \in \mathbb{R}$ and note that $u(\varphi)H(\theta)u(\varphi)^{-1}$ and $H(\theta + \varphi)$ are both analytic in B_α and they are clearly equal if $\theta \in \mathbb{R}$. ∎

Before turning to some examples, we note that there is an operator version of dilation analytic potentials, but that the related class $\mathscr{C}_\alpha$ is contained in $\mathscr{F}_\alpha$ (see Problem 73).

Example 1 Let V be a central potential given by a real-valued function $V(r)$. Then, for real θ, $V(\theta)$ is multiplication by the function $V(e^\theta r)$. It is now easy to see (Problem 74) that if $V(r)$ has an analytic continuation $V(z)$ to a sector $\{z\,|\,|\arg z| < \alpha\}$ with

$$\lim_{\substack{|z| \to \infty \\ |\arg z| < \beta}} V(z) = 0$$

and

$$\sup_{0 < |\varphi| < \beta}\int_{|r|,\,|r'| \le 1} |V(e^{i\varphi}r)|\,|V(e^{i\varphi}r')|\,|r - r'|^{-2}\,d^3r\,d^3r' < \infty$$

for each $\beta < \alpha$, then V is dilation analytic. In fact, for each $\theta \in B_\alpha$, $V(\theta)$ is multiplication by $V(e^\theta r)$ which is in $R + (L^\infty)_\varepsilon$. In particular, the Coulomb potential $V(r) = r^{-1}$ is in $\mathscr{F}_\infty$ and the Yukawa potential $V(r) = e^{-\mu r}/r$, $\mu > 0$, is in $\overline{\mathscr{F}}_{\pi/2}$.

Example 2 Dilation analytic potentials need not be "local," i.e., they do not have to be multiplication operators. For example, let $\psi \in L^2$ be an analytic vector for the infinitesimal generator of $u(\theta)$. Thus $\{u(\theta)\psi\}_{\theta \in \mathbb{R}}$ has an analytic continuation to a strip $\{\theta \,|\, |\operatorname{Im} \theta| < \alpha\}$ (Problem 75). Let V be the rank-one operator $(\psi, \cdot)\psi$. Then V is dilation analytic; in fact $V(\theta) = (\psi(\bar\theta), \cdot)\psi(\theta)$.

Our ultimate goal in this section is to analyze the spectrum of the Hamiltonian H on $L^2(\mathbb{R}^{3N-3})$ obtained by removing the center of mass from $\tilde{H} = -\sum_{i=1}^N (2\mu_i)^{-1}\Delta_i + \sum_{i<j} V_{ij}(r_i - r_j)$ where each V_{ij} is an operator on $L^2(\mathbb{R}^3)$ that is dilation analytic. We begin by considering the two-body case and then use the method of the two-body case in the general situation supplementing it with the Weinberg–Van Winter equations and Ichinose's lemma.

Theorem XIII.36 Let $H_0 = -\Delta$ on $L^2(\mathbb{R}^3)$ and let $V \in \mathscr{F}_\alpha$ for some $\alpha > 0$. Suppose that $\theta \in B_\alpha$ and let $H(\theta) = e^{-2\theta}H_0 + V(\theta)$. Then:

(a) $\sigma(H(\theta))$ is only dependent on $\operatorname{Im} \theta$.
(b) $\sigma(H(\theta))$ consists of $\{e^{-2\theta}\lambda \,|\, \lambda \in [0, \infty)\} \cup \sigma_d(\theta)$ where σ_d is a set whose only possible limit point is zero. Each $\mu \in \sigma_d(\theta)$ is an eigenvalue of finite multiplicity.
(c) $\sigma_d(\bar\theta) = \overline{\sigma_d(\theta)}$.
(d) If $0 < \operatorname{Im} \theta < \tfrac{1}{2}\pi$, then $\sigma_d(\theta) \subset \mathbb{R} \cup \{\mu \,|\, -2\operatorname{Im} \theta < \arg \mu < 0\}$ and $\mathbb{R} \cap \sigma_d(\theta) = \sigma_{pp}(H(0))\setminus\{0\}$. Moreover, if $0 < \operatorname{Im} \varphi < \operatorname{Im} \theta < \tfrac{1}{2}\pi$, then $\sigma_d(\varphi) \subset \sigma_d(\theta)$.
(e) $\sigma_{\text{sing}}(H(0)) = \varnothing$.

Proof (a) and (c) follow from Proposition 1 which says that $H(\theta)$ and $H(\theta + \varphi)$ are unitarily equivalent if $\varphi \in \mathbb{R}$ and $H(\theta)^* = H(\bar\theta)$. Corollary 2 to Theorem XIII.14 implies that $\sigma_{\text{ess}}(e^{2\theta}H(\theta)) = \sigma_{\text{ess}}(H_0) = [0, \infty)$. This proves (b) except that a priori $\sigma_d(\theta)$ could have limit points anywhere in $e^{-2\theta}\mathbb{R}_+$. This leaves (d) and (e). Fix θ_0 and suppose that $E \in \sigma_d(\theta_0)$. Since $H(\theta)$ is an analytic family of type (B), Theorem XII.13 implies that for θ near θ_0, $H(\theta)$ has eigenvalues near E and that these nearby eigenvalues are given by all the branches of one or more functions $f_1(\theta), \ldots, f_n(\theta)$ analytic near θ_0 with at worst algebraic branch points near θ_0. On the other hand, if φ is real, $H(\theta_0)$ and $H(\theta_0 + \varphi)$ are unitarily equivalent, so the only eigenvalue of $H(\theta_0 + \varphi)$ near E is E. We conclude that $f_i(\theta) = E$ if θ is near θ_0 and $\theta - \theta_0$ is real. But analyticity then implies that $f_i(\theta) = E$ for the entire region where f is defined! Thus we have proven that $\sigma_d(\theta)$ is locally constant in the sense that if $\lambda \in \sigma_d(\theta_0)$, then it is in $\sigma_d(\theta)$ for all θ near θ_0. On the other hand, if $\theta_n \to \theta$ and $\lambda \in \sigma_d(\theta_n)$ for each n, then $\lambda \in \sigma(H(\theta))$ since $H(\theta_n) \to H(\theta)$ in norm resolvent sense. Therefore a simple connectedness argument (Problem 76) proves the following statement: If $\gamma(t)$ ($0 \le t \le 1$) is a curve in B_α and $\lambda \in \sigma_d(\gamma(0))$, then either $\lambda \in \sigma_d(\gamma(1))$ or $\lambda \in \sigma_{\text{ess}}(H(\gamma(t)))$ for some $t \in (0, 1]$. In particular, if $\lambda \notin \{\mu | -2 \operatorname{Im} \theta \le \arg \mu \le 0\}$ and if $0 < \operatorname{Im} \theta < \tfrac{1}{2}\pi$, then $\lambda \in \sigma_d(\theta)$ if and only if $\lambda \in \sigma_d(0)$ for we can take the curve $\gamma(t) = t\theta$. Thus

$$\sigma_d(\theta) \cap \{\mu | 0 < \arg \mu < 2\pi - 2 \operatorname{Im} \theta\} = \sigma_{\text{pp}}(H) \cap (-\infty, 0)$$

Similarly, if $0 < \operatorname{Im} \varphi < \operatorname{Im} \theta$, then $\sigma_d(\varphi) \subset \sigma_d(\theta)$. This shows that $\sigma_d(\theta)$ cannot have any limit points other than zero. To complete the proof of (d), we need only show that $\sigma_d(\theta) \cap (0, \infty) = \sigma_{\text{pp}}(H) \cap (0, \infty)$. This requires an additional idea which is also needed for the proof of (e). Let

$$N_\alpha = \{\psi \in L^2(\mathbb{R}^3) \,|\, u(\theta)\psi \text{ has an analytic continuation from } \mathbb{R} \text{ to } B_\alpha\}$$

N_α is precisely the set of analytic vectors ψ for the infinitesimal generator D of $u(\theta)$ for which $\sum_{t=0}^\infty t^n \|D^n \psi\|/n!$ has radius of convergence at least α. Let $\psi \in N_\alpha$ and let $\psi(\theta)$ denote the continuation of $u(\theta)\psi$. Consider the function $f(z, \theta) = (\psi(\bar\theta), (H(\theta) - z)^{-1} \psi(\theta))$. For each fixed $\theta \in B_\alpha$, $f(z, \theta)$ is analytic in z for $z \in \mathbb{C} \backslash \sigma(H(\theta))$ and meromorphic in $\mathbb{C} \backslash \sigma_{\text{ess}}(\theta)$. On the other hand, let us fix z with $\operatorname{Im} z > 0$. Then $f(z, \theta)$ is analytic in θ in the region

$$R_z = \{\theta \,|\, -\min\{\alpha, \tfrac{1}{2} \arg z\} < \operatorname{Im} \theta < \min\{\alpha, \tfrac{1}{2}\pi\}\}.$$

Since for $\varphi \in \mathbb{R}$,

$$f(z, \varphi) = (u(\varphi)\psi, (H(\varphi) - z))^{-1} u(\varphi)\psi) = (u(\varphi)\psi, u(\varphi)(H(0) - z)^{-1}\psi)$$
$$= f(z, 0)$$

we conclude by analyticity that $f(z, \cdot)$ is constant in R_z. In particular, if $0 < \operatorname{Im} \theta_0 < \min\{\alpha, \tfrac{1}{2}\pi\}$, then $f(z, \theta_0)$ provides an analytic continuation of $f(z, 0)$ from $\mathbb{C} \backslash \sigma(H(0))$ to $\mathbb{C} \backslash \sigma(H(\theta_0))$ (see Figure XIII.5). In particular, if $\psi \in N_\alpha$, then $(\psi, (H - z)^{-1}\psi)$ has continuous boundary values as $z \to \mu + i0$ if $\mu \in \mathbb{R} \backslash \sigma_d(\theta_0)$. Since N_α is dense, and $\sigma_d(\theta_0) \cap \mathbb{R}$ is discrete, we conclude that $\sigma_{\text{sing}}(H(0)) = \varnothing$ by appealing to Theorem XIII.20.

Finally, we return to the proof that $\sigma_d(\theta) \cap (0, \infty) = \sigma_{\text{pp}}(H(0)) \cap (0, \infty)$ if $0 < \operatorname{Im} \theta < \min\{\alpha, \tfrac{1}{2}\pi\}$. Since $H(0)$ is self-adjoint, the functional calculus immediately implies that $\operatorname{s-lim}_{\varepsilon \downarrow 0} i\varepsilon(H - x - i\varepsilon)^{-1} = P_{\{x\}}$ for any $x \in \mathbb{R}$.

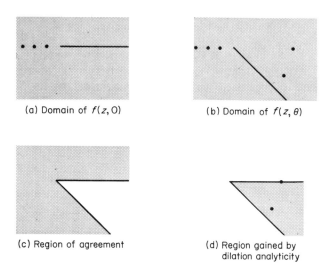

FIGURE XIII.5 Analytic continuation of matrix elements of the resolvent.

Suppose that $x \notin \sigma_d(\theta) \cap (0, \infty)$. Then, by the above argument, $(\psi, (H - z)^{-1}\psi)$ has an analytic continuation from $\{z \mid \operatorname{Im} z > 0\}$ to a neighborhood of x if $\psi \in N_\alpha$. In particular, $\lim_{\varepsilon \downarrow 0} i\varepsilon(\psi, (H - x - i\varepsilon)^{-1}\psi) = 0$ so $P_{\{x\}}\psi = 0$. Since N_α is dense, $P_{\{x\}} = 0$, i.e., $x \notin \sigma_{\mathrm{pp}}(H(0)) \cap (0, \infty)$. Conversely, let $x \in \sigma_d(\theta) \cap (0, \infty)$. Let η be an eigenvector for $H(\theta)$ with eigenvalue x. Then $(\varphi, (H(\theta) - z)^{-1}\eta)$ has a pole at $z = x$ for some φ. Since $N_{2\alpha}$ is dense, we can find $\varphi_n, \eta_n \in N_{2\alpha}$ so that $\varphi_n \to \varphi$, $\eta_n \to \eta$. For $z \notin \sigma(H(\theta))$, $(\varphi_n, (H(\theta) - z)^{-1}\eta_n)$ converges to $(\varphi, (H(\theta) - z)^{-1}\eta)$, so, by the argument principle (which relates the number of poles of a meromorphic function inside a contour to the integral of f'/f on the contour), we conclude that for all large n, $(\varphi_n, (H(\theta) - z)^{-1}\eta_n)$ has a pole at $z = x$. Let $\psi = \eta_n(-\theta)$, $\kappa = \varphi_n(-\theta)$. Then $(\psi, (H - z)^{-1}\kappa) = (\varphi_n, (H(\theta) - z)^{-1}\eta_n)$ if $\operatorname{Im} z > 0$ so we conclude that $(\psi, (H - z)^{-1}\kappa)$ has a pole at $z = x$. Thus $(\psi, P_{\{x\}}\kappa) \neq 0$ and therefore $x \in \sigma_{\mathrm{pp}}(H)$. ∎

In the above proof a critical role was played by the fact that we could explicitly find $\sigma_{\mathrm{ess}}(H(\theta))$ on account of Weyl's theorem. In the N-body case, we have to replace Weyl's theorem with a theorem of the same type as the HVZ theorem. There is one important difference between what we need now to analyze $H(\theta)$ and the analysis of Section 5—namely the potentials $V_{ij}(\theta)$ are not self-adjoint. As preparation for an N-body analogue to Theorem XIII.36, we therefore prove:

Proposition 2 Let $\tilde{H} = \tilde{H}_0 + \sum_{1 \leq i < j = N} \tilde{V}_{ij}$ where:

(i) $\tilde{H}_0 = \sum_{i=1}^{N} (2\mu_i)^{-1} \Delta_i$.
(ii) Under the decomposition induced by writing $\mathbb{R}^{3N} = \mathbb{R}^3 \times \mathbb{R}^{3N-3}$ with $\mathbb{R}^3$ corresponding to the coordinate $r_i - r_j$, $\tilde{V}_{ij} = V_{ij} \otimes I$.
(iii) Each V_{ij} is a relatively form compact perturbation of $-\Delta$ on $L^2(\mathbb{R}^3)$, but is not necessarily symmetric.

Let H be $\tilde{H}$ with its center of mass removed. For each cluster $C \subset \{1, \ldots, N\}$, define $H(C_i)$ according to the prescription in Section XI.5. For each cluster decomposition $D = \{C_1, \ldots, C_k\}$ call the family of numbers $\{E_1 + \cdots + E_k \mid E_i \in \sigma_{\text{disc}}(H(C_i))\}$ the set of **D-thresholds**, written Σ_D. Let Σ denote the union of the Σ_D over all D with at least two clusters. Then

$$\sigma_{\text{ess}}(H) \subset \bigcup_{\lambda \in \Sigma} (\lambda + [0, \infty))$$

Proof We shall only indicate the changes necessary in the argument in Section 5; the details are left to the reader (Problem 77). If $N = 2$, then $\Sigma = \{0\}$ and the conclusion of the theorem is part of Weyl's essential spectrum theorem. Thus we can suppose that the theorem holds for all M-body systems with $M \leq N - 1$ and prove it for N-body systems. The first step is to prove that

$$\sigma(H_D) \subset \bigcup_{\lambda \in \Sigma} (\lambda + [0, \infty)) \tag{65}$$

for any D with at least two clusters. To see this, write $D = \{C_1, \ldots, C_k\}$ so that $H_D = \sum_{i=1}^{k} H(C_i) + T_D$. Now, each V_{ij} is a form-bounded perturbation of H_0 with relative bound zero so each $H(C_i)$ is a strictly m-sectorial operator with arbitrarily small opening angle. Moreover, under the natural decomposition $\mathcal{H} \equiv L^2(\mathbb{R}^{3N-3}) = \mathcal{H}_D \otimes \mathcal{H}_{C_1} \otimes \cdots \otimes \mathcal{H}_{C_k}$ each summand of $\sum_{i=1}^{k} H(C_i) + T_D$ acts on a different factor in the tensor product. As a consequence, Ichinose's lemma (Theorem XIII.35 and its corollary) is applicable and $\sigma(H_D) = \sum_{i=1}^{k} \sigma(H(C_i)) + \sigma(T_D)$. Using the fact that $\sigma(H(C_i)) = \sigma_{\text{disc}}(H(C_i)) \cup \sigma_{\text{ess}}(H(C_i))$ and the inductive hypothesis, it is easy to prove (65).

At this point one can mimic the proof of the HVZ theorem fairly closely. Since $\sum V_{ij}$ is a form-bounded perturbation of relative bound zero, the perturbation series for $G(E) = (H_0 + \sum V_{ij} - E)^{-1}$ converges in norm if E is very negative. Writing $R(E) = (H_0 - E)^{+1/2} G(E)(H_0 - E)^{1/2}$, we can resum $R(E) = D_R(E) + I_S(E) R(E)$ as in Section 5. Both $D_R(E)$ and $I_S(E)$ can be expressed in terms of resolvents for $(H_D - E)^{-1}$ where D has at least two clusters. By our analysis above, they both have analytic continuations to $\mathbb{C} \backslash S$ where $S = \bigcup_{\lambda \in \Sigma} (\lambda + [0, \infty))$. This step requires one to use the induction hypothesis to conclude that S is closed. Finally, one shows that each connected diagram is compact, uses this to prove $I_S(E)$ is compact, and employs the analytic Fredholm theorem to complete the proof. ∎

At this point it is easy to prove the main result of this section. For ease of statement, we first define:

Definition Let $V_{ij} \in \mathscr{F}_\alpha$ for each pair $1 \le i < j \le N$ and let $\tilde{V}_{ij} = V_{ij} \otimes I$ on $L^2(\mathbb{R}^{3N})$ in terms of the decomposition $\mathbb{R}^{3N} = \mathbb{R}^3 \times \mathbb{R}^{3N-3}$ where the first coordinate is $r_i - r_j$. Let $\tilde{H}_0 = -\sum_{i=1}^N (2\mu_i)^{-1} \Delta_i$ and let $\tilde{H} = \tilde{H}_0 + \sum \tilde{V}_{ij}$ For $\theta \in B_\alpha$, let $\tilde{H}(\theta) = e^{-2\theta}\tilde{H}_0 + \sum \tilde{V}_{ij}(\theta)$. Let $H(\theta)$ and H denote $\tilde{H}(\theta)$, $\tilde{H}$ with their center of mass motion removed. For every cluster decomposition $D = \{C_1, \ldots, C_k\}$, let $\Sigma_D(\theta) \equiv \{E_1 + \cdots + E_k \mid E_i \in \sigma_{\text{disc}}(H_{C_i}(\theta))\}$ and let $\Sigma(\theta) = \bigcup_{\#(D) \ge 2} \Sigma_D(\theta)$. Finally we let $\Sigma_{\min} = \min\{\lambda \mid \lambda \in \Sigma(0)\}$.

Theorem XIII.37 Let H be an N-body Hamiltonian on $L^2(\mathbb{R}^{3N-3})$ with dilation analytic pair potentials of the type described in the last definition. Then:

(a) $\sigma(H(\theta))$ and $\Sigma(\theta)$ depend only on Im θ.
(b) $\sigma(H(\theta)) = \sigma_{\text{ess}}(\theta) \cup \sigma_{\text{d}}(\theta)$ where the essential spectrum $\sigma_{\text{ess}}(\theta) = \{\mu + e^{-2\theta}\lambda \mid \mu \in \Sigma(\theta), \lambda \in [0, \infty)\}$ and where $\sigma_{\text{d}}(\theta)$ is the discrete spectrum of $\sigma(H(\theta))$.
(c) $\sigma_{\text{d}}(\theta) = \overline{\sigma_{\text{d}}(\theta)}$; $\Sigma(\bar{\theta}) = \overline{\Sigma(\theta)}$.
(d1) If $0 < \text{Im } \theta < \min\{\alpha, \tfrac{1}{2}\pi\}$, then

$$\Sigma(\theta) \subset \mathbb{R} \cup \{\Sigma_{\min} + \mu \mid -2 \text{ Im } \theta < \arg \mu < 0\}$$

and $\mathbb{R} \cap \Sigma(\theta) = \Sigma(0)$. Moreover, if $0 < \text{Im } \varphi < \text{Im } \theta < \tfrac{1}{2}\pi$, then $\Sigma(\varphi) \subset \Sigma(\theta)$.

(d2) If $0 < \text{Im } \theta < \tfrac{1}{2}\pi$, then $\sigma_{\text{d}}(\theta) \subset \mathbb{R} \cup \{\Sigma_{\min} + \mu \mid -2 \text{ Im } \theta < \arg \mu < 0\}$ and $\mathbb{R} \cap \sigma_{\text{d}}(\theta) = \sigma_{\text{pp}}(H) \backslash \Sigma(0)$. Moreover, if $0 < \text{Im } \varphi < \text{Im } \theta < \tfrac{1}{2}\pi$, then $\sigma_{\text{d}}(\varphi) \subset \sigma_{\text{d}}(\theta)$.

(e) $\sigma_{\text{sing}}(H) = \varnothing$.

Proof Since the proof directly follows that of Theorem XIII.36, we leave the details to the reader (Problems 78, 79). (a) and (c) follow from $U(\varphi)H(\theta)U(\varphi)^{-1} = H(\theta + \varphi)$ and $H(\bar{\theta}) = H(\theta)^*$, and the containment $\sigma_{\text{ess}}(\theta) \subset \{\mu + e^{-2\theta}\lambda \mid \mu \in \Sigma(\theta), \lambda \in [0, \infty)\}$ follows from Proposition 2. The containment $\{\mu + e^{-2\theta}\lambda \mid \mu \in \Sigma(\theta), \lambda \in [0, \infty)\} \subset \sigma_{\text{ess}}(\theta)$ requires a separate argument (Problem 78). (d1) for N-body systems follows from (d2) for $(N-1)$-body systems so only (d2) need be proven. This is accomplished, as in Theorem XIII.36 in two parts. First, one uses the perturbation theory of discrete spectrum to show that $\sigma_d(\theta)$ is locally constant and uses this to prove all of (d2) but $[\Sigma_{\min}, \infty) \cap \sigma_d(\theta) = \sigma_{pp}(H)\backslash(\Sigma(0) \cup (-\infty, \Sigma_{\min}))$. This last fact is proven by analytic continuation of matrix elements $(\psi, (H - z)^{-1}\psi)$ for $\psi \in N_\alpha$. (e) follows as a by-product of this last development. ∎

Corollary An N-body quantum Hamiltonian with center of mass motion removed and with Coulomb or Yukawa potentials has an empty continuous singular spectrum.

Eigenvalues of $H(\theta)$ in $\sigma_d(\theta)\backslash\sigma_d(0)$ are called **resonance eigenvalues** of H. We have already discussed their significance in Section XII.6. Points in $\Sigma(\theta)\backslash\Sigma$ are called **complex thresholds** or **resonance thresholds**.

XIII.11 Properties of eigenfunctions

With the exception of the question of proving that $\sigma_{pp} = \sigma_{\text{disc}}$ (which we turn to in Section 13), we have completed our analysis of the different kinds of spectra of Schrödinger operators on infinite space with two body potentials going to zero at ∞. We turn now to a set of problems that are not spectral analysis per se but which are closely related. In this section we concentrate on regularity properties and falloff of eigenfunctions. There is a close connection between these two sets of properties for as we have seen (see Sections IX.2, 3) smoothness of a function ψ is closely connected to decrease properties of the Fourier transform $\hat\psi$. We shall state all our theorems below in terms of $\psi(x)$, the "x-space function," but in proofs $\hat\psi(p)$, "the p-space function" will often enter. We first discuss smoothness of $\psi(x)$, then the falloff of ψ in the sense that $\psi \in D(\exp(a|x|))$ in certain circumstances. Finally, we combine the two sets of ideas to prove pointwise bounds $|\psi(x)| \leq C \exp(-a|x|)$.

To state our smoothness results we introduce spaces of **uniformly Hölder continuous functions**:

Definition Let $0 < \theta < 1$. Then we say $f \in C_\theta(\mathbb{R}^n)$ if $f: \mathbb{R}^n \to \mathbb{C}$ is bounded and obeys

$$|f(x) - f(y)| \leq C|x - y|^\theta$$

for some fixed C and all $x, y \in \mathbb{R}^n$. We set
$$\|f\|_{(\theta)} \equiv \|f\|_\infty + \sup_{x \neq y} |x-y|^{-\theta} |f(x) - f(y)|$$
We say $f \in C_\theta^1(\mathbb{R}^n)$ if f is bounded, C^1, and for each $j = 1, \ldots, n$, $\partial f/\partial x_j \in C_\theta(\mathbb{R}^n)$. We set
$$\|f\|_{\theta, 1} = \|f\|_\infty + \sum_{j=1}^n \|D_j f\|_{(\theta)}$$

The connection of Hölder continuity and the Fourier transform is illustrated by:

Lemma 1 Let $(1 + k^2)^{\beta/2} \hat{f}(k) \in L^1(\mathbb{R}^n)$. Then:
(a) If $0 < \beta < 1$, then $f \in C_\beta$ and
$$\|f\|_{(\beta)} \leq c\|(1+k^2)^{\beta/2} \hat{f}(\cdot)\|_1$$
(b) If $1 < \beta < 2$, then $f \in C_{\beta-1}^1$ and
$$\|f\|_{\beta-1, 1} \leq c_n \|(1+k^2)^{\beta/2} \hat{f}\|_1$$

Proof (a) For each real s, $|e^{is} - 1| \leq 2$ and $|e^{is} - 1| \leq |\int_0^s e^{it} \, dt| \leq s$, so $|e^{is} - 1| \leq s^\beta 2^{1-\beta}$. Thus
$$|e^{ik \cdot x} - e^{ik \cdot y}| = |e^{ik \cdot (x-y)} - 1|$$
$$\leq 2^{1-\beta} |k|^\beta |x-y|^\beta$$
so that
$$|f(x) - f(y)| \leq (2\pi)^{-n/2} \int |\hat{f}(k)| |e^{ik \cdot x} - e^{ik \cdot y}| \, dk$$
$$\leq \left[2^{1-\beta} (2\pi)^{-n/2} \int |\hat{f}(k)| |k|^\beta \, dk\right] |x-y|^\beta$$
$$\leq 2^{1-\beta} (2\pi)^{-n/2} \|(1+k^2)^{\beta/2} \hat{f}\|_1 |x-y|^\beta$$

Since $\|f\|_\infty \leq \|\hat{f}\|_1$, (a) follows.

The proof of (b) is similar to (a) and is left to the reader (Problem 80). ∎

We can now see what sort of smoothness result to expect by considering $H = -\Delta + V$ on $\mathbb{R}^3$ when $V \in L^2 + L^\infty$. Then $D(H) = D(-\Delta)$ by Kato's theorem. But if $\psi \in D(-\Delta)$, then $(1+k^2)\hat{\psi} \in L^2$ so for any $\beta < \frac{1}{2}$, $(1+k^2)^{\beta/2} \hat{\psi} \in L^1$ since $(1+k^2)^{-(1-\beta/2)}$ is then in $L^2(\mathbb{R}^3)$. We conclude that

$D(H) \subset C_\theta$ if $\theta < \frac{1}{2}$. For N-body Hamiltonians with two-body potentials, it is reasonable to expect that eigenfunctions should have no worse singularities than those produced by the two-body potentials in the two-body case. To state a sharp result, we introduce a special class of potentials.

Definition Let n be given and suppose $\sigma \geq 1$ and $\sigma > \frac{1}{2}n$. We say that a real-valued measurable function V on $\mathbb{R}^n$ is in class $M_\sigma^{(n)}$ if and only if

$$\hat{V} \in L^{\sigma'} + L^1$$

where $\sigma' = (1 - \sigma^{-1})^{-1}$ is the dual index to σ.

Example 1 Let $V(r) = |r|^{-1}$ on $\mathbb{R}^3$ be the Coulomb potential. Then $\hat{V}(k) = c|k|^{-2} \in L^{\sigma'} + L^1$ if $\sigma' > \frac{3}{2}$ so $V \in M_\sigma^{(3)}$ if $\sigma < 3$. The same conclusion holds for the Yukawa potential $V(r) = e^{-\mu|r|}/|r|$.

Proposition

(a) If $\sigma \geq 2$ and $\sigma > \frac{1}{2}n$ and if $V \in M_\sigma^{(n)}$, then V is $-\Delta$-bounded with relative bound zero.
(b) If $\sigma \geq 1$ and $\sigma > \frac{1}{2}n$ and if $V \in M_\sigma^{(n)}$, then V is $-\Delta$ form-bounded with relative bound zero.

Proof (a) By the Plancherel theorem, we need only prove that $\|(p^2 + a)^{-1}\hat{V} * \psi\|_2 \leq C_a \|\psi\|_2$ for all $\psi \in L^2$ where C_a is independent of ψ and can be chosen arbitrarily small by making a large. This follows from Young's and Hölder's inequalities. The proof of (b) is similar. ∎

The following theorem gives regularity properties of eigenfunctions since any eigenfunction of $\tilde{H}$ is clearly in $C^\infty(\tilde{H})$. Since $C^\infty(\tilde{H})$ is left invariant by $e^{-it\tilde{H}}$, it also implies regularity properties of solutions of the time-dependent Schrödinger equation with sufficiently regular initial data.

Theorem XIII.38 (the Kato–Simon theorem) Let

$$\tilde{H} = -\sum_{i=1}^{N} (2\mu_i)^{-1}\Delta_i + \sum_{i<j} V_{ij}(r_i - r_j)$$

as an operator on $L^2(\mathbb{R}^{Nn})$ where a point in $\mathbb{R}^{Nn}$ is written $\langle r_1, \ldots, r_N \rangle$ with $r_i \in \mathbb{R}^n$. Suppose that $\sigma \geq 1$, $\sigma > \frac{1}{2}n$ and that each V_{ij} is in $M_\sigma^{(n)}$. Then:

(a) If $\psi \in C^\infty(\tilde{H})$, then $\psi \in C_\theta$ for any $\theta < \min\{1, 2 - \sigma^{-1}n\}$.
(b) If $\sigma > n$, and $\psi \in C^\infty(\tilde{H})$, then $\psi \in C_\theta^1$ for any $\theta < 1 - \sigma^{-1}n$.

Moreover, the embeddings are continuous in the sense that given θ, there exist m and C with $\|\psi\|_{(\theta)} \leq C[\|\tilde{H}^m \psi\| + \|\psi\|]$ for all $\psi \in C^\infty(\tilde{H})$. Identical results hold if $\tilde{H}$ is replaced with H, the operator on $L^2(\mathbb{R}^{(N-1)n})$, obtained by removing the center of mass motion from $\tilde{H}$.

Proof We consider only the case where $\sigma \geq 2$, thus avoiding the need for quadratic form techniques. The general $\sigma \geq 1$ case is discussed in the reference quoted in the Notes. We also leave it to the reader to prove that the embeddings are continuous (Problem 82).

Let $H_0 = -\sum_{i=1}^{N}(2\mu_i)^{-1}\Delta_i$ and let $t_0(k)$ be the function $\sum_{i=1}^{N}(2\mu_i)^{-1}k_i^2$ on $\mathbb{R}^{Nn}$. We first claim that

$$\|(t_0(k)+1)^{-\beta}\hat{V}_{ij} * \eta\|_p \leq C\|\eta\|_p \tag{66}$$

for any $p \in [1, 2]$, $\eta \in L^p(\mathbb{R}^{Nn})$, and β with $\beta > (2\sigma)^{-1}n$. In (66), C is independent of η and p (it is dependent only on V_{ij}, t_0, and β) and $\hat{V}_{ij}$ is the Fourier transform in all Nn variables; thus

$$(\hat{V}_{ij} * \eta)(k_1, \ldots, k_N)$$
$$= \int (2\pi)^{(Nn-n)/2} f(\ell) \eta(k_1, \ldots, k_i - \ell, \ldots, k_j + \ell, \ldots, k_N)\, d^n\ell$$

where f is the $\mathbb{R}^n$ Fourier transform of V_{ij}. To prove (66) we first notice that since $p \leq 2$ and $\sigma \geq 2$, Hölder's and Young's inequalities imply that

$$\|(k^2+1)^{-\beta}f * \varphi\|_p \leq C_2 \|\varphi\|_p \tag{67}$$

if $\varphi \in L^p(\mathbb{R}^n)$ since $\beta > (2\sigma)^{-1}n$ tells us that $(k^2+1)^{-\beta} \in L^\sigma$. From (67) we immediately conclude that

$$\|[(k_i-k_j)^2+1]^{-\beta}\hat{V}_{ij} * \eta\|_p \leq C_1 \|\eta\|_p$$

for all $\eta \in L^p(\mathbb{R}^{Nn})$ by raising both sides of (67) to the pth power and integrating over dummy variables. Since $[(k_i-k_j)^2+1]^\beta[t_0(k)+1]^{-\beta}$ is bounded, (66) holds.

Next suppose that $p \in [1, 2]$, $\beta > (2\sigma)^{-1}n$, and we know that $\psi \in D(\tilde{H})$ with $\hat{\psi}$ and $\mathscr{F}(\tilde{H}\psi)$ in L^p. Then we claim that $(t_0(k)+1)^{(1-\beta)}\hat{\psi} \in L^p$. For we can write

$$\hat{\psi} = \mathscr{F}[(H_0+I)^{-1}(\tilde{H}+I-V)\psi]$$
$$= (t_0(k)+1)^{-1}(\mathscr{F}(\tilde{H}\psi)+\hat{\psi}) - (2\pi)^{-Nn/2}(t_0(k)+1)^{-1}\sum_{i<j}\hat{V}_{ij} * \hat{\psi}$$

so that (66) implies that $(t_0(k)+1)^{(1-\beta)}\hat{\psi} \in L^p$.

Now we can complete the proof. Since $\sigma > \frac{1}{2}n$, choose β so that $(2\sigma)^{-1}n < \beta < 1$. Suppose that $\psi \in C^\infty(\tilde{H})$. Then by the above,

$$(k^2 + 1)^{1-\beta} \mathscr{F}(\tilde{H}^m \psi) \in L^2$$

for all m. Since $(k^2 + 1)^{-(1-\beta)} \in L^r$ for all $r \in (r_0, \infty)$ and some $r_0 < \infty$, we conclude that $\mathscr{F}(\tilde{H}^m \psi) \in L^q$ as long as $q^{-1} - r_0^{-1} < \frac{1}{2}$ and $q \geq 1$. By repeating this argument j times we see that $\mathscr{F}(\tilde{H}^m \psi) \in L^q$ as long as $q^{-1} - jr_0^{-1} < \frac{1}{2}$. Choosing j suitably, we conclude that $\mathscr{F}(\tilde{H}\psi)$ and $\hat{\psi}$ are in L^1. Invoking the argument in the preceding paragraph once more we see that $(k^2+1)^{1-\beta}\hat{\psi} \in L^1$ so long as $\beta > (2\sigma)^{-1}n$ and $\psi \in C^\infty(\tilde{H})$. Lemma 1 now completes the proof. ∎

Corollary Let H be an atomic Hamiltonian. Then any $\psi \in C^\infty(H)$ is a C^∞ function on $\mathbb{R}^{3N-3} \setminus \{\langle r_1, \ldots, r_{N-1} \rangle \mid$ some $r_i = 0$ or some $r_i = r_j\}$ and is uniformly Hölder continuous of order θ for any $\theta < 1$ on all of $\mathbb{R}^{3N-3}$.

Proof The first statement follows from the technique of the elliptic regularity theorem (Problem 83); the second from Theorem XIII.38. ∎

* * *

The second topic we discuss in this section is that of the exponential falloff of eigenfunctions in the discrete spectrum. We first present L^2 results in various forms and then pointwise bounds. We begin with an L^2 result which is easy to prove and later give a sharper L^2 result which requires some kinematical preliminaries.

Theorem XIII.39 Let $\tilde{H} = -\sum_{i=1}^N (2\mu_i)^{-1}\Delta_i + \sum_{i<j} V_{ij}(r_i - r_j)$ on $L^2(\mathbb{R}^{3N})$ and let H be the operator on $L^2(\mathbb{R}^{3N-3})$ obtained by removing the center of mass motion from $\tilde{H}$. Suppose that each V_{ij} is a form-bounded perturbation of $-\Delta$ on $\mathbb{R}^3$ with relative bound 0. Let $E \in \sigma_{\text{disc}}(H)$ and let $H\psi = E\psi$. Then $\psi \in D(e^{ar})$ for some $a > 0$ where $r = (\sum_{j=1}^{3N-3} x_j^2)^{1/2}$ in terms of some set of coordinates for $\mathbb{R}^{3N-3}$.

Proof Since

$$e^{ar} \leq \exp(a\sqrt{3N-3} \max |x_j|) \leq \sum_{j=1}^{3N-3} \exp(a|x_j|\sqrt{3N-3})$$

we need only show that ψ is an analytic vector for each position operator x_j. We can write $H = H_0 + V$ with $H_0 = t(p)$ where $t(v) = \sum_{i<j} a_{ij} v_i v_j$ and p is the $(3N-3)$-tuple of operators with $p_j = -i\,\partial/\partial x_j$ and $\{a_{ij}\}$ is positive definite.

The idea is to employ the dilation analytic techniques of Section 10 but with a different group. Fix j and let $W(\alpha) = e^{i\alpha x_j}$. Then
$$H_0(\alpha) \equiv W(\alpha)H_0 W(\alpha)^{-1} = t(p_1, \ldots, p_j - \alpha, \ldots, p_{3N-3})$$
and
$$V(\alpha) \equiv W(\alpha)VW(\alpha)^{-1} = V$$
Thus both $H_0(\alpha)$ and $V(\alpha)$ have analytic continuations to the entire α plane. Moreover, it is easy to see that for each fixed α, $\tfrac{1}{2}H_0 \le \operatorname{Re} H_0(\alpha) + c_\alpha$ where c_α is an α-dependent constant. It follows that $H(\alpha) \equiv H_0(\alpha) + V(\alpha)$ is an entire analytic family of type (B).

The theory of Section XII.2 is therefore applicable, so there is an a and functions $f_1, \ldots, f_k$ in $\{\alpha \,|\, |\alpha| < a\}$ with at worst algebraic singularities at $\alpha = 0$ so that the branches of $f_1(\alpha), \ldots, f_k(\alpha)$ are all the eigenvalues of $H(\alpha)$ near E. Moreover, there exists a projection valued analytic function $P(\alpha)$ on $\{\alpha \,|\, |\alpha| < a\}$ so that Ran $P(\alpha)$ is the set of eigenvectors for $H(\alpha)$ with eigenvalues $f_1(\alpha), \ldots, f_k(\alpha)$. Now, for α real, $H(\alpha) = W(\alpha)HW(\alpha)^{-1}$ so for such α, $f_1(\alpha) = \cdots = f_k(\alpha) = E$ and $W(\alpha)P(0)W(\alpha)^{-1} = P(\alpha)$. It follows by analyticity that f_i is always equal to E and that
$$W(\alpha_0)P(\alpha)W(\alpha_0)^{-1} = P(\alpha + \alpha_0)$$
if α_0 is real and $|\alpha| < a$, $|\alpha + \alpha_0| < a$. The fact that ψ is an analytic vector for x_j now follows from the proposition below. ∎

Proposition (O'Connor's lemma) Let $W(\alpha) = e^{i\alpha A}$ be a one-parameter unitary group and let D be a connected region in $\mathbb{C}$ with $0 \in D$. Suppose that a projection-valued analytic function $P(\alpha)$ is given on D with $P(0)$ of finite rank so that
$$W(\alpha_0)P(\alpha)W(\alpha_0)^{-1} = P(\alpha + \alpha_0)$$
for all pairs $\langle \alpha, \alpha_0 \rangle$ with α_0 real and with $\alpha, \alpha + \alpha_0 \in D$. Let $\psi \in \operatorname{Ran} P(0)$. Then the function $\psi(\alpha) \equiv W(\alpha)\psi$ has an analytic continuation from $D \cap \mathbb{R}$ to D. In particular, ψ is an analytic vector for A.

Proof Let $\mathscr{A}_D$ denote the set of vectors φ for which $\varphi(\alpha) \equiv W(\alpha)\varphi$ has an analytic continuation from $D \cap \mathbb{R}$ to D. Then $\mathscr{A}_D$ is dense in the underlying Hilbert space (see Section X.6) so $P(0)[\mathscr{A}_D]$ is dense in Ran $P(0)$. Since Ran $P(0)$ is finite dimensional, $P(0)[\mathscr{A}_D] = \operatorname{Ran} P(0)$. Thus we need only prove that $P(0)[\mathscr{A}_D] \subset \mathscr{A}_D$. Let $\varphi \in \mathscr{A}_D$ and let $\eta(\alpha) = P(\alpha)\varphi(\alpha)$ for $\alpha \in D$. Clearly $\eta(\alpha)$ is analytic and for $\alpha \in D \cap \mathbb{R}$
$$\eta(\alpha) = P(\alpha)W(\alpha)\varphi = W(\alpha)(P(0)\varphi)$$
so $P(0)\varphi \in \mathscr{A}_D$. ∎

To obtain better results on how big the constant a in the statement $\psi \in D(e^{ar})$ can be, it is necessary to introduce some kinematics connected with the meaning of r.

Definition Given particles with masses $\mu_1, \ldots, \mu_N$ at points $\mathbf{r}_1, \ldots, \mathbf{r}_N$ we define the **total mass** $M = \sum \mu_i$, the **center of mass** $\mathbf{R} = M^{-1} \sum_{i=1}^{N} \mu_i \mathbf{r}_i$, and the **radius of gyration** $r = [M^{-1} \sum \mu_i(\mathbf{r}_i - \mathbf{R})^2]^{1/2}$.

Definition Given an N-body free quantum Hamiltonian $\tilde{H}_0$ on $L^2(\mathbb{R}^{3N})$, a set of coordinates $\zeta_1, \ldots, \zeta_{N-1}$ orthogonal to $\mathbf{R}$ is called a **set of normal coordinates** if and only if

$$\tilde{H}_0 = -(2M)^{-1}\left[\Delta_\mathbf{R} + \sum_{i=1}^{N} \Delta_{\zeta_i}\right]$$

To see that normal coordinates exist, pass first to a set of Jacobi coordinates, $\xi_1, \ldots, \xi_{N-1}$ so that

$$\tilde{H}_0 = (-2M)^{-1}\Delta_\mathbf{R} + \sum (-2M_j)^{-1}\Delta_{\xi_j}$$

and let $\zeta_j = (M_j/M)^{1/2}\xi_j$.

Proposition

(a) The radius of gyration r is a function of the coordinates orthogonal to $\mathbf{R}$.
(b) If $\zeta_1, \ldots, \zeta_{N-1}$ is a set of normal coordinates, then

$$r = \left(\sum_{i=1}^{N-1} |\zeta_i|^2\right)^{1/2}$$

Proof We first notice that

$$Mr^2 + MR^2 = \sum_{i=1}^{N} \mu_i r_i^2$$

Let $\eta_i = (\mu_i/M)^{1/2}r_i$ so that

$$\tilde{H}_0 = -(2M)^{-1}\sum_{i=1}^{N}\Delta_{\eta_i} \tag{68a}$$

and

$$Mr^2 + MR^2 = M\sum_{i=1}^{N}\eta_i^2 \tag{68b}$$

By (68a), the coordinate change from $\langle \eta_1, \ldots, \eta_N \rangle$ to $\langle \mathbf{R}, \zeta_1, \ldots, \zeta_{N-1} \rangle$ is orthogonal if $\langle \zeta_1, \ldots, \zeta_{N-1} \rangle$ is a set of normal coordinates, and thus by (68b),

$$r^2 + R^2 = R^2 + \sum_{i=1}^{N-1} \zeta_i^2 \quad \blacksquare$$

Theorem XIII.40 (O'Connor–Combes–Thomas theorem) Let $\tilde{H} = \tilde{H}_0 + \sum_{i<j} V_{ij}(r_i - r_j)$ be an N-body Hamiltonian on $L^2(\mathbb{R}^{3N})$ with each $V_{ij} \in R + (L^\infty)_\varepsilon$. Let H be the operator on $L^2(\mathbb{R}^{3N-3})$ obtained from $\tilde{H}$ by removing the center of mass motion. Let $\Sigma = \inf \sigma_{\text{ess}}(H)$ and let ψ be an eigenfunction of H with eigenvalue $E < \Sigma$. Then $\psi \in D(e^{ar})$, where r is the radius of gyration, whenever $a^2 < 2M(\Sigma - E)$.

Proof We outline the proof and leave the details to the reader. Let $\zeta_1, \ldots, \zeta_{N-1}$ be a set of normal coordinates and define $W(\alpha) = \exp(i(\alpha_1 \cdot \zeta_1 + \cdots + \alpha_{N-1} \cdot \zeta_{N-1}))$ for $\alpha = \langle \alpha_1, \ldots, \alpha_{N-1} \rangle \in \mathbb{C}^{3N-3}$. Suppose we can prove that $\psi \in D(W(\alpha))$ if $|\alpha| \equiv (|\alpha_1|^2 + \cdots + |\alpha_{N-1}|^2)^{1/2} < \sqrt{2M(\Sigma - E)}$. Then a simple geometric argument (Problem 86a) shows that $\psi \in D(e^{ar})$ if $a < \sqrt{2M(\Sigma - E)}$.

It is thus sufficient to prove that $W(\alpha)\psi$ defined for $\alpha \in \mathbb{R}^{3N-3}$ has a continuation into the region $\{\alpha \,|\, |\text{Im } \alpha|^2 < 2M(\Sigma - E)\}$. Let

$$H_0(\alpha) = -(2M)^{-1} \sum_{i=1}^{N-1} (\nabla_{\zeta_i} - i\alpha_i)^2$$

and let $H(\alpha) = H_0(\alpha) + V$. Then, an analysis similar to that given in Section 10 (Problem 86b) shows that

$$\sigma_{\text{ess}}(H(\alpha)) = \bigcup_D \{\lambda + \mu \,|\, \lambda \in \Sigma_D(\alpha), \mu \in K_D\}$$

where $\Sigma_D(\alpha)$ are the thresholds associated with a cluster decomposition D and $K_D = \{t_D(p - \alpha) \,|\, p \in \mathbb{R}^{3N-3}\}$ where t_D is the kinetic energy between the clusters in D. By induction, it is easy to show (Problem 86c) that

$$\sigma_{\text{ess}}(H(\alpha)) \subset \{\lambda \,|\, \text{Re } \lambda \geq \Sigma - (2M)^{-1}(\text{Im } \alpha)^2\}$$

Thus if $(\text{Im } \alpha)^2 < \sqrt{2M(\Sigma - E)}$, E remains disjoint from $\sigma_{\text{ess}}(H(\alpha))$. Thus, by following the arguments of Section 10, E is an eigenvalue of $H(\alpha)$ for all such α. The proof is completed (Problem 86d) by mimicking the proof of Theorem XIII.39 and employing O'Connor's lemma. $\blacksquare$

If the potentials are dilation analytic, one can also prove exponential falloff for eigenfunctions associated with eigenvalues in the continuum. This result will have applications in Section 13.

XIII.11 Properties of eigenfunctions

Theorem XIII.41 Let $\tilde{H} = \tilde{H}_0 + \sum_{i<j} V_{ij}$ where each multiplication operator V_{ij} is a dilation analytic potential in some $\mathscr{F}_\beta$. Let E be a non-threshold eigenvalue for H with eigenfunction ψ. Then:

(a) $\psi \in D(e^{ar})$ for some $a > 0$.
(b) If $U(\theta)$ is the family of dilations, and $\beta_0 = \min\{\beta, \tfrac{1}{2}\pi\}$, then $U(\theta)\psi \equiv \psi(\theta)$ has a continuation into the strip $\{\theta \,|\, |\mathrm{Im}\,\theta| < \beta_0\}$ and for each θ in the strip, $\psi(\theta) \in D(e^{ar})$ for some $a > 0$.
(c) If E is larger than the largest threshold, then the β_0 of (b) may be chosen to be $\min\{\beta, \pi\}$.
(d) If each V_{ij} has the property that $V_{ij}(\theta)$ has a continuation into the strip $\{\theta\,|\,|\mathrm{Im}\,\theta| \leq \beta\}$ analytic in the interior, continuous on the whole strip, and if $\beta < \tfrac{1}{2}\pi$ (or if E is larger than the largest threshold, if $\beta < \pi$), then $\psi(\theta)$ has a continuation to that strip, analytic in the interior, continuous on the whole strip, with $\psi(\theta) \in D(e^{ar})$ for some $a > 0$.

Proof By the analysis of Section 10, E is an eigenvalue of $H(\theta)$ for all θ in the strip $\{\theta\,|\,|\mathrm{Im}\,\theta| < \beta_0\}$; or in case (d), $\{\theta\,|\,|\mathrm{Im}\,\theta| \leq \beta_0\}$. Let $P(\theta)$ be the associated projection, defined as a spectral projection if θ is real and by the method of Section XII.2 if $\mathrm{Im}\,\theta \neq 0$. Clearly, $P(\theta)$ is analytic in the regions $\{\theta\,|\,0 < |\mathrm{Im}\,\theta| < \beta_0\}$ and by the arguments in Section 10, $P(\theta)$ is continuous in $\{\theta\,|\,0 \leq |\mathrm{Im}\,\theta| < \beta_0\}$. Moreover $P(\theta)$ is self-adjoint if θ is real. By the Schwartz reflection principle, $P(\theta)$ is analytic in the whole strip $\{\theta\,|\,|\mathrm{Im}\,\theta| < \beta_0\}$. The analyticity properties of $\psi(\theta)$ now follow from O'Connor's lemma.

If $\mathrm{Im}\,\theta \neq 0$, E is a discrete eigenvalue so we can prove that $\psi(\theta) \in D(e^{ar})$ for some $a > 0$ by following the proof of Theorem XIII.39. Let us suppose that $\psi(\theta_0) \in D(\exp(a_0 e^{\theta_0} r))$ with $|\mathrm{Im}\,\theta_0| < \tfrac{1}{4}\pi$. We will show that $\psi \in D(e^{a_0 r})$. Let $\varepsilon > 0$ be given, let $\eta \in L^2(\mathbb{R}^{3N-3})$, and define

$$F(\theta) = (\eta, \exp(-\varepsilon r^2 e^{2\theta})\exp(a_0 e^\theta r)\psi(\theta))$$

Clearly $F(\theta)$ is analytic in the strip $\{\theta\,|\,|\mathrm{Im}\,\theta| < |\mathrm{Im}\,\theta_0| + \delta\}$ for suitable δ. Moreover $F(\theta)$ is uniformly bounded on the strip and for $\mathrm{Im}\,\theta = \pm|\mathrm{Im}\,\theta_0|$, we have $|F(\theta)| \leq \|\eta\|\,\|\exp(a_0 e^{\theta_0} r)\psi(\theta_0)\|$. By the maximum principle, this last inequality holds for all θ; in particular

$$|(\eta, e^{-\varepsilon r^2} e^{a_0 r}\psi)| \leq C\|\eta\|$$

for all η and ε. It follows that $\psi \in D(\exp(a_0 r))$. ∎

As with Theorem XIII.39, Theorem XIII.41 has a sharp form where an estimate on a_0 is given (see the Notes).

We finally turn to proving pointwise exponential bounds:

Theorem XIII.42 Let H be an N-body Hamiltonian on $L^2(\mathbb{R}^{3N-3})$ with potentials in $M_\sigma^{(3)}$ for some $\sigma > \frac{3}{2}$. Suppose that $\psi \in C^\infty(H)$ and for some a and all n, $H^n\psi \in D(e^{ar})$ where r is the radius of gyration. Then, for each ε, there is a constant C_ε with

$$|\psi(\zeta)| \leq C_\varepsilon e^{-(a-\varepsilon)r}$$

for all ζ. In particular, if the hypotheses of Theorem XIII.40 hold and, in addition, each $V_{ij} \in M_\sigma^{(3)}$, then

$$|\psi(\zeta)| \leq C_a e^{-ar}$$

for any $a < \sqrt{2M(\Sigma - E)}$.

Proof The proof is an application of the Fourier transform, especially of the ideas behind the Paley–Wiener theorem. Pass to normal coordinates so that $r = (\sum_{i=1}^{N-1} |\zeta_i|^2)^{1/2}$. If $\varphi \in D(e^{ar})$, then for any $k \in \mathbb{C}^{3N-3}$ with $|\text{Im } k| < a$, $\varphi_k \equiv e^{-ik\cdot\zeta}\varphi(\zeta)$ is in $L^1(\mathbb{R}^{3N-3})$ because $\exp(-br) \in L^2$ for any $b > 0$. Moreover, the map $k \mapsto \varphi_k$ is analytic in k as an L^1-valued function. Thus the function

$$\hat{\varphi}(k) = (2\pi)^{-(3N-3)/2} \int \varphi_k(\zeta)\, d\zeta$$

is analytic in the tube $\{k \mid |\text{Im } k| < a\}$. By the Plancherel theorem, for any $b < a$, there is a constant C_b with

$$\int_{k \in \mathbb{R}^{3N-3}} |\hat{\varphi}(k + i\kappa)|^2\, d^{3N-3}k < C_b \tag{69}$$

if $|\kappa| \leq b$. At this point, we follow the proof of Theorem XIII.38. By analytic continuation, $\hat{\varphi}(k)$ obeys the equation

$$(2M)^{-1}k^2\hat{\varphi}(k) + (2\pi)^{-(3N-3)/2}\int \hat{V}(\ell)\hat{\varphi}(k-\ell)\, d^{3N-3}\ell = \widehat{H\varphi}(k)$$

for all k with $|\text{Im } k| < a$ whenever φ and $H\varphi$ are in $D(e^{ar})$. By following the proof of Theorem XIII.38 and using (69) one shows (Problem 88) that for any $b < a$, there is constant D_b with

$$\int |\hat{\psi}(k + i\kappa)|\, d^{3N-3}\kappa < D_b \tag{70}$$

if $|\kappa| \leq b$ and if $\psi, H\psi, \ldots, H^m\psi \in D(\exp(ar))$ for suitable m. Since $\hat\psi(\cdot + i\kappa)$ is the Fourier transform of $e^{\kappa\cdot\zeta}\psi(\zeta)$, (70) implies that

$$|e^{\kappa\cdot\zeta}\psi(\zeta)| \leq D_b(2\pi)^{-(3N-3)/2}$$

if $|\kappa| \leq b$. Taking a sup over such κ we see that

$$|e^{br}\psi(\zeta)| \leq D_b(2\pi)^{-(3N-3)/2} \quad \blacksquare$$

Corollary Let H be an atomic Hamiltonian. Then any eigenfunction ψ corresponding to a point in the discrete spectrum obeys an estimate of the form

$$|\psi(\zeta)| \leq C_a e^{-ar}$$

for some positive a.

XIII.12 Nondegeneracy of the ground state

Our main result in this section will concern a self-adjoint operator H that is bounded below and has an eigenvalue as the lowest point in its spectrum. Under certain conditions, we shall show that the eigenspace corresponding to that eigenvalue is one dimensional and that the corresponding eigenvector is a strictly positive function in a particular realization of the underlying Hilbert space as an L^2 space. The eigenvalue is called the **ground state energy** and the eigenvector the **ground state**. As is the case with so much spectral analysis, the method of verifying the conditions in applications is a perturbative technique. At first sight this is surprising since we shall be interested in Schrödinger operators and $H_0 = -\Delta$ does not possess any eigenfunctions! It will thus be important that all our results will be in terms of conclusions of the form: "Either H has no eigenvalue below the essential spectrum or it has a nondegenerate ground state energy with associated eigenvector strictly positive." Finally, we note that, as usual, it is easier to study the bounded operators e^{-tH} and $(H + 1)^{-1}$ rather than H. If E_0 is the ground state energy for H, then e^{-tE_0} is the *largest* eigenvalue of e^{-tH}. We begin by studying a bounded operator A and its largest eigenvalue.

Definition Let $\langle M, \mu \rangle$ be a σ-finite measure space. $f \in L^2(M, d\mu)$ is called **positive** if f is nonnegative a.e. and is not the zero function. f is called **strictly positive** if $f(x) > 0$ a.e.. A bounded operator A on L^2 is called **positivity preserving** if Af is positive whenever f is positive. A is called **positivity**

improving if Af is strictly positive whenever f is positive. Finally, A is called **ergodic** if and only if it is positivity preserving and for any $u, v \in L^2$ that are both positive there is some $n > 0$ for which $(u, A^n v) \neq 0$.

We remark that by definition the zero function is not positive; so if A is positivity preserving, then Af is not the zero function for any positive function f. Moreover, by the following reformulation, every positivity improving map is ergodic: A function $g \in L^2(M, d\mu)$ is strictly positive if and only if $(f, g) > 0$ for all positive functions f. Thus, a bounded operator A on $L^2(M, d\mu)$ is positivity improving if and only if $(u, Av) > 0$ for all positive functions $u, v \in L^2(M, d\mu)$.

Example 1 Let $A = (-\Delta + 1)^{-1}$ on $L^2(\mathbb{R}^3)$. Equation (IX.30) states that

$$(f, Ag) = \int \overline{f(x)} g(y) \frac{e^{-|x-y|}}{4\pi |x-y|} d^3x \, d^3y$$

Since $(4\pi|x-y|)^{-1} e^{-|x-y|}$ is strictly positive, A is positivity improving. Similarly, the explicit formula for $e^{t\Delta}$ on $L^2(\mathbb{R}^n)$ shows that it is also positivity improving.

Example 2 Let $\langle M, \mu \rangle$ be a probability measure space, i.e., $\mu(M) = 1$ and let $T: M \to M$ be a measure preserving map. Let $(Af)(x) = f(Tx)$. A is a unitary operator that is clearly positive preserving. It is never positivity improving. A is ergodic if and only if T is ergodic in the sense defined in Section II.5 (Problem 90).

Example 3 (second quantized operators on a Fock space) We saw in the proof of Theorem X.61 (Section X.9) that $\Gamma(e^{-tA})$ is positivity preserving as an operator on Q space whenever A is a positive self-adjoint operator on the one-particle space that commutes with the distinguished complex conjugation. We shall examine later the question of when it is ergodic or positivity improving.

The main abstract theorem of this section is the following:

Theorem XIII.43 Let A be a bounded, positive operator on $L^2(M, d\mu)$. Suppose that A is positivity preserving and that $\|A\|$ is an eigenvalue. Then the following are equivalent:

(a) $\|A\|$ is a simple eigenvalue and the associated eigenvector is strictly positive.

(b) A is ergodic.
(c) $L^\infty(M) \cup \{A\}$ acts irreducibly, i.e., no nontrivial closed subspace is left invariant by both A and every bounded multiplication operator.

Proof We prove (a) $\Rightarrow$ (b) $\Rightarrow$ (c) $\Rightarrow$ (a).

(a) $\Rightarrow$ (b) Let $B = A/\|A\|$ and let $\{P_\Omega\}$ be the spectral projections for B. Since $x^n \to 0$ if $0 \le x < 1$ and $x^n \to 1$ if $x = 1$, the functional calculus implies that

$$\operatorname*{s-lim}_{n \to \infty} B^n = P_{\{1\}}$$

By hypothesis, $P_{\{1\}} = (\psi, \cdot)\psi$ where ψ is strictly positive. Thus for any positive u, v,

$$\lim_{n \to \infty} (u, B^n v) = (u, \psi)(\psi, v) > 0$$

As a result, for some n, $(u, A^n v) = \|A\|^n (u, B^n v) > 0$.

(b) $\Rightarrow$ (c) Suppose (c) does not hold. Let S be a nontrivial subspace left invariant by $L^\infty(M) \cup \{A\}$. Let $f \in S$ and let $h = \bar{f}/|f| \in L^\infty(M)$. Then $|f| = hf \in S$. Similarly, if $g \in S^\perp$, then $|g| \in S^\perp$. Choose $f \in S$, $g \in S^\perp$, $f \ne 0 \ne g$. Since A leaves S invariant, $A^n |f| \in S$, so

$$(|g|, A^n |f|) = 0$$

for all n. Thus A is not ergodic.

(c) $\Rightarrow$ (a) By hypothesis, $\|A\|$ is an eigenvalue. Let ψ be an eigenvector of A with eigenvalue $\|A\|$ and suppose that ψ is real valued. Since $|\psi| \pm \psi \ge 0$, we know that $A(|\psi| \pm \psi) \ge 0$ so

$$|A\psi| \le A|\psi|$$

Thus

$$(|\psi|, A|\psi|) \ge (|\psi|, |A\psi|) \ge (\psi, A\psi) = \|A\| \|\psi\|^2$$

As a result, $A|\psi| = \|A\| |\psi|$ so $|\psi|$ is also an eigenvector. We shall show that $|\psi|$ is strictly positive. Let $S = \{f \in L^2(M, d\mu) \mid f\bar\psi = 0 \text{ a.e.}\}$. S is clearly a subspace and moreover it is left invariant by $L^\infty(M)$. Let $S_+ = \{g \in S \mid g \ge 0\}$. Then for $f \in S_+$, $(Af, |\psi|) = (f, A|\psi|) = \|A\|(f, |\psi|) = 0$. Since Af is positive, $(Af)\bar\psi = 0$ a.e., that is, $Af \in S_+$. Thus, A leaves S_+ invariant. Since $S = S_+ - S_+ + i(S_+ - S_+)$, A leaves S invariant. Thus by hypothesis (c), $S = \{0\}$ or $S = \mathscr{H}$. Since $\psi \notin S$ and $\psi \ne 0$, $S = \{0\}$, from which it follows that $|\psi| > 0$ a.e.

Thus we know that any real eigenvector with eigenvalue $\|A\|$ is nonzero a.e. and that also $A|\psi| = \|A\| |\psi|$. Thus $|\psi| - \psi$ is either an eigenvector with eigenvalue $\|A\|$ or it is identically 0 so that $|\psi| - \psi$ is either almost everywhere vanishing or almost everywhere nonvanishing. This means that every real eigenvector with eigenvalue $\|A\|$ is a.e. strictly positive or a.e. strictly negative. If A had two real eigenvectors with eigenvalue $\|A\|$, it would have two orthogonal eigenvectors which were both a.e. positive. Since this is impossible, we conclude that A has only one real eigenvector of eigenvalue $\|A\|$ and that this eigenvector is strictly positive.

Finally, let ψ be an arbitrary eigenvector with eigenvalue $\|A\|$. Since A takes positive functions into positive functions, it takes real functions into real functions, so that $\text{Re}(A\psi) = A(\text{Re }\psi)$. Thus Re ψ and Im ψ are eigenvectors with eigenvalue $\|A\|$. It follows that ψ is a complex multiple of the unique real eigenvector. ∎

To apply this to the lowest eigenvalue of a self-adjoint operator, we can use either the resolvent or the semigroup generated by the operator. We first note:

Proposition Let H be a self-adjoint operator that is bounded from below. Let $E = \inf \sigma(H)$. Then e^{-tH} is positivity preserving for *all* $t > 0$ if and only if $(H - \lambda)^{-1}$ is positivity preserving for *all* $\lambda < E$.

Proof The proposition follows from the formulas

$$(H - \lambda)^{-1}\varphi = \int_0^\infty e^{\lambda t} e^{-Ht}\varphi \, dt \qquad (\lambda < E)$$

$$e^{-tH}\varphi = \lim_{n \to \infty} \left(1 + \frac{tH}{n}\right)^{-n} \varphi \qquad (t > 0) \tag{71}$$

These formulas, which are familiar from the theory of semigroups, (Section X.8) can be proven easily in the self-adjoint case by the functional calculus.

Theorem XIII.44 Let H be a self-adjoint operator that is bounded from below on $L^2(M, d\mu)$. Suppose that e^{-tH} is positivity preserving for all $t > 0$ and that $E = \inf \sigma(H)$ is an eigenvalue. Then the following are equivalent:

(a) E is a simple eigenvalue with a strictly positive eigenvector.
(b) $(H - \lambda)^{-1}$ is ergodic for some $\lambda < E$.
(c) e^{-tH} is ergodic for some $t > 0$.
(d) $(H - \lambda)^{-1}$ is positivity improving for all $\lambda < E$.
(e) e^{-tH} is positivity improving for all $t > 0$.

Proof By Theorem XIII.43, (a), (b), (c) are equivalent and clearly (d) ⇒ (b) and (e) ⇒ (c). We complete the proof by showing that (c) ⇒ (d) and (e).

(c) ⇒ (d) Let u and v be positive. Since some e^{-tH} is ergodic, $(u, e^{-sH}v) > 0$ for some $s > 0$. But $(u, e^{-sH}v)$ is continuous so $(u, e^{-sH}v) > 0$ for s in some interval. Thus

$$(u, (H - \lambda)^{-1}v) = \int_0^\infty e^{\lambda s}(u, e^{-sH}v)\, ds > 0$$

(c) ⇒ (e) Let u, v be positive and let $B = \{t > 0 \mid (u, e^{-tH}v) > 0\}$. Since B is nonempty and $(u, e^{-tH}v)$ is analytic in a neighborhood of the positive real axis, $(0, \infty)\setminus B$ can have only 0 as a limit point. In particular B contains arbitrarily small numbers. Thus, if we can show that $t > s$ and $s \in B$ implies $t \in B$, we can conclude that $B = (0, \infty)$. Fix $s \in B$. Then $(u, e^{-sH}v) > 0$ so that $u(\cdot)(e^{-sH}v)(\cdot)$ is not identically zero. Let $w = \min\{u, e^{-sH}v\}$. Then $w(m)$ is not identically zero. Since e^{-tH} is positivity preserving,

$$(u, e^{-\tau H}(e^{-sH}v)) \geq (u, e^{-\tau H}w) = (e^{-\tau H}u, w)$$
$$\geq (e^{-\tau H}w, w) = \|e^{-\tau H/2}w\|^2 > 0$$

We have used the fact that w is positive and that $e^{-\tau H/2}$ is positivity preserving to conclude that $e^{-\tau H/2}w \neq 0$. Thus $s \in B$ and $\tau > 0$ imply $s + \tau \in B$. ∎

Example 3, revisited Let A be an operator on a complex Hilbert space $\mathscr{H}$ that commutes with a distinguished complex conjugation and which obeys $A \geq cI$ for some $c > 0$. Let $H = d\Gamma(A)$ be the second quantization of A on $\mathscr{F}_s(\mathscr{H})$ thought of as an operator on $L^2(Q, d\mu)$ (see Section X.7). Then $H\Omega_0 = 0$ while $A \upharpoonright \{\Omega_0\}^\perp \geq cI$. Thus H has a nondegenerate strictly positive ground state. We conclude that $\Gamma(e^{-tA}) = e^{-tH}$ is positivity improving for all $t > 0$.

In applications of Theorem XIII.44, the following perturbation theorem is very useful. There is another perturbation result based on the resolvent rather than the semigroup which is also useful (see Problems 91, 92).

Theorem XIII.45 Let H and H_0 be semibounded, self-adjoint operators on $L^2(M, d\mu)$ where $\langle M, \mu \rangle$ is a σ-finite measure space. Suppose that there exists a sequence of bounded multiplication operators V_n so that $H_0 + V_n$ converges to H in strong resolvent sense and so that $H - V_n$ converges to H_0 in strong resolvent sense. Suppose, moreover, that $H - V_n$ and $H_0 + V_n$ are uniformly bounded from below. Then:

(a) e^{-tH} is positivity preserving if and only if e^{-tH_0} is positivity preserving.
(b) $\{e^{-tH}\} \cup L^\infty(M, d\mu)$ acts irreducibly on $L^2(M, d\mu)$ if and only if $\{e^{-tH_0}\} \cup L^\infty(M, d\mu)$ acts irreducibly on $L^2(M, d\mu)$.

Proof By the Trotter product formula (Theorem VIII.31) and the continuity of the functional calculus (Theorem VIII.20),

$$e^{-tH} = \underset{n\to\infty}{\text{s-lim}} \left(\underset{m\to\infty}{\text{s-lim}} \; [e^{-tH_0/m} e^{-tV_n/m}]^m \right) \quad (72a)$$

$$e^{-tH_0} = \underset{n\to\infty}{\text{s-lim}} \left(\underset{m\to\infty}{\text{s-lim}} \; [e^{-tH/m} e^{+tV_n/m}]^m \right) \quad (72b)$$

Since $e^{\pm tV_n/m}$ is positivity preserving, we see that (a) holds. Moreover, (72) and the fact that $e^{\pm tV_n/m} \in L^\infty(M)$ imply that any subspace left invariant by e^{-tH_0} and $L^\infty(M)$ is left invariant by e^{-tH}. ∎

We can now apply Theorem XIII.44 to Schrödinger operators and to spatially cut-off $P(\varphi)_2$ Hamiltonians.

Theorem XIII.46 Let H be the Hamiltonian of an N-body Schrödinger system with center of mass motion removed. Suppose that the potentials V_{ij} are in $R + (L^\infty)_\varepsilon$. Then, if H has an eigenvalue at the bottom of its spectrum, the eigenvalue is nondegenerate and the corresponding eigenfunction is strictly positive.

Proof By Theorems XIII.43 and XIII.44, we need only prove that e^{-tH} is positivity preserving and that $\{e^{-tH}\} \cup L^\infty(\mathbb{R}^{3N-3})$ acts irreducibly. By Example 1 and the proof (b) ⇒ (c) in Theorem XIII.43, we know these facts for e^{-tH_0}. Let $V_{ij}^{(n)}$ be defined by

$$V_{ij}^{(n)}(x) = \begin{cases} V_{ij}(x) & \text{if } |V_{ij}(x)| \leq n \\ n & \text{if } V_{ij}(x) > n \\ -n & \text{if } V_{ij}(x) < -n \end{cases}$$

It is easy to prove that $V_{ij} - V_{ij}^{(n)} \to 0$ in the Rollnik norm. By Theorem VIII.25c, $H_0 + \sum V_{ij}^{(n)} \to H$ and $H - \sum V_{ij}^{(n)} \to H_0$ in norm resolvent sense. Thus, Theorem XIII.45 is applicable and the proof is completed. ∎

Corollary Let H satisfy the hypothesis of Theorem XIII.46 and suppose that $H\psi = E\psi$ where $E = \inf \sigma(H)$. Let U be any positivity preserving unitary operator commuting with H. Then $U\psi = \psi$.

Proof Since U commutes with H, $H(U\psi) = E(U\psi)$. But E is nondegenerate so $U\psi = \alpha\psi$. Since U is unitary, $|\alpha| = 1$ and since U is positivity preserving and ψ is positive, $\alpha = 1$. ∎

For example, if the V_{ij} are all spherically symmetric and H has an eigenvalue at the bottom of its spectrum, that eigenvector is left invariant by rotations and reflections.

We warn the reader that Theorem XIII.46 deals with the lowest eigenvalue of H as an operator on all of $L^2(\mathbb{R}^{3N-3})$. If H commutes with a family $\mathscr{P}$ of permutations of the coordinates and we restrict to a subspace of $L^2(\mathbb{R}^{3N-3})$ with a certain permutation symmetry, then the lowest eigenvalue of the restricted operator need not be nondegenerate. In fact, by the corollary, unless we look at the subspace left *pointwise* invariant by $\mathscr{P}$, the ground state of the restricted operator is an excited state of the unrestricted operator. This implies the statement that *ground states of Boson systems are nondegenerate while ground states of Fermi systems may be degenerate*.

Theorem XIII.47 Let $V \in L^2_{\text{loc}}(\mathbb{R}^m)$ be positive and suppose $\lim_{|x|\to\infty} V(x) = \infty$. Then $-\Delta + V$ has a nondegenerate strictly positive ground state.

Proof We shall see in Section 14 that $-\Delta + V$ has purely discrete spectrum so the bottom of its spectrum is certainly an eigenvalue. The proof of the remaining conclusions follows closely that of Theorem XIII.46 with one detail different. Let $V_n = \min\{V, n\}$. Then $-\Delta + V_n$, $-\Delta + V$, $-\Delta$, and $-\Delta + (V - V_n)$ are all essentially self-adjoint on $C_0^\infty(\mathbb{R}^m)$ by Theorem X.28. Moreover, for any $\psi \in C_0^\infty(\mathbb{R}^m)$, $V_n \psi \to V\psi$ in L^2. Thus by Theorem VIII.25a we have the necessary strong resolvent convergence. ∎

If V is allowed to be very singular, then $-\Delta + V$, defined as a sum of quadratic forms, can have a degenerate ground state (see Example 2 in Section 13 and Problem 95). A reference for the following result is given in the Notes.

Theorem XIII.48 Let G be a closed subset of $\mathbb{R}^n$ of measure zero and suppose that $V \in L^1_{\text{loc}}(\mathbb{R}^n \setminus G)$ with $V \geq 0$. Let $H = -\Delta + V$ be defined as the sum of quadratic forms. Then:

(a) If $\mathbb{R}^n \setminus G$ is connected and $\inf \sigma(H)$ is an eigenvalue, then it is a simple eigenvalue and the corresponding eigenfunction is strictly positive.
(b) If $\mathbb{R}^n \setminus G$ is disconnected, V can be chosen so that H has a degenerate ground state.

Finally, we turn to existence and uniqueness of ground states in the framework of hypercontractive semigroups (see Section X.9):

XIII: SPECTRAL ANALYSIS

Theorem XIII.49 Let H_0 generate a positivity improving hypercontractive semigroup on a probability measure space $\langle M, \mu \rangle$. Suppose that V is a (not necessarily bounded) multiplication operator with V and e^{-V} in $\bigcap_{p<\infty} L^p(M, d\mu)$. Let $H = H_0 + V$. Then:

(a) $E = \inf \sigma(H)$ is an eigenvalue.
(b) E is nondegenerate.
(c) The ground state eigenfunction is strictly positive.

In particular, (a)–(c) hold for the spatially cutoff $P(\varphi)_2$ Hamiltonians.

Proof We shall first prove (b) and (c) assuming (a) and then prove (a). Let V_n be defined by

$$V_n(q) = \begin{cases} V(q) & \text{if } |V(q)| \leq n \\ n & \text{if } V(q) > n \\ -n & \text{if } V(q) < -n \end{cases}$$

Since $V_n \to V$ in each L^p ($p < \infty$) and $\sup_n \{\|e^{-V_n}\|_p, \|e^{-(V-V_n)}\|_p\} < \infty$, Theorem X.60 says that $H_0 + V_n \to H_0 + V$ and $H_0 + V - V_n \to H_0$ in norm resolvent sense. Since e^{-tH_0} is a positivity improving semigroup, Theorem XIII.45 is applicable and thus (b) and (c) hold if (a) is true.

To prove (a), we proceed as follows. Let t be chosen so that $A = \exp(-tH)$ is bounded from L^2 to L^4. We need only prove that $\|A\|$ is an eigenvalue of A. A **finite partition** of M is a family $\{S_1, \ldots, S_n\} = \alpha$ of measurable subsets of M that are mutually disjoint, and whose union is M. If α and β are partitions, we write $\beta > \alpha$ if each set in β is a subset of a set in α. For each partition α, define the projection

$$P_\alpha = \sum_{S \in \alpha} \mu(S)^{-1}(\chi_S, \cdot)\chi_S$$

and let $A_\alpha = P_\alpha A P_\alpha$. We first prove the following properties of $\{P_\alpha\}$:

(i) s-$\lim_\alpha P_\alpha = 1$.
(ii) P_α is positivity preserving.
(iii) P_α is a contraction on each $L^p(M, d\mu)$.

To prove (i), we note that for any simple function ψ, $P_\alpha \psi = \psi$ for α sufficiently large. Since the simple functions are dense, (i) follows. (ii) is obvious, and (iii) follows from (ii), $P_\alpha 1 = 1$, and Theorem X.55a. Thus $\{A_\alpha\}$ obeys:

(iv) s-$\lim_\alpha A_\alpha = A$.
(v) $\|A\| = \lim_\alpha \|A_\alpha\|$.
(vi) There is a K with $\|A_\alpha \psi\|_4 \leq K\|A_\alpha\| \|\psi\|_2$ for all $\psi \in L^2$ and all α.

(iv) follows from (i). (vi) follows from (iii) and the fact that $\|A\psi\|_4 \leq \tilde{K}\|\psi\|_2$ for some $\tilde{K}$ and the bound $\|A_\alpha\| \geq$ (A1, 1). To prove (v), notice first that $\|A_\alpha\| \leq \|A\|$ since $\|P_\alpha\| = 1$. On the other hand, by (iv), $\|A\| \leq \underline{\lim}\|A_\alpha\|$. Thus $\lim\|A_\alpha\|$ exists and equals $\|A\|$.

Fix a partition $\alpha = \{S_1, \ldots, S_n\}$. Then Ran P_α is naturally isomorphic to $\mathbb{C}^n$ via the map $\sum c_i \chi_{S_i} \mapsto (c_1, \ldots, c_n)$. View $\mathbb{C}^n$ as $L^2(\{1, \ldots, n\}, dv)$ with $v(i) = \mu(S_i)$. Then A_α is positivity preserving on $\mathbb{C}^n$. Since A_α is a positive definite finite-dimensional matrix, $\|A_\alpha\|$ is an eigenvalue, so we can find $\varphi_\alpha \in$ Ran P_α with $A_\alpha \varphi_\alpha = \|A_\alpha\|\varphi_\alpha$. By the argument in the proof of Theorem XIII.43, $\psi_\alpha = |\varphi_\alpha|$ is an eigenvector. Normalize ψ_α so that $\|\psi_\alpha\|_2 = 1$. Then, on the one hand, $\|\psi_\alpha\|_4 \leq K$, and on the other $1 = \|\psi_\alpha\|_2 \leq \|\psi_\alpha\|_1^{1/3} \|\psi_\alpha\|_4^{2/3}$ by Hölder's inequality. Thus $\|\psi_\alpha\|_1 \geq K^{-2}$. Let ψ be a weak limit point of the net $\{\psi_\alpha\}$, and let $\{\psi_\beta\}$ be a subnet converging to ψ. Then for any $\varphi \in L^2$,

$$(\varphi, A\psi) = (A\varphi, \psi) = \lim_\beta (A\varphi, \psi_\beta) = \lim_\beta \|A_\beta\|(\varphi, \psi_\beta) = \|A\|(\varphi, \psi)$$

so $A\psi = \|A\|\psi$. Moreover, since $\psi_\beta \geq 0$, $\|\psi_\alpha\|_1 = (1, \psi_\alpha) \geq K^{-2}$. Thus $(1, \psi) \geq K^{-2}$. In particular, $\psi \neq 0$. Thus $\|A\|$ is an eigenvalue. ∎

Appendix 1 to Section XIII.12 The Beurling–Deny criteria

There is an especially simple criterion for a positive self-adjoint operator H on $L^2(M, d\mu)$ to generate a positivity preserving semigroup.

Theorem XIII.50 (first Beurling–Deny criterion) Let $H \geq 0$ be a self-adjoint operator on $L^2(M, d\mu)$. Extend $(\psi, H\psi)$ to all of L^2 by setting it equal to infinity when $\psi \notin Q(H)$. Then the following are equivalent:

(a) e^{-tH} is positivity preserving for all $t > 0$.
(b) $(|u|, H|u|) \leq (u, Hu)$ for all $u \in L^2$.
(c) e^{-tH} is reality preserving and

$$(u_+, Hu_+) \leq (u, Hu)$$

for all real-valued $u \in L^2$. Here $u_+(x) \equiv \max\{u(x), 0\}$.
(d) e^{-tH} is reality preserving and

$$(u_+, Hu_+) + (u_-, Hu_-) \leq (u, Hu)$$

for all real-valued $u \in L^2$. Here $u_- = u_+ - u$.

Proof We shall prove that (a) ⇒ (b) ⇒ (a). The proofs that (a) ⇒ (d) and (c) ⇒ (a) are similar and left to the reader (Problem 96).

(a) ⇒ (b) With our convention on extending $(\cdot, H \cdot)$ to L^2,

$$(u, Hu) = \lim_{t \to 0} t^{-1}[(u, (1 - e^{-tH})u)] \tag{73}$$

By hypothesis (a), $|e^{-tH}u| \leq e^{-tH}|u|$, so

$$(u, e^{-tH}u) = \|e^{-tH/2}u\|^2 \leq \|e^{-tH/2}|u|\|^2 = (|u|, e^{-tH}|u|)$$

Thus, since $(u, u) = (|u|, |u|)$, we have that

$$(u, (e^{-tH} - 1)u) \leq (|u|, (e^{-tH} - 1)|u|)$$

so that (b) follows from (73).

(b) ⇒ (a) Suppose that u is a positive function and that $a > 0$. Let $w = (H + a)^{-1}u$ and define $Q(\varphi) = (\varphi, H\varphi) + a(\varphi, \varphi)$ on all of L^2. One first notes that for $\varphi \in D(H)$ and any ψ,

$$Q(\varphi + \psi) = Q(\varphi) + Q(\psi) + 2\operatorname{Re}((H + a)\varphi, \psi)$$

Thus, if $\operatorname{Re} v \geq 0$,

$$Q(w + v) = Q(w) + Q(v) + 2\operatorname{Re}(u, v) \geq Q(w) + Q(v)$$

Put differently, if $\varphi \equiv w + v$ obeys $\operatorname{Re} \varphi \geq \operatorname{Re} w$, then $Q(\varphi) \geq Q(w)$ with equality only if $\varphi = w$. By hypothesis (b), $Q(|w|) \leq Q(w)$. Since $\operatorname{Re}|w| \geq \operatorname{Re} w$, we see that $|w| = w$, that is, $w \geq 0$. We have thus seen that $(H + a)^{-1}$ is positivity preserving for any $a > 0$. Therefore, by the proposition preceding Theorem XIII.44, the semigroup is positivity preserving. ∎

Example 1 Criterion (d) implies that a finite-dimensional positive definite matrix generates a positivity preserving semigroup if and only if its off-diagonal elements are all negative. This is quite easy to prove directly in a more general form (Problem 97).

Example 2 One can see directly, without appealing to the Trotter product formula, that if V is a positive multiplication operator and H_0 generates a positivity preserving semigroup, then so does $H_0 + V$ defined as a sum of forms. For, when $Q(H) = Q(H_0)$, criterion (b) holds for H_0 if and only if it holds for H.

Example 3 It is easy to see (Problem 99) that the Dirichlet and Neumann Laplacians defined in Section 15 obey criterion (b) and thus generate positivity preserving semigroups; equivalently, the integral kernels of their resolvents ("Green's functions") are positive functions.

There is a slightly more subtle criterion for e^{-tH} to be a contraction on all the L^p spaces as well as positivity preserving.

Theorem XIII.51 (second Beurling–Deny criterion) Let $H \geq 0$ be a self-adjoint operator that generates a positivity preserving semigroup on $L^2(M, d\mu)$. Then the following are equivalent:

(a) e^{-tH} is a contraction on each $L^p(M, d\mu)$ for all p and all $t \geq 0$.
(b) e^{-tH} is a contraction on $L^\infty(M, d\mu)$ for all $t > 0$.
(c) Let $(f \wedge 1)(x) \equiv \min\{f(x), 1\}$. For each $f \geq 0$,

$$((f \wedge 1), H(f \wedge 1)) \leq (f, Hf)$$

(d) Let F be an arbitrary function from $\mathbb{C}$ to $\mathbb{C}$ obeying $|F(x)| \leq |x|$ and $|F(x) - F(y)| \leq |x - y|$. Then

$$(F(f), HF(f)) \leq (f, Hf)$$

for all $f \in L^2$.

Proof We shall show that (c) $\Rightarrow$ (b) and (a) $\Rightarrow$ (d). The implication (b) $\Rightarrow$ (a) follows by a simple modification of the duality interpolation argument used in the proof of Theorem X.55a, and (d) $\Rightarrow$ (c) follows if we note that $F(z) = \min\{\text{Re } z, 1\}$ obeys the restrictions on F.

(c) $\Rightarrow$ (b) Fix $u \in L^2$ with $0 \leq u \leq 1$. Define

$$\psi(v) = (v, Hv) + \|u - v\|^2 = (v, (H + 1)v) + \|u\|^2 - 2 \text{ Re}(u, v)$$

Let $R = (H + 1)^{-1}$. Then $\psi(Ru) = \|u\|^2 - (u, Ru)$ and

$$((Ru - v), (H + 1)(Ru - v)) = (u, Ru) + (v, (H + 1)v) - 2 \text{ Re}(u, v)$$

so we have

$$\psi(v) = \psi(Ru) + ((Ru - v), (H + 1)(Ru - v))$$

This says that $\psi(v)$ takes its minimum value when and only when $v = Ru$. Now, since $u \leq 1$,

$$|(u - (v \wedge 1))(x)| \leq |(u - v)(x)|$$

and, by hypothesis $(v \wedge 1, H(v \wedge 1)) \leq (v, Hv)$ if $v \geq 0$. Thus

$$\psi(Ru \wedge 1) \leq \psi(Ru)$$

Since Ru minimizes ψ, we have that $Ru \wedge 1 = Ru$, i.e., $Ru \leq 1$. Thus R is a contraction on L^∞. Similarly, $(1 + \varepsilon H)^{-1}$ is a contraction on L^∞ for all ε so that $e^{-tH} = \lim(1 + tH/n)^{-n}$ is a contraction on L^∞.

$(a) \Rightarrow (d)$ By (73), it clearly suffices to show that

$$(F(f), (1 - e^{-tH})F(f)) \leq (f, (1 - e^{-tH})f)$$

for all $t > 0$. Let K be a finite-dimensional subspace of $L^2(M, d\mu)$ of functions of the form $\sum_{i=1}^n \alpha_i \chi_{A_i}$ where $\{A_i\}_{i=1}^n$ are disjoint sets of finite measure. Under the order $K' > K$ if $K \subset K'$, the set of such K's is a directed set. If $P(K)$ is the projection onto K, s-$\lim_K P(K) = 1$, so it suffices to prove that

$$(F(P(K)f), (1 - e^{-tH})F(P(K)f)) \leq (P(K)f, (1 - e^{-tH})P(K)f)$$

for each K. Let

$$b_{ij} = (\chi_{A_i}, (1 - e^{-tH})\chi_{A_j})$$

Then we need only prove that

$$\sum_{i,j} \overline{F(\alpha_i)} F(\alpha_j) b_{ij} \leq \sum_{i,j} \bar{\alpha}_i \alpha_j b_{ij} \tag{74}$$

Let $\lambda_i = (\chi_{A_i}, \chi_{A_i})$ and $a_{ij} = (\chi_{A_i}, e^{-tH}\chi_{A_j})$. Then $a_{ij} \geq 0$ and since $\|e^{-tH}\chi_{A_j}\|_1 \leq \lambda_j$, we have

$$\sum_i a_{ij} \leq \lambda_j$$

Let $m_j = \lambda_j - \sum_i a_{ij} \geq 0$. Then $b_{ij} = \lambda_i \delta_{ij} - a_{ij}$, so

$$\sum_{i,j} \bar{z}_i z_j b_{ij} = \sum_{i<j} a_{ij} |z_i - z_j|^2 + \sum_j m_j |z_j|^2$$

From this representation, (74) follows. ∎

Example 3, revisited Using criterion (c), it is easily seen that Neumann and Dirichlet Laplacians generate contraction semigroups on L^p (Problem 99).

Appendix 2 to XIII.12 **The Levy–Khintchine formula**

In order to apply the theorems of Section 12 one needs methods of proving that semigroups are positivity preserving. If the generator is of the form $-\Delta + V(x)$, then, as we have seen, the generated semigroup is positivity preserving since the explicit representation of $e^{t\Delta}$ on $L^2(\mathbb{R}^n)$ shows that it is positivity preserving and the Trotter product formula allows us to conclude from this that $e^{t(\Delta-V)}$ is positivity preserving also. It is a natural question to ask for what other functions $F(-i\nabla)$ the operator $F(-i\nabla) + V(x)$ generates a positivity preserving semigroup. Certainly, if

$$F(-i\nabla) = -\sum_{i,j=1}^{n} a_{ij}\frac{\partial}{\partial x_i}\frac{\partial}{\partial x_j} + c$$

where $\{a_{ij}\}$ is a strictly positive definite matrix, the answer is yes since by change of variables $F(-i\nabla)$ can be reduced to $-\Delta + c$. However, it is not clear a priori which other functions F are allowed.

Let F be a continuous complex-valued function on $\mathbb{R}^n$ whose real part is bounded from below. As in Section IX.7, we define

$$\widehat{F(-i\nabla)\varphi} = F(p)\hat{\varphi}$$

Because of the semiboundedness of the real part of F, $F(-i\nabla)$ generates a strongly continuous semigroup on $L^2(\mathbb{R}^n)$. For V's that are not too nasty, Theorem X.50 shows that $F(-i\nabla) + V(x)$ also generates a strongly continuous semigroup and the Trotter product formula (Theorem X.51) reduces the question of whether $\exp(-t(F(-i\nabla) + V(x)))$ is positivity perserving to the same question for $\exp(-tF(-i\nabla))$. The point of this appendix is to characterize in different ways the set of F's so that this is true.

First, set $G(x) = e^{-tF(x)}$ and let f and g be positive functions in $\mathscr{S}(\mathbb{R}^n)$. Then,

$$(f, G(-i\nabla)g) = \widehat{(G(p)\hat{g})}(\bar{f})$$
$$= (2\pi)^{-n/2}(\check{G} * g)(\bar{f})$$
$$= (2\pi)^{-n/2}(\check{G} * (g * \tilde{\bar{f}}))(0)$$

Thus, if $\check{G}$ is a polynomially bounded positive measure, then $(f, G(-i\nabla)g) \geq 0$ for such f and g. Thus, by the Bochner–Schwartz theorem (Theorem IX.10), if G is positive definite, then $G(-i\nabla)$ is positivity preserving. Conversely, if $G(-i\nabla)$ is positivity preserving and we set $g_y(x) = f(x + y)$, then

$$(2\pi)^{-n/2}(\check{G} * f * \tilde{\bar{f}})(y) = (2\pi)^{-n/2}(\check{G} * g_y * \tilde{\bar{f}})(0)$$
$$= (f, G(-i\nabla)g_y)$$
$$\geq 0$$

Since $\tilde{G}$ is tempered, the left-hand side is a polynomially bounded function (Theorem IX.4a). Thus by Bochner's theorem, $\mathscr{F}(\tilde{G} * f * \tilde{f}) = (2\pi)^n |\hat{f}|^2 G$ is a positive definite function. If we now let $f(x) = j_\varepsilon(x)$, an approximate identity, then as $\varepsilon \downarrow 0$, $|\hat{j}_\varepsilon|^2 G \to G$ uniformly on compact sets, so G is positive definite.

This proves the first of the characterizations given by the following theorem. The second equivalence gives a characterization in terms of F itself.

Theorem XIII.52 Let $F(x)$ be a complex-valued function on $\mathbb{R}^n$ whose real part is bounded below. Then the following are equivalent:

(a) $e^{-tF(-i\nabla)}$ is a positivity preserving semigroup.
(b) For each $t > 0$, $e^{-tF(x)}$ is a positive definite distribution in the sense of Bochner.
(c) $\overline{F(x)} = F(-x)$ and

$$\sum_{i,j=1}^m F(\mathbf{x}_i - \mathbf{x}_j)\bar{z}_i z_j \leq 0 \tag{75}$$

for all $\mathbf{x}_1, \ldots, \mathbf{x}_m \in \mathbb{R}^n$ and $\mathbf{z} \in \mathbb{C}^m$ satisfying $\sum_{i=1}^m z_i = 0$.

Proof We have already proven that (a) and (b) are equivalent. In order to prove that (b) and (c) are equivalent it is sufficient to show that if $A = \{a_{ij}\}$ is an $m \times m$ matrix and $M(t)$ is the matrix with entries $M(t)_{ij} = e^{ta_{ij}}$, then $M(t)$ is positive definite for all $t \geq 0$ if and only if A is **conditionally positive definite**, i.e., $(A\zeta, \zeta) \geq 0$ for all $\zeta \in \mathbb{C}^m$ satisfying $\sum_{i=1}^m \zeta_i = 0$. So, suppose that $M(t)$ is positive definite for $t \geq 0$ and let $\sum_{i=1}^m \zeta_i = 0$. Then $(\zeta, M(t)\zeta) = 0$ at $t = 0$ and $(\zeta, M(t)\zeta) \geq 0$ for $t \geq 0$. Therefore,

$$(\zeta, A\zeta) = \frac{d}{dt}(\zeta, M(t)\zeta)\bigg|_{t=0} \geq 0$$

so A is conditionally positive definite.

Conversely, suppose that A is conditionally positive definite. Let e be the vector whose components all equal $1/\sqrt{m}$ and let P be the projection onto the orthogonal complement of e. The hypothesis on A is that PAP is positive definite. Thus, if we define $\tilde{a}_{ij} = (PAP)_{ij}$ and write

$$A = PAP + (1-P)A(1-P) + PA(1-P) + (1-P)AP$$

then one can easily check that

$$a_{ij} = \tilde{a}_{ij} + \bar{b}_i + b_j$$

for some vector b. Therefore, $M(t)_{ij} = \tilde{M}(t)_{ij} C(t)_{ij}$ where $\tilde{M}(t)_{ij} = e^{t\tilde{a}_{ij}}$ and $C(t)_{ij} = e^{t\bar{b}_i} e^{tb_j}$. Now, using the lemma below inductively, we have that $\{e^{D_{ij}}\}$

Appendix 2 to XIII.12 The Levy–Khintchine formula 215

is positive definite if $\{D_{ij}\}$ is positive definite. Applying this result we see that $\tilde{M}(t)_{ij}$ is positive definite. $C(t)_{ij}$ is obviously positive definite, so using the lemma again we conclude that $\{M(t)_{ij}\}$ is positive definite. ∎

The proof of the theorem is completed by the following lemma.

Lemma Let $\{D_{ij}\}$ and $\{F_{ij}\}$ be positive definite matrices. Then $\{D_{ij}F_{ij}\}$ is positive definite.

Proof Let $\{\mu_k, d_k\}$ and $\{\lambda_\ell, f_\ell\}$ be the eigenvalues and corresponding eigenfunctions of $\{D_{ij}\}$ and $\{F_{ij}\}$ respectively. If $\{e_i\}$ is the standard basis in $\mathbb{C}^n$, then

$$D_{ij}F_{ij} = \sum_{\ell,k=1}^{n} \mu_k \lambda_\ell (e_i, d_k)(e_i, f_\ell)(d_k, e_j)(f_\ell, e_j)$$

Now, fix k and ℓ and let $s_i = (e_i, d_k)(e_i, f_\ell)$. Then $\{s_i \bar{s}_j\}$ is positive definite. Therefore, $\{D_{ij}F_{ij}\}$ is positive definite since it a sum (with positive coefficients) of positive definite matrices. ∎

Functions that satisfy (75) are called **conditionally negative definite**. Although we have characterized the functions in which we are interested, it is very important to have other characterizations since, first, (75) is not easy to check, and secondly, it is hard to extract other necessary conditions from (75). Two further characterizations are given by the following theorem.

Theorem XIII.53 Let $F(x)$ be a complex-valued function on $\mathbb{R}^n$ whose real part is bounded below. Then the following are equivalent:

(a) $e^{-tF(-i\nabla)}$ is a positivity preserving semigroup.
(b) For each $a \in \mathbb{C}^n$, $-D^*DF$ is a positive definite distribution, where $D = \sum_{i=1}^{n} a_i \partial_i$ and $D^* = -\sum_{i=1}^{n} \bar{a}_i \partial_i$.
(c) (the Levy–Khintchine formula) There exists a positive finite measure ν on $\mathbb{R}^n$, with $\nu(\{0\}) = 0$, a positive definite matrix $A = \{a_{ij}\}$, a real vector $\boldsymbol{\beta}$, and a real number α so that

$$F(\mathbf{x}) = \alpha + i\boldsymbol{\beta} \cdot \mathbf{x} + \mathbf{x} \cdot A\mathbf{x} - \int_{\mathbb{R}^n} \left[e^{i\mathbf{x}\cdot\mathbf{y}} - 1 - \frac{i\mathbf{x}\cdot\mathbf{y}}{1+y^2} \right] \frac{1+y^2}{y^2} d\nu(\mathbf{y}) \quad (76)$$

Proof We shall show that (75) is equivalent to (b) and to (c). So, suppose that (75) holds. Then a simple approximation argument shows that

$$F(\bar{f} * f) = \int F(x-y)f(x)\overline{f(y)}\,dx\,dy \leq 0$$

216   XIII: SPECTRAL ANALYSIS

for any $f \in C_0^\infty(\mathbb{R}^n)$ that satisfies $\int f(x)\, dx = 0$. Now, if $D = \sum a_i\, \partial_i$, then $\int Df(x)\, dx = 0$ for all $f \in C_0^\infty(\mathbb{R}^n)$, so

$$0 \leq -F(\widetilde{Df} * Df) = -F(D^*D(\tilde{f} * f))$$
$$= -D^*DF(\tilde{f} * f)$$

Thus, $-D^*DF$ is a positive definite distribution.

We now show that (b) implies (c). Although the idea is simple, this is the lengthiest part of the proof. First, there are some complications with the multidimensional case; and secondly, there are technical complications caused by the fact that a priori F is only an ordinary distribution. Since we want to deal with the Fourier transform of F, we introduce the space $\mathscr{L}$ of Fourier transforms of functions in C_0^∞ with the topology $g_\alpha \to g$ in $\mathscr{L}$ if and only if $\check{g}_\alpha \to \check{g}$ in C_0^∞. $\mathscr{L}$ has an intrinsic description on account of the Paley–Wiener theorem (Theorem IX.11), but that will not concern us. Let G, "the Fourier transform of F," be defined as the linear functional on $\mathscr{L}$ given by

$$(G, g) = (F, \hat{g})$$

Choose a function $\alpha \in \mathscr{L}$ so that $\alpha(x) = 1 + O(x^3)$ at $x = 0$. To see that such an α exists, first choose any $\beta \in \mathscr{L}$ with $\beta(0) = 1$. Then

$$\beta(x) = 1 + \sum_i b_i x_i + \sum_{i,j} a_{ij} x_i x_j + O(x^3)$$

Let $p(x)$ be the polynomial $p(x) = 1 + \sum c_i x_i + \sum_{i,j} d_{ij} x_i x_j$ with $c_i = -b_i$; $d_{ij} = -a_{ij} + b_i b_j$. Then $\alpha(x) = p(x)\beta(x)$ lies in $\mathscr{L}$ since C_0^∞ is closed under the application of differential operators, and $\alpha(x) = 1 + O(x^3)$.

Our goal will be to prove that there is a tempered positive measure σ and real numbers a_0, b_i ($i = 1, \ldots, n$), $c_{ij}(i, j = 1, \ldots, n)$ so that

$$(2\pi)^{n/2} F(x) = -\int_{|\lambda|>0} [e^{i\lambda \cdot x} - \alpha(\lambda)(1 + i\lambda \cdot x)]\, d\sigma(\lambda)$$
$$+ a_0 + i\sum_{j=1}^n b_j x_j + \sum_{i,j=1}^n c_{ij} x_i x_j \qquad (77)$$

with $\{c_{ij}\}$ a positive definite matrix and

$$\int_{0<|\lambda|\leq 1} |\lambda|^2\, d\sigma + \int_{|\lambda|\geq 1} d\sigma < \infty$$

Appendix 2 to XIII.12 The Levy-Khintchine formula   217

We leave the rewriting of (77) into the form (76) to the reader (Problem 100). Letting G be the Fourier transform of F, (77) is clearly equivalent to

$$(G, \varphi) = -\int_{|\lambda|>0} [\varphi(\lambda) - \alpha(\lambda)(\varphi(0) + \lambda \cdot \nabla\varphi(0))] \, d\sigma(\lambda)$$
$$+ a_0 \varphi(0) + \sum_{j=1}^{n} b_j (\partial_j \varphi)(0) - \sum_{i,j=1}^{n} c_{ij} (\partial_i \partial_j \varphi)(0) \qquad (78)$$

for all $\varphi \in \mathscr{L}$. Suppose that we can prove that

$$(G, \varphi) = -\int_{|\lambda| \geq 0} \varphi(\lambda) \, d\sigma(\lambda) - \sum_{i,j=1}^{n} c_{ij} (\partial_i \partial_j \varphi)(0) \qquad (79)$$

for all φ that vanish to order one at $\lambda = 0$. Then (78) follows by applying (79) to $\varphi(\lambda) - \alpha(\lambda)[\varphi(0) + \lambda \cdot \nabla\varphi(0)]$. Next we note that the right-hand side of (79) defines a continuous functional on $\{\varphi \in \mathscr{L} \mid \varphi(0) = 0 = \partial_j \varphi(0), j = 1, \ldots, n\}$ as long as

$$\int_{0 < |\lambda| \leq 1} |\lambda|^2 \, d\sigma + \int_{|\lambda| \geq 1} d\sigma < \infty$$

(Problem 101) and that the sums of functions of the form $\lambda_i \lambda_j \psi$ with $\psi \in \mathscr{L}$ are the same as those $\varphi \in \mathscr{L}$ satisfying $\varphi(0) = 0 = \partial_j \varphi(0)$ (Problem 102).

Thus, to prove (76) it suffices to prove that there is a measure $d\sigma$ with the required bound, and a positive definite matrix c_{ij} so that (79) holds for all φ of the form $\lambda_i \lambda_j \psi$ with $\psi \in \mathscr{L}$. We first claim that there is a signed measure dv_{ij} on $\mathbb{R}^n$ so that

$$(G, \lambda_i \lambda_j \psi) = -\int \psi(\lambda) \, dv_{ij}(\lambda) \qquad (80)$$

for all $\psi \in \mathscr{L}$. For $-(\lambda_i \pm \lambda_j)^2 G$ defines a tempered positive measure by hypothesis (b), Bochner's theorem, and

$$\lambda_i \lambda_j = \tfrac{1}{4}[(\lambda_i + \lambda_j)^2 - (\lambda_i - \lambda_j)^2]$$

Since

$$(G, \lambda_i \lambda_j (\lambda_k \lambda_\ell \eta)) = (G, \lambda_k \lambda_\ell (\lambda_i \lambda_j \eta))$$

we see that the measures dv_{ij} obey the condition

$$\lambda_k \lambda_\ell \, dv_{ij} = \lambda_i \lambda_j \, dv_{k\ell} \qquad (81)$$

We can therefore define a measure $d\sigma$ on $\mathbb{R}^n \setminus \{0\}$ which is tempered at ∞ so that $\int_{0 < |\lambda| < 1} |\lambda|^2 \, d|\sigma| < \infty$ as follows: Let $\Omega_{ij} = \{\lambda \mid \lambda_i \lambda_j \neq 0\}$. On Ω_{ij}, let

$d\mu = (\lambda_i \lambda_j)^{-1} dv_{ij}$. By (81) this measure is a well-defined measure on $\bigcup \Omega_{ij} = \mathbb{R}^n \setminus \{0\}$. In particular, by (81), $\mu\{\lambda \mid \lambda_i \lambda_j = 0; \lambda \neq 0\} = 0$, so by (80),

$$(G, \lambda_i \lambda_j \psi) = -\int_{|\lambda|>0} (\lambda_i \lambda_j \psi(\lambda)) \, d\mu - v_{ij}(\{0\})\psi(0)$$

Now, setting $\varphi = \lambda_i \lambda_j \psi$, $\psi(0) = (\partial_i \partial_j \varphi)(0)$ and moreover, $(\partial_k \partial_\ell \varphi)(0) = 0$ unless $\{k, \ell\} = \{i, j\}$. Thus, (79) is proven with $c_{ij} = v_{ij}(\{0\})$.

All that remains is to prove that $d\sigma$ is positive with $\int_{|\lambda|>1} d\sigma < \infty$ and that $\{c_{ij}\}$ is positive definite. To see that σ is positive, we note that by hypothesis, $\lambda_i^2 \, d\sigma$ is positive for each i and so $\lambda^2 \, d\sigma$ is positive. Moreover, for any $\zeta \in \mathbb{C}^n$, $\sum \bar\zeta_i \zeta_j c_{ij} = v_\zeta(\{0\}) \geq 0$ where v_ζ is the Fourier transform of $-D_\zeta^* D_\zeta F$ with $D_\zeta = \sum_{i=1}^n \zeta_i \partial_i$. Finally, $\int_{|\lambda|>1} d\sigma < \infty$ by an argument (Problem 103) based on the formal relation $F(0) = -\int_{|\lambda|>0} (1 - \alpha(\lambda)) \, d\sigma + a_0$. We have therefore proven that (b) implies (c).

It remains to show that the Levy–Khintchine formula implies (75). First, we claim that

$$\left| e^{ixy} - 1 - \frac{ixy}{1+y^2} \right| \leq C \left[\frac{y^2}{1+y^2} \right](1 + x^2) \tag{82}$$

For, when $|y| \geq 1$, the left-hand side is bounded by $2 + \tfrac{1}{2}x$ and when $|y| \leq 1$ by $C|xy|^2 + |xy|(y^2/(1+y^2))$. From (82) any F of the form (76) is a tempered distribution satisfying

$$|F(x)| \leq C(1 + |x|^2) \tag{83}$$

and by a direct calculation each distribution $-(D^*D)F$ is positive definite. Since F is tempered, it has a Fourier transform $\hat{F}$. Moreover, as in the proof that (b) implies (c), the Fourier transform of $\partial_i \partial_j F$ is a signed measure dv_{ij} with $dv_\zeta = \sum \bar\zeta_i \zeta_j \, dv_{ij}$ a positive measure for each $\zeta \in \mathbb{C}^n$. Thus, for any functions $f_i(k)$ $(i = 1, \ldots, n)$,

$$\sum_{i,j} \int \overline{f_i(k)} f_j(k) \, dv_{ij}(k) \geq 0$$

Now suppose that $\int f(x) \, dx = 0$. Then $\hat f(0) = 0$ so, by a simple argument, $\hat f(k) = \sum k_i f_i(k)$ with $f_i(k) \in \mathcal{S}$. It follows that

$$\int F(x-y)\overline{f(x)}f(y) \, dx \, dy = -\sum_{i,j} \int \overline{f_i(k)} f_j(k) \, dv_{ij}(k) \leq 0 \quad \blacksquare$$

Example 1 We now answer the question of which self-adjoint constant coefficient partial differential operators generate positivity preserving semigroups. Let $F(p)$ be a polynomial in n variables that is bounded below, and suppose that $\exp(-tF(-i\mathbf{V}))$ is positivity preserving. Then by Theorem XIII.52, F is conditionally negative definite and by (83), F must have order less than or equal to two. Furthermore, the self-adjointness condition $\overline{F(p)} = F(p)$ and the condition $F(p) = F(-p)$ (from (75)) imply that the first-order terms are absent. Thus, $F(p) = \sum a_{ij} p_i p_j + c$. Finally, for $\zeta \in \mathbb{C}^n$ and $D_\zeta = \sum_{i=1}^{n} \zeta_i \partial_i$, we have

$$(D_\zeta^* D_\zeta F)(0) = -\sum a_{ij} \bar{\zeta}_i \zeta_j$$

Since $D_\zeta^* D_\zeta F$ is negative definite by part (b) of Theorem XIII.53, $\{a_{ij}\}$ must be positive definite. Thus, the *only* constant coefficient self-adjoint partial differential operators that generate positivity preserving semigroups are of the form

$$F(-i\mathbf{V}) = -\sum_{i=1}^{n} a_{ij} \partial_i \partial_j + c$$

where $\{a_{ij}\}$ is positive definite. On the other hand, each of these may be reduced by change of variables to $-\Delta + c$ so they all do generate positivity preserving semigroups.

From this example it is clear that if we want to find new conditionally negative functions (that satisfy $\overline{F(p)} = F(p)$) we have to look further than polynomials. To do this, it is useful to simplify condition (b) of Theorem XIII.53.

Theorem XIII.54 Suppose that F is a spherically symmetric, polynomially bounded continuous function satisfying $F(0) = 0$. Suppose further that ΔF is positive definite. Then F is conditionally negative definite.

Proof Let $G = -\hat{F}$. Thus $k^2 G$ is a polynomially bounded positive measure. So G extends from $\mathscr{S}(\mathbb{R}^n)$ to a set including all functions of the form $k^2 f(k)$ where f is continuous and $k^n f(k) \to 0$ as $|k| \to \infty$. In particular, if $g \in \mathscr{S}$, then

$$k_i k_j k_\ell g = k^2 [(k_i k_j k_\ell) k^{-2}] g$$

has this form and thus as in the proof of Theorem XIII.53, there is a measure μ on $\mathbb{R}^n \setminus \{0\}$ so that

$$G(k_i k_j k_\ell g) = \int k_i k_j k_\ell g(k) \, d\mu$$

for all g in $\mathscr{S}$. Moreover, $\int_{|k|\leq 1} k^2 \, d\mu(k) < \infty$. Thus, as in the proof of Theorem XIII.53,

$$G(f) = \int_{|\lambda|>0} [\varphi(\lambda) - \alpha(\lambda)(\varphi(0) + \lambda \cdot \nabla\varphi(0) + \tfrac{1}{2}\sum \lambda_i \lambda_j \, \partial_i \partial_j \varphi(0))] \, d\mu$$

$$- a_0 \varphi(0) - \sum_{j=1}^{n} b_j \, \partial_j \varphi(0) + \sum_{i,j=1}^{n} c_{ij}(\partial_i \, \partial_j \varphi)(0)$$

where $\alpha \in C_0^\infty$ is identically 1 near 0. Since $\int k^2 \, d\mu(k) < \infty$, we can eliminate the $\lambda_i \lambda_j (\partial_i \, \partial_j \varphi)(0)$ term in the integral and absorb it into the c_{ij} term. Thus

$$(2\pi)^{n/2} F(x) = -\int_{|\lambda|>0} [e^{i\lambda x} - \alpha(\lambda)(1 + \lambda \cdot x)] \, d\mu(\lambda) \tag{84}$$

$$+ \alpha_0 + i\beta \cdot x + \sum_{i,j} \gamma_{ij} x_i x_j$$

so to prove that F is conditionally negative definite, we need only prove that γ_{ij} is positive definite. Since F is rotationally invariant and α can be chosen rotationally invariant, γ_{ij} must equal $\gamma \delta_{ij}$ so we need only prove that $\gamma > 0$. But the Fourier transform of ΔF is $d\nu = \lambda^2 \, d\mu + 2\delta(\lambda)\mathrm{Tr}(\gamma_{ij})$ so $n\gamma = \mathrm{Tr}(\gamma_{ij}) = \tfrac{1}{2}\nu(\{0\})$ is positive. ∎

Example 2 We claim that

$$F_1(p) = |p|^\alpha; \quad 1 \leq \alpha \leq 2 \text{ if } n = 1, \quad 0 \leq \alpha \leq 2 \text{ if } n > 1$$
$$F_2(p) = \sqrt{p^2 + m^2}; \quad m > 0$$

are conditionally positive definite. For we can compute ΔF in both cases and find:

$$\Delta F_1(p) = \alpha(n + \alpha - 2)|p|^{\alpha-2}$$
$$\Delta F_2(p) = (n-1)(p^2 + m^2)^{-1/2} + m^2(p^2 + m^2)^{-3/2}$$

So we need only prove that $|p|^{\alpha-2}$, $(p^2 + m^2)^{-1/2}$ and $(p^2 + m^2)^{-3/2}$ are positive definite. Now, p^2 is conditionally positive definite, so e^{-tp^2} is positive definite. Since, for $\alpha < 2$,

$$|p|^{\alpha-2} = c_\alpha \int_0^\infty t^{-\alpha/2} e^{-tp^2} \, dt$$

$|p|^{\alpha-2}$ is positive definite as the integral of positive definite functions convergent as an integral of distributions. Similarly, for $\beta > 0$,

$$(p^2 + m^2)^{-\beta} = d_\beta \int_0^\infty t^{\beta-1} e^{-t(p^2+m^2)} \, dt$$

so $(p^2 + m^2)^{-\beta}$ is positive definite for $\beta > 0$.

The choice for the free Hamiltonian $h_0 = (p^2 + m^2)^{1/2} - m$ is of some interest since it is the quantum analogue of the energy of a relativistic particle. It should describe spinless particles (e.g., pions) in the region where relativity is of some importance but where particle creation phenomena (field theoretic effects) are not yet important.

Our last topic is an aside from the main motivation of this appendix, but we include it because of its importance to probability theory.

Definition A probability measure μ on $\mathbb{R}^n$ is called **infinitely divisible** if for any n, there is a probability measure v_n with $\mu = v_n * v_n * \cdots * v_n$ (n times). The Fourier transforms of probability measures are called **characteristic functions**. A characteristic function $f(x)$ is called **infinitely divisible** if for each n, there is a characteristic function $f_n(x)$ so that $f(x) = (f_n(x))^n$.

What we would like to do is to characterize those functions that are infinitely divisible characteristic functions. Let F be continuous. First, notice that if $e^{-tF(p)}$ is positive definite for all $t > 0$, then $e^{-F(p)}$ is an infinitely divisible characteristic function. But the converse is also true. For suppose that $e^{-F(p)}$ is infinitely divisible. Then, there is a positive definite function g_2 so that $g_2(p)^2 = e^{-F(p)}$. Clearly, $g_2(p) = \pm\exp(-\frac{1}{2}F(p))$ for all p. Since g_2 and F are continuous and $g_2(0) = 1 = +\exp(-\frac{1}{2}F(0))$, the plus sign always occurs. In this way we see that $\exp(-2^{-n}F)$ is positive definite for all n. Since products and limits of positive definite functions are positive definite, we conclude that $\exp(-tF(p))$ is positive definite for all t. This remark combined with the theorems that we have already proven allow us to characterize infinitely divisible characteristic functions.

Theorem XIII.55 (the Levy–Khintchine theorem) Every infinitely divisible characteristic function G is of the form $G = e^{-F}$ for some F satisfying (76).

Proof Let G_m be the positive definite function with $(G_m)^m = G$. Then $G_m(0) = 1$ and since G is continuous, $G(x) \neq 0$ for $|x| < \varepsilon$ for suitable ε. Thus $G_m(x) \to 1$ for $|x| < \varepsilon$ as $m \to \infty$. By the lemma below, $G_m(x) \to 1$ for all x so that G is never zero. By a fundamental topological property of $\mathbb{R}^n$, any nonvanishing complex-valued continuous function on $\mathbb{R}^n$ has a unique continuous logarithm once it is fixed at one point. Thus, there is a unique continuous function F with $G = e^{-F}$ and $F(0) = 0$. Since G is bounded, the real part of F is bounded below. By the remark before the theorem, $\exp(-tF(x))$ is positive definite for all $t > 0$, so by Theorem XIII.52, $e^{-tF(-i\nabla)}$ is a positivity preserving semigroup. Thus, by Theorem XIII.53, there is a finite positive measure so that the Levy–Khintchine formula (76) holds. ∎

Lemma Let G_m be a sequence of positive definite fuctions with $G_m(0) = 1$. Suppose that as $m \to \infty$, $G_m(x) \to 1$ for x in some neighborhood of zero. Then $G_m(x) \to 1$ for all x.

Proof By hypothesis, there exist measures μ_m on $\mathbb{R}^n$ with

$$G_m(x) = \int e^{ixy} \, d\mu_m(y)$$

By hypothesis, each μ_m has mass 1. Moreover, an elementary use of Fubini's theorem shows that

$$(2a)^{-n} \int_{|x_i| \leq a} [1 - G_m(x)] \, dx = \int \left(1 - \prod_{i=1}^{n} \frac{\sin a y_i}{a y_i}\right) d\mu_m(y)$$

$$\geq C \mu_m\{y \,|\, y_i \geq a^{-1} \text{ for some } i\}$$

where $C = \min_{y \geq 1} (1 - y^{-1} \sin y) > 0$. Thus, by the hypothesis on G_m, for some a, $\mu_m\{y \,|\, y_i \geq a^{-1} \text{ for some } i\} \to 0$ as $m \to \infty$. View the μ_m as measures on $\dot{\mathbb{R}}^n$, the one-point compactification of $\mathbb{R}^n$. Let μ_∞ be any weak limit point of the μ_m; such limit points exist since $\mathcal{M}_{+,1}(\dot{\mathbb{R}}^n)$ is compact. Since $\mu_m\{y\,|\,|y| \geq a^{-1}\} \to 0$ as $m \to \infty$ for some a, we can conclude that μ_∞ is a measure on $\mathbb{R}^n$, i.e., $\mu_\infty(\{\infty\}) = 0$ and that $\int f(y) \, d\mu_{m(i)} \to \int f(y) \, d\mu_\infty$ for *any* continuous bounded function on $\mathbb{R}^n$ and not just those going to a constant at ∞. In particular, $\int e^{ixy} \, d\mu_\infty(y) = 1$ for $|x|$ small. Let Q be the family of points in this set of small x, all of whose coordinates are rational. It is easy to see that $\bigcap_{x \in Q} \{y\,|\,e^{ixy} = 1\} = \{0\}$ so we conclude that $\mu_\infty = \delta_0$, the point mass at zero. Since every limit point of μ_m is δ_0, $\mu_m \to \delta_0$ weakly, and as above $\int f(y) \, d\mu_m(y) \to f(0)$ for any continuous bounded f. In particular, $G_m(x) \to 1$ for all x. ∎

XIII.13 Absence of positive eigenvalues

We have thus far avoided discussing the question of whether Schrödinger operators can have eigenvalues embedded in the continuous spectrum. This question turns out to be very difficult; the results that do exist generally require more detailed assumptions than for any other spectral problem. Moreover, the problem is especially frustrating since one has good physical reasons, which we discuss below, for expecting that such eigenvalues should not occur. Let us consider the intuition behind a simple case. Consider a potential $V(x)$ that goes to zero as $|x| \to \infty$. Classically, the only way to

XIII.13 Absence of positive eigenvalues

prevent a particle of positive energy from reaching infinity is to put up a barrier, i.e., trap it by energy conservation. But quantum mechanically, there is tunneling through barriers, so one would guess that a positive energy bound state cannot occur. This guess is wrong!

Example 1 (the Wigner–Von Neumann potential) To find a potential V so that there is a square integrable ψ with $(-\Delta + V)\psi = \psi$, we try to guess ψ. If u is spherically symmetric and $u(r) = r\psi(r)$, then $-u'' + Vu = u$ or $V = 1 + u''u^{-1}$. For V to go to zero at infinity, $u''u^{-1}$ has to go to -1 at infinity, which suggests that we try the ansatz $u(r) = (\sin r)w(r)$. Then

$$V(r) = w''(r)w(r)^{-1} + 2(\cot r)w'(r)w(r)^{-1}$$

For V to be nonsingular, we need w' to vanish whenever $\sin r$ does. Thus w must behave something like

$$g(r) = 2r - \sin 2r = 4\int_0^r \sin^2 x \, dx \tag{85}$$

Of course, if we take $w = g$, then u will not be square integrable. But if we take $w = (1 + g(r)^2)^{-1}$, then w will be square integrable and $w' = -2g'g \times (1 + g^2)^{-2}$ will have zeros in the right places. Making this choice of w, we compute V and find

$$V(r) = [1 + g(r)^2]^{-2}(-32 \sin r)[g(r)^3 \cos r - 3g(r)^2 \sin^3 r \\ + g(r)\cos r + \sin^3 r]$$

where $g(r)$ is given by (85). This rather complicated V has the property that $-\Delta + V$ has an eigenvalue at $+1$ with eigenvector

$$\psi(r) = (r^{-1} \sin r)(1 + g(r)^2)^{-1}$$

even though V is bounded and $V \to 0$ at infinity so that $[0, \infty) \subset \sigma(-\Delta + V)$ by Weyl's essential spectrum theorem. Notice that V has the asymptotic behavior

$$V(r) = -8 \sin 2r/r + O(r^{-2})$$

That is, V oscillates and falls off slowly at ∞. We shall see that both these features are critical for positive eigenvalues to occur. Further discussion of this kind of example can be found in the second and third appendices to Section XI.8.

The physicist confident in the prediction that positive eigenvalues cannot occur must confront Example 1 and answer the question: Why is the intuition wrong? The critical aspect of V is its oscillations. As in Example 1 in the

Appendix to Section X.1, we have used the fact that quantum-mechanical particles are reflected off bumps. The potential V is constructed in such a way that for a particular standing wave pattern these reflections occur coherently. This interpretation suggests that the occurrence of positive energy eigenvalues is a very unusual thing depending on detailed "coherence" properties in the potential. It also indicates that it will be difficult to formulate simple general conditions that exclude the possibility of such bound states.

Example 2 Let $V(x) = ||x| - 1|^{-1}$ on $\mathbb{R}^3$. V is not even in $L^1_{\text{loc}}(\mathbb{R}^3)$, but $Q(V) \cap Q(-\Delta)$ is dense and so $-\Delta + V$ can be defined as a sum of quadratic forms. This operator will have positive energy eigenvalues for a simple reason: Any $\psi \in Q(V) \cap Q(-\Delta)$ will vanish on the sphere of radius 1 in a suitable sense and so $-\Delta + V$ will leave $\mathcal{H}_1 = L^2(\{x \mid |x| \leq 1\})$ invariant. Thus $\mathcal{H}$ will decompose into a direct sum $\mathcal{H}_1 \oplus \mathcal{H}_2$ so that $H = H_1 \oplus H_2$ and H_1 will have purely discrete spectrum. The proofs of these assertions are left to the reader (Problem 104).

This example is rather artificial, but it does show that the question of positive eigenvalues is connected with the question of whether $-\Delta + V$ can have eigenvectors with compact support. We return to this aspect of the question below and in the appendix.

Example 3 Let V be a spherical square well on $\mathbb{R}^3$, $V(x) = -c$ if $|x| \leq 1$, and $V(x) = 0$ if $|x| > 1$. We seek solutions of $(-\Delta + V)\psi = 0$. If $\psi(x) = |x|^{-1} u(|x|) Y_{\ell m}(\hat{x})$ where $\hat{x}$ is the unit vector in direction x and $Y_{\ell m}$ is a spherical harmonic, then u obeys

$$-u'' + \ell(\ell + 1) r^{-2} u + V(r) u = 0 \tag{86}$$

The solutions of (86) in the region $r > 1$ where $V = 0$ are $r^{-\ell}$ and $r^{\ell+1}$. If the solution of (86) that is regular at $r = 0$ is pure $r^{-\ell}$ in the region $r > 1$ (and this will occur as c is varied) and $\ell \geq 1$, then $(-\Delta + V)\psi = 0$ will have square integrable solutions.

This example illustrates that zero energy eigenvalues, and more generally threshold eigenvalues in n-body systems, are perfectly normal occurrences that we cannot hope to eliminate.

Example 4 Consider the helium atom Hamiltonian

$$H(\beta) = -\Delta_1 - \Delta_2 - 2|r_1|^{-1} - 2|r_2|^{-1} + \beta |r_1 - r_2|^{-1}$$

which we have discussed extensively in Section XII.6. When $\beta = 0$, we found a high multiplicity of eigenvalues in the continuum. This was because states with the proper energy to decay could not decay because decay would involve the transfer of energy from one particle to another and such transfer cannot occur when $\beta = 0$ because the two particles do not interact with each other. Even when $\beta \neq 0$, there remained eigenvalues embedded in the continuum because of an underlying symmetry, namely, "naturalness" of parity. Notice that all these embedded eigenvalues occur at negative energies.

The moral of the above example is that it should be especially difficult to prove that n-body Schrödinger operators have no embedded eigenvalues at negative energies. In the above cases, there is a simple reason that such eigenvalues occur. But given an n-body system, how does one express the fact that there are no "hidden" symmetries that could produce embedded eigenvalues?

With the examples in mind, it is clear why the problem at hand is so difficult. One has an intuition that positive energy eigenvalues do not occur in "reasonable" situations, but it is clear that such eigenvalues can occur due to "unreasonable" subtleties. We shall present four rather different approaches to the problem. The first is for the two-body central case. In this case, the Schrödinger equation can be formally reduced to an ordinary differential equation, so it is fairly easy to obtain results. We do not attempt to give the strongest conditions involving local singularities.

Theorem XIII.56 Let V be a spherically symmetric function on $\mathbb{R}^n$ with

$$\int_a^\infty |V(r)| \, dr < \infty \tag{87}$$

for some $a > 0$. Suppose that $V \in L^2_{\text{loc}}(\mathbb{R}^n \backslash 0)$ and let H be any self-adjoint extension of $-\Delta + V$ on $C_0^\infty(\mathbb{R}^n \backslash 0)$ that commutes with rotations. Then H has no strictly positive eigenvalues.

Proof Suppose that $(-\Delta + V)\psi = E\psi$. Expand ψ in a partial wave (spherical harmonic) expansion. Then some partial wave component is nonzero, and so we have a distributional solution of

$$-u'' + cr^{-2}u + Vu = Eu \tag{88}$$

on $(0, \infty)$ that is square integrable. Here c is a constant depending on the partial wave index. As in Section XI.8, since $E > 0$, we can find two independent Jost solutions of the integral equation associated to (88) because of the condition (87). It is not difficult (Problem 105) to show that any solution of (88) must be a linear combination of these two Jost solutions. Since no such linear combination is square integrable at infinity, u must be identically zero. ∎

While the potential of Example 1 fails to obey (87), it is precisely of the form discussed in the appendix to Section XI.8 where we constructed Jost solutions for potentials of the form

$$V(x) = \sum_{i=1}^{N} c_i r^{-1} \sin(\alpha_i r) + O(r^{-1-\varepsilon})$$

for any $k^2 \neq 0, \frac{1}{4}\alpha_1^2, \ldots, \frac{1}{4}\alpha_N^2$. Thus, by the above argument, we see that the Wigner–Von Neumann potential can have a positive eigenvalue only at energy one, the energy at which it does have such a bound state.

Our next result will allow potentials that are not spherically symmetric. One important element of the method involves a "unique continuation theorem for Schrödinger operators." To see how such a result enters suppose that $V \in C_0^\infty(\mathbb{R}^3)$ with supp $V \subset \{x \mid |x| < R\}$. If $(-\Delta + V)u = Eu$, then u obeys $-\Delta u = Eu$ on $\{x \mid |x| > R\} = \Omega$. As above, we can expand u in a spherical harmonic expansion in Ω, and so conclude that $u = 0$ on Ω. To conclude that u is identically zero we need:

Theorem XIII.57 Let V be a function on $\mathbb{R}^n$ so that:

(i) Multiplication by V is $-\Delta$-bounded with relative bound less than 1.
(ii) There is a closed set S of measure zero so that $\mathbb{R}^n \backslash S$ is connected and so that V is bounded on any compact subset of $\mathbb{R}^n \backslash S$.

Let $H = -\Delta + V$ and suppose that $Hu = Eu$ for some E and $u \in L^2$. Suppose that u vanishes on some open subset of $\mathbb{R}^n$. Then u is identically zero.

As the proof in the appendix shows, this result is really a local result dealing with distributional solutions of $(-\Delta + V)u = Eu$ which are locally in $D(-\Delta)$. The reason for hypothesis (i) is just to ensure that eigenfunctions are locally in $D(-\Delta)$.

The above proves that $\sigma_p(-\Delta + V) \cap (0, \infty) = \emptyset$ for $V \in C_0^\infty$. This is a special case of the following general result:

Theorem XIII.58 (Kato–Agmon–Simon theorem) Suppose that V is a potential satisfying the hypothesis of Theorem XIII.57 and that, in addition, $V = V_1 + V_2$ where:

(i) V_1 is bounded outside some ball $\{x \mid |x| < R_0\}$, and $|x| V_1(x) \to 0$ as $|x| \to \infty$.

XIII.13 Absence of positive eigenvalues

(ii) V_2 is bounded outside some ball, $\{x \mid |x| < R_0\}$, and $V_2(x) \to 0$ as $|x| \to \infty$.

(iii) If we consider V_2 as a map $r \mapsto V_2(r, \cdot)$ from $(0, \infty)$ to $L^\infty(S^{n-1})$ where S^{n-1} is the $(n-1)$-dimensional sphere, then for $|x| > R_0$, V_2 is differentiable as an L^∞-valued function and $\overline{\lim}_{r \to \infty} r \, \partial V_2/\partial r \leq 0$.

Then, $H = -\Delta + V$ has no strictly positive eigenvalues.

Proof Consider the alternative hypotheses:

(ii′) $V_2(x) < 0$ for $|x| > R_0$ replaces $V_2(x) \to 0$.
(iii′) $\partial V_2(x)/\partial r \leq -r^{-1} V_2$ for $|x| > R_0$ replaces $\lim_{r \to \infty} r \, \partial V_2/\partial r \leq 0$.

Suppose that we can show that $-\Delta + V$ has no strictly positive eigenvalues assuming (i), (ii′), (iii′). Given V_2 obeying (ii) and (iii), the function $\tilde{V}_2 = V_2 - \varepsilon$ obeys (ii′) and (iii′), at least if R_0 is taken sufficiently large. Applying the putative (i)–(iii′) result to $-\Delta + V_1 + \tilde{V}_2$ we see that $-\Delta + V$ has no eigenvalues in (ε, ∞). Since ε is arbitrary, we can complete the proof of the theorem. Thus we suppose that (i), (ii′), (iii′) hold.

Suppose that ψ is a real eigenfunction for $-\Delta + V$ with eigenvalue $E > 0$. Define a function w from $(0, \infty)$ to $L^2(S^{n-1}, d\Omega)$ by

$$w(r, \Omega) = r^{(n-1)/2} \psi(r\Omega) \tag{89}$$

so that

$$\int_0^\infty \|w(r)\|^2_{L^2(S^{n-1}, d\Omega)} \, dr < \infty$$

From now on we drop the subscript on the norm. More generally, given any $\varphi \in L^2(\mathbb{R}^n)$, define $w_\varphi \in L^2((0, \infty); L^2(S^{n-1}, d\Omega), dr)$ by (89) with ψ replaced by φ. If $\varphi \in C_0^\infty(\mathbb{R}^n)$, then

$$w_{H\varphi} = -w_\varphi'' - r^{-2} B w_\varphi + \tfrac{1}{4}(n-1)(n-3) r^{-2} w_\varphi + V w_\varphi \tag{90}$$

where B is the Laplace–Beltrami operator on $L^2(S^{n-1})$ described in Example 4 of the Appendix to Section X.1. (90) can be proven by using a spherical harmonic expansion as in that example. Below, we only need to know that $-B$ is a positive operator.

In terms of the function w of (89), define for $r > R_0$,

$$F(r) = (w', w') + r^{-2}(w, Bw) + (w, (E - V_2(r))w) \tag{91}$$

where all inner products are in $L^2(S^{n-1}, d\Omega)$. The objects occurring in (91) are all well defined, for one first notes that for $\varphi \in C_0^\infty$,

$$\int_0^\infty [(\dot{w}_\varphi, \dot{w}_\varphi) - r^{-2}(w_\varphi, B w_\varphi)] \, dr = \|\nabla \varphi\|^2_{L^2(\mathbb{R}^n)}$$

where $\dot{w}_\varphi = r^{(n-1)/2}(d/dr)(r^{-(n-1)/2}w_\varphi)$. From this estimate it follows that w'_φ and (w_φ, Bw_φ) are defined almost everywhere on $(0, \infty)$ if $\varphi \in Q(-\Delta)$.

Choose $R_1 > R_0$ so that

$$|rV_1(r)| + \tfrac{1}{4}(n-1)(n-3)r^{-1} < k \equiv \sqrt{E}$$

for $r > R_1$. Such an R_1 exists by hypothesis (i). We claim that almost everywhere

$$F(r) \geq r^{-1}r_1 F(r_1), \qquad r > r_1 > R_1 \tag{92}$$

Let us first give a formal proof of (92). Formally, if $r > R_1$,

$$\frac{d}{dr}(rF(r)) = 2r(w', w'' + r^{-2}Bw + (E - V_2)w)$$

$$+ \|w'\|^2 - r^{-2}(w, Bw) + E\|w\|^2 - (w, (rV_2)'w)$$

$$\geq 2(w', (rV_1 + \tfrac{1}{4}(n-1)(n-3)r^{-1})w) + \|w'\|^2 + E\|w\|^2$$

$$\geq 0$$

where the first inequality follows from the (formal) eigenfunction relation and the facts that $-B$ is positive and that $(rV_2)' = rV'_2 + V_2 \leq 0$ by hypothesis (iii'). The final inequality comes from the choice of R_1. Of course, (92) comes from integrating this inequality.

Now we give a rigorous proof of (92). Let $\varphi \in C_0^\infty$ and define F_φ by (91) with w_φ replacing w. The differentiations done formally above on F are legitimate on F_φ and imply that

$$\frac{d}{dr}(rF_\varphi) \geq 2r(w'_\varphi, (Ew_\varphi - w_{H\varphi}))$$

for $r > R$. Therefore, if $r > r_1 > R_1$, we have

$$rF_\varphi(r) \geq r_1 F_\varphi(r_1) + \int_{r_1}^r 2t(w'_\varphi(t), Ew_\varphi(t) - w_{H\varphi}(t))\,dt$$

Now pick $\varphi_n \in C_0^\infty$ so that $\varphi_n \to \psi$ and $-\Delta\varphi_n \to -\Delta\psi$ (such a choice is possible since C_0^∞ is a core for $-\Delta$). Then $H\varphi_n \to H\psi = E\psi$. Thus, $w_{\varphi_n} \to w$, $w_{H\varphi_n} \to Ew$, and $w'_{\varphi_n} \to w'$ in $L^2((0, \infty); L^2(S^{n-1}, d\Omega), dr)$ and, after passing to a subsequence, $F_{\varphi_n} \to F$ pointwise a.e. This proves (92).

From (92) we deduce that $F(r) \leq 0$ for $r > R_1$ because if this were not so, then the left-hand side of (91) would not be integrable. But it is integrable since $\psi \in D(-\Delta)$ (Problem 110). We now want to prove that $w = 0$ for $r > R_2$ for suitable R_2. This will imply that ψ vanishes outside some sphere and so, by the unique continuation theorem, ψ is identically zero. In our

XIII.13 Absence of positive eigenvalues

considerations below, we shall only make "formal" computations, leaving it to the reader to supply the arguments analogous to the F_φ argument we have just given.

For any $m \geq 0$, let $w_m = r^m w$ and let

$$G(m, r) = \|w'_m\|^2 + (k^2 - k^2 R_1 r^{-1} + m(m+1)r^{-2})\|w_m\|^2$$
$$+ r^{-2}(w_m, Bw_m) - (w_m, V_2(r)w_m)$$

Notice first that w_m obeys

$$w''_m - 2mr^{-1}w'_m + r^{-2}[m(m+1) - \tfrac{1}{4}(n-1)(n-3) + B]w_m + (k^2 - V)w_m$$
$$= 0$$

Then, using the inequality

$$-(r^2 V_2)' = -r^1 V_2 - r(rV_2)' \geq 0$$

one easily computes that for $r > R_0$,

$$\frac{d}{dr}(r^2 G(m, r)) \geq 2r\left[(2m+1)\|w'_m\|^2 + k^2\left(1 - \frac{R_1}{2r}\right)\|w_m\|^2\right.$$
$$\left. + (w'_m, (rV_1 + \tfrac{1}{4}(n-1)(n-3)r^{-1} - k^2 R_1)w_m)\right]$$

Thus, for $r > R_1$,

$$\frac{d}{dr}(r^2 G(m, r)) \geq 2r\left[(2m+1)\|w'_m\|^2 + \frac{1}{2}k^2\|w_m\|^2\right.$$
$$\left. - (k + k^2 R_1)\|w'_m\|\,\|w_m\|\right]$$

It follows that for some m_0, $r^2 G(m, r)$ is monotone increasing on (R_1, ∞) if $m > m_0$.

Suppose now that $w(r_0) \neq 0$ for some $r_0 > R_1$. Writing

$$G(m, r) = r^{2m}[\|w' + mr^{-1}w\|^2 + (k^2 - k^2 R_1 r^{-1} + m(m+1)r^{-2})\|w\|^2$$
$$- (w, V_2 w) + r^{-2}(w, Bw)] \tag{93}$$

we see that $G(m, r_0) > 0$ for m sufficiently large. Combining this with the monotonicity proven above, we conclude: If $w(r_0) \neq 0$ for some $r_0 > R_1$, then for some M (depending on r_0) $G(m, r) > 0$ for all $m > M$, $r > r_0$.

Now we can complete the proof. Given r_0 with $w(r_0) \neq 0$ and $r_0 > R_1$, choose $m_0 > M$ and then $R_2 > r_0$ so that for $r > R_2$,

$$-k^2 R_1 r^{-1} + m_0(2m_0 + 1)r^{-2} < 0$$

Since $\int_{R_2}^{\infty} \|w\|^2 \, dr < \infty$, $\|w\|$ is not strictly monotone increasing on all of $[R_2, \infty)$ and so there is some $r_1 > R_2$ so that

$$\frac{d}{dr}\|w\|^2 \bigg|_{r=r_1} = 2(w', w) \leq 0$$

In particular, at $r = r_1$,

$$\|w' + mr^{-1}w\|^2 \leq \|w'\|^2 + m^2 r^{-2}\|w\|^2$$

so by (93)

$$0 < r_1^{-2m_0} G(m_0, r_1) \leq \|w'\|^2 + k^2\|w\|^2 - (w, V_2 w) + r^{-2}(w, Bw) = F(r_1)$$

Given (92), this implies that $\int \|w\|^2 \, dr = \infty$ or $\|w\|^2 = 0$, so $w(r_0) = 0$ for all $r_0 > R_1$. By unique continuation (Theorem XIII.57), ψ is identically zero. ∎

Corollary Suppose that V satisfies the hypotheses of Theorem XIII.57. If V has compact support or if V is repulsive near ∞ ($\partial V/\partial r \leq 0$), then there are no bound states of positive energy.

Notice that because of its asymptotic behavior, the potential of Example 1 fails both the $rV_1 \to 0$ condition and the $r \, \partial V_2/\partial r$ condition.

The third method for controlling positive energy bound states that we shall discuss involves the virial theorem. This theorem asserts that if ψ is an eigenfunction of $H = -\Delta + V$, then under suitable hypotheses on V:

$$2(\psi, (-\Delta)\psi) = (\psi, \mathbf{r} \cdot \nabla V \psi) \tag{94}$$

(94) is clearly a useful tool in studying bound states. For example, if $\mathbf{r} \cdot \nabla V \leq 0$ (which is an expression of the fact that V is repulsive), then clearly H can have no bound states since then the right-hand side of (94) would always be negative and the left strictly positive. Formally, (94) comes from the following argument. Let

$$D = \frac{in}{2} + i \sum_{i=1}^{n} x_i \frac{\partial}{\partial x_i}$$

Then

$$i[D, H] = 2(-\Delta) - \mathbf{r} \cdot \nabla V$$

so that if ψ is an eigenfunction with eigenvalue E,

$$2(\psi, -\Delta \psi) - (\psi, \mathbf{r} \cdot \nabla V) = (\psi, i[D, H]\psi)$$
$$= -iE\{(\psi, D\psi) - (\psi, D\psi)\}$$
$$= 0$$

since H is Hermitian. This argument is only formal since D is an unbounded operator and ψ need not be in its domain. In fact, there are modifications of the potential in Example 1 which have eigenfunctions that are not in the domain of D (Problem 107)! The key to our proof of the virial theorem is the realization that the group generated by D is the group of dilations discussed in Section 10. For $a > 0$, define the unitary family

$$(U_a \psi)(x) \equiv (e^{-iD \log a}\psi)(x) = a^{n/2}\psi(ax)$$

and set

$$V_a(x) = V(ax)$$

so that $V_a = U_a V U_a^{-1}$ for a multiplication operator V.

Theorem XIII.59 (the virial theorem)   Suppose that V is a multiplication operator on $L^2(\mathbb{R}^n)$ and that:

(i) V is $-\Delta$-bounded with relative bound less than one.
(ii) There is a multiplication operator W on $L^2(\mathbb{R}^n)$ with $D(W) \supset D(-\Delta)$ so that for all $\psi \in D(-\Delta)$,

$$(a-1)^{-1}(V_a - V)\psi \to W\psi \quad \text{as} \quad a \to 1 \tag{95}$$

Then if $\psi \in D(-\Delta)$ and $-\Delta\psi + V\psi = E\psi$, we have

$$2(\psi, -\Delta\psi) = (\psi, W\psi) = 2(\psi, (E - V)\psi) \tag{96}$$

Proof   Since $U_a(-\Delta)U_a^{-1} = a^{-2}(-\Delta)$, we have

$$(-\Delta + a^2 V_a)\psi_a = a^2 E \psi_a$$

as well as $(-\Delta + V)\psi = E\psi$. These two relations imply that

$$E(a^2 - 1)(\psi_a, \psi) = ((-\Delta + a^2 V_a)\psi_a, \psi) - (\psi_a, (-\Delta + V)\psi)$$
$$= a^2(V_a\psi_a, \psi) - (\psi_a, V\psi)$$

since $-\Delta$ is symmetric. Thus

$$(a+1)(\psi_a, V_a\psi) + (a-1)^{-1}(\psi_a, (V_a - V)\psi) = E(a+1)(\psi_a, \psi)$$

Taking the limit as $a \to 1$ yields (96). ∎

We remark that formally W is just $\mathbf{r} \cdot \nabla V$. In fact, since $C_0^\infty(\mathbb{R}^n) \subset D(-\Delta)$, if (95) holds, W is given by $\mathbf{r} \cdot \nabla V$ where the derivatives are interpreted in the sense of distributions. Secondly, (95) is typically proven in the following way: Suppose that $(a-1)^{-1}(V_a - V)$ converges pointwise to a function W and that there is a multiplication operator $\widetilde{W}$ with $D(\widetilde{W}) \supset D(-\Delta)$ so that

$$|(a-1)^{-1}(V_a - V)| \leq \widetilde{W}$$

pointwise a.e. for a close to one. Then $D(W) \supset D(-\Delta)$ and the dominated convergence theorem allows one to conclude (95). ∎

Theorem XIII.60 Let V be a real-valued function that is $-\Delta$-bounded with relative bound less than one. Then $-\Delta + V$ has no positive eigenvalues if at least *one* of the following conditions holds:

(i) V satisfies the hypotheses of Theorem XIII.59 and V is repulsive (that is, $V(ar) \leq V(r)$ for all r and all $a > 1$).
(ii) V is homogeneous of degree $-\alpha$ where $0 < \alpha < 2$ (that is, $V(ar) = a^{-\alpha}V(r)$).
(iii) V satisfies the hypotheses of Theorem XIII.59 and for some $b > 0$ we have

$$-\Delta - \tfrac{1}{2}(1+b)W - bV \geq 0 \qquad (97)$$

Proof If (ii) holds, then it is easy to check that the hypotheses of Theorem XIII.59 hold. So, in all three cases let W be as in Theorem XIII.59. Then

$$(\psi, -\Delta\psi) = \tfrac{1}{2}(\psi, W\psi)$$

for any eigenfunction ψ. In case (i), $W(x) \leq 0$ so $\psi = 0$ since $H_0 \geq 0$ and $\mathrm{Ker}(H_0) = \{0\}$. In case (ii), $W = -\alpha V$, so by (96),

$$2E(\psi, \psi) = (2-\alpha)(\psi, V\psi) = -\alpha^{-1}(2-\alpha)(\psi, W\psi)$$
$$= -2\alpha^{-1}(2-\alpha)(\psi, H_0\psi)$$

Since $2 - \alpha > 0$ and $-\Delta \geq 0$, we conclude that $E < 0$. Finally, in case (iii), for any eigenfunction ψ, we have

$$-bE(\psi, \psi) = -b(\psi, (H_0 + V)\psi) + (1+b)(\psi, (H_0 - \tfrac{1}{2}W)\psi)$$
$$= (\psi, [H_0 - \tfrac{1}{2}(1+b)W - bV]\psi)$$

so that (97) implies that $E \leq 0$. ∎

We emphasize that the potentials in this theorem are not required to go to zero at infinity, so the theorem can be applied to many-body Schrödinger operators.

Example 5 (Coulomb systems, including atomic Hamiltonians) Suppose that we have an n-body system with two body forces that are all of the form $a_{ij}|r_i - r_j|^{-1}$. Then V is homogeneous of degree -1, so by (ii) of Theorem XIII.60, $H = -\Delta + V$ has no positive eigenvalues.

Example 6 Let us illustrate (iii) of Theorem XIII.60 by a simple example. Let $V \in C_0^\infty(\mathbb{R}^n)$ with $n \geq 3$. We have seen in Section 3 that $-\Delta + \lambda V$ has no negative eigenvalues so long as λ is sufficiently small. What about positive eigenvalues? The smoothness techniques of Section 7 imply that there are no positive eigenvalues either. And, Theorem XIII.58 says that there are no positive eigenvalues for any value of λ. An alternative proof is available for small λ, since $-\mathbf{r} \cdot \nabla V - V$ is also in $C_0^\infty(\mathbb{R}^n)$. Thus, for λ small $-\Delta - \lambda(\mathbf{r} \cdot \nabla V + V)$ has no negative eigenvalues and so is a positive operator. It follows from part (iii) of Theorem XIII.60 that $-\Delta + \lambda V$ has no positive eigenvalues for λ small.

Theorem XIII.60 does not exhaust the information available from the virial theorem. For example, one can deal with sums of potentials of the type considered there (Problem 108). For some potentials one can prove the absence of eigenvalues in some interval (a, ∞) with $a > 0$.

Example 7 Let

$$H = -\sum_{j=1}^{n} \Delta_j + \sum_{j=1}^{n} V_j(\mathbf{r}_j) + \sum_{i<j} V_{ij}(\mathbf{r}_i - \mathbf{r}_j)$$

on $L^2(\mathbb{R}^{3n})$ where $V_j(\mathbf{r}_j) = -b_j r_j^{-1} \exp(-a_j r_j)$ and

$$V_{ij}(\mathbf{r}) = b_{ij} r^{-1} \exp(-a_{ij} r)$$

with all a and b positive constants. Let $U(r) = r^{-1}(\exp(-ar))$. Then $r \, dU/dr = -U - a \exp(-ar)$ so that if $V = \sum_{j=1}^{n} V_j + \sum_{i<j} V_{ij}$, then

$$\mathbf{x} \cdot \nabla V + V \leq \sum_{j=1}^{n} a_j b_j$$

It follows that, if $(-\Delta + V)\psi = E\psi$, then

$$E(\psi, \psi) = -(\psi, H_0 \psi) + (\psi, (\mathbf{x} \cdot \nabla V + V)\psi)$$

$$\leq \sum_{j=1}^{n} a_j b_j (\psi, \psi)$$

so that H has no eigenvalues in $(\sum_{j=1}^{n} a_j b_j, \infty)$.

There is a connection between virial theorem methods and the result of Theorem XIII.58. For example, if V is negative for all x and homogeneous of degree $-\alpha$ with $\alpha \leq 1$, then near infinity the conditions (ii') and (iii') of the proof of Theorem XIII.58 hold, so we can conclude that $-\Delta + V$ has no positive eigenvalues by either method. The distinction between the types of results is that Theorem XIII.58 requires only information at infinity, while Theorem XIII.60 requires global hypotheses.

There is also a connection between virial theorem methods and our final method. The role played by dilations in the above suggests the dilation analytic methods of Section 10 might be relevant to the problem of positive energy bound states. To see that they are, let us first reconsider the case of Coulomb potentials:

Example 5, revisited Let H be the Hamiltonian of an n-body system whose two body potentials are all homogeneous of some fixed degree $-\alpha$ where $0 < \alpha < 1$. Then the potentials are all dilation analytic, in fact,

$$U(\theta)VU(\theta)^{-1} = e^{-\alpha\theta}V$$

where $U(\theta)$ is the group of dilations of Section 10. Suppose that E_0 is the largest threshold of $H_0 + V$ (we will eventually prove that $E_0 = 0$) and that $E > E_0$ is an eigenvalue of H. By our general analysis of dilation analytic Hamiltonians, $H(\theta)$ will have E as a discrete eigenvalue so long as $0 < \text{Im } \theta < \pi$. Writing

$$H(\theta) = e^{-2\theta}(H_0 - e^{-(\alpha-2)(\theta-\theta_0)}V)$$

with $\theta_0 = i\pi/(2 - \alpha)$, we note that

$$H(\theta_0) = e^{-2\theta_0}(H_0 - V)$$

so that any eigenvalues of $H(\theta_0)$ must have argument $2i\theta_0 \neq 0$ or $\pi + 2i\theta_0 \neq 0$. Thus E cannot be an eigenvalue of $H(\theta_0)$ and so of H (see Figure XIII.6a). The condition $\alpha < 1$ is needed to ensure that $\text{Im } \theta_0 < \pi$.

Now by induction we claim that $E_0 = 0$, for this is true in the two-body system and clearly if it is true in all k-body systems with $k < n$, then by the above argument all k-body systems with $k < n$ have no positive eigenvalues and so the n-body system has no positive thresholds. Notice that by the above argument we can actually prove that $H(\theta)$ has no eigenvalues in $\{E \mid 0 \leq \arg E < 2 \text{ Im } \theta\}$, for $0 > \text{Im } \theta > -\pi$.

Now let V be a Coulomb potential. Let V_α be the potential obtained by replacing $|r_i - r_j|^{-1}$ by $|r_i - r_j|^{-\alpha}$. By analytic perturbation theory, if $H_0 + V$ has a positive eigenvalue, then for, say $\theta = -i\pi/3$, $H_0(\theta) + V_\alpha(\theta)$ will have an eigenvalue near the positive real axis for $\alpha - 1$ small. But, by general principles, this eigenvalue cannot have small negative argument for α real. By the above argument, for α real and less than 1, it cannot have argument zero or small and positive. This contradiction shows that $H_0 + V$ cannot have any positive eigenvalues (see Figure XIII.6b).

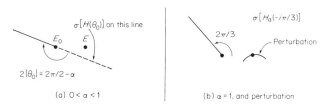

FIGURE XIII.6

Example 5 (revisited, alternative argument) There is another way of obtaining the results we have just discussed that allows $0 < \alpha < 2$ and which also requires all potentials to be homogeneous of the same degree. $H_0(\theta) = e^{-2\theta}H_0$, $V(\theta) = e^{-\alpha\theta}V$. Thus for any $\varphi \in Q(H_0)$, $(\varphi, H_0(\theta)\varphi)$ has argument $2\pi - 2\,\mathrm{Im}\,\theta$ while $(\varphi, V(\theta)\varphi)$ has argument $2\pi - \alpha\,\mathrm{Im}\,\theta$ or $\pi - \alpha\,\mathrm{Im}\,\theta$. Thus, since $\alpha < 2$, $(\varphi, H(\theta)\varphi)$ has argument between $\pi - \alpha\,\mathrm{Im}\,\theta$ and $2\pi - \alpha\,\mathrm{Im}\,\theta$ for $\mathrm{Im}\,\theta$ small and positive. In particular, $(\varphi, H(\theta)\varphi)$ cannot have argument zero and so $H(\theta)\varphi$ cannot be $E\varphi$ with $E > 0$. Since $H(\theta)$ cannot have positive eigenvalues, neither can H.

There is one especially interesting feature of the above argument: There is a distinction between positive and negative eigenvalues, as there must be due to Example 4; the distinction occurs at $\mathrm{Im}\,\theta = \pm\tfrac{1}{2}\pi$. For $|\mathrm{Im}\,\theta| < \tfrac{1}{2}\pi$, any eigenvalue of H at a nonthreshold point is a discrete eigenvalue of $H(\theta)$. As $\mathrm{Im}\,\theta$ approaches $\pm\tfrac{1}{2}\pi$, the continuous spectrum approaches the negative eigenvalues, so we cannot assert that the eigenvalues remain for $\mathrm{Im}\,\theta > \tfrac{1}{2}\pi$ (and in particular, for $\theta = \theta_0$). But positive eigenvalues stay away from the continuous spectrum for $|\mathrm{Im}\,\theta| \in (0, \pi)$. This distinction suggests that dilation analytic potentials allowing a continuation up to $\mathrm{Im}\,\theta = \tfrac{1}{2}\pi$ might be special. We therefore define:

Definition We say that a "potential" V lies in $\overline{\mathscr{F}}_{\pi/2}$ if and only if:
(1) V is a quadratic form with form domain $D(-\Delta)$.
(2) $(-\Delta + 1)^{-1/2}[U(\theta)VU(\theta)^{-1}](-\Delta + 1)^{-1/2}$ defined as an operator for $\theta \in \mathbb{R}$ is the restriction to the real axis of a compact operator-valued function that is analytic in the strip $|\mathrm{Im}\,\theta| < \tfrac{1}{2}\pi$ and norm continuous in the strip $|\mathrm{Im}\,\theta| \leq \tfrac{1}{2}\pi$.

As we shall see, $\overline{\mathscr{F}}_{\pi/2}$ includes the Coulomb and Yukawa potentials. We shall prove presently that any N-body system all of whose two-body potentials lie in $\overline{\mathscr{F}}_{\pi/2}$ has no positive eigenvalues. To do this, we need a complex variables theorem related to Theorem XII.18.

Lemma (Carlson's theorem) Suppose that f is a complex-valued function defined and continuous in $\{z \mid \operatorname{Re} z \geq 0\}$ and analytic in $\{z \mid \operatorname{Re} z > 0\}$. Suppose that

$$|f(z)| \leq Me^{A|z|} \qquad \text{all } z \text{ with } \operatorname{Re} z \geq 0 \qquad (98a)$$

$$|f(iy)| \leq Me^{-B|y|} \qquad \text{all } y \in \mathbb{R} \qquad (98b)$$

where $B > 0$. Then f is identically zero.

Proof Let us first prove a slightly strengthened version of the maximum principle called the **Phragmen–Lindelöf principle**: Suppose that g is analytic in the wedge

$$N = \{re^{i\theta} \mid r > 0, \alpha < \theta < \beta\}$$

and continuous in $\bar{N}$ and that $\beta - \alpha < \pi$. Then if $|g(z)| \leq C_1 \exp(C_2|z|)$ in $\bar{N}$, we have

$$|g(z)| \leq \sup_{y \geq 0} \{|g(ye^{i\alpha})|, |g(ye^{i\beta})|\} \equiv D$$

To prove this, we first suppose without loss that $\beta < \frac{1}{2}\pi\mu^{-1}$, $\alpha > -\frac{1}{2}\pi\mu^{-1}$, where $\mu > 1$. Let

$$g_\varepsilon(z) = g(z) \exp(-\varepsilon z^\mu)$$

Then $g_\varepsilon \to 0$ as $|z| \to \infty$ in $\bar{N}$, so, by the ordinary maximum principle, $g_\varepsilon(z)$ takes its maximum on ∂N. Thus

$$|g_\varepsilon(z)| \leq D$$

for any ε. Taking $\varepsilon \downarrow 0$, we see that $|g(z)| \leq D$.

Now suppose that f obeys (98). Let

$$h(z) = f(z)\exp(-iBz - Az)$$

Then $|h(z)| \leq M$ if $|\arg z| = 0$ or $\frac{1}{2}\pi$ and $|h(z)| \leq M \exp((A+B)|z|)$ for all z. Thus, by the Phragmen–Lindelöf principle, $|h(z)| \leq M$ if $0 \leq \arg z \leq \frac{1}{2}\pi$. It follows that

$$|f(z)| \leq M \exp(A|z|\cos\theta - B|z||\sin\theta|); \qquad \theta = \arg z \qquad (99)$$

for $0 \leq \theta \leq \frac{1}{2}\pi$ and similarly for $0 \geq \theta \geq -\frac{1}{2}\pi$.

Now consider the function $g(x) = f(ix)$ for x real and let $\hat{g}$ be its Fourier transform. By (99),

$$|g(x - iy)| \leq M \exp(A|y|), \qquad y > 0$$

so a Paley–Wiener argument implies that $\hat{g}(k) = 0$ if $k < -A$. Since
$$|g(x)| \leq M \exp(-\beta |x|)$$
Paley–Wiener arguments also imply that $\hat{g}$ is analytic in a neighborhood of the real axis. It follows that g is identically zero so f is zero also. ∎

Theorem XIII.61 (Balslev–Simon theorem) Let H be an N-body Schrödinger operator on $L^2(\mathbb{R}^{nN-n})$ obtained by removing the center of mass motion from
$$\tilde{H} = -\sum_{i=1}^{N} (2m_i)^{-1}\Delta_i + \sum_{i=j} V_{ij}(r_i - r_j)$$
where each V_{ij} is a form in $\mathcal{F}_{\pi/2}$ which is local (i.e., the form of a multiplication operator) and symmetric. Then H has no eigenvalues in $(0, \infty)$.

Proof As in Example 5 (revisited), if we can prove that H has no eigenvalues in $(0, \infty)$ under the hypothesis that H has no positive thresholds, then by an induction argument, we can prove that H has no positive eigenvalues.

If H has no positive thresholds and $E > 0$ is an eigenvalue, then E is also an eigenvalue of $H(\theta)$ for $|\text{Im } \theta| \leq \frac{1}{2}\pi$ and the projection
$$P(\theta) = -(2\pi i)^{-1} \int_{|\lambda - E| = \varepsilon} (H(\theta) - \lambda)^{-1} d\lambda$$
defined for $|\text{Im } \theta| > 0$, is analytic in $|\text{Im } \theta| < \frac{1}{2}\pi$ and equal to the spectral projection $P_{\{E\}}(H)$ for $\theta = 0$. Thus, by O'Connor's lemma (the proposition following Theorem XIII.39), any eigenvector ψ with $H\psi = E\psi$ has the property that $U(\theta)\psi$ extends to a function analytic in $\{\theta \,|\, |\text{Im } \theta| < \frac{1}{2}\pi\} \equiv N$, continuous in $\bar{N}$. Moreover, since E is an isolated eigenvalue of $H(\pm i\frac{1}{2}\pi)$, by the arguments of Section 11, $U(\pm i\frac{1}{2}\pi)\psi$ falls off exponentially in the sense that for some $b > 0$, $e^{br}U(\pm i\frac{1}{2}\pi)\psi \in L^2$.

Heuristically, the argument would proceed as follows: Intuitively, the statement that ψ is dilation analytic means that $\psi(x)$, thought of as a function of $r = |x|$ and $\Omega = x/|x|$, is analytic in r for fixed Ω in the region $\text{Re } r > 0$ with
$$e^{-\frac{1}{2}(nN-n)i\theta}(U(\theta)\psi)(r, \Omega) = \psi(re^{i\theta}, \Omega)$$
$\psi \in L^2$ implies that $\psi(z, \Omega)$ is more or less bounded and $U(\pm i\frac{1}{2}\pi)\psi \in D(e^{br})$ implies that $\psi(z, \Omega)$ falls off exponentially if $\arg z = \pm \frac{1}{2}\pi$. Thus $\psi = 0$ by

Carlson's theorem. This argument does not go through because all our information is L^2, but it is the intuition behind the argument below.

Fix $g \in C_0^\infty(\mathbb{R}^{nN-n}\setminus\{0\})$, a function with support away from zero; say supp $g \subset \{x \mid |x| > R\}$. For $z \in (0, \infty)$, define

$$F(z) = z^{nN/2} \int \overline{g(x)} \psi(zx) \, d^{nN-n}x$$

$$= z^{n/2}(g, U(\ln z)\psi)$$

Using this formula, we can extend F to a function analytic in $\{z \mid \text{Re } z > 0\}$, continuous in $\{z \mid \text{Re } z \geq 0\}$.

Since $U(\theta)$ is unitary for θ real,

$$|F(z)| \leq |z|^{n/2} \|g\| \sup_{|y| \leq \pi/2} \|U(iy)\psi\|$$

for all z with Re $z \geq 0$. Moreover, for β real and positive

$$|F(i\beta)| = \left| \beta^{nN/2} \int \overline{g(x)} \left[U\left(i\frac{\pi}{2}\right)\psi \right](\beta x) \, dx \right|$$

$$= \beta^{n/2} \int_{|x| \geq \beta R} \overline{g(\beta^{-1}x)} \left[U\left(\frac{\pi}{2}i\right)\psi \right](x) \, dx$$

$$\leq e^{-\beta Rb} |\beta|^{nN/2} \|g\| \left\| e^{br} U\left(\frac{\pi}{2}i\right)\psi \right\|$$

and similarly for $F(-i\beta)$. Thus, by Carlson's theorem, $F = 0$ so $(g, \psi) = 0$. Since the functions $g \in C_0^\infty(\mathbb{R}^{nN-n}\setminus\{0\})$ are dense, $\psi = 0$. ∎

Example 7, revisited The potential $V(r) = e^{-ar}/r$ is in $\overline{\mathscr{F}}_{\pi/2}$ with $V_\theta(r) = e^{-\theta}r^{-1}\exp(-ae^\theta r)$. Thus the Hamiltonians of Example 7 have no eigenvalues in $(0, \infty)$. This is an improvement of the virial theorem result. However, there are cases where the virial method gives information while Theorem XIII.61 gives none. For example, if $V(r) = be^{-ar}$ with $b > 0$, then $-\Delta + V$ has no positive eigenvalues by the virial theorem, but V is *not* in $\overline{\mathscr{F}}_{\pi/2}$ since $\exp(-ae^\theta r)$ is not a relatively compact multiplication operator, so $V_\theta(r)$ cannot have a *norm* continuous extension to $\{\theta \mid |\text{Im } \theta| = \tfrac{1}{2}\pi\}$.

Example 8 If each V_{ij} is homogeneous of degree β_{ij} with $0 < \beta_{ij} < 2$. Then $\tilde{H} = -\sum_{i=1}^n \Delta_i + \sum_{i<j} V_{ij}$ has the property that after its center of mass motion is removed, it has no positive energy bound states. This result is not obtainable with virial theorem methods although it is reasonable from the virial theorem point of view.

In the above we have concentrated on Schrödinger operators, but the ideas are applicable in other situations. In our study of acoustical scattering by the Lax–Phillips method (Section XI.11), we needed the following result:

Theorem XIII.62 Let σ and λ be real C^2 functions on $\mathbb{R}^3$ so that $\sigma(x) > 0$, $\lambda(x) > 0$, and such that $\sigma_0 - \sigma(x)$ and $\lambda_0 - \lambda(x)$ have compact support for some constants σ_0 and λ_0. Let H be the self-adjoint operator on $L^2(\mathbb{R}^3)$ obtaining by closing the quadratic form

$$q(f,f) = \int_{\mathbb{R}^3} |\nabla(\lambda f)|^2 \sigma \, dx$$

on $C_0^\infty(\mathbb{R}^3)$. Then H has no eigenvalues.

Proof Clearly, for some $a > 0$, $\sigma(x) \geq a$ for all x. Thus, for $f \in D(H)$,

$$(Hf, f) = q(f, f) \geq a \int_{\mathbb{R}^3} |\nabla(\lambda f)|^2 \, dx \geq 0$$

Thus H cannot have any negative eigenvalues and if $Hf = 0$, then $\lambda f = 0$, which implies that $f = 0$, so zero cannot be an eigenvalue either. It remains to eliminate strictly positive eigenvalues.

Suppose that $E > 0$ and that $Hf = Ef$. Let B_R be a ball of radius R containing the supports of $\sigma_0 - \sigma(x)$ and $\lambda_0 - \lambda(x)$. Then, on $\Omega = \mathbb{R}^3 \backslash B_R$, f obeys

$$-\Delta f = (\lambda_0^2 \sigma_0)^{-1} E f$$

and so f vanishes on Ω by the argument preceding Theorem XIII.57. Thus, to conclude that f is zero we need a unique continuation theorem for H. But since $Hf = Ef$, we have that

$$(\Delta f)(x) = \alpha(x) f(x) + \beta(x) \cdot \nabla f(x)$$

for suitable bounded functions $\alpha(x)$ and $\beta(x)$, so that

$$|(\Delta f)(x)| \leq M(|f(x)| + |\nabla f(x)|)$$

By a slight extension of the methods of the Appendix (see Problem 113), there is a unique continuation theorem for such an inequality. ∎

Appendix to XIII.13 Unique continuation theorems for Schrödinger operators

In this appendix we shall prove Theorem XIII.57. In fact, we shall prove the following stronger theorem:

Theorem XIII.63 Let $u \in H^2_{\text{loc}}$, that is, $\varphi u \in D(-\Delta)$ for each $\varphi \in C_0^\infty(\mathbb{R}^n)$. Let D be an open connected set in $\mathbb{R}^n$ and suppose that

$$|\Delta u(x)| \leq M|u(x)|$$

almost everywhere in D. Then, if u vanishes in the neighborhood of a single point $x_0 \in D$, u is identically zero in D.

The proof of this result, which we give only when $n = 3$ (see Problems 111, 112 for general n), relies on the following beautiful estimate for a C^∞ function f with support in $\{x \,|\, 0 < |x| < R\}$:

$$\int |x|^\alpha |f(x)|^2 \, d^3x \leq \tfrac{4}{3} R^4 \int |x|^\alpha |\Delta f|^2 \, d^3x \tag{100}$$

for *all* real α. We shall prove (100) by using the "partial wave expansion" of f,

$$f(x) = \sum_{\ell, m} f_{\ell m}(|x|) Y_{\ell m}(\Omega_x)$$

described explicitly below. Thus, we begin by proving an analogue of (100) for each partial wave.

Lemma 1 Let f be a C^∞ function with support in $(0, 1)$. For each $\ell = 0, 1, 2, \ldots$, define

$$g_\ell = -f'' + \frac{\ell(\ell+1)}{x^2} f$$

Then for *any* real α

$$\int_0^1 x^\alpha |f(x)|^2 \, dx \leq \tfrac{4}{3}(2\ell + 1)^{-2} \int_0^1 x^\alpha |g_\ell(x)|^2 \, dx$$

Proof Fix ℓ and let D denote the formal differential operator $-d^2/dx^2 + \ell(\ell+1)/x^2$. Then, as a formal operator, $Dh_+ = Dh_- = 0$ where $h_+(x) = x^{\ell+1}$ and $h_-(x) = x^{-\ell}$. Now, since f vanishes near zero and one, we can freely integrate by parts the expressions $\int f \, Dh_\pm = 0$ to conclude that

$$\int_0^1 g_\ell(x) h_\pm(x) \, dx = 0 \tag{101}$$

We claim that, setting $g = g_\ell$,

$$f(x) = [h_+(x) A(x) + h_-(x) B(x)]/(2\ell + 1) \tag{102a}$$

$$A(x) = \int_x^1 g(y) h_-(y) \, dy; \quad B(x) = \int_0^x g(y) h_+(y) \, dy \tag{102b}$$

Appendix to XIII.13 Unique continuation theorems

(102) is a standard variation of parameters formula from ordinary differential equations which can be proven as follows: Let $F(x)$ denote the right-hand side of (102a). Then $DF = g$ since $h_+ h'_- - h'_+ h_- = 2\ell + 1$, and by (101) F has support strictly in $(0, 1)$. Since $D(F - f) = 0$, and $F - f$ is zero near 0, $F - f = 0$ by the uniqueness of solutions of ordinary differential equations.

Next, we claim that for any β,

$$|A(x)x^\beta| \leq \int_0^1 |x^\beta h_-(x)g(x)|\, dx \tag{103a}$$

$$|B(x)x^\beta| \leq \int_0^1 |x^\beta h_+(x)g(x)|\, dx \tag{103b}$$

Suppose $\beta > 0$. Then,

$$|A(x)x^\beta| \leq x^\beta \int_x^1 y^{-\beta} |y^\beta h_-(y)g(y)|\, dy$$

$$\leq \int_x^1 y^\beta |h_-(y)g(y)|\, dy$$

If $\beta \leq 0$, we use (101) to rewrite A so that

$$|A(x)x^\beta| \leq x^\beta \int_0^x y^{-\beta} |y^\beta h_-(y)g(y)|\, dy$$

$$\leq \int_0^x |y^\beta h_-(y)g(y)|\, dy$$

This proves (103a); the proof of (103b) is similar.

As a result,

$$H(x) \equiv [x^{\alpha-2}|f(x)|^2]$$

$$\leq (2\ell + 1)^{-2}(|x^{\frac{1}{2}\alpha+\ell}A(x)| + |x^{\frac{1}{2}\alpha-\ell-1}B(x)|)^2$$

$$\leq \frac{4}{(2\ell+1)^2}\left(\int_0^1 |x^{\alpha/2}g(x)|\, dx\right)^2$$

$$\leq \frac{4}{(2\ell+1)^2} \int_0^1 x^\alpha |g(x)|^2\, dx$$

In particular

$$\int_0^1 x^\alpha |f(x)|^2\, dx = \int_0^1 x^2 H(x)\, dx \leq \frac{4}{3(2\ell+1)^2} \int_0^1 x^\alpha |g(x)|^2\, dx \quad \blacksquare$$

Lemma 2 Let h be in $C_0^\infty(\mathbb{R}^3)$ with support in $\{x \mid 0 < |x| < 1\}$. Then, for any real α,

$$\int |x|^\alpha |h(x)|^2 \, d^3x \leq \tfrac{4}{3} \int |x|^\alpha |\Delta h(x)|^2 \, d^3x$$

Proof We use the spherical harmonic decomposition described in Example 4 of the Appendix to Section X.1. There is a distinguished orthonormal basis $\{Y_{\ell m}(\Omega)\}$, $\ell = 0, 1, \ldots$; $m = -\ell, -\ell + 1, \ldots, \ell - 1, \ell$, of $L^2(S^2)$ where S^2 is the unit sphere in $\mathbb{R}^3$. Given $f \in L^2(\mathbb{R}^3)$, we define

$$f_{\ell, m}(r) = r \int f(r\Omega) \overline{Y_{\ell, m}(\Omega)} \, d\Omega$$

Then, for $f \in C_0^\infty$,

$$rf(r\Omega) = \sum_{\ell, m} f_{\ell m}(r) Y_{\ell m}(\Omega)$$

$$-r(\Delta f)(r\Omega) = \sum_{\ell, m} (D_\ell f_{\ell m})(r) Y_{\ell m}(\Omega)$$

where $D_\ell = -d^2/dr^2 + \ell(\ell + 1)r^{-2}$. Thus, using the orthonormality of the $Y_{\ell m}$, if h has support in $\{x \mid 0 < |x| < 1\}$,

$$\int |x|^\alpha |h(x)|^2 \, d^3x = \sum_{\ell, m} \int r^\alpha |h_{\ell m}(r)|^2 \, dr$$

$$\leq \sum_{\ell, m} \tfrac{4}{3}(2\ell + 1)^{-2} \int r^\alpha |D_\ell h_{\ell m}(r)|^2 \, dr$$

$$\leq \tfrac{4}{3} \int |x|^\alpha |\Delta h(x)|^2 \, d^3x$$

where we have used Lemma 1. ∎

Lemma 3 Suppose that $h \in L^2(\mathbb{R}^3, d^3x)$ lies in the domain of $-\Delta$ and that h vanishes outside some compact subset of $\{x \mid 0 < |x - a| < R\}$ for some $a \in \mathbb{R}^3$; $R > 0$. Then

$$\int |x - a|^\alpha |h(x)|^2 \, d^3x \leq \tfrac{4}{3}R^4 \int |x - a|^\alpha |\Delta h(x)|^2 \, d^3x \tag{104}$$

for *any* real α.

Proof Suppose first that $h \in C_0^\infty(\mathbb{R}^3)$. Then (104) follows from Lemma 2 after the change of variables $y = (x - a)/R$. Knowing (104) for all $h \in C_0^\infty(\mathbb{R}^3)$ with the required support property yields (104) for all $h \in D(-\Delta)$ by a simple approximate identity argument. ∎

Proof of Theorem XIII.63 For each $x \in D$, define

$$R_x = \min\{(\tfrac{128}{3}M^2)^{-1/4}, \tfrac{1}{2} \operatorname{dist}(x, \partial D)\}$$

Appendix to XIII.13 Unique continuation theorems

We claim that, if $u(x)$ vanishes for x near y and obeys $|\Delta u(x)| \leq M|u(x)|$, then $u(x)$ vanishes in $\{x \mid |x-y| \leq R_y\}$. Accepting this claim for the moment, let us prove that $u \equiv 0$ if u vanishes near some $x_0 \in D$. Given $y \in D$, choose a smooth curve γ in D joining x_0 to y. Suppose that γ has length L. Since γ is compact, $\text{dist}(x, \partial D)$ is bounded away from zero on γ so R_x is bounded below on γ, say by R_0. Choose an integer n so that $\frac{1}{2}R_0 n \geq L > \frac{1}{2}R_0(n-1)$, and $x_1, \ldots, x_{n-1}$, $x_n = y$ on γ so that the length of γ between x_i and x_{i-1} $(i=1,\ldots,n-1)$ is $\leq \frac{1}{2}R_0$. In particular, $|x_i - x_{i-1}| \leq \frac{1}{2}R_0$. Since u vanishes near x_0, by our claim, it vanishes if $|x - x_0| < R_0$ and so near x_1. Repeating this argument, we see that $u(y) = 0$, so $u \equiv 0$ in D.

We now turn to the proof of the claim. Fix y and let χ be a function in C_0^∞ that is 1 if $|x-y| \leq R_y$ and 0 if $|x-y| \geq \frac{3}{2}R_y$. Then χu has support in $\{x \mid 0 < |x-y| < 2R_y\}$ so, by Lemma 3, for any $\beta > 0$,

$$\int_{|x-y| \leq R_y} |x-y|^{-\beta} |u(x)|^2 \, dx$$

$$\leq \int |x-y|^{-\beta} |(\chi u)(x)|^2 \, dx$$

$$\leq \tfrac{4}{3}(2R_y)^4 \int_{|x-y| \leq 2R_y} |x-y|^{-\beta} |\Delta(\chi u)(x)|^2 \, dx$$

$$\leq \tfrac{64}{3} M^2 R_y^4 \int_{|x-y| \leq R_y} |x-y|^{-\beta} |u(x)|^2 \, dx + C R_y^{-\beta}$$

$$\leq \tfrac{1}{2} \int_{|x-y| \leq R_y} |x-y|^{-\beta} |u(x)|^2 \, dy + C R_y^{-\beta}$$

where

$$C = \tfrac{64}{3} R_y^4 \int_{R_y \leq |x-y| \leq 2R_y} |\Delta(\chi u)|^2 \, dx$$

is finite since $\chi u \in D(-\Delta)$. Therefore, if $R_1 < R_y$,

$$\int_{|x-y| \leq R_1} |u(x)|^2 \, dx \leq R_1^\beta \int_{|x-y| < R_y} |x-y|^{-\beta} |u(x)|^2 \, dx$$

$$\leq 2C(R_1/R_y)^\beta$$

Letting β go to infinity, we see that $\int_{|x-y| \leq R_1} |u(x)|^2 \, dx = 0$. This proves our claim, and thus the result. ∎

We are now ready to restate and prove Theorem XIII.57.

Theorem XIII.57 Let V be a function on $\mathbb{R}^n$ so that:

(i) Multiplication by V is $-\Delta$-bounded with relative bound less than 1.
(ii) There is a closed set S of measure zero so that $\mathbb{R}^n \backslash S$ is connected and so that V is bounded on any compact subset of $\mathbb{R}^n \backslash S$.

Let $H = -\Delta + V$ and suppose that $Hu = Eu$ for some E and $u \in L^2$. Suppose that u vanishes on an open subset of $\mathbb{R}^n$. Then u is identically zero.

Proof Suppose that u vanishes near $x_0 \in \mathbb{R}^n \backslash S$. By choosing a neighborhood of a curve between x_0 and some $y \in \mathbb{R}^n \backslash S$ we can find an open connected set $D \subset \mathbb{R}^n \backslash S$, with $x_0, y \in D$, so that $\bar{D}$ is compact in $\mathbb{R}^n \backslash S$. Since V is bounded on D, say by M_0, and $u \in D(-\Delta + V) = D(-\Delta)$,

$$|-\Delta u| \leq (M_0 + |E|)|u|$$

on D, so u vanishes in D by Theorem XIII.63. It follows that u is zero on $\mathbb{R}^n \backslash S$. ∎

XIII.14 Compactness criteria and operators with compact resolvent

In most of this chapter we have studied the continuous spectrum of self-adjoint operators, especially Schrödinger operators with potentials that decay at infinity. In this section, by contrast, we concentrate on criteria for proving that an operator has purely discrete spectrum. First, we show that a semibounded self-adjoint operator A has entirely discrete spectrum if and only if its resolvent is a compact operator. To apply this result, we need criteria that guarantee that certain subsets of L^2 are compact; these criteria are supplied by results of Rellich and Riesz. We then apply these techniques to provide extensions of Theorem XIII.16 which shows that Schrödinger operators with potentials that grow at infinity have purely discrete spectrum.

These ideas are intimately connected with various compact embedding theorems for Sobolev spaces which are required in various applications to partial differential equations and, in particular, in the Lax–Phillips theory (Section XI.11). In the middle part of this section we describe these embedding theorems.

As a further application, we show that the Hamiltonian H associated to a statistical mechanical system in a box has purely discrete spectrum and that $e^{-\beta H}$ is trace class. We conclude the section by giving criteria that guarantee that part of the spectrum of a self-adjoint operator is discrete.

Theorem XIII.64 Let A be a self-adjoint operator that is bounded from below. Then the following are equivalent:

(i) $(A - \mu)^{-1}$ is compact for some $\mu \in \rho(A)$.
(ii) $(A - \mu)^{-1}$ is compact for all $\mu \in \rho(A)$.
(iii) $\{\psi \in D(A) \mid \|\psi\| \leq 1; \|A\psi\| \leq b\}$ is compact for all b.
(iv) $\{\psi \in Q(A) \mid \|\psi\| \leq 1, (\psi, A\psi) \leq b\}$ is compact for all b.
(v) There exists a complete orthonormal basis $\{\varphi_n\}_{n=1}^{\infty}$ in $D(A)$ so that $A\varphi_n = \mu_n \varphi_n$ with $\mu_1 \leq \mu_2 \leq \cdots$ and $\mu_n \to \infty$.
(vi) $\mu_n(A) \to \infty$ where $\mu_n(\cdot)$ is given by the min–max principle.

Proof We shall prove (i) $\Leftrightarrow$ (ii); (v) $\Leftrightarrow$ (vi); (v) $\Rightarrow$ (iv) $\Rightarrow$ (iii) $\Rightarrow$ (i) $\Rightarrow$ (v).

(i) $\Leftrightarrow$ (ii) is trivial in one direction and follows in the other by the first resolvent formula and the fact that the compact operators are an ideal.

(v) $\Leftrightarrow$ (vi) follows directly from the min–max principle, Theorem XIII.1.

(v) $\Rightarrow$ (iv) Let
$$F_b = \{\psi \in Q(A) \mid \|\psi\| \leq 1, (\psi, A\psi) \leq b\}$$
We will prove that F_b is closed and that, for any ε, it can be covered by finitely many open balls of radius ε. It easily follows from this that it is compact (Problem 114). Let $\psi_n \in F_b$ with $\psi_n \to \psi$. Then clearly $\|\psi\| \leq 1$, so we need only show that $\psi \in Q(A)$ and $(\psi, A\psi) \leq b$. It is clearly sufficient to prove that for any $\lambda < \infty$, $(\psi, AP_{(-\infty, \lambda)} \psi) \leq b$ where P_Ω is the family of spectral projections for A. But, since $AP_{(-\infty, \lambda)}$ is bounded,
$$(\psi, AP_{(-\infty, \lambda)} \psi) = \lim_{n \to \infty} (\psi_n, AP_{(-\infty, \lambda)} \psi_n)$$
$$\leq \lim_{n \to \infty} (\psi_n, A\psi_n) \leq b$$

Thus F_b is closed. Now let $-a$ be a lower bound for A so that $B = A + a \geq 0$ and $\psi \in F_b$ implies $(\psi, B\psi) \leq b + a$. Given ε, choose N so that $\mu_N + a \geq 2\varepsilon^{-1}(b + a)$. Then $(\psi, B\psi) \leq b + a$ and hypothesis (v) imply that
$$\sum_{n \geq N} |(\psi, \varphi_n)|^2 \leq \tfrac{1}{2}\varepsilon$$

Thus any $\psi \in F_b$ is within $\sqrt{\varepsilon/2}$ of a point of the unit ball in the subspace generated by $\{\varphi_1, \ldots, \varphi_{N-1}\}$. Since this ball can be covered by finitely many $\sqrt{\varepsilon/2}$-balls, F_b can be covered by finitely many ε-balls.

(iv) $\Rightarrow$ (iii) The set
$$D_b = \{\psi \in D(A) \mid \|\psi\| \leq 1, \|A\psi\| \leq b\}$$
is closed by an argument similar to that proving F_b closed. By the Schwarz inequality, $D_b \subset F_b$, so D_b is compact if F_b is compact.

(iii) ⇒ (i) Without loss of generality, we may assume that A is positive. Let $M = \{\psi = (A+1)^{-1}\varphi \mid \|\varphi\| \leq 1\}$. Suppose that $\psi \in M$ and $\varphi = (A+1)\psi$. Then

$$\|\psi\| \leq \|(A+1)^{-1}\| \|\varphi\| \leq \|\varphi\| \leq 1$$

and

$$\|A\psi\| \leq \|A(A+1)^{-1}\| \|\varphi\| \leq \|\varphi\| \leq 1$$

so $\psi \in D_1$. Thus, $\bar{M} \subset \bar{D}_1$ so that D_1 being compact implies that M is precompact, i.e., that $(A+1)^{-1}$ is a compact operator.

(i) ⇒ (v) Since (i) ⇒ (ii), we may suppose that $(A+a)^{-1}$ is compact where a is real and $-a$ is a strict lower bound for A. Since $(A+a)^{-1}$ is compact, the Hilbert–Schmidt theorem (Theorem VI.16) assures us that there exists an orthonormal basis $\{\varphi_n\}$ with $(A+a)^{-1}\varphi_n = \lambda_n \varphi_n$ and $\lambda_n \to 0$. Since $(A+a)^{-1}$ is positive, $\lambda_n \geq 0$, so we can renumber the λ's and φ's to ensure that $\lambda_1 \geq \lambda_2 \geq \cdots$. Taking $\mu_n = \lambda_n^{-1} - a$, (v) follows. ∎

On account of the criteria (iii), (iv) of Theorem XIII.64, it is important to have criteria that assure us that a subset S of $L^2(\mathbb{R}^n)$ is compact. To see the ideas involved, consider first the case where $\mathcal{H} = L^2(-\pi, \pi)$ and A is the self-adjoint extension of $i\,d/dx \upharpoonright C_0^\infty(-\pi, \pi)$ with boundary conditions $\varphi(-\pi) = \varphi(\pi)$. Let

$$S = \{\psi \in L^2(-\pi, \pi) \mid \|\psi\| \leq 1, \|A\psi\| \leq 1\}$$

Expanding each $\psi \in S$ in its Fourier series, we can map S isometrically onto

$$\hat{S} = \{\{a_n\} \in \ell_2(-\infty, \infty) \mid \sum |a_n|^2 \leq 1, \sum n^2 |a_n|^2 \leq 1\}$$

The second condition ensures that the tails of sequences in $\hat{S}$ are uniformly small and using this it is easy to show that $\hat{S}$ is compact (Problem 115), which proves that S is compact. This situation is reminiscent of Ascoli's theorem where compactness of support, uniform boundedness, and uniform equicontinuity combined to guarantee compactness of a set of functions. In this case the uniform equicontinuity is replaced by the smoothness condition, $\|A\psi\| \leq 1$ for all $\psi \in S$, which is equivalent to a decay condition on the Fourier coefficients of ψ. If we now try to extend this idea to $L^2(\mathbb{R})$, it is clear that the smoothness condition itself is not enough. For given any $f \in C_0^\infty(-\pi, \pi)$, the set of translates $f_n(x) = f(x + 2n\pi)$ cannot have any convergent subsequence. The difficulty is that these functions are not uniformly small for large x. So this suggests that to get a compact set in $L^2(\mathbb{R})$ we should require both the functions and their Fourier transforms to be "uniformly small" at ∞. This is just the hypothesis of Rellich's criterion below.

XIII.14 Operators with compact resolvent

We give a short proof of Rellich's criterion based on Weyl's theorem and Theorem XIII.64. A longer proof which follows the Fourier series intuition we have just discussed is outlined in Problem 116.

Definition Let F be a function on $\mathbb{R}^n$ that is measurable and nonnegative almost everywhere. We say that $F \to \infty$ if and only if, for every $N > 0$, there is an R_N so that $F(x) \geq N$ for almost all x with $|x| \geq R_N$.

Theorem XIII.65 (Rellich's criterion) Let F and G be two functions on $\mathbb{R}^n$ so that $F \to \infty$ and $G \to \infty$. Then

$$S = \left\{ \psi \mid \int |\psi(x)|^2 \, dx \leq 1, \int F(x) |\psi(x)|^2 \, dx \leq 1, \int G(p) |\hat{\psi}(p)|^2 \, dp \leq 1 \right\}$$

is a compact subset of $L^2(\mathbb{R}^n)$.

Proof We use the basic idea behind the proof of Theorem XIII.16. By replacing F by $\min\{F(x), x^2\}$ and similarly changing G, we can suppose without loss that $F(x) \leq x^2$ and $G(p) \leq p^2$. Thus we can define a self-adjoint operator $A = G(p) + F(x)$ as a densely defined form sum on $Q(G(p)) \cap Q(F(x))$. By the proof in Theorem XIII.64, S is closed. Further, since

$$S \subset \{\psi \in L^2 \mid \|\psi\| \leq 1, (\psi, A\psi) \leq 2\}$$

it suffices to show that $\mu_m(A) \to \infty$ as $m \to \infty$ by the equivalence (iv) $\Leftrightarrow$ (vi) in Theorem XIII.64.

Now let V be a bounded function of compact support. Then, we claim that $V(x)(G(p) + 1)^{-1}$ is compact: For $V(x)(\varepsilon p^n + G(p) + 1)^{-1}$ is Hilbert–Schmidt since $(\varepsilon p^n + G(p) + 1)^{-1}$ is in L^2. Moreover, since $G \to \infty$, $(\varepsilon p^n + G(p) + 1)^{-1} \to (G(p) + 1)^{-1}$ in L^∞ as $\varepsilon \downarrow 0$, so $V(x)(G(p) + 1)^{-1}$ is a norm limit of Hilbert–Schmidt operators.

From Weyl's theorem (Theorem XIII.14), we see that $G(p) + V(x)$ has discrete spectrum in $(-\infty, 0)$ so that, in particular, $\mu_m(G(p) + V(x)) \geq -1$ for m sufficiently large.

Given $\alpha > 0$, define V_α by

$$V_\alpha(x) = \min\{F(x), \alpha + 1\} - \alpha - 1$$

V_α has compact support since $F \to \infty$. Since

$$\mu_m(A) \geq \mu_m(G(p) + V_\alpha(x)) + \alpha + 1$$

we see that $\mu_m(A) \geq \alpha$ for m sufficiently large. Since α is arbitrary, $\mu_m(A) \to \infty$ as $m \to \infty$. ∎

Theorem XIII.66 (M. Riesz's criterion)   Let $p < \infty$. Let $S \subset L^p(\mathbb{R}^n)_1$, the unit ball in L^p. A necessary and sufficient condition that the norm closure of S be norm compact is that:

(1) $f \to 0$ in L^p sense at infinity uniformly in S, i.e., for any ε, there is a bounded set $K \subset \mathbb{R}^n$ so that

$$\int_{\mathbb{R}^n \setminus K} |f(x)|^p \, dx \leq \varepsilon^p \tag{105}$$

for all $f \in S$;

and

(2) $f(\cdot - y) \to f$ uniformly in S as $y \to 0$, i.e., for any ε, there is a δ so that $f \in S$ and $|y| < \delta$ imply that

$$\int_{\mathbb{R}^n} |f(x - y) - f(x)|^p \, dx \leq \varepsilon^p \tag{106}$$

Proof   Suppose first that $\bar{S}$ is compact. Fix $\alpha > 0$. Find $f_1, \ldots, f_n$ so that the $(\alpha/3)$-balls about the f_i cover S. Then find K and δ so that (105) and (106) hold for $f = f_1, \ldots, f_n$ and $\varepsilon = \alpha/3$. This can be done since for any $g \in L^p$, $\lim_{K \uparrow \mathbb{R}^n} \int_{\mathbb{R}^n \setminus K} |g|^p \, dx = 0$ and $g(\cdot - y) \to g$. By an $\alpha/3$ argument, (105) and (106) hold with $\varepsilon = \alpha$.

Conversely, suppose that S obeys (1) and (2). For any compact $\Omega \subset \mathbb{R}^n$ and finite numbers α, β,

$$T(\Omega, \alpha, \beta) = \{f \in C_0^\infty(\mathbb{R}^n) \,|\, \operatorname{supp} f \subset \Omega, \|f\|_\infty \leq \alpha, \|\nabla f\|_\infty \leq \beta\}$$

is precompact in $C(\Omega)$ by Ascoli's theorem, and so a fortiori precompact in $L^p(\mathbb{R}^n)$. Thus, it suffices, given ε, to find Ω, α and β so that given $f \in S$, there is a $g \in T(\Omega, \alpha, \beta)$ with $\|f - g\|_p < \varepsilon$. For since T can be covered by finitely many ε-balls, we can conclude that S can be covered by finitely many 2ε-balls.

Given ε, find K and δ so that for $f \in S$ and $|y| \leq \delta$:

$$\int_{\mathbb{R}^n \setminus K} |f(x)|^p \, dx \leq (\varepsilon/4)^p$$

$$\|f(\cdot - y) - f\|_p \leq \varepsilon/4$$

Let η be any C^∞ function supported in $\{y \,|\, |y| < \delta\}$ with $\int \eta(x) \, dx = 1$ and let χ be the characteristic function of $K' = \{y \,|\, \operatorname{dist}(y, K) < \delta\}$. Let $\Omega = \{y \,|\, \operatorname{dist}(y, K) \leq 2\delta\}$, $\alpha = \|\eta\|_q$, and $\beta = \|\nabla \eta\|_q$, where $q = p(p-1)^{-1}$. Given $f \in S$, let $g = \eta * (\chi f)$. We claim that

$$\|g - f\|_p \leq \varepsilon \tag{107}$$

Moreover, $g \in T(\Omega, \alpha, \beta)$, for it is evident that supp $g \subset \Omega$ and the norm conditions follow from Young's inequality.

Thus, we need only prove (107) to see that $\bar{S}$ is compact. Now, by choice of K and the definition of K', if $|y| < \delta$,

$$\int_{\mathbb{R}^n \setminus K'} |f(x-y)|^p \, dx \le \int_{\mathbb{R}^n \setminus K} |f(x)|^p \, dx \le (\varepsilon/4)^p$$

so

$$\|(\chi f)(\cdot - y) - \chi f\|_p \le \|f(\cdot - y) - f\|_p + \|(1-\chi)f\|_p$$
$$+ \|[1 - \chi(\cdot - y)]f(\cdot - y)\|_p$$
$$\le 3\varepsilon/4$$

Thus,

$$\|\eta * \chi f - \chi f\|_p \le \int \eta(y) \|\chi f(\cdot - y) - \chi f\|_p \, dy \le 3\varepsilon/4$$

so

$$\|g - f\|_p \le \|g - \chi f\|_p + \|(1-\chi)f\|_p \le \varepsilon \quad \blacksquare$$

Using these last two theorems, we can completely analyze the spectrum of Schrödinger operators with potentials going to infinity at infinity. The following result strengthens Theorem XIII.16.

Theorem XIII.67 Let $V \in L^1_{\text{loc}}(\mathbb{R}^n)$ be bounded from below and suppose that $V \to \infty$. Then $H = -\Delta + V$ defined as a sum of quadratic forms is an operator with compact resolvent. In particular, H has purely discrete spectrum and a complete set of eigenfunctions.

Proof By Theorem XIII.64, we need only show that

$$F_H \equiv \{\psi \in Q(H) \mid \|\psi\| \le 1, (\psi, H\psi) \le b\}$$

is compact. Since it is closed by the arguments of Theorem XIII.64, we need only prove it is contained in a compact set. But, since $-\Delta$ and V are positive definite, $(\psi, H\psi) \le b$ implies that $(\psi, -\Delta \psi) \le b$ and $(\psi, V\psi) \le b$ so

$$F_H \subset \left\{ \psi \mid \|\psi\| \le 1, \int p^2 |\hat{\psi}(p)|^2 \, dp \le b, \int V(x)|\psi(x)|^2 \, dx \le b \right\}$$

and this set is compact by Rellich's criterion. $\blacksquare$

To extend this last result to the case where V has some negative singularities, we first prove a simple perturbation result.

Theorem XIII.68 Let A be a semibounded self-adjoint operator with compact resolvent. Let b be a symmetric form-bounded perturbation of A with relative form bound strictly smaller than 1. Let $C = A + b$ defined as a sum of forms. Then C has compact resolvent.

Proof By hypothesis,
$$|b(\psi, \psi)| \leq \alpha(\psi, A\psi) + \beta(\psi, \psi)$$
with $\alpha < 1$. Then for any $\psi \in Q(C) = Q(A)$,
$$(\psi, C\psi) \geq (1 - \alpha)(\psi, A\psi) - \beta(\psi, \psi)$$
It follows from the form version of the min-max principle (Theorem XIII.2) that
$$\mu_n(C) \geq (1 - \alpha)\mu_n(A) - \beta$$
Since $\mu_n(A) \to \infty$, we conclude that $\mu_n(C) \to \infty$, so C has a compact resolvent by criterion (vi) of Theorem XIII.64. ∎

Theorem XIII.69 Let $n \geq 3$. Let $V = V_1 + V_2$ where $V_2 \in L^{n/2}(\mathbb{R}^n) + L^\infty(\mathbb{R}^n)$, $V_1 > 0$, $V_1 \to \infty$, and $V_1 \in L^1_{\text{loc}}(\mathbb{R}^n)$. Then $H = -\Delta + V_1 + V_2$, defined as a sum of quadratic forms, is an operator with compact resolvent.

Proof By a form version of Strichartz theorem (Problem 119), V_2 is $-\Delta$-form bounded with relative bound zero. So, since V_1 is positive, V_2 is also $(-\Delta + V_1)$-form bounded with relative bound zero. The result now follows from Theorems XIII.67 and XIII.68. ∎

As a result of the above considerations, we have a complete qualitative description of the spectrum of $-\Delta + V$ when $V \to \infty$. What more can we ask for in a general framework? The actual eigenvalues $\lambda_n(V) \equiv \mu_n(-\Delta + V)$, and corresponding eigenfunctions $\psi_n(x)$, are obviously dependent on detailed information about V and change as V changes locally, but one can hope that the qualitative behavior of $\lambda_n(V)$ as $n \to \infty$ and $\psi_n(x)$ as $x \to \infty$ depends only on the qualitative behavior of V as $x \to \infty$. For the case of eigenvalues, we shall study this problem in Section 15, but we can use the methods of Section 11 to study the behavior of $\psi_n(x)$ as $x \to \infty$.

XIII.14 Operators with compact resolvent

Theorem XIII.70 Suppose that H obeys the hypotheses of Theorem XIII.69. Then any eigenfunction of H obeys $\psi(x) \in D(e^{a|x|})$ for any $a > 0$. If $V \geq 0$, then for any $a > 0$, there is a C with

$$|\psi(x)| \leq Ce^{-a|x|}$$

Proof As in the proof of Theorem XIII.39, we need only prove that ψ is an entire vector for $W(\alpha) = e^{i\alpha x_j}$ for each j to conclude that $\psi \in D(\exp(a|x|))$ for all $a > 0$. As in that theorem

$$W(\alpha) H W(\alpha)^{-1} = -\sum_{i \neq j} \partial_i^2 + (i\,\partial_j + \alpha)^2 + V \tag{108}$$

for α real. The right-hand side of (108) defines an entire analytic family of type (B) (see Section XII.2) for all complex α. Since $(H(\alpha) - \lambda)^{-1}$ is compact for α real, it is compact for all α. Moreover, as in the proof of the Theorem XIII.39, $H(\alpha)$ has locally constant spectrum. Using the compactness of the resolvent it is easy to see that $H(\alpha)$ has spectrum independent of α (Problem 121), so by O'Connor's lemma, all its eigenfunctions are entire in α. This proves the first assertion.

Now suppose that V is positive. Fix k and for $a \in \mathbb{R}$, let

$$H_0(a) = (i\,\partial_k - ia)^2 - \sum_{j \neq k} \partial_j^2 = e^{ax_k} H_0 e^{-ax_k}$$

Then $\exp(-tH_0(a))$ has the kernel

$$(4\pi t)^{-n/2} e^{ax_k} e^{-|x-y|^2/4t} e^{-ay_k}$$

and so is positivity preserving. Moreover, since this kernel is an L^2 function of $x - y$, Young's inequality implies that for any $t > 0$,

$$\|e^{-tH_0(a)}\psi\|_\infty \leq C_{a,t} \|\psi\|_2 \tag{109}$$

Now, let V_n be a monotone increasing sequence of bounded multiplication operators with $V_n \uparrow V$. Then $H_0(a) + V_n$ converges to $H_0(a) + V$ in strong resolvent sense (by a monotone convergence theorem for forms) so

$$e^{-t(H_0(a)+V)} = \underset{n\to\infty}{\text{s-lim}}\ \underset{m\to\infty}{\text{s-lim}}\ [e^{-tH_0(a)/m} e^{-tV_n/m}]^m$$

Since $e^{-tH_0(a)/m}$ is positivity preserving and $0 \leq e^{-tV_n/m} \leq 1$, we conclude that

$$|e^{-t(H_0(a)+V)}f| \leq e^{-tH_0(a)}|f|$$

pointwise a.e. But by (109), $\exp(-tH_0(a))|f| \in L^\infty$. Thus for any $f \in L^2$, $e^{-t(H_0(a)+V)}f \in L^\infty$; and, in particular, eigenfunctions of $H_0(a) + V$ are in L^∞. But, if ψ is an eigenfunction of $H_0 + V$, then $H_0(a) + V$ has $e^{ax_k}\psi$ as an eigenfunction. Thus $e^{ax_k}\psi \in L^\infty$ for all a, which proves the second result. ∎

The eigenfunctions of $-\Delta + x^2$ are Gaussians times products of Hermite polynomials, so the previous result on falloff faster than any exponential is far from optimal. One expects stronger results to hold if V goes to infinity fast enough. First, one can simply generalize the Combes–Thomas method a little further:

Theorem XIII.71 Let V have form of Theorem XIII.69. Let $W(x) = \int_0^{|x|} \sqrt{V_0(s)}\, ds$ where $V_0(r) = \text{ess inf}_{|x|=r} V_1(x)$. Then for any eigenfunction ψ of $-\Delta + V$, $\psi \in D(\exp(\tfrac{1}{2}\beta W(x)))$ for any $\beta < \sqrt{5} - 1$.

Proof Let $U(\alpha) = \exp(i\alpha W(x))$. Then,
$$U(\alpha)(-\Delta)U(\alpha)^{-1} = (i\nabla + \alpha\nabla W)^2$$
$$= -\Delta + \alpha^2(\nabla W)^2 + \alpha(\nabla W)\cdot(i\nabla) + \alpha(i\nabla)\cdot(\nabla W)$$

But $|\nabla W| = V_0^{1/2} \le V_1^{1/2}$, so
$$\alpha^2(\nabla W)^2 \le \alpha^2(-\Delta + V_2 + C) + \alpha^2(\nabla W)^2 \le \alpha^2(-\Delta + V + \text{const})$$
and
$$\pm[\alpha(\nabla W)\cdot(i\nabla) + \alpha(i\nabla)\cdot\nabla W] \le (|\alpha| + \varepsilon)(-\Delta + V + \text{const})$$
so $U(\alpha)(-\Delta + V)U(\alpha)^{-1}$ extends to an analytic family in the region $|\alpha|^2 + |\alpha| < 1$. The result now follows by mimicking the method of proof of the last theorem. ∎

By totally different methods, one can prove the following sharper result (see the references in the Notes for a proof):

Theorem XIII.72 Let V be a positive C^∞ function on $\mathbb{R}^n$ and let $H = -\Delta + V$. Suppose that ψ is an eigenfunction for H. Then:

(a) If $V(x) \ge c|x|^{2n} - d$ for some c and d, then for every $\varepsilon > 0$, there is a D with
$$|\psi(x)| \le D\exp(-(n+1)^{-1}(c-\varepsilon)^{1/2}|x|^{n+1})$$
for all x.

(b) If $V(x) \le c|x|^{2n} + d$ for some c and d and ψ is the ground state wave function then, for any $\varepsilon > 0$, there is an $E > 0$ with
$$\psi(x) \ge E\exp(-(n+1)^{-1}(c+\varepsilon)^{1/2}|x|^{n+1})$$

The results of these preceding two theorems can be understood in terms of the WKB approximation,

$$\psi \sim \exp\left(-\int \sqrt{V-E}\, dx\right)$$

* * *

We turn now to compact embedding theorems for local Sobolev spaces. There are at least two natural candidates:

Definition Let $\Lambda \subset \mathbb{R}^n$ be an open set. $H^m(\Lambda)$ is the set of functions $f \in L^2(\Lambda)$ whose distributional derivatives $D^\alpha f$ are in $L^2(\Lambda)$ for all α with $|\alpha| \leq m$. $H^m(\Lambda)$ is a Hilbert space under the norm

$$\|f\|_m = \left(\sum_{|\alpha| \leq m} \|D^\alpha f\|_2^2\right)^{1/2}$$

$H_0^m(\Lambda)$ is defined to be the completion of $C_0^\infty(\Lambda)$ in the norm $\|\cdot\|_m$. $H^m(\Lambda)$ and $H_0^m(\Lambda)$ are called **local Sobolev spaces**.

In general, $H_0^m(\Lambda)$ is a proper subset of $H^m(\Lambda)$.

Example 1 Let $\Lambda = (0, 1) \subset \mathbb{R}$. Let $g(x) = e^x$ and let $f \in C_0^\infty(\Lambda)$. In the natural inner product on $H^1(\Lambda)$,

$$(g, f)_1 = \left(\left(-\frac{d^2}{dx^2} + 1\right)g, f\right)_{L^2} = 0$$

Thus $H_0^1(\Lambda) \subset \{g\}^\perp$ so that $H_0^1(\Lambda) \neq H^1(\Lambda)$.

The reader familiar with the theory of self-adjoint extensions (Section X.1) will notice a connection between Example 1 and the fact that $-d^2/dx^2$ is not essentially self-adjoint on $C_0^\infty(0, 1)$. In fact, using the fact that $-\Delta$ is never essentially self-adjoint on $C_0^\infty(\Lambda)$ for bounded Λ, one can show that $H_0^1(\Lambda)$ is always different from $H^1(\Lambda)$ for bounded Λ. Notice that $H_0^m(\mathbb{R}^n) = H^m(\mathbb{R}^n) = D((-\Delta)^{m/2})$.

There are many embedding relations for these Sobolev spaces.

Proposition Let Λ and Λ' be open sets with $\Lambda \subset \Lambda'$ and suppose that $m \geq j \geq 0$. Then:

(a) $H_0^m(\Lambda) \subset H^m(\Lambda)$.
(b) $H^m(\Lambda) \subset H^j(\Lambda)$.

(c) $H_0^m(\Lambda) \subset H_0^j(\Lambda)$.
(d) Let $f \in H(\Lambda')$. Then $f \restriction \Lambda$, the restriction of f to Λ, lies in $H^m(\Lambda)$.
(e) If $f \in H^m(\Lambda)$ and $f = 0$ on $\Lambda \setminus C$ where C is a compact subset of Λ, then $f \in H_0^m(\Lambda)$.
(f) If f is in $D((-\Delta)^{m/2})$ where $-\Delta$ denotes the operator on $L^2(\mathbb{R}^n)$, then $f \restriction \Lambda \in H^m(\Lambda)$.

All inclusions are norm nonincreasing.

Proof (a), (b), and (d) follow immediately from the definitions. (c) comes by using the norm inequality implicit in (b) and taking completions of $C_0^\infty(\mathbb{R}^n)$. (e) is an elementary "approximation of the identity" argument, and (f) follows by using the Fourier transform to estimate each term in $\|f\|_{H^m(\mathbb{R}^n)}$ in terms of $\|(-\Delta)^{m/2}f\|_{L^2}$ and $\|f\|_{L^2}$. ∎

We now prove two easy compactness results.

Theorem XIII.73 Let A be the Friedrichs extension of $-\Delta$ on $C_0^\infty(\Lambda)$ (the Dirichlet Laplacian). Suppose that Ω is bounded. Then:

(a) $H_0^m(\Lambda) \subset D(A^{m/2})$ and the norms are equivalent.
(b) If $m > j$, then the unit ball of $H_0^m(\Lambda)$ is a compact subset of $H_0^j(\Lambda)$.

Proof Let $f \in C_0^\infty(\Lambda)$. Then $\|f\|_m$ and $\|(-\Delta)^{m/2}f\|$ are equivalent norms since we can think of f as an element of L^2 in which case the derivatives of f obey $\mathscr{F}(D^\alpha f) = (ik)^\alpha \hat{f}$ and

$$k^{2m} \leq C_1 \sum_{|\alpha| \leq m} (k^\alpha)^2 \leq C_2(k^{2m} + 1)$$

Now clearly $C_0^\infty(\Lambda) \subset D(A^n)$ for all n and $A^n f = (-\Delta)^n f$, so that

$$\|A^{m/2}f\|^2 = (f, (-\Delta)^m f) = (f, A^m f) = \|A^{m/2}f\|^2$$

As a result, the completion of C_0^∞ in the $\|\cdot\|_m$ norm is clearly in $D(A^{m/2})$.

Next, we claim that A is an operator with compact resolvent. To show that $F_b \equiv \{\psi \in Q(A) \mid \|\psi\| \leq 1, (\psi, A\psi) \leq b\}$ is compact in $L^2(\Lambda)$, it clearly suffices to prove that it is contained in a compact subset of $L^2(\mathbb{R}^n)$. But, as above $Q(A) \subset Q(-\Delta)$ with $(\psi, A\psi) = (\psi, -\Delta\psi)$ where $-\Delta$ means the usual operator on $L^2(\mathbb{R}^n)$. Let F be any function that is 1 on Λ with $F \to \infty$. Then

$$F_b \subset \{\psi \in L^2(\mathbb{R}^n) \mid \|\psi\| \leq 1, (\psi, -\Delta\psi) \leq b, (\psi, F\psi) \leq 1\}$$

so by Theorem XIII.65, F_b is contained in a compact subset of $L^2(\mathbb{R}^n)$.

Since $(A+1)^{-1}$ is compact from $L^2(\Lambda)$ to $L^2(\Lambda)$, it is compact from $D(A^{m/2})$ to $D(A^{m/2})$ for any m because the unitary map $(A+1)^{-m/2}$ from $L^2(\Lambda)$ to $D(A^{m/2})$ commutes with $(A+1)^{-1}$. Since $(A+1)^{-1}$ is compact, so is $(A+1)^{-1/2}$; thus, the unit ball of $D(A^{(m+1)/2})$ is compact in $D(A^{m/2})$. Since $H_0^j(\Lambda)$ is a subset of $D(A^{j/2})$, the compactness result on the H_0^j follows. ∎

We warn the reader that although $H_0^1(\Lambda) = D(A^{1/2})$ by the definition of A, $H_0^m(\Lambda)$ are unequal to $D(A^{m/2})$ for $m > 1$.

Corollary Let Ω be a bounded open set in $\mathbb{R}^n$. The Dirichlet Laplacian $-\Delta_D^\Omega$ has compact resolvent.

Proof By Theorem XIII.73,
$$\{\psi \in Q(-\Delta_D^\Omega) \mid (\psi, -\Delta_D^\Omega \psi) \leq 1 \text{ and } \|\psi\| \leq 1\}$$
is compact in $L^2(\Omega)$, so by Theorem XIII.64, $-\Delta_D^\Omega$ has compact resolvent. ∎

Lemma Let Λ be a bounded open set and let Λ' be open with $\bar{\Lambda} \subset \Lambda'$. Then there exists $\eta \in C_0^\infty(\Lambda')$ with $\eta \equiv 1$ on Λ.

Proof For any $x \in \bar{\Lambda}$, find an $r_x > 0$ so that the ball of radius $2r_x$ about x lies in Λ'. By compactness, we can cover $\bar{\Lambda}$ by a finite number of balls $B_1, \ldots, B_m$ with radii r_{x_i}. In the usual way find $\eta_i \in C_0^\infty(\mathbb{R}^n)$ with $\eta_i \equiv 1$ on B_i and supp η_i in the ball of radius $2r_{x_i}$ about x_i. Now, just define $\eta = 1 - \prod_{i=1}^m (1 - \eta_i)$. ∎

Theorem XIII.74 Let Λ be a bounded open set and let Λ' be open with $\bar{\Lambda} \subset \Lambda'$. Let $m > j$. Then the map of $H^m(\Lambda')$ into $H^j(\Lambda)$ by restriction is compact.

Proof Pick $\eta \in C_0^\infty(\Lambda')$ with $\eta \equiv 1$ on Λ. Given a sequence $\{f_k\}$ in $H^m(\Lambda')$ with $\|f_k\|_m \leq 1$ let $g_k \equiv \eta f_k$. Then, it is easy to see that $\|g_k\|_m \leq c$ for some fixed c. Moreover, each g_k lies in $H_0^m(\Lambda')$. As a result, we can find a subsequence $g_{k(i)}$ converging in $H_0^j(\Lambda')$. But then $g_{k(i)} \restriction \Lambda \equiv f_{k(i)} \restriction \Lambda$ converges in $H^j(\Lambda)$. ∎

Corollary 1 Let $-\Delta$ denote the Laplacian on $\mathbb{R}^m$ and let u_n be a sequence of functions in $Q(-\Delta)$ satisfying $\|\nabla u_n\| \leq c$, $\|u_n\| \leq c$ for some c. Then, given $R > 0$, u_n has a subsequence converging in the norm $(\int_{|x| \leq R} |v(x)|^2 \, dx)^{1/2}$. Similarly, given $u_n \in D(-\Delta)$ with $\|\nabla u_n\| \leq c$, $\|-\Delta u_n\| \leq c$, then u_n has a subsequence converging in the seminorm $(\int_{|x| \leq R} |\nabla v(x)|^2 \, dx)^{1/2}$.

Proof Choose $\Lambda = \{x \mid |x| \leq R\}$ and $\Lambda' = \mathbb{R}^m$. Since $Q(-\Delta) = H^1(\mathbb{R}^m)$ and $\|u\|_{H^1(\mathbb{R}^m)} \leq \|\nabla u\|_{L^2} + \|u\|_{L^2}$, the first part follows from Theorem XIII.74. The second part follows from the first if we note that $\|\nabla \partial_i u\| \leq \|\Delta u\|$ since we can pass to the Fourier transform and use the fact that $\sum_j (k_j k_i)^2 \leq k^4$. ∎

The above corollary is needed in the study of scattering by inhomogeneous media using the Lax–Phillips method (Section XI.11). For scattering off an obstacle with Dirichlet boundary conditions, one needs instead the following corollary. The proof, which is a little longer, is outlined in Problem 124.

Corollary 2 Let K be a compact subset of $\mathbb{R}^m$ and let $-\Delta_D$ denote the Dirichlet Laplacian on $L^2(\mathbb{R}^m \setminus K)$. Let u_n be a sequence of functions in $Q(-\Delta_D)$ with $(u_n, (-\Delta_D)u_n) \leq c^2$, $\|u_n\| \leq c$ for some c. Then, given $R > 0$, $\{u_n\}$ has a subsequence converging in the norm $(\int_{\{x \mid |x| \leq R\} \setminus K} |v(x)|^2 \, dx)^{1/2}$. Similarly, given $u_n \in D(-\Delta_D)$ with $\|-\Delta_D u_n\| \leq c$ and $(u_n, (-\Delta_D)u_n) \leq c^2$, then, if $m \geq 3$, $\{u_n\}$ has a subsequence converging in the seminorm

$$\left(\int_{\{x \mid |x| \leq R\} \setminus K} |\nabla v(x)|^2 \, dx \right)^{1/2}$$

The case of Neumann boundary conditions is more subtle.

Example 2 Let $\Gamma_1, \ldots, \Gamma_n, \ldots$ be an infinite sequences of disjoint open balls with radii decreasing so fast that $\Lambda = \bigcup \Gamma_i$ is bounded. Let f_i be the function that is that positive multiple of the characteristic function of Γ_i with $\|f_i\| = 1$ in $L^2(\Lambda)$. Then the f_i are an orthonormal set in each $H^m(\Lambda)$ so that $H^m(\Lambda)$ is not compactly embedded in $H^j(\Lambda)$ for any m or j.

While this example is somewhat artificial since Λ is not connected, it is reasonably clear that one should be able to join the Γ's with extremely narrow "corridors" without destroying the noncompactness of the embeddings. In general compactness results for the embedding $H^m(\Lambda) \subset H^j(\Lambda)$ depend on very subtle geometric properties of Λ. We shall settle for stating one of the stronger results.

Definition Let Λ be a bounded open set in $\mathbb{R}^m$ and let $\partial \Lambda \equiv \bar{\Lambda} \setminus \Lambda$ be its topological boundary. Λ is said to have the **segment property** if $\partial \Lambda$ has a finite open covering $\{O_i\}$ and corresponding nonzero vectors $\{y_i\}$ so that for $0 < t < 1$, $x + ty_i$ is in Λ if $x \in \bar{\Lambda} \cap O_i$.

Let us illustrate the definition by some examples in $\mathbb{R}^2$. First, it is clear that if $\partial \Lambda$ is a smooth curve, then Λ has the segment property. Also, if Λ is bounded by finitely many smooth curves that meet at nonzero angles, then Λ will have the segment property. The arcs in Figure XIII.7 can even meet in

FIGURE XIII.7 A domain with the segment property.

cusps and retain the segment property as long as the cusps are not too wiggly (Figure XIII.8). The canonical example of a nice region without the segment property is a disk with a line segment removed (see Figure XIII.9).

References for the following compact embedding theorem are given in the notes.

(a) A smooth cusp (b) A wiggly cusp

FIGURE XIII.8

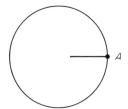

FIGURE XIII.9 The point A destroys the segment property.

Theorem XIII.75 Let Λ be a bounded open set having the segment property. Then $H^m(\Lambda)$ is compactly embedded in $H^j(\Lambda)$ for $m > j$.

By following the proof of the corollary to Theorem XIII.73, we immediately conclude:

Corollary 1 Let Ω be a bounded open region in $\mathbb{R}^m$ obeying the segment property. The Neumann Laplacian $-\Delta_N^\Omega$ has compact resolvent.

Corollary 2 Let K be a compact subset of $\mathbb{R}^m$ and suppose that $\mathbb{R}^m\setminus K$ obeys the segment property in the sense that it holds for $\mathscr{B}_{R_0}\setminus K$ where $\mathscr{B}_{R_0}$ is some open ball containing K. Let $-\Delta_N$ denote the Neumann Laplacian on $L^2(\mathbb{R}^m\setminus K)$ and suppose that $u_n \in Q(-\Delta_N)$ is a sequence of functions so that $\|u_n\| \le c$ and $(u_n, (-\Delta_N)u_n) \le c^2$ for some c. Then for each $R > 0$, $\{u_n\}$ has a subsequence that converges in the norm $(\int_{\{x \mid |x| \le R\}\setminus K} |v(x)|^2 \, dx)^{1/2}$. Similarly, if $u_n \in D(-\Delta_N)$ and $\|-\Delta_N u_n\| \le c$, $(u_n, (-\Delta_N)u_n) \le c^2$, then $\{u_n\}$ has a subsequence converging in the seminorm $(\int_{\{x \mid |x| \le R\}\setminus K} |\nabla v(x)|^2 \, dx)^{1/2}$.

The reader is asked to deduce Corollary 2 from Theorem XIII.75 in Problem 124.

* * *

Another situation where operators with compact resolvent enter naturally is in the study of quantum statistical mechanical systems in boxes. Let Ω be a bounded open subset of $\mathbb{R}^3$. For an N-particle system, let $\Omega_N = \Omega \times \cdots \times \Omega \subset \mathbb{R}^{3N}$ and let $\mathscr{H} = L^2(\Omega_N, d^{3N}x)$. If one wishes to deal with particles obeying Bose–Einstein statistics or Fermi–Dirac statistics, one needs to take the Hilbert space to be $\mathscr{H}_s$, the space of all symmetric functions, or $\mathscr{H}_a$, the space of antisymmetric functions. All the results that we discuss here go through without change. Let H_0 denote the Friedrichs extension of $-\Delta$ on $C_0^\infty(\Omega_N)$. H_0 is the total kinetic energy. Let W be an infinitesimally small form perturbation of $-\Delta$ on $\mathbb{R}^3$, say $W \in R$, the Rollnik class, and let

$$V = \sum_{1 \le i < j \le N} W(\mathbf{r}_i - \mathbf{r}_j)$$

as a multiplication operator on $\mathscr{H}$. Then it is not hard to see that V is an infinitesimal form perturbation of H_0 (Problem 122). The first step in studying statistical mechanical systems is:

Theorem XIII.76 The above operator $H = H_0 + V$ has compact resolvent. Moreover, for any $\beta > 0$, $e^{-\beta H}$ is trace class.

Proof Since V is a form bounded perturbation with relative bound zero, it is easy to see that, for any $\varepsilon > 0$, there is a c with

$$(1 - \varepsilon)(H_0 - c) \le H \le (1 + \varepsilon)(H_0 + c)$$

so that

$$(1 - \varepsilon)[\mu_n(H_0) - c] \leq \mu_n(H) \leq (1 + \varepsilon)(\mu_n(H_0) + c) \quad (110)$$

From (110) and Theorem XIII.64, we see that H has compact resolvent if and only if H_0 does. Notice that $e^{-\beta A}$ is trace class if and only if $\sum_{n=1}^{\infty} e^{-\beta \mu_n(A)} < \infty$. Thus, again using (110), $e^{-\beta H}$ is trace class for all positive β if and only if $e^{-\beta H_0}$ is trace class for all positive β. So we have reduced the proof to the case $V = 0$. Make the Ω_N dependence of H_0 explicit by writing $H_0(\Omega_N)$. Since, by definition of the Friedrichs extension, $C_0^{\infty}(\Omega_N)$ is a form core for $H_0(\Omega_N)$, and since $C_0^{\infty}(\Omega) \subset C_0^{\infty}(\Omega')$ if $\Omega \subset \Omega'$, by the min-max principle we see that

$$\mu_n(H_0(\Omega)) \geq \mu_n(H_0(\Omega'))$$

if $\Omega \subset \Omega'$. Any Ω_N can be inbedded in a box $\Omega' \subset \mathbb{R}^{3N}$, so we need just show

$$\sum_n \exp(-\beta \mu_n[H_0(\Omega')]) < \infty \quad (111)$$

If Ω' has sides $\ell_1, \ldots, \ell_{3N}$, its eigenvalues are just

$$E_{m_1, \ldots, m_{3N}} \equiv \sum_{i=1}^{3N} m_i^2 (\pi/\ell_i)^2$$

where $m_1, \ldots, m_{3N}$ are arbitrary positive integers. Let $a = \min\{(\pi/\ell_i)^2 \mid i = 1, \ldots, 3N\}$. Then $E_{m_1, \ldots, m_{3N}} \geq a \sum_{i=1}^{3N} m_i^2 \geq a \sum_{i=1}^{3N} m_i$ so that

$$\sum_n \exp(-\beta \mu_n[H_0(\Omega')]) \leq \sum_{m_i \geq 1} \exp\left(-\beta a \sum_{i=1}^{3N} m_i\right)$$

$$= \left[\sum_{m=1}^{\infty} \exp(-\beta a m)\right]^{3N} < \infty$$

since the geometric series is convergent. This proves (111) and the theorem. ∎

The above theorem is just the start of the discussion of quantum statistical mechanics. For further discussion, see the reference in the Notes.

What we have discussed thus far in this section relates compactness and discreteness of the entire spectrum. We conclude by giving conditions on a self-adjoint operator guaranteeing that a part of its spectrum is discrete.

Theorem XIII.77 Let A be a self-adjoint operator. Then the part of spectrum of A in (a, b) is purely discrete if and only if $f(A)$ is compact for every continuous function f with $\mathrm{supp}\, f \subset (a, b)$.

Proof Let supp $f \subset [c, d]$, $a < c \leq d < b$. If A has purely discrete spectrum in (a, b), then there are only finitely many points $\lambda_1, \ldots, \lambda_n \in \sigma(A) \cap [c, d]$ since no point in $[c, d]$ is a limit point of $\sigma(A)$. Moreover, each spectral projection $P(\lambda_i)$ is finite dimensional. Thus

$$f(A) = \sum_{i=1}^n f(\lambda_i) P(\lambda_i)$$

is an operator vanishing on the orthogonal complement of a finite-dimensional space and so it is compact.

Conversely, let each $f(A)$ with supp $f \subset (a, b)$ be compact. Let $[c, d] \subset (a, b)$. Let f be a positive continuous function with $0 \leq f \leq 1$, $f(x) = 1$ if $c \leq x \leq d$ and supp $f \subset (a, b)$. Then $P([c, d])$, the spectral projection for $[c, d]$ obeys $P([c, d]) f(A) = P([c, d])$ so it is compact. Therefore, it has finite-dimensional range so that any $\lambda \in [c, d]$ is either in $\rho(A)$ or $\sigma_{\text{disc}}(A)$. ∎

Two important corollaries of Theorem XIII.77 are:

Corollary 1 Let A_n be a sequence of self-adjoint operators with purely discrete spectrum in an interval (a, b). Let A_n converge to A in norm resolvent sense. Then A has purely discrete spectrum in (a, b).

Proof By Theorem VIII.20, if f is continuous with support in (a, b), then $f(A_n) \to f(A)$ in norm. Since a norm limit of compact operators is compact, the theorem follows from Theorem XIII.77. ∎

Corollary 2 Let A_n be a sequence of self-adjoint operators with $\sigma_{\text{ess}}(A_n) = [\Sigma_n, \infty)$. Suppose that $A_n \to A$ in norm resolvent sense. Then Σ_n converges as $n \to \infty$ to some Σ (possibly ∞) and $\sigma_{\text{ess}}(A) = [\Sigma, \infty)$.

Proof By passing to a subsequence, we may suppose that $\Sigma_n \to \Sigma$. By Corollary 1, $\sigma_{\text{ess}}(A) \subset [\Sigma, \infty)$ and by Theorem VIII.23, $(\Sigma, \infty) \subset \sigma_{\text{ess}}(A)$. Thus $\sigma_{\text{ess}}(A) = [\Sigma, \infty)$. In particular, inf $\sigma_{\text{ess}}(A)$ is the only possible limit point of the original sequence Σ_n so Σ_n converges. ∎

XIII.15 The asymptotic distribution of eigenvalues

We know a great deal more about the forces which produce the vibrations of sound than about those which produce the vibrations of light. To find out the different tunes sent out by a vibrating system is a problem which may or may not be solvable in certain special cases, but it would baffle the most skilful mathematician to solve the inverse problem and to find out the shape of a bell by means of the sounds which it is capable of sending out. And this is the problem which ultimately spectroscopy hopes to solve in the case of light. In the meantime we must welcome with delight even the smallest step in the desired direction.   A. Schuster, 1882

XIII.15 The asymptotic distribution of eigenvalues

In this section we describe a method for obtaining qualitative information about point spectra. This method has its roots in classical mathematical physics where one wants to know the asymptotic behavior as $\lambda \to \infty$ of the function $N(\lambda) = \dim P_{(-\infty, \lambda)}(A)$ where A is a self-adjoint operator that arises in a boundary value problem. The most famous example of such a result is Weyl's proof that for $-\Delta$ with Dirichlet boundary conditions in a bounded region $\Omega \subset \mathbb{R}^m$, $N(\lambda)$ is asymptotically equal to $C_m \lambda^{m/2}$ times the volume of Ω, where C_m is the m-dependent constant

$$C_m = [m 2^{m-1} \pi^{m/2} \Gamma(\tfrac{1}{2}m)]^{-1}$$

with $\Gamma(\cdot)$ the Euler gamma function; explicitly

$$\Gamma(\tfrac{1}{2}m) = \begin{cases} (k-1)! & (m = 2k) \\ 2^{-2k}(2k)!\, \pi^{1/2}/k! & (m = 2k+1) \end{cases}$$

where k is an integer.

The method we develop will also be used to answer quantum-mechanical questions such as: If $-\Delta + V$ has compact resolvent, how rapidly does $\dim P_{(-\infty, E)}(-\Delta + V)$ grow as $E \to \infty$? If V is a potential going to zero at infinity but so slowly that $\dim P_{(-\infty, 0)}(-\Delta + V) = \infty$, how rapidly does $\dim P_{(-\infty, E)}(-\Delta + V)$ grow as $E \uparrow 0$? If W is a short-range potential, what is the rate of growth of $\dim P_{(-\infty, 0)}(-\Delta + \lambda W)$ as $\lambda \to \infty$? Elements of this technique have also proven useful in studying the pressure in quantum statistical mechanics, the Fock space energy per unit volume in constructive field theory, and the validity of the Thomas–Fermi model in atomic physics.

There is associated with these problems an intuition that is critical for understanding the structure of the machinery we shall develop. Consider first Weyl's result. The constant C_m is not really as unpleasant or unnatural as it looks. If τ_m is the volume of the unit ball in $\mathbb{R}^m$, then $C_m = \tau_m/(2\pi)^m$. Thus the asymptotic value of $N(\lambda)$ is

$$N(\lambda) \sim (\tau_m \lambda^{m/2})(\text{vol } \Omega)/(2\pi)^m$$

so in some sense $N(\lambda)$ appears to be associated with a volume in $\mathbb{R}^{2m}$ (instead of $\mathbb{R}^m$ as one might guess directly from the formula $C_m(\text{vol } \Omega)$). In fact, since $\tau_m \lambda^{m/2} = \text{vol}\{p \in \mathbb{R}^m \mid p^2 < \lambda\}$, we see that

$$N(\lambda) \sim \text{vol}\{\langle x, p\rangle \mid x \in \Omega, p^2 < \lambda\}/(2\pi)^m$$

Thus $(2\pi)^m N(\lambda)$ is asymptotic to the volume of the region of phase space where a classical particle of mass $m = \tfrac{1}{2}$, moving freely within Ω (with, for example, elastic collisions with the wall), has energy $E_{\text{class}} \leq \lambda$. In units with $\hbar = 1$, $-\Delta_D$, the Laplacian with Dirichlet boundary conditions in Ω, is just the quantum energy corresponding to the classical free particle in Ω. Thus, we see that as $\lambda \to \infty$, the number of quantum states is given by a classical phase space volume divided by $(2\pi)^m$. This is in agreement with the "old quantum theory" Bohr–Sommerfeld quantization condition that each quantum state is associated with a volume of h^m in phase space since $h \equiv 2\pi\hbar = 2\pi$ in our units. It is an interesting historical accident that Weyl's result was approximately contemporary to the work of Bohr and Sommerfeld although the connection was not completely realized until some years afterward. The results we shall prove for the quantum questions mentioned above will all have the interpretation of a phase space volume divided by $(2\pi)^n$. Thus, for example, for W short range,

$$(2\pi)^{-m} \operatorname{vol}\{\langle p, x\rangle \,|\, p^2 + \lambda W(x) < 0\} = \frac{\tau_m}{(2\pi)^m} \int_{\{x \,|\, W(x) < 0\}} \lambda^{m/2} |W(x)|^{m/2} \, d^m x$$

which is asymptotically the expression which we shall obtain for $\dim P_{(-\infty, 0)}(-\Delta + \lambda W)$.

The idea of the method that we shall use to study the above problems is quite simple and is connected to the phase space intuition just described. For cubes, the eigenvalues of the Laplacian with Dirichlet boundary conditions $(-\Delta_D)$ or Neumann boundary conditions $(-\Delta_N)$ can be computed explicitly and the asymptotic formula of Weyl can be directly verified (Proposition 2); the content of the formula in this case is essentially the statement that the volume of a large ball in $\mathbb{R}^m$ is approximately the number of points in it whose coordinates are integral. When Ω is a union of cubes (together with any common boundaries) the formula of Weyl is much subtler but the fact that phase space volumes "are local in x space," i.e., that according to Weyl, $N(\lambda; \Omega_1 \cup \Omega_2)$ is asymptotically equal to $N(\lambda; \Omega_1) + N(\lambda; \Omega_2)$ suggests that we try to decouple the cubes in Ω along their common boundary and use the fact that we know how to control individual cubes. Here a rather remarkable fact about boundary conditions helps us. There are two operators that we can compare with $-\Delta_D^\Omega$. One operator has added Dirichlet boundary conditions on the common boundaries, call it $-\tilde{\Delta}_D$, and the other has added Neumann boundary conditions, $-\tilde{\Delta}_N$. There is decoupling in $-\tilde{\Delta}_D$ in the sense that $-\tilde{\Delta}_D = \bigoplus_\alpha (-\Delta_D^{\Omega_\alpha})$ (and similarly for N) where Ω is the union of cubes Ω_α together with common boundary (Theorem XIII.79). Moreover one can sandwich $N(\lambda; -\Delta_D^\Omega)$ between $N(\lambda; -\tilde{\Delta}_D)$ and $N(\lambda; -\tilde{\Delta}_N)$, i.e., $N(\lambda; -\tilde{\Delta}_D) \leq N(\lambda; -\Delta_D^\Omega) \leq N(\lambda; N\tilde{\Delta}_N)$. We shall refer to this combined method of decoupling and monotonicity as **Dirichlet–Neumann bracketing**.

XIII.15 The asymptotic distribution of eigenvalues

We begin our formal discussion with a study of Dirichlet and Neumann boundary conditions.

Definition Let Ω be an open region of $\mathbb{R}^m$ with connected components $\Omega_1, \ldots$ (finite or infinite). The **Dirichlet Laplacian for Ω**, $-\Delta_D^\Omega$, is the unique self-adjoint operator on $L^2(\Omega, d^m x)$ whose quadratic form is the closure of the form $q(f, g) = \int \overline{\nabla f} \cdot \nabla g \, d^m x$ with domain $C_0^\infty(\Omega)$. The **Neumann Laplacian for Ω**, $-\Delta_N^\Omega$, is the unique self-adjoint operator on $L^2(\Omega, d^m x)$ whose quadratic form is $q(f, g) = \int \overline{\nabla f} \cdot \nabla g \, d^m x$ on the domain

$$H^1(\Omega) = \{f \in L^2(\Omega) \mid \nabla f \in L^2(\mathbb{R}^m)\}$$

where by ∇f we mean the distributional gradient, i.e.,

$$\int (\nabla f) \varphi \, dx = -\int f(\nabla \varphi) \, dx$$

for all $\varphi \in C_0^\infty(\mathbb{R}^m)$. When a fixed Ω is intended, we shall drop the superscript Ω.

There are various alternative descriptions of $-\Delta_D^\Omega$ and $-\Delta_N^\Omega$. Clearly, $-\Delta_D^\Omega$ is the Friedrichs extension by definition, and therefore $Q(-\Delta_D^\Omega)$ is the space $H_0^1(\Omega)$ defined in the preceding section. It follows from our discussion there that

$$\{f \in H^1(\Omega) \mid \operatorname{supp} f \text{ is compact in } \Omega\}$$

is contained in $Q(-\Delta_D^\Omega)$. For any Ω, one also has the following description: Let D denote the *operator* closure of the gradient on $C_0^\infty(\Omega)$. Then $-\Delta_D^\Omega = D^*D$ while $-\Delta_N^\Omega = DD^*$. For nice regions, in particular for those obeying the segment condition,

$$\{f \in H^1(\Omega) \mid f \text{ is } C^\infty \text{ up to } \partial\Omega\}$$

is a form core for $-\Delta_N^\Omega$. If Ω has a nice boundary, $-\Delta_D$ has an *operator* core functions that are C^∞ up to $\partial\Omega$ and that vanish on $\partial\Omega$, and $-\Delta_N$ has an *operator* core functions that are C^∞ up to $\partial\Omega$ and whose normal derivatives vanish on $\partial\Omega$. It is this last description that is responsible for the names "Dirichlet" and "Neumann" after the classical boundary value problems of potential theory. For the case of cubes, we shall need this last description.

Proposition 1 Let Ω be a cube in $\mathbb{R}^m$. Then:

(a) $D_D \equiv \{f \mid f \text{ is } C^\infty \text{ up to } \partial\Omega \text{ with } f \upharpoonright \partial\Omega = 0\}$ is an operator core for $-\Delta_D$ and for $f \in D_D$,

$$-\Delta_D f = -\sum_{i=1}^{m} \frac{\partial^2 f}{\partial x_i^2}$$

(b) $D_N \equiv \{f \mid f \text{ is } C^\infty \text{ up to } \partial\Omega \text{ with } \partial f/\partial n \upharpoonright \partial\Omega = 0\}$ is an operator core for $-\Delta_N$ and for such f,

$$-\Delta_N f = -\sum_{i=1}^{m} \frac{\partial^2 f}{\partial x_i^2}$$

Here $\partial/\partial n$ denotes the normal derivative at the boundary.

Proof (a) Without loss of generality, let $\Omega = (-1, 1)^m$. Let A denote the operator $-\Delta$ with domain D_D. We wish to show that $-\Delta_D = \bar{A}$. A is clearly symmetric, and by separation of variables (multiple Fourier series) we can find a complete orthonormal basis of eigenfunctions $\{\varphi_n\}$ for A. If $A\varphi_n = \lambda_n \varphi_n$, it is easy to see that $\varphi \in D(\bar{A})$ if and only if $\sum \lambda_n^2 |(\varphi_n, \varphi)|^2 < \infty$; and from this we conclude that $\bar{A}$ is self-adjoint.

Clearly, $C_0^\infty(\Omega) \subset D(\bar{A}) \subset Q(\bar{A})$ and as quadratic forms $A \upharpoonright C_0^\infty(\Omega) \times C_0^\infty(\Omega) = -\Delta_D \upharpoonright C_0^\infty(\Omega) \times C_0^\infty(\Omega)$, so it is sufficient to prove that $Q(\bar{A}) \subset Q(-\Delta_D)$. Since $D_D = D(A)$ is a form core for $\bar{A}$ as a quadratic form, we need only prove that $D_D \subset Q(-\Delta_D)$. Thus it suffices to find for each $f \in D_D$ a sequence $f_n \in C_0^\infty(\Omega)$ with $\|f_n - f\| + \|\nabla f_n - \nabla f\| \to 0$. Given $f \in D_D$, let $g_n(x) = f((1 + n^{-1})x)$ if $|x_i| \leq (1 + n^{-1})^{-1}$ and zero otherwise. Then $g_n(x)$ is continuous and piecewise C^1 with bounded gradient. Moreover, $g_n \to f$ and $\nabla g_n \to \nabla f$ in L^2. Let $j_\delta(x)$ be an approximate identity. Then $g_n * j_\delta \to g_n$ and $\nabla(g_n * j_\delta) \to \nabla g_n$ as $\delta \to 0$. Since $g_n * j_\delta \in C_0^\infty(\Omega)$ for δ small, we can find a suitable sequence $\{f_n\}$.

(b) Assume that $\Omega = (-1, 1)^m$. Let B denote the operator $-\Delta$ with domain D_N. As in part (a), B is easily seen to be essentially self-adjoint by looking at its eigenfunctions. Moreover, since $D(B) \subset H^1(\Omega)$ and $(f, Bf) = \int |\nabla f|^2 d^2x$ for $f \in D(B)$ (by the boundary condition), we have $Q(\bar{B}) \subset Q(-\Delta_N)$, so that it is sufficient to prove that $Q(-\Delta_N) \subset Q(\bar{B})$. To show this, it is enough to prove that $H^1(\Omega) \subset Q(\bar{B})$.

Let f be in $H^1(\Omega)$. We first claim that if g and ∇g are continuous up to the boundary and $g(\pm 1, x_2, \ldots, x_m) = 0$, then

$$(\partial_1 f, g) = -(f, \partial_1 g) \tag{112}$$

Suppose first that g vanishes on all of $\partial\Omega$. Then, as in the proof of (a) we can find $g_n \in C_0^\infty(\Omega)$ so that $g_n \to g$, $\nabla g_n \to g$ so that (112) follows since it holds for $g = g_n$. Now, let η_n be a sequence of C^∞ functions on Ω depending only on $x_2, \ldots, x_m$ with compact support in

$$\{x \mid |x_2| \leq 1 - n^{-1}, \ldots, |x_m| \leq 1 - n^{-1}\}$$

such that $\eta_n(x) \uparrow 1$ as $n \to \infty$. Then (112) holds for $g\eta_n$ and thus for g also since $\partial_1(g\eta_n) = \eta_n \, \partial_1 g$.

Let $\{\psi_n\}$ be the eigenfunctions for B given by (115) below. By replacing $\sin(n_1 \pi x/2)$ by $\cos(n_1 \pi x/2)$ when n_1 is odd and $\cos(n_1 \pi x/2)$ by $\sin(n_1 \pi x/2)$ when n_1 is even, we can find an orthonormal family $\{\varphi_n\}_{n_1 \geq 1}$ so that $\partial_1 \varphi_n = \pm(\pi/2)n_1 \psi_n$. The functions φ_n and $\nabla \varphi_n$ are continuous up to the boundary and $\varphi_n(\pm 1, x_2, \ldots, x_m) = 0$. Using (112), we find

$$\sum_n |(f, \psi_n)|^2 n_1^2 = \left(\frac{2}{\pi}\right)^2 \sum_n |(\partial_1 f, \varphi_n)|^2$$

$$\leq \left(\frac{2}{\pi}\right)^2 \|\partial_1 f\|_2^2$$

since the φ_n are orthonormal. Doing a similar procedure in each variable, we conclude that

$$\sum_n (1 + n^2)|(f, \psi_n)|^2 \leq \|f\|_2^2 + \left(\frac{2}{\pi}\right)^2 \|\nabla f\|_2^2$$

so $f \in Q(\bar{B})$. ∎

At first sight it seems surprising that the operator $-\Delta_N^\Omega$ which we have defined in a way that makes no mention of $\partial\Omega$ or normal derivatives should obey Neumann boundary conditions. On an intuitive level, one can understand this in several ways. Since $-\Delta_N^\Omega = DD^*$, for f to be in the domain of $-\Delta_N^\Omega$, we need D^*f to be in the domain of D. And, intuitively, the domain of D consists of functions that vanish on $\partial\Omega$.

The second intuition is clearest in one dimension so let $\Omega = (a, b)$. The operator C defined by the Neumann quadratic form should be a self-adjoint extension of $-d^2/dx^2$ on $C_0^\infty(a, b)$ satisfying two properties:

(i) For any function f that is C^∞ on $[a, b]$, there should be a sequence $f_n \in D(C)$ so that $f_n \to f$ in H^1.
(ii) For any $f \in D(C)$, $(f, Cf) = \|df/dx\|^2$.

We could not impose Dirichlet conditions on C because that would require us to make the f_n vanish at a and b, which would require that $d(f_n - f)/dx$ be large near the end points. This would also rule out periodic boundary conditions that link the two end points. However, condition (i) does not rule out boundary conditions of the form $df/dx + \alpha f = 0$. Condition (ii) forces α to be zero (Problem 127).

The main advantage of the above characterization is that it provides explicit eigenvalues and eigenvectors of $-\Delta_D$ and $-\Delta_N$. To label them we denote the nonnegative integers by $\mathbb{Z}_+$ and the strictly positive integers as $\mathbb{Z}_{++}$, so $\mathbb{Z}_+ = \mathbb{Z}_{++} \cup \{0\}$. Then the eigenvectors and eigenvalues of $-\Delta_D$ in $(-a, a)^m$ are labeled by $\mathbb{Z}_{++}^m$ with eigenvectors

$$\Phi_{n;a}(x) = a^{-m/2} \prod_{i=1}^{m} \varphi_{n_i}(x_i/a) \tag{113}$$

where

$$\varphi_k(x) = \cos(k\pi x/2); \quad k = 1, 3, 5, \ldots$$
$$\varphi_k(x) = \sin(k\pi x/2); \quad k = 2, 4, 6, \ldots$$

and eigenvalues

$$E_n(a) = \left(\frac{\pi}{2a}\right)^2 \sum_{i=1}^{m} n_i^2 \tag{114}$$

For $-\Delta_N$ in $(-a, a)^m$, the eigenvectors are labeled by $\mathbb{Z}_+^m$ with eigenvectors

$$\Psi_{n;a}(x) = a^{-m/2} \prod_{i=1}^{m} \psi_{n_i}(x_i/a) \tag{115}$$

where

$$\psi_k(x) = \sin(k\pi x/2); \quad k = 1, 3, 5, \ldots$$
$$\psi_k(x) = \cos(k\pi x/2); \quad k = 2, 4, 6, \ldots$$
$$\psi_k(x) = \tfrac{1}{2}\sqrt{2}; \quad k = 0$$

and with eigenvalues still given by (114). Because of this explicit listing, one can prove Weyl's theorem directly for cubes.

Proposition 2 Let $N_D(a, \lambda)$ (respectively, $N_N(a, \lambda)$) denote the dimension of the spectral projection $P_{[0, \lambda)}$ for $-\Delta_D$ (respectively, $-\Delta_N$) on $(-a, a)^m$.

XIII.15 The asymptotic distribution of eigenvalues

Then for all a, λ, we have

$$|N_D(a, \lambda) - \tau_m(2a/2\pi)^m \lambda^{m/2}| \leq C(1 + (a^2\lambda)^{(m-1)/2}) \quad (116)$$

and

$$|N_N(a, \lambda) - \tau_m(2a/2\pi)^m \lambda^{m/2}| \leq C(1 + (a^2\lambda)^{(m-1)/2}) \quad (117)$$

where τ_m is the volume of the unit ball in $\mathbb{R}^m$ and C is a suitable constant.

Proof We begin with two remarks. The first explains why the result is true. On account of (114) N_D (respectively, N_N) is the number of points of $\mathbb{Z}_{++}^m$ (respectively, $\mathbb{Z}_+^m$) within a ball of radius $(2a/\pi)\lambda^{1/2}$. For large λ, this should be approximately the volume of an "octant" of this sphere, i.e., $2^{-m}\tau_m((2a/\pi)\lambda^{1/2})^m$. The error should be a "surface term," i.e., of order $(a^2\lambda)^{(m-1)/2}$. The second remark is to note that $-\Delta_D$ in $(-1, 1)^m$ is unitarily equivalent under scaling to $a^2(-\Delta_D)$ in $(-a, a)^m$, so $N_D(a, \lambda) = N_D(1, a^2\lambda)$. As a result, we need only prove the estimates when $a = 1$.

Let S_λ denote the ball of radius $(2/\pi)\lambda^{1/2}$. For each $n \in \mathbb{Z}_{++}^m \cap S_\lambda$, the unit cube $\{x \mid n_i - 1 \leq x_i \leq n_i\}$ is contained in the upper octant of S_λ, so

$$N_D(\lambda) \leq \tau_m(2a/2\pi)^m \lambda^{m/2} \quad (118)$$

Since the upper octant of S_λ is covered by the unit cubes $\{x \mid n_i \leq x_i < n_i + 1\}$ as n runs through $\mathbb{Z}_+^m \cap S_\lambda$, we have

$$\tau_m(2a/2\pi)^m \lambda^{m/2} \leq N_N(\lambda) \quad (119)$$

Finally, it is clear that

$$N_N(\lambda) - N_D(\lambda) = \#\{S_\lambda \cap (\mathbb{Z}_+^m \setminus \mathbb{Z}_{++}^m)\} = \bigcup_{k=0}^{m-1} M_k(\lambda)$$

where

$$M_k(\lambda) = \{n \in \mathbb{Z}_+^m \cap S_\lambda \mid \text{exactly } k \text{ of the } n_i \text{ are nonzero}\}$$

(see Figure XIII.10 in case $m = 3$). As in our bound on N_D, $\#(M_k(\lambda))$ is dominated by the k-dimensional volume of $\binom{m}{k}$ octants of $(2/\pi)\lambda^{1/2}$-balls in k space. Therefore,

$$\sum_{k=1}^{m-1} M_k(\lambda) \leq \text{const}\left(\sum_{k=0}^{m-1} \lambda^{k/2}\right) \leq C(1 + \lambda^{(m-1)/2})$$

Thus

$$N_N(\lambda) - N_D(\lambda) \leq C(1 + \lambda^{(m-1)/2}) \quad (120)$$

(118)–(120) clearly imply (116) and (117). ∎

Dirichlet and Neumann Laplacians are useful for analyzing general eigenvalue problems because of two critical properties that occur when a Dirichlet or Neumann surface is added. Consider disjoint Ω_1 and Ω_2 and $\Omega = (\overline{\Omega_1 \cup \Omega_2})^{\text{int}}$ as shown in Figure XIII.11. We shall show below that: (1) Dirichlet and Neumann surfaces decouple (Proposition 3); (2) Dirichlet surfaces raise energies, and Neumann surfaces lower energies (Proposition 4). To be precise about the decoupling we make some preliminary remarks

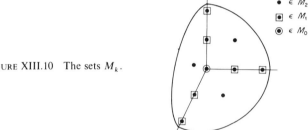

FIGURE XIII.10 The sets M_k.

about direct sums of self-adjoint operators. Let $\mathscr{H} = \mathscr{H}_1 \oplus \mathscr{H}_2$. Let A_1 be a self-adjoint operator on $\mathscr{H}_1$, and A_2 a self-adjoint operator on $\mathscr{H}_2$. Let A be the operator with domain $D(A) = \{\langle \varphi, \psi \rangle \mid \varphi \in D(A_1), \psi \in D(A_2)\}$ with $A \langle \varphi, \psi \rangle = \langle A_1 \varphi, A_2 \psi \rangle$. We shall write $A = A_1 \oplus A_2$. There are several properties of $A_1 \oplus A_2$ whose proofs (which use the spectral theorem and/or the fundamental criterion) we leave to the reader (Problem 133):

(1) $A_1 \oplus A_2$ is self-adjoint.
(2) If D_1 is a core for A_1 and D_2 is a core for A_2, then

$$D_1 \oplus D_2 \equiv \{\langle \varphi, \psi \rangle \mid \varphi \in D_1, \psi \in D_2\}$$

is a core for $A_1 \oplus A_2$.
(3) $Q(A_1 \oplus A_2) = Q(A_1) \oplus Q(A_2)$; and if $\langle \varphi, \psi \rangle \in Q(A_1) \oplus Q(A_2)$, then

$$(\langle \varphi, \psi \rangle, (A_1 \oplus A_2)\langle \varphi, \psi \rangle) = (\varphi, A_1 \varphi) + (\psi, A_2 \psi)$$

(4) For any Borel subset $\Omega \subset \mathbb{R}$,

$$P_\Omega(A_1 \oplus A_2) = P_\Omega(A_1) \oplus P_\Omega(A_2)$$

(5) If $N(\lambda, A) = \dim P_{(-\infty, \lambda)}(A)$, then

$$N(\lambda, A_1 \oplus A_2) = N(\lambda, A_1) + N(\lambda, A_2)$$

XIII.15 The asymptotic distribution of eigenvalues

Proposition 3 Let Ω_1 and Ω_2 be disjoint open sets so that $L^2(\Omega_1 \cup \Omega_2) = L^2(\Omega_1) \oplus L^2(\Omega_2)$. Under this decomposition

$$-\Delta_D^{\Omega_1 \cup \Omega_2} = -\Delta_D^{\Omega_1} \oplus -\Delta_D^{\Omega_2}$$
$$-\Delta_N^{\Omega_1 \cup \Omega_2} = -\Delta_N^{\Omega_1} \oplus -\Delta_N^{\Omega_2}$$

Proof We consider the Dirichlet case; the Neumann case is similar. Given $f \in C_0^\infty(\Omega_1 \cup \Omega_2)$, we let $f_i = f \restriction \Omega_i$. Then $f_i \in C_0^\infty(\Omega_i)$ and clearly

$$\int_{\Omega_1 \cup \Omega_2} \overline{\nabla f} \cdot \nabla g \, d^m x = \int_{\Omega_1} \overline{\nabla f_1} \cdot \nabla g_1 \, d^m x + \int_{\Omega_2} \overline{\nabla f_2} \cdot \nabla g_2 \, d^m x$$

Since this relation clearly extends to the closures of the quadratic forms, the quadratic forms are equal so the result follows. ∎

As an example of the use of this proposition we have:

Corollary Let $N_D(\Omega, \lambda)$ (respectively, $N_N(\Omega, \lambda)$) be the dimension of the spectral projection $P_{[0, \lambda)}$ for $-\Delta_D^\Omega$ (respectively, $-\Delta_N^\Omega$). Then if $\Omega_1, \ldots, \Omega_k$ are disjoint,

$$N_D\left(\bigcup_{i=1}^k \Omega_i, \lambda\right) = \sum_{i=1}^k N_D(\Omega_i, \lambda)$$
$$N_N\left(\bigcup_{i=1}^k \Omega_i, \lambda\right) = \sum_{i=1}^k N_N(\Omega_i, \lambda)$$

To state the monotonicity properties, we extend slightly the definition of $A \leq B$ given in Section 2.

Definition Let A and B be self-adjoint operators that are nonnegative where A is defined on a dense subset of a Hilbert space $\mathcal{H}$ and B is defined on a dense subset of a Hilbert subspace $\mathcal{H}_1 \subseteq \mathcal{H}$. We write $0 \leq A \leq B$ if and only if

(i) $Q(A) \supset Q(B)$.
(ii) For any $\psi \in Q(B)$,

$$0 \leq (\psi, A\psi) \leq (\psi, B\psi)$$

The point of this definition is that on account of the min–max principle (in the form of Theorem XIII.2), we immediately have

Lemma If $0 \leq A \leq B$, then:

(a) $\dim P_{[0, \lambda]}(A) \geq \dim P_{[0, \lambda]}(B)$, all $\lambda > 0$.
(b) $\mu_n(A) \leq \mu_n(B)$ for all n where μ_n is given by the min–max principle.

Proof Since $Q(A)$ contains more trial functions,

$$\min_{\substack{\varphi \in Q(A) \\ \varphi \perp \psi_1, \ldots, \psi_n}} (\varphi, A\varphi) \leq \min_{\substack{\varphi \in Q(B) \\ \varphi \perp \psi_1, \ldots, \psi_n}} (\varphi, A\varphi)$$

$$\leq \min_{\substack{\varphi \in Q(B) \\ \varphi \perp \psi_1, \ldots, \psi_n}} (\varphi, B\varphi)$$

so that (b) holds. (a) follows from (b) and the min–max principle. ∎

Proposition 4

(a) If $\Omega \subset \Omega'$, then $0 \leq -\Delta_D^{\Omega'} \leq -\Delta_D^{\Omega}$.
(b) For any Ω, $0 \leq -\Delta_N^{\Omega} \leq -\Delta_D^{\Omega}$.
(c) Let Ω_1, Ω_2 be disjoint open subsets of an open set Ω so that $(\overline{\Omega_1 \cup \Omega_2})^{\text{int}} = \Omega$, and $\Omega \backslash \Omega_1 \cup \Omega_2$ has measure 0 (see Figure XIII.11).

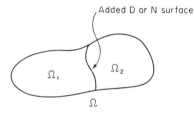

FIGURE XIII.11 Adding a surface.

Then

$$0 \leq -\Delta_D^{\Omega} \leq -\Delta_D^{\Omega_1 \cup \Omega_2}$$

$$0 \leq -\Delta_N^{\Omega_1 \cup \Omega_2} \leq -\Delta_N^{\Omega}$$

Proof (a) This statement is to be interpreted in the sense that any $f \in L^2(\Omega)$ is viewed as an element of $L^2(\Omega')$ by setting it equal to zero in $\Omega' \backslash \Omega$. With this definition, $C_0^\infty(\Omega) \subset C_0^\infty(\Omega')$ and $-\Delta_D^{\Omega'} \upharpoonright C_0^\infty(\Omega) \times C_0^\infty(\Omega) = -\Delta_D^{\Omega}$, as quadratic forms, so (a) follows.

(b) Since $C_0^\infty(\Omega) \subset H^1(\Omega)$, this is immediate.

(c) Clearly $C_0^\infty(\Omega_1 \cup \Omega_2) \subset C_0^\infty(\Omega)$ since $\Omega \supset \Omega_1 \cup \Omega_2$; (indeed, the first part of (c) is a special case of (a)). On the other hand, if $f \in H^1(\Omega)$, its restriction to $\Omega_1 \cup \Omega_2$ is clearly in $H^1(\Omega_1) \oplus H^1(\Omega_2)$. Moreover, since $\Omega \backslash \Omega_1 \cup \Omega_2$ has measure zero

$$\int_\Omega |\nabla f|^2 \, dx = \int_{\Omega_1 \cup \Omega_2} |\nabla f|^2 \, d^3 x \quad \blacksquare$$

The main point about the inequalities in (c) is that by adding an extra Dirichlet "surface" $\Omega \backslash \Omega_1 \cup \Omega_2$, we add an extra requirement on the wave functions (they must vanish on the surface) and so the eigenvalues go up. But it is by *removing* a Neumann "surface" that we add an extra requirement because the functions must be "smooth" across the surface after we remove the boundary condition.

As a first application, we prove the result of Weyl relating the asymptotic density of eigenvalues of $-\Delta_D^\Omega$ and the volume of Ω when Ω is sufficiently nice.

Definition A standard 2^{-n} **cube** in $\mathbb{R}^m$ is a cube of the form

$$\left[\frac{a_1}{2^n}, \frac{a_1 + 1}{2^n}\right) \times \cdots \times \left[\frac{a_m}{2^n}, \frac{a_m + 1}{2^n}\right)$$

in $\mathbb{R}^m$ with $a_1, \ldots, a_m$ integers. Given a set $\Omega \subset \mathbb{R}^m$, we let $W_n^-(\Omega)$ be the volume of those standard 2^{-n} cubes contained in Ω and $W_n^+(\Omega)$ be the volume of standard 2^{-n} cubes that intersect Ω. Thus if Ω is Lebesgue measurable,

$$W_n^-(\Omega) \leq W_{n+1}^-(\Omega) \leq \mu(\Omega) \leq W_{n+1}^+(\Omega) \leq W_n^+(\Omega) \tag{121}$$

The limit $\lim W_n^-(\Omega) = W_\infty^-(\Omega)$ (respectively, $\lim W_n^+(\Omega) = W_\infty^+(\Omega)$) is called the **inner** (respectively, **outer**) **Jordan content** of Ω. If $W_\infty^+(\Omega) = W_\infty^-(\Omega)$, then we say that Ω is a **contented set** and $W(\Omega) = W_\infty^\pm(\Omega)$ is called its **content**.

Notice that by (121) if Ω is both contented and Lebesgue measurable, then $W(\Omega) = \mu(\Omega)$. One can also prove that any contented set is Lebesgue measurable (Problem 128).

Theorem XIII.78 Let Ω be a bounded open set in $\mathbb{R}^m$. Let $N_D(\Omega, \lambda)$ be the dimension of the range of the spectral projection $P_{[0, \lambda)}$ for $-\Delta_D^\Omega$. Then, if Ω is contented,

$$\lim_{\lambda \to \infty} N_D(\Omega, \lambda)/\lambda^{m/2} = \frac{\tau_m}{(2\pi)^m} W(\Omega)$$

Proof We shall show that for any n,

$$\varlimsup_{\lambda \to \infty} N_D(\Omega, \lambda)/\lambda^{m/2} \leq (2\pi)^{-m} \tau_m W_n^+(\Omega) \tag{122a}$$

$$\varliminf_{\lambda \to \infty} N_D(\Omega, \lambda)/\lambda^{m/2} \geq (2\pi)^{-m} \tau_m W_n^-(\Omega) \tag{122b}$$

which implies the result.

Let $\Omega_n^\pm$ denote the union of cubes whose volume enters in the definition of $W_n^\pm(\Omega)$, and let $\{C_{n,\alpha}^\pm\}$ be the interiors of the actual cubes themselves, so that $\bar{\Omega}_n^\pm = \bigcup_\alpha \bar{C}_{n,\alpha}^\pm$. Now, by (a) and (c) of Proposition 4,

$$-\Delta_D^\Omega \leq -\Delta_D^{\Omega_n^-} \leq -\Delta_D^{\bigcup C_{n,\alpha}^-} = \bigoplus_\alpha -\Delta_D^{C_{n,\alpha}^-}$$

where we have used Proposition 3 in the last step. Thus, by Proposition 4,

$$N_D(\Omega, \lambda) \geq \sum_\alpha N_D(C_{n,\alpha}^-, \lambda)$$
$$= (\#\alpha) N_D(2^{-n-1}, \lambda)$$
$$= W_n^-(\Omega) 2^{nm} N_D(2^{-n-1}, \lambda)$$

By Proposition 2,

$$\lim_{\lambda \to \infty} N_D(2^{-n-1}, \lambda) \lambda^{-m/2} = \tau_m (2\pi)^{-m} 2^{-nm}$$

so (122b) holds. Similarly, by (a)–(c) of Proposition 4,

$$-\Delta_D^\Omega \geq -\Delta_D^{\Omega_n^+} \geq -\Delta_N^{\Omega_n^+} \geq -\Delta_N^{\bigcup C_{n,\alpha}^+} = \bigoplus -\Delta_N^{C_{n,\alpha}^+}$$

and (122a) follows by mimicking the steps above. ∎

The structure of the Dirichlet–Neumann bracketing technique is now clear. One uses the monotonicity under addition of Dirichlet or Neumann boundaries and their decoupling properties to reduce a problem to one about cubes which is then solved by Proposition 2. That phase volumes enter is a result of the fact that they enter in (116) and (117). Thus, for example, we have:

Theorem XIII.79 Let V be a continuous function on $\mathbb{R}^m$ that has compact support. Let $N(\lambda)$ be the dimension of the spectral projection $P_{(-\infty, 0)}$ for $-\Delta + \lambda V$. Then

$$\lim_{\lambda \to \infty} N(\lambda)/\lambda^{m/2} = \tau_m (2\pi)^{-m} \int_{\{x | V(x) \leq 0\}} (-V(x))^{m/2} d^m x$$

XIII.15 The asymptotic distribution of eigenvalues

Proof For each n, let $\{C_{n,\alpha}\}_\alpha$ be a listing of all the standard 2^{-n}-cubes. Define V_n^+ (respectively, V_n^-) to be that function piecewise constant over each $C_{n,\alpha}$ with value $\max\{V(x)\,|\,x \in C_{n,\alpha}\}$ (respectively, $\min\{V(x)\,|\,x \in C_{n,\alpha}\}$). We shall prove that for each n,

$$\varlimsup_{\lambda \to \infty} N(\lambda)/\lambda^{m/2} \leq \tau_m(2\pi)^{-m} \int_{\{x\,|\,V_n^-(x) \leq 0\}} (-V_n^-(x))^{m/2}\, d^m x \qquad (123\text{a})$$

and

$$\varliminf_{\lambda \to \infty} N(\lambda)/\lambda^{m/2} \geq \tau_m(2\pi)^{-m} \int_{\{x\,|\,V_n^+(x) \leq 0\}} (-V_n^+(x))^{m/2}\, d^m x \qquad (123\text{b})$$

Since

$$\lim_{n \to \infty} \int_{\{x\,|\,V_n^\pm(x) \leq 0\}} (-V_n^\pm(x))^{m/2}\, d^m x = \int_{\{x\,|\,V(x) \leq 0\}} (-V(x))^{m/2}\, d^m x$$

on account of the continuity of V, (123) implies the theorem.

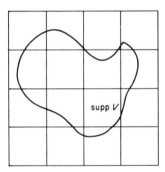

FIGURE XIII.12

Suppose that supp V is in a cube $(-k, k)^m$ with k integral. Let $-\Delta_+^n$ (respectively, $-\Delta_-^n$) be the Laplacian on all of $\mathbb{R}^m$ but with Dirichlet (respectively, Neumann) boundary conditions on the boundaries of all the 2^{-n} cubes in $(-k, k)^m$ (Figure XIII.12). Now, since $V_n^- \leq V \leq V_n^+$ and $-\Delta_-^n \leq -\Delta \leq -\Delta_+^n$, by Proposition 4, we have that

$$-\Delta_-^n + V_n^- \leq -\Delta + V \leq -\Delta_+^n + V_n^+$$

so with the obvious meaning for $N_n^\pm(\lambda)$, to prove (123) we need only prove that

$$\lim_{\lambda \to \infty} N_n^\pm(\lambda)/\lambda^{m/2} = \tau_m(2\pi)^{-m} \int_{V_n^\pm(x) \leq 0} (-V_n^\pm(x))^{m/2}\, dx \qquad (124)$$

But, by the decoupling of Neumann and Dirichlet boundaries, $-\Delta_{\pm}^n + \lambda V_n^{\pm}$ is a direct sum of operators of the form $-\Delta_D^{\Omega} + \lambda c$ or $-\Delta_N^{\Omega} + \lambda c$ for constants c and cubes Ω and one positive operator given by the Laplacian on $\mathbb{R}^m \setminus (-k, k)^m$. Since for any positive operator A,

$$\dim((\operatorname{Ran} P_{(-\infty, 0)}(A + \lambda c)) = \dim(\operatorname{Ran} P_{[0, -\lambda c)}(A))$$

(124) follows from Proposition 2. ∎

The method of proof used to demonstrate the limit theorems does not extend to V's that are not at least locally bounded. However, one can extend the limit theorems by approximating nonsmooth V's by V's in C_0^{∞}. The critical element in this approximation argument is the existence of bounds on $N(V)$ that have the right coupling constant behavior, such as the Cwikel–Lieb–Rosenbljum bound (Theorem XIII.12).

Theorem XIII.80 For any $V \in L^{m/2}(\mathbb{R}^m)$, $m \geq 3$, one has

$$\lim_{\lambda \to \infty} N(\lambda V)/\lambda^{m/2} = \tau_m (2\pi)^{-m/2} \int_{V \leq 0} (-V(x))^{m/2} \, d^m x$$

Proof Let A be an arbitrary self-adjoint operator and let $\tilde{N}(A) = \dim(E_{(-\infty, 0)}(A))$ where $E_\Omega(A)$ is the spectral family for A. We first claim that

$$\tilde{N}(A + B) \leq \tilde{N}(A) + \tilde{N}(B) \qquad (125)$$

for any self-adjoint operators A and B that are bounded below with $Q(A) \cap Q(B)$ dense and $A + B$ the form sum. (125) follows from the min–max principle: If $\tilde{N}(A)$, $\tilde{N}(B) < \infty$, let $\psi_1, \ldots, \psi_{\tilde{N}(A)}$ be a basis for $E_{(-\infty, 0)}(A)$ and $\psi_{\tilde{N}(A)+1}, \ldots, \psi_{\tilde{N}(A)+\tilde{N}(B)}$ a basis for $E_{(-\infty, 0)}(B)$. If $\varphi \in Q(A) \cap Q(B)$ is in $[\psi_1, \ldots, \psi_{\tilde{N}(A)+\tilde{N}(B)}]^{\perp}$, then $(\varphi, A\varphi) \geq 0$ and $(\varphi, B\varphi) \geq 0$, so $\mu_{\tilde{N}(A)+\tilde{N}(B)+1}(A + B) \geq 0$ by the min–max principle. This proves (125).

Fix $\varepsilon > 0$, small. Choose $V_k \in C_0^{\infty}$ so that $V_k \to V$ in $L^{m/2}(\mathbb{R}^m)$. Then

$$-\Delta + \lambda V = [-(1 - \varepsilon)\Delta + \lambda V_k] + [\varepsilon(-\Delta) + \lambda(V - V_k)]$$

so that by (125) and $\tilde{N}(-\alpha \Delta + V) = N(\alpha^{-1} V)$,

$$N(\lambda V) \leq N((1 - \varepsilon)^{-1} \lambda V_k) + N(\varepsilon^{-1} \lambda (V - V_k))$$
$$\leq N((1 - \varepsilon)^{-1} \lambda V_k) + c\varepsilon^{-m/2} \lambda^{m/2} \| V - V_k \|_{m/2}^{m/2}$$

using (10). By the limit theorem just proven for $V_k \in C_0^\infty$,

$$\varlimsup_{\lambda \to \infty} N(\lambda V)/\lambda^{m/2} \leq (1-\varepsilon)^{-m/2}(2\pi)^{-m/2}\tau_m \int_{V_k \leq 0} (-V_k)^{m/2} d^m x$$
$$+ c\varepsilon^{-m/2}\|V - V_k\|_{m/2}^{m/2}$$

Taking $k \to \infty$ and then $\varepsilon \to 0$, we find that

$$\varlimsup_{\lambda \to \infty} N(\lambda V)/\lambda^{m/2} \leq (2\pi)^{-m/2}\tau_m \int_{V \leq 0} (-V)^{m/2} d^m x \qquad (126)$$

Similarly,

$$-\Delta + \lambda V_k = [-(1-\varepsilon)\Delta + \lambda V] + [\varepsilon(-\Delta) + \lambda(V_k - V)]$$

so using the limit theorem and (10) as above,

$$(2\pi)^{-m/2}\tau_m \int_{V_k \leq 0} (-V_k)^{m/2} d^m x \leq \left[\varliminf_{\lambda \to \infty} N(\lambda V)/\lambda^{m/2}\right](1-\varepsilon)^{-m/2}$$
$$+ c\varepsilon^{-m/2}\|V - V_k\|_{m/2}^{m/2}$$

Again taking $k \to \infty$ and then $\varepsilon \to 0$, we obtain

$$(2\pi)^{-m/2}\tau_m \int_{V \leq 0} (-V)^{m/2} d^m x \leq \varliminf_{\lambda \to \infty} N(\lambda V)/\lambda^{m/2} \qquad (127)$$

(126) and (127) imply the theorem. ∎

In our next result, which answers a question raised in Section 14, we make stronger assumptions on V than are absolutely necessary.

Theorem XIII.81 Let V be a measurable function on $\mathbb{R}^m$ ($m \geq 2$) obeying

$$c_1(|x|^\beta - 1) \leq V(x) \leq c_2(|x|^\beta + 1) \qquad (128)$$
$$|V(x) - V(y)| \leq c_3[\max\{|x|, |y|\}]^{\beta-1}|x - y| \qquad (129)$$

for some $\beta > 1$ and suitable constants $c_1, c_2, c_3 > 0$. Let $N(E) = \dim \operatorname{Ran} P_{(-\infty, E)}$ for $-\Delta + V$ and let

$$g(E) = \frac{\tau_m}{(2\pi)^m} \int_{\{x | V(x) \leq E\}} (E - V(x))^{m/2} d^m x \qquad (130)$$

Then

$$\lim_{E \to \infty} N(E)/g(E) = 1$$

Proof Let $\{\Omega_\alpha\}$ be a list of all unit cubes with vertices at integral lattice points. Let V^+ (respectively, V^-) be that piecewise constant function with value $\sup_{\Omega_\alpha} V(x)$ (respectively, $\inf_{\Omega_\alpha} V(x)$) in the cube Ω_α. Let $-\Delta^+$ (respectively, $-\Delta^-$) be the Laplacian with Dirichlet (respectively, Neumann) boundary conditions on the sides of the cubes Ω_α. Let $N_\pm(E)$ be dim Ran $P_{(-\infty, E)}$ for $-\Delta^\pm + V^\pm$ and let $g_\pm(E)$ be given by (130) with V replaced by $V^\pm$. Then, by Dirichlet–Neumann bracketing,

$$N_-(E) \geq N(E) \geq N_+(E) \tag{131}$$

By (128) it is easy to see that

$$c_3(E^\gamma - 1) \leq g_+(E) \leq g(E) \leq g_-(E) \leq c_4(E^\gamma + 1) \tag{132}$$

where $\gamma = \tfrac{1}{2}m + m\beta^{-1}$. Moreover, by (129) and the bound

$$|(E-a)^{m/2} - (E-b)^{m/2}| \leq \frac{m}{2} \max[(E-a)^{\frac{1}{2}m-1}, (E-b)^{\frac{1}{2}m-1}]|b-a|$$

we have that

$$|g_-(E) - g_+(E)| \leq c_5(E^{\gamma - \beta^{-1}} + 1) \tag{133}$$

By (131)–(133), the theorem follows if we can prove that

$$\lim_{E \to \infty} N_\pm(E)/g_\pm(E) = 1 \tag{134}$$

By the decoupling nature of Dirichlet and Neumann boundary conditions

$$N_\pm(E) = \sum_{\{\alpha | V_\pm \restriction \Omega_\alpha \leq E\}} \eta_\pm(E - V_\pm \restriction \Omega_\alpha)$$

where $\eta_+(\lambda)$ (respectively, $\eta_-(\lambda)$) is dim(Ran $P_{[0, \lambda)}$) for $-\Delta$ in the unit cube with Dirichlet (respectively, Neumann) boundary conditions. By Proposition 2,

$$\left|\eta_\pm(\lambda) - \frac{\tau_m}{(2\pi)^m}\lambda^{m/2}\right| \leq c_6(1 + \lambda^{(m-1)/2})$$

Since

$$\sum_{\{\alpha | V_\pm \restriction \Omega_\alpha \leq E\}} \frac{\tau_m}{(2\pi)^m}(E - V_\pm \restriction \Omega_\alpha)^{m/2} = g_\pm(E)$$

we need only show that the error term

$$\varepsilon_\pm(E) \equiv \sum_{\{\alpha | V_\pm \restriction \Omega_\alpha \leq E\}} [1 + (E - V_\pm \restriction \Omega_\alpha)^{(m-1)/2}]$$

is small compared to $g_\pm(E)$. But clearly

$$\varepsilon_\pm(E) \leq (1 + E^{(m-1)/2})\,\text{vol}\{x \mid V_\pm(x) \leq E\}$$

By (128), this volume is bounded by $\text{const}(1 + E^{1/\beta})^m$ so

$$\varepsilon_\pm(E) \leq c_6(1 + E^{\gamma - \frac{1}{2}})$$

and therefore by (132),

$$\varepsilon_\pm(E)/g_\pm(E) \to 0$$

This proves (134) and thus completes the proof of the theorem. ∎

With some simple modifications, the theorem and proof extend to the case $m = 1$.

By methods similar to the above, one may prove the following result (Problem 131) which expresses how quickly $\dim \text{Ran } P_{(-\infty, E]}$ diverges for those Schrödinger operators for which we proved $\dim \text{Ran } P_{(-\infty, 0]} = \infty$ in Section 3.

Theorem XIII.82 Let V be a measurable function on $\mathbb{R}^m$ that obeys

$$-c_1(|x| + 1)^{-\beta} \leq V(x) \leq -c_2(|x| + 1)^{-\beta}$$

$$|V(x) - V(y)| \leq c_3[\min\{|x|, |y|\} + 1]^{-\beta - 1}|x - y|$$

for some $\beta < 2$ and $c_1, c_2, c_3 > 0$. Let $N(E) = \dim \text{Ran } P_{(-\infty, -E]}$ for $-\Delta + V$ and let

$$g(E) = \frac{\tau_m}{(2\pi)^m} \int_{\{x \mid V(x) < -E\}} (-E - V(x))^{m/2}\, d^m x$$

Then

$$\lim_{E \uparrow 0} N(E)/g(E) = 1$$

We also note (Problem 132) that the asymptotic behavior of $N(E)$ is unchanged if we replace $-\Delta + V$ by $-\Delta + V + W$ where W is such that $-\Delta + \lambda W$ has only finitely many negative bound states for any λ. For example, this holds if W is in C_0^∞ (Problem 20) or in $L^{m/2}$ with $m \geq 3$.

By using methods quite different from those discussed thus far in this section, one can sometimes obtain more detailed information on the asymptotics of eigenvalues. Typical of further developments is the following:

Theorem XIII.82.5 Let H_0 be the operator $-d^2/dx^2$ on $L^2(0, 1)$ with the boundary conditions $u(0) = u(1) = 0$. Let $E_n(V)$ be the nth eigenvalue of $H_0 + V$. Then, for $V \in L^\infty(0, 1)$:

$$\lim_{n \to \infty} \left\{ E_n(V) - (n\pi)^2 - \int_0^1 V(x)\,dx \right\} = 0$$

Proof Fix $V \in L^\infty$, let $E_n(a) = E_n(aV)$, and let $\psi_n(a)$ be the eigenvector for $H(a) \equiv H_0 + aV$ with eigenvalue $E_n(a)$. Since $H(a)$ has only simple eigenvalues, the functions $E_n(a)$ are real analytic by the results of Section XII.2 and, in particular, by the explicit formulas of Section XII.1,

$$\frac{dE_n}{da}(a) = (\psi_n(a), V\psi_n(a))$$

$$\frac{d^2 E_n}{da^2}(a) = -(V\psi_n(a), R_n(a) V\psi_n(a))$$

where $R_n(a)$ is the operator $(I - P_n(a))(H(a) - E_n(a))^{-1}$ with P_n the projection onto ψ_n. From the first formula, $|E_n(1) - E_n(0)| \le \|V\|_\infty$ and, in particular, since $E_n(0) = (n\pi)^2$, we have that

$$|E_n(a) - E_{n-1}(a)| \ge (2n - 1)\pi^2 - 2\|V\|_\infty a$$

It follows that for some N_0, and all $a \in (0, 1)$

$$R_n(a) = \max(|E_n(a) - E_{n-1}(a)|^{-1}, |E_n(a) - E_{n+1}(a)|^{-1})$$
$$\le cn^{-1}$$

for all $n \ge N_0$. Thus, for such a and n, $d^2 E_n(a)/da^2 \le c\|V\|_\infty^2 n^{-1}$ so, by Taylor's theorem with remainder,

$$\left| E_n(1) - (n\pi)^2 - \frac{dE_n}{da}(0) \right| \le \tfrac{1}{2} c \|V\|_\infty^2 n^{-1}$$

Since $\psi_n(0) = \sqrt{2} \sin(n\pi x)$, we conclude that at $a = 0$, $dE_n/da = \tilde{V}(0) - \tfrac{1}{2}\tilde{V}(2n) - \tfrac{1}{2}\tilde{V}(-2n)$ where $\tilde{V}(m) = \int_0^1 e^{-im\pi x} V(x)\,dx$. By the Riemann–Lebesgue lemma, $\tilde{V}(m) \to 0$ as $m \to \infty$, so $dE_n/da \to \tilde{V}(0)$ as $n \to \infty$. ∎

Notice that the error $E_n(V) - (n\pi)^2 - \int_0^1 V(x)\,dx$ is the sum of two terms, one of which is $O(n^{-1})$ and the other of which is Re $\tilde{V}(2n)$. For smooth V's, this term is also at least $O(n^{-1})$ but, in general, $\sup_{n \geq m} \tilde{V}(2n)$ can go to zero arbitrarily slowly. The above results depend critically on the one-dimensional nature of the problem, for it is only in one dimension that the distance between eigenvalues of H_0 diverges as n goes to infinity.

XIII.16 Schrödinger operators with periodic potentials

In this section we study Schrödinger operators $-\Delta + V$ where V is a periodic function. That is, we assume that for some basis $\{a_i\}_{i=1}^n \in \mathbb{R}^n$, V satisfies

$$V(\mathbf{x} + \mathbf{a}_i) = V(\mathbf{x}) \tag{135}$$

As we shall discuss, these operators are important in solid state physics.

We have already seen that the spectral properties of Schrödinger operators are highly dependent on the behavior of V at infinity. Basically, we have studied three distinct classes of Schrödinger operators. The class whose spectral properties were easiest to establish were those with $V(x) \to \infty$ as $x \to \infty$; this class had empty essential spectrum (Theorem XIII.16). The next simplest class consisted of the "one-body Schrödinger operators" where $V(x) \to 0$ as $x \to \infty$, at least in some "average sense" (such as $V(x) \in L^p(\mathbb{R}^n, dx)$ for some $p < \infty$); under fairly general hypotheses these operators have $\sigma_{\text{ess}} = [0, \infty)$ (see Theorem XIII.15) and empty singular continuous spectrum (Theorem XIII.33). The third class is made up of the "N-body Schrödinger operators" for which $V(x) \to 0$ as $x \to \infty$ in "most" directions (i.e., those directions in which all "coordinate" differences $|\mathbf{r}_i - \mathbf{r}_j| \to \infty$) but for which V did not have a limit in tubes about those spatial directions where $\mathbf{r}_i = \mathbf{r}_j$ (some i, j). These operators were much harder to analyze; we saw that under fairly general circumstances $\sigma_{\text{ess}} = [\Sigma, \infty)$ where Σ was a "computable" number (Theorem XIII.17) but were only able to prove $\sigma_{\text{sing}} = \emptyset$ under specialized hypotheses (Theorems XIII.27, XIII.29, and XIII.36). We see therefore that spectral properties are very sensitive to the behavior of V at infinity. Since V's obeying (135) do not have a limit as $x \to \infty$ in any direction one might expect the analysis of periodic Schrödinger operators to be difficult.

The property that allows one to analyze $H = -\Delta + V$ when V is periodic is that H has a large symmetry group. For letting

$$(U(\mathbf{t})\psi)(\mathbf{x}) = \psi(\mathbf{x} + \sum_{i=1}^{n} t_i \mathbf{a}_i)$$

where $t \in \mathbb{Z}^n$, we see that (formally)

$$U(t)H = HU(t) \tag{136}$$

One can in fact prove that $U(t)e^{-iHs} = e^{-iHs}U(t)$ (Problem 135). A part of the analysis of H is then a special case of general symmetry arguments which are the subject of Chapter XVI. In this sense our discussion here is premature. We emphasize to the reader that the constant fiber direct integrals described below are an example of a construction from Chapter XVI (with most of the essential features) and that the fact that periodic Schrödinger operators have a direct integral decomposition is a direct consequence of (136). We remark that historically the essentials of the decomposition were discovered both by mathematicians (Floquet) and physicists (Bloch) who did not realize they were speaking group theory. We too shall not explicitly use the connection with the symmetry group here but will develop the theory directly.

Let $\mathcal{H}'$ be a separable Hilbert space and $\langle M, \mu \rangle$ a σ-finite measure space. In Section II.1, we constructed the Hilbert space $L^2(M, d\mu; \mathcal{H}')$ of square integrable $\mathcal{H}'$-valued functions. Notice that if μ is a sum of point measures at a finite set of points $m_1, \ldots, m_k$, then any $f \in L^2(M, d\mu; \mathcal{H}')$ is determined by the k-tuple $\langle f(m_1), \ldots, f(m_k) \rangle$ so $L^2(M, d\mu; \mathcal{H}')$ is isomorphic to the direct sum $\bigoplus_{i=1}^{m} \mathcal{H}'$. In some sense then, $L^2(M, d\mu; \mathcal{H}')$ for more general μ is a kind of "continuous direct sum" but with identical summands. We shall thus call $\mathcal{H} \equiv L^2(M, d\mu; \mathcal{H}')$ a **constant fiber direct integral** and write

$$\mathcal{H} = \int_M^{\oplus} \mathcal{H}' \, d\mu$$

It may seem silly to give an old familiar object a strange new name, but the new name is intended to convey a new emphasis on the "fibers" $\mathcal{H}'$ rather than the points of M. A particular class of operators on $\mathcal{H}$ will concern us. A function $A(\cdot)$ from M to $\mathcal{L}(\mathcal{H}')$ is called measurable if and only if for each $\varphi, \psi \in \mathcal{H}'$, $(\varphi, A(\cdot)\psi)$ is measurable. $L^{\infty}(M, d\mu; \mathcal{L}(\mathcal{H}'))$ denotes the space of (equivalence class of a.e. equal) measurable functions from M to $\mathcal{L}(\mathcal{H}')$ with

$$\|A\|_{\infty} \equiv \operatorname{ess\,sup} \|A(m)\|_{\mathcal{L}(\mathcal{H}')} < \infty$$

XIII.16 Schrödinger operators with periodic potentials

Definition A bounded operator A on $\mathcal{H} = \int_M^\oplus \mathcal{H}' \, d\mu$ is said to be **decomposed** by the direct integral decomposition if and only if there is a function $A(\cdot)$ in $L^\infty(M, d\mu; \mathscr{L}(\mathcal{H}'))$ so that for all $\psi \in \mathcal{H}$,

$$(A\psi)(m) = A(m)\psi(m) \tag{137}$$

We then call A **decomposable** and write

$$A = \int_M^\oplus A(m) \, d\mu(m)$$

The $A(m)$ are called the **fibers** of A.

We first note that every $A(\cdot)$ in $L^\infty(M, d\mu; \mathcal{H}')$ is associated with some decomposable operator:

Theorem XIII.83 If $A(\cdot) \in L^\infty(M, d\mu; \mathscr{L}(\mathcal{H}'))$, then there is a unique decomposable operator $A \in \mathscr{L}(\mathcal{H})$ so that (137) holds. Moreover $\|A\|_{\mathscr{L}(\mathcal{H})} = \|A(\cdot)\|_\infty$.

Proof Uniqueness is obvious. We must only show that (137) takes measurable square integrable $\mathcal{H}'$-valued functions ψ into measurable square integrable $\mathcal{H}'$-valued functions and that the operator A so defined is bounded with norm $\|A(\cdot)\|_\infty$. Let $\psi \in L^2(M, d\mu; \mathcal{H}')$. Let $\{\eta_k\}_{k=1}^\infty$ be an orthonormal basis for $\mathcal{H}'$. Then $A(m)\psi(m) = \sum_{k=1}^\infty (\eta_k, \psi(m))A(m)\eta_k$, a.e. in m since $A(\cdot)$ is a.e. a bounded operator. Now, by definition of measurability for $A(\cdot)$, $A(m)\eta_k$ is weakly measurable, so for any $N < \infty$, $\varphi_N(m) \equiv \sum_{k=1}^N (\eta_k, \psi(m))A(m)\eta_k$ is strongly measurable (Theorem IV.22). Moreover,

$$\int \|\varphi_N(m)\|^2 \, d\mu = \int \left\| A(m) \sum_{k=1}^N (\eta_k, \psi(m))\eta_k \right\|^2 d\mu$$

$$\leq \|A(\cdot)\|_\infty^2 \int \left\| \sum_{k=1}^N (\eta_k, \psi(m))\eta_k \right\|^2 d\mu$$

$$\leq \|A(\cdot)\|_\infty^2 \|\psi\|^2 \tag{138}$$

A similar computation shows that that φ_N is Cauchy in $\mathcal{H}$. Thus it has a limit $\varphi \in L^2(M, d\mu; \mathcal{H}')$. But for almost all $m \in M$, $\varphi_N(m)$ converges to $A(m)\psi(m)$ in $\mathcal{H}'$. It follows that $A(\cdot)\psi(\cdot) \in L^2(M, d\mu; \mathcal{H}')$. By (138)

$$\|A(\cdot)\psi(\cdot)\| \leq \|A\|_\infty \|\psi\|$$

so A is bounded and $\|A\|_{\mathscr{L}(\mathcal{H})} \leq \|A(\cdot)\|_\infty$.

To prove the converse inequality, let $\{\beta_k\}_{k=1}^\infty$ be a dense subset of the unit ball in $\mathscr{H}'$ and let $f \in L^1(M, d\mu)$. We may decompose f as $f = gh$, with $g, h \in L^2$ and $\|g\|_2^2 = \|h\|_2^2 = \|f\|_1$. Fix k, ℓ and let $\psi = \bar{g}\beta_k$ and $\varphi = h\beta_\ell$. Then

$$\left| \int f(m)(\beta_k, A(m)\beta_\ell) \, d\mu \right| = |(\psi, A\varphi)| \le \|A\| \|\psi\| \|\varphi\|$$

$$= \|A\| \|\beta_k\| \|\beta_\ell\| \int |f(m)| \, d\mu$$

Since $L^\infty(M)$ is the dual of $L^1(M)$, it follows that

$$|(\beta_k, A(m)\beta_\ell)| \le \|\beta_k\| \|\beta_\ell\| \|A\|_{\mathscr{L}(\mathscr{H})}$$

a.e. in m. It follows that $\|A(\cdot)\|_\infty \le \|A\|_{\mathscr{L}(\mathscr{H})}$. ∎

The above theorem sets up an isometric isomorphism of $L^\infty(M, d\mu; \mathscr{L}(\mathscr{H}'))$ and the decomposable operators on $\int_M^\oplus \mathscr{H}' \, d\mu$. Both these spaces are algebras in a natural way, and it is easy to see that the algebraic structure is preserved. $L^\infty(M, d\mu; \mathbb{C})$ is the natural subalgebra of $L^\infty(M, d\mu; \mathscr{L}(\mathscr{H}'))$ corresponding to those decomposable operators whose fibers are all multiples of the identity.

Theorem XIII.84 Let $\mathscr{H} = \int_M^\oplus \mathscr{H}' \, d\mu$ where $\langle M, \mu \rangle$ is a separable σ-finite measure space and $\mathscr{H}'$ is separable. Let $\mathscr{A}$ be the algebra of decomposable operators whose fibers are all multiples of the identity. Then $A \in \mathscr{L}(\mathscr{H})$ is decomposable if and only if A commutes with each operator in $\mathscr{A}$.

Proof It is obvious that any decomposable A commutes with all the operators in $\mathscr{A}$, so we need only prove the converse. Since μ is σ-finite, we can find a strictly positive $F \in L^1$ so that $d\nu = F \, d\mu$ has unit mass. Let $\tilde{\mathscr{H}} = \int_M^\oplus \mathscr{H}' \, d\nu$. Then the map $U: \mathscr{H} \to \tilde{\mathscr{H}}$ by $Ug = F^{-1/2}g$ is unitary and $U\mathscr{A}U^{-1} = \tilde{\mathscr{A}}$. Moreover, A is decomposable if and only if UAU^{-1} is decomposable. As a result, we suppose without loss that $\int d\mu = 1$.

Suppose that A in $\mathscr{L}(\mathscr{H})$ commutes with any operator in $\mathscr{A}$. Choose an orthonormal basis $\{\eta_k\}_{k=1}^\infty$ for $\mathscr{H}'$ and let F_k be the element of $\mathscr{H}$ with $F_k(x) = \eta_k$ for all x. The F_k are orthonormal since $\int d\mu = 1$. Moreover, any $\psi \in \mathscr{H}$ has an expansion $\psi = \sum_{k=1}^\infty f_k(x)F_k$ with each $f_k \in L^2(M, d\mu; \mathbb{C})$ and $\|\psi\|^2 = \sum_k \|f_k\|^2$ (see Problem 12 of Chapter II). Define functions $a_{km}(x)$ by $AF_k = \sum_{m=1}^\infty a_{km}(x)F_m$. Choose a countable dense set D in $\mathscr{H}'$ of vectors of the form $\sum_{k=1}^{N} \alpha_k \eta_k = \varphi$ and set $A(x)\varphi = \sum_{k,m} \alpha_k a_{km}(x)\eta_m$. Then, for any $f \in L^\infty(M, d\mu; \mathbb{C})$,

$$A(f\varphi) = f(A\varphi) = \sum_k^N f\alpha_k AF_k$$
$$= \sum_{k,m} f\alpha_k a_{km}(\cdot) F_m$$

since $f \mathbb{1} \in \mathscr{A}$. Thus

$$\int |f(x)|^2 \sum_m \left| \sum_k \alpha_k a_{km}(x) \right|^2 d\mu(x) \leq \|A\|^2 \left(\int |f(x)|^2 \right) \sum_k |\alpha_k|^2$$

It follows that, for almost all x and all $\varphi \in D$,

$$\|A(x)\varphi\| \leq \|A\| \|\varphi\|$$

so $A(x)$ can be extended to an operator on $\mathscr{L}(\mathscr{H}')$ and $A(\cdot) \in L^\infty$. Let B be the corresponding decomposable operator. Let $\psi \in \mathscr{H}$ have the form $\psi = \sum_{k=1}^{N} f_k(x) F_k$ with each $f_k \in L^\infty$. Then

$$(A\psi)(x) = \sum_{k=1}^{N} f_k(x)(AF_k(x)) = \sum_{k=1}^{N} f_k(x)(A(x)\eta_k) = A(x) \sum_{k=1}^{N} f_k(x)\eta_k$$
$$= (B\psi)(x)$$

Since such ψ's are dense, $A = B$. ∎

The construction we use below depends basically on the fact that the $U(t)$ generate an algebra that is isomorphic to the algebra $\mathscr{A}$ for a suitable constant fiber direct integral decomposition of $\mathscr{H} = L^2(\mathbb{R}^n, dx)$.

Since $-\Delta + V$ is unbounded, we need to discuss unbounded decomposable self-adjoint operators.

Definition A function $A(\cdot)$ from a measure space M to the (not necessarily bounded) self-adjoint operators on a Hilbert space $\mathscr{H}'$ is called **measurable** if and only if the function $(A(\cdot) + i)^{-1}$ is measurable. Given such a function, we define an operator A on $\mathscr{H} = \int_M^\oplus \mathscr{H}'$ with domain

$$D(A) = \left\{ \psi \in \mathscr{H} \,\middle|\, \psi(m) \in D(A(m)) \text{ a.e.}; \int_M \|A(m)\psi(m)\|_{\mathscr{H}'}^2 \, d\mu(m) < \infty \right\}$$

by

$$(A\psi)(m) = A(m)\psi(m)$$

We write $A = \int_M^\oplus A(m) \, d\mu$.

The properties of such operators are summarized by:

Theorem XIII.85 Let $A = \int_M^\oplus A(m) \, d\mu$ where $A(\cdot)$ is measurable and $A(m)$ is self-adjoint for each m. Then:

(a) The operator A is self-adjoint.
(b) A self-adjoint operator A on $\mathcal{H}$ has the form $\int_M^\oplus A(m) \, d\mu$ if and only if $(A + i)^{-1}$ is a bounded decomposable operator.
(c) For any bounded Borel function F on $\mathbb{R}$,

$$F(A) = \int_M^\oplus F(A(m)) \, d\mu \tag{139}$$

(d) $\lambda \in \sigma(A)$ if and only if for all $\varepsilon > 0$,

$$\mu(\{m \mid \sigma(A(m)) \cap (\lambda - \varepsilon, \lambda + \varepsilon) \neq \varnothing\}) > 0$$

(e) λ is an eigenvalue of A if and only if

$$\mu(\{m \mid \lambda \text{ is an eigenvalue of } A(m)\}) > 0$$

(f) If each $A(m)$ has purely absolutely continuous spectrum, then so does A.
(g) Suppose that $B = \int_M^\oplus B(m) \, d\mu(m)$ with each $B(m)$ self-adjoint. If B is A-bounded with A-bound a, then a.e. $B(m)$ is $A(m)$-bounded with $A(m)$-bound $a(m) \leq a$. If $a < 1$, then

$$A + B = \int_M^\oplus (A(m) + B(m)) \, d\mu \tag{140}$$

is self-adjoint on $D(A)$.

Proof (a) We first note that A is symmetric, so by the fundamental criterion, we need only prove that $\text{Ran}(A \pm i) = \mathcal{H}$. Let $C(m) = (A(m) + i)^{-1}$. By hypothesis, $C(m)$ is measurable and $\|C(m)\| \leq 1$, so we can define $C = \int_M^\oplus C(m) \, d\mu$. Let $\psi = C\eta$ for $\eta \in \mathcal{H}$. Then, a.e., $\psi(m) \in \text{Ran } C(m) = D(A(m))$ and

$$\|A(m)\psi(m)\| = \|A(m)C(m)\eta(m)\| \leq \|\eta(m)\| \in L^2(d\mu)$$

so $\psi \in D(A)$. Moreover $(A + i)\psi = \eta$ so $\text{Ran}(A + i) = \mathcal{H}$. Similarly, since $(A(m) - i)^{-1} = C(m)^*$ is weakly measurable, $\text{Ran}(A - i) = \mathcal{H}$.

XIII.16 Schrödinger operators with periodic potentials

(b) We leave this to the reader (Problem 136).

(c) Let us sketch the argument leaving the details to the reader (Problem 136). By the argument in (a), for any λ with $\operatorname{Im} \lambda \neq 0$,

$$(A - \lambda)^{-1} = \int_M^\oplus (A(m) - \lambda)^{-1} \, d\mu(m)$$

Since $e^{iAt} = \lim_{n \to \infty} (1 - (itA/n))^{-n}$ (by the functional calculus), one sees, employing the dominated convergence theorem, that

$$e^{iAt} = \int_M^\oplus e^{itA(m)} \, d\mu$$

If $F \in \mathscr{S}(\mathbb{R})$, (139) follows by use of the Fourier transform. By a suitable limiting argument, (139) holds for arbitrary F.

(d) A particular case of (139) is

$$P_{(a,b)}(A) = \int_M^\oplus P_{(a,b)}(A(m)) \, d\mu$$

Now (d) follows by noting that $\lambda \in \sigma(A)$ if and only if $P_{(\lambda - \varepsilon, \lambda + \varepsilon)}(A) \neq 0$ for all $\varepsilon > 0$ and that $\int_M^\oplus T(m) \, d\mu = 0$ if and only if $T(m) = 0$ a.e.

(e) The proof is similar to (d) using

$$P_{\{\lambda\}}(A) = \int_M^\oplus P_{\{\lambda\}}(A(m)) \, d\mu$$

(f) Let $\psi \in \mathscr{H}$ and let dv be the spectral measure for A associated to ψ. Let dv_m be the spectral measure for $A(m)$ associated to $\psi(m)$. Then

$$dv = \int_M (dv_m) \, d\mu(m)$$

in the sense that

$$\int_\mathbb{R} F(x) \, dv = \int_M \left(\int_\mathbb{R} F(x) \, dv_m \right) d\mu(m) \tag{141}$$

(141) follows immediately from (139). Now, if each $A(m)$ has purely absolutely continuous spectrum, then

$$dv_m(x) = g_m(x) \, dx$$

for some $g_m \in L^1(\mathbb{R}, dx)$ with $\int g_m(x) \, dx = \|\psi(m)\|_{\mathscr{H}'}^2$. Thus

$$g(x) = \int g_m(x) \, d\mu(m)$$

is in $L^1(\mathbb{R}, dx)$ and, by (141),
$$dv = g(x)\, dx$$
It follows that $\psi \in \mathcal{H}_{ac}$ for A, so that A has purely absolutely continuous spectrum.

(g) If $\|B\psi\| \leq a\|A\psi\| + b\|\psi\|$, then $\|B(A + ik)^{-1}\| \leq a + bk^{-1}$ for any positive integer k. Therefore,
$$\|B(m)(A(m) + ik)^{-1}\| \leq a + bk^{-1}$$
a.e. so $B(m)$ is $A(m)$-bounded with bound $a(m) \leq a$. (140) is immediate. ∎

Part (f) of this last theorem says that a sufficient condition for $A = \int_M^\oplus A(m)\, d\mu(m)$ to have purely absolutely continuous spectrum is that each $A(m)$ have purely absolutely continuous spectrum. But this is certainly not necessary. In fact, A can have purely absolutely continuous spectrum even though each $A(m)$ has purely discrete spectrum! The following theorem illustrates the phenomenon.

Theorem XIII.86 Let $\langle M, d\mu \rangle$ be $[0, 1]$ with Lebesgue measure. Let $\mathcal{H}'$ be a fixed separable infinite-dimensional space and let $A = \int_{[0,\,1]}^\oplus A(m)\, d\mu(m)$ with each $A(m)$ self-adjoint. Suppose we are given $\mathcal{H}'$-valued functions $\{\psi_n(\cdot)\}_{n=1}^\infty$ on $[0, 1]$, real analytic on $(0, 1)$, continuous on $[0, 1]$, and complex-valued functions $E_n(\cdot)$, analytic in a neighborhood of $[0, 1]$, so that:
(i)   No $E_n(\cdot)$ is constant.
(ii) $A(m)\psi_n(m) = E_n(m)\psi_n(m)$ for all $m \in [0, 1]$; $n = 1, 2, \ldots$.
(iii) For each m, the set $\{\psi_n(m)\}_{n=1}^\infty$ is a complete orthonormal basis for $\mathcal{H}'$.

Then A has purely absolutely continuous spectrum.

Proof Let
$$\mathcal{H}_n = \{\psi \in \mathcal{H}\,|\,\psi(m) = f(m)\psi_n(m);\, f \in L^2(M;\, d\mu)\}$$
Then the $\mathcal{H}_n$ are closed subspaces that are mutually orthogonal and $\mathcal{H} = \oplus\, \mathcal{H}_n$ since any $\psi \in \mathcal{H}$ has an expansion (Problem 134):
$$\psi = \sum_{n=1}^\infty (\psi_n(m), \psi(m))\psi_n(m)$$

Moreover, each $\mathcal{H}_n$ lies in $D(A)$ with $A[\mathcal{H}_n] \subset \mathcal{H}_n$. Consider the unitary map $U_n: \mathcal{H}_n \to L^2([0, 1], dx)$, given by $U_n(f(m)\psi_n(m)) = f(m)$. Then $A_n \equiv U_n A U_n^{-1}$ is given by
$$(A_n f)(m) = E_n(m)f(m) \qquad (142)$$

We need only show that each A_n has purely absolutely continuous spectrum. Since $E_n(\cdot)$ is analytic in a neighborhood of $[0, 1]$ and nonconstant, dE_n/dm has only finitely many zeros in $(0, 1)$, say at $m_1, \ldots, m_{k-1}$. Let $m_0 = 0$ and $m_k = 1$. Then

$$L^2([0, 1], dx) = \bigoplus_{j=1}^{k} L^2((m_{j-1}, m_j), dx)$$

A_n leaves each summand invariant and acts on the summand by (142). On each interval (m_{j-1}, m_j), $E_n(\cdot)$ is strictly monotone, either increasing or decreasing. Consider the case where it is increasing. Define $\alpha: (E_n(m_{j-1}), E_n(m_j)) \to (m_{j-1}, m_j)$ by $E_n(\alpha(\lambda)) = \lambda$. Then α is differentiable and the Stieltjes measure $d\alpha$ is absolutely continuous with respect to dx. In fact,

$$d\alpha = \left[\left(\frac{dE}{dm}\right)\bigg|_{m=\alpha(\lambda)}\right]^{-1} d\lambda$$

Let U be the unitary operator from $L^2((m_{j-1}, m_j), dx)$ to $L^2((E_n(m_{j-1}), E_n(m_j)), d\lambda)$ given by

$$(Uf)(\lambda) = \left(\frac{d\alpha}{d\lambda}\right)^{+1/2} f(\alpha(\lambda))$$

Then

$$(UA_n U^{-1})g(\lambda) = \lambda g(\lambda)$$

We have thus explicitly constructed a spectral representation for $A_n \upharpoonright L^2([m_{j-1}, m_j], dx)$ for which the spectral measure $d\alpha$ is Lebesgue measure. It follows that each A_n, and thus A, has purely absolutely continuous spectrum. ∎

We turn now to an analysis of Schrödinger operators with periodic potentials. We first consider the case of one dimension with V piecewise continuous where differential equation methods are available and then the case of higher dimension and more general V.

To motivate our analysis, suppose that $V \in C_0^\infty(\mathbb{R})$ with bounded derivatives so that $-d^2/dx^2 + V$ takes $\mathscr{S}(\mathbb{R})$ into itself. If $f \in \mathscr{S}(\mathbb{R})$, then

$$\left[\left(-\frac{d^2}{dx^2} + V\right)f\right]^{\wedge}(p) = p^2 \hat{f}(p) + (2\pi)^{-1/2} \int \hat{V}(p - p') \hat{f}(p') \, dp' \quad (143)$$

where the integral in (143) is a formal symbol for the convolution of the distribution $\hat{V}$ and the function $\hat{f}$. Now, let us suppose that V has period 2π. Then V has a uniformly convergent Fourier series (see Theorem II.8):

$$V(x) = \sum_{n=-\infty}^{\infty} \tilde{V}_n e^{inx} \qquad (144)$$

where

$$\tilde{V}_n = \int_{-\pi}^{\pi} \tilde{V}(x) e^{-inx} \frac{dx}{2\pi}$$

(144) suggests that

$$(2\pi)^{-\frac{1}{2}} \hat{V}(p) = \sum_{n=-\infty}^{\infty} \tilde{V}_n \delta(p-n) \qquad (145)$$

since putting (145) formally into the Fourier inversion formula yields (144). In fact, one can prove (145) as follows: If $f \in \mathcal{S}(\mathbb{R})$, then the uniform convergence of (144) implies that

$$\int f(x) V(x)\, dx = (2\pi)^{1/2} \sum_{n=-\infty}^{\infty} \tilde{V}_n \hat{f}(n)$$

from which (145) follows if the sum is viewed as convergent in the weak $(\sigma(\mathcal{S}', \mathcal{S}))$ topology on $\mathcal{S}'$.

Now that we have analyzed Fourier transforms of periodic tempered distributions, we can use this analysis to rewrite (143) as

$$\left[\left(-\frac{d^2}{dx^2} + V\right) f\right]^\wedge (p) = p^2 \hat{f}(p) + \sum_{n=-\infty}^{\infty} \tilde{V}_n \hat{f}(p-n)$$

Thus, if $H = -d^2/dx^2 + V$, then $\widehat{Hf}(p)$ depends only on the values $\hat{f}(p-n)$; $n \in \mathbb{Z}$. We have therefore proven:

Theorem XIII.87 (direct integral decomposition of periodic Schrödinger operators—p-space version in one dimension) Let $\mathcal{H}' = \ell_2$ and let $\mathcal{H} = \int_{(-1/2, 1/2]}^{\oplus} \mathcal{H}'\, dx$. For $q \in (-\frac{1}{2}, \frac{1}{2}]$, let

$$(H(q)g)_j = (q+j)^2 g_j + \sum_{n=-\infty}^{\infty} \tilde{V}_n g_{j-n}$$

where $\tilde{V}_n$ are the Fourier series coefficients of some $V \in C^\infty(\mathbb{R})$ with period 2π. Map $L^2(\mathbb{R}, dx)$ to $\mathcal{H}$ by

$$[(Uf)(q)]_j = \hat{f}(q+j)$$

Let $H = -d^2/dx^2 + V$ on $L^2(\mathbb{R})$. Then,

$$UHU^{-1} = \int_{(-1/2,\,1/2]}^{\oplus} H(q)\, dq$$

One can get quite far in the analysis of H by using this p-space version of the direct integral decomposition. In fact, this will be our main tool in the multidimensional case. However, in the one-dimensional case, the x-space translation of Theorem XIII.87 gives a little more information. While we could use Theorem XIII.87 directly to write down the x-space version, we shall give an independent proof, using Theorem XIII.87 merely for the following motivation. In case $V = 0$ the operator $H(q)$ has eigenvalues $(q+j)^2$ and eigenfunctions that are basically the Fourier transforms of the functions $e^{i(q+j)x}$. This suggests that somehow $H(q)$ is related to the operator $-d^2/dx^2$ on $L^2([0, 2\pi], dx)$ but with the boundary conditions

$$\psi(2\pi) = e^{2\pi i q}\psi(0), \qquad \psi'(2\pi) = e^{2\pi i q}\psi'(0)$$

Lemma Let $\mathcal{H}' = L^2([0, 2\pi], dx)$. Let

$$\mathcal{H} = \int_{[0,\,2\pi)}^{\oplus} \mathcal{H}' \frac{d\theta}{2\pi} \tag{146}$$

Then $U: L^2(\mathbb{R}, dx) \to \mathcal{H}$ given by

$$(Uf)_\theta(x) = \sum_{n=-\infty}^{\infty} e^{-i\theta n} f(x + 2\pi n) \tag{147}$$

for θ and x in $[0, 2\pi)$, is well defined for $f \in \mathcal{S}(\mathbb{R})$ and uniquely extendable to a unitary operator. Moreover,

$$U\left(-\frac{d^2}{dx^2}\right)U^{-1} = \int_{[0,\,2\pi)}^{\oplus} \left(-\frac{d^2}{dx^2}\right)_\theta \frac{d\theta}{2\pi} \tag{148}$$

where $(-d^2/dx^2)_\theta$ is the operator $-d^2/dx^2$ on $L^2([0, 2\pi])$ with the boundary conditions

$$\psi(2\pi) = e^{i\theta}\psi(0), \qquad \psi'(2\pi) = e^{i\theta}\psi'(0)$$

Proof For $f \in \mathscr{S}$, the sum (147) is clearly convergent. To prove that Uf is in $\mathscr{H}$, we compute that for $f \in \mathscr{S}$,

$$\int_0^{2\pi} \left(\int_0^{2\pi} \left| \sum_{n=-\infty}^{\infty} e^{-in\theta} f(x + 2\pi n) \right|^2 dx \right) \frac{d\theta}{2\pi}$$

$$= \int_0^{2\pi} \left[\left(\sum_{n=-\infty}^{\infty} \sum_{j=-\infty}^{\infty} \overline{f(x + 2\pi n)} f(x + 2\pi j) \right) \int_0^{2\pi} e^{-i(j-n)\theta} \frac{d\theta}{2\pi} \right] dx$$

$$= \int_0^{2\pi} \left(\sum_{n=-\infty}^{\infty} |f(x + 2\pi n)|^2 \right) dx = \int_{-\infty}^{\infty} |f(x)|^2 \, dx$$

where we have used the Fubini and Plancherel theorems. Thus we see that U is well defined and has a unique extension to an isometry. To see that U is onto $\mathscr{H}$, we compute U^*. For $g \in \mathscr{H}$, we define, for $0 \leq x \leq 2\pi$, $n \in \mathbb{Z}$,

$$(U^*g)(x + 2\pi n) = \int_0^{2\pi} e^{in\theta} g_\theta(x) \frac{d\theta}{2\pi} \tag{149}$$

A direct computation shows that this is indeed the formula for the adjoint of U. Moreover,

$$\|U^*g\|_2^2 = \int_{-\infty}^{\infty} |(U^*g)(y)|^2 \, dy$$

$$= \int_0^{2\pi} \left(\sum_{n=-\infty}^{\infty} |(U^*g)(2\pi n + x)|^2 \right) dx$$

$$= \int_0^{2\pi} \left(\sum_{n=-\infty}^{\infty} \left| \int e^{in\theta} g_\theta(x) \frac{d\theta}{2\pi} \right|^2 \right) dx$$

$$= \int_0^{2\pi} \left(\int_0^{2\pi} |g_\theta(x)|^2 \frac{d\theta}{2\pi} \right) dx$$

$$= \|g\|^2$$

In the next to the last step we have used the Parseval relation for Fourier series.

To verify (148), let A be the operator on the right-hand side of (148). We shall show that if $f \in \mathscr{S}(\mathbb{R})$, then $Uf \in D(A)$ and $U(-f'') = A(Uf)$. Since $-d^2/dx^2$ is essentially self-adjoint on $\mathscr{S}$ and A is self-adjoint, (148) will follow. So, suppose $f \in \mathscr{S}(\mathbb{R}^n)$. Then Uf is given by the convergent sum (147) so that Uf is C^∞ on $(0, 2\pi)$ with $(Uf)'_\theta(x) = (Uf')_\theta(x)$ and similarly for higher derivatives. Moreover, it is clear that

$$(Uf)_\theta(2\pi) = \sum_{n=-\infty}^{\infty} e^{-i\theta n} f(2\pi(n+1))$$

$$= \sum_{n=-\infty}^{\infty} e^{-i\theta(n-1)} f(2\pi n) = e^{i\theta}(Uf)_\theta(0)$$

Similarly, $(Uf)'_\theta(2\pi) = e^{i\theta}(Uf_\theta)'(0)$. Thus, for each θ, $(Uf)_\theta \in D((-d^2/dx^2)_\theta)$ and

$$\left(-\frac{d^2}{dx^2}\right)_\theta (Uf) = U(-f'')_\theta$$

We conclude that $Uf \in D(A)$ and $A(Uf) = U(-f'')$. This proves (148). ∎

Theorem XIII.88 (direct integral decomposition of periodic Schrödinger operators—x-space version in one dimension)   Let V be a bounded measurable function on $\mathbb{R}$ with period 2π. For $\theta \in [0, 2\pi)$, let

$$H(\theta) = \left(-\frac{d^2}{dx^2}\right)_\theta + V(x)$$

as an operator on $L^2[0, 2\pi]$. Let U be given by (147). Then, under the decomposition (146),

$$U\left(-\frac{d^2}{dx^2} + V\right)U^{-1} = \int_{[0, 2\pi)}^{\oplus} H(\theta) \frac{d\theta}{2\pi} \qquad (150)$$

Proof   Let V be the θ-independent operator acting on the fiber $\mathscr{H}' = L^2([0, 2\pi), dx)$ by

$$(V_\theta f)(x) = V(x)f(x), \qquad 0 \le x \le 2\pi$$

(150) follows from Theorem XIII.85g and the lemma if we can prove that

$$UVU^{-1} = \int_{[0, 2\pi)}^{\oplus} V_\theta \frac{d\theta}{2\pi} \qquad (151)$$

By (147), for $f \in \mathscr{S}$,

$$(UVf)_\theta(x) = \sum_{n=-\infty}^{\infty} e^{-in\theta} V(x + 2\pi n) f(x + 2\pi n)$$

$$= V(x) \sum_{n=-\infty}^{\infty} e^{-in\theta} f(x + 2\pi n)$$

$$= V_\theta (Uf)_\theta(x)$$

since V is periodic. This proves (151) and so (150). ∎

292   XIII: SPECTRAL ANALYSIS

As a result, to analyze $-d^2/dx^2 + V$ with V periodic, we need only analyze $(-d^2/dx^2)_\theta + V$ for each θ. As a preliminary, we note:

Lemma

(a) For each $\theta \in [0, 2\pi)$, $(-d^2/dx^2)_\theta$ has compact resolvent.
(b) For $\theta = 0$, $\exp(-t(-d^2/dx^2)_{\theta=0})$ is a positivity improving semigroup (see Section 12).
(c) $[(-d^2/dx^2)_\theta + 1]^{-1}$ is an analytic operator-valued function of θ in a neighborhood of $[0, 2\pi)$.

Proof We shall later prove the analogue of this lemma in the multidimensional case by using general arguments that could be used here. However, it is easy to obtain explicit formulas for $K_\theta \equiv [(-d^2/dx^2)_\theta + 1]^{-1}$. Let $f \in C_0^\infty(0, 2\pi)$. Let K be the inverse of $-d^2/dx^2 + 1$ defined on all of $L^2(\mathbb{R})$. Let $g = Kf$. By our arguments in Section IX.7, K is an integral operator with kernel $G(x - y)$ where $\hat{G}(p) = (2\pi)^{-1/2}(p^2 + 1)^{-1}$. A direct computation of G is possible (Problem 137) and one finds

$$g(x) \equiv Kf(x) = \tfrac{1}{2}\int e^{-|x-y|} f(y)\, dy \tag{152}$$

Now, both Kf and $K_\theta f$ solve the differential equation $-u''(x) + u(x) = f(x)$ on $(0, 2\pi)$. It follows that their difference $v = K_\theta f - Kf$ obeys $-v'' + v = 0$ so that

$$(K_\theta f)(x) = g(x) + ae^x + be^{-x}$$

Since $K_\theta f \in D((-d/dx^2)_\theta)$, a and b must be chosen so that $K_\theta f$ obeys the boundary conditions

$$u(2\pi) = e^{i\theta} u(0), \qquad u'(2\pi) = e^{i\theta} u'(0) \tag{153}$$

Direct computation using (152) shows that

$$(K_\theta f)(x) = \int_0^{2\pi} G_\theta(x, y) f(y)\, dy$$

$$G_\theta(x, y) = \tfrac{1}{2} e^{-|x-y|} + \alpha(\theta) e^{x-y} + \beta(\theta) e^{y-x} \tag{154}$$

$$\alpha(\theta) = \tfrac{1}{2}(e^{2\pi - i\theta} - 1)^{-1}$$

$$\beta(\theta) = \tfrac{1}{2}(e^{2\pi + i\theta} - 1)^{-1}$$

One can read the properties claimed for $(-d^2/dx^2)_\theta$ directly from (154). Since $G_\theta(x, y)$ is bounded in x, y for each fixed θ,

$$\int_0^{2\pi} \int_0^{2\pi} |G_\theta(x, y)|^2 \, dx \, dy < \infty$$

so K_θ is Hilbert–Schmidt and so compact, proving (a). By direct examination, the kernel $G_{\theta=0}(x, y)$ is strictly positive. A similar computation proves that $[(-d^2/dx^2)_{\theta=0} + a]^{-1}$ has a strictly positive kernel for any $a > 0$ and so by Theorem XIII.44 and the preceding proposition, $\exp(-t(-d^2/dx^2)_{\theta=0})$ is a positivity improving semigroup. Finally, to prove (c), we note that the formulas (154) allow us to define a Hilbert–Schmidt operator K_θ for any θ with $|\text{Im } \theta| < 2\pi$ and that $\theta \to K_\theta$ is clearly analytic in θ. ∎

It may seem striking at first sight that $K_\theta - K_{\theta'}$ is a rank two operator for any θ, θ', but, in fact, this is just a reflection of the fact that $-d^2/dx^2 \upharpoonright C_0^\infty(0, 2\pi)$ has deficiency indices $\langle 2, 2 \rangle$ so that K_θ is completely determined in a θ-independent way on the closure of the space $(-d^2/dx^2 + 1)[C_0^\infty(0, 2\pi)]$ which has codimension 2.

An analysis similar to that above shows that $((-d^2/dx^2)_\theta + a)^{-1}$ is analytic in the region $|\text{Im } \theta| < 2\pi\sqrt{a}$ so that the map $\theta \mapsto (-d^2/dx^2)_\theta$ can be extended to an entire analytic family. This family is neither type (A) nor type (B).

Armed with the lemma, we are prepared for a complete analysis of the operators

$$H(\theta) = \left(-\frac{d^2}{dx^2}\right)_\theta + V \tag{155}$$

Theorem XIII.89 Suppose that V is piecewise continuous and periodic of period 2π. Then:

(a) $H(\theta)$ has purely discrete spectrum and is real analytic in θ.
(b) $H(\theta)$ and $H(2\pi - \theta)$ are antiunitarily equivalent under ordinary complex conjugation. In particular, their eigenvalues are identical and their eigenfunctions are complex conjugates.
(c) For $\theta \in (0, \pi)$, or in $(\pi, 2\pi)$, $H(\theta)$ has only nondegenerate eigenvalues.
(d) Let $E_n(\theta)$ $(n = 1, 2, \ldots; 0 \leq \theta \leq \pi)$ denote the nth eigenvalue of $H(\theta)$. Then each $E_n(\cdot)$ is analytic in $(0, \pi)$ and continuous at $\theta = 0$ and π.

(e) For n odd (respectively, even) $E_n(\theta)$ is strictly monotone increasing (respectively, decreasing) as θ increases from 0 to π. In particular,

$$E_1(0) < E_1(\pi) \leq E_2(\pi) < E_2(0) \leq \cdots \leq E_{2n-1}(0) < E_{2n-1}(\pi) \leq E_{2n}(\pi)$$
$$< E_{2n}(0) \leq \cdots$$

See Figure XIII.13.

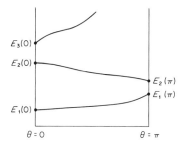

FIGURE XIII.13 Bands in one-dimensional Schrödinger operators.

(f) One can choose the eigenvectors $\psi_n(\theta)$ so that they are analytic in θ for $\theta \in (0, \pi) \cup (\pi, 2\pi)$, continuous at π and 0 (with $\psi_n(0) = \psi_n(2\pi)$).

Proof (a) This follows directly from the lemma and the basic perturbation Theorems XII.11 and XIII.64.

(b) When $V = 0$, this is a simple consequence of the definition of $(-d^2/dx^2)_\theta$. Since $\overline{V\psi} = V\bar\psi$, the results hold for general V.

(c) If E is an eigenvalue of $H(\theta)$, $\theta \in (0, \pi)$, then $-u'' + Vu = Eu$ has a solution obeying the boundary condition (153). So $\bar u$ is a solution obeying a distinct boundary condition. Since $-u'' + Vu = Eu$ has only two linearly independent solutions and not all of them obey (153), at most one can.

(d) Consider $E_1(0)$. This is a simple eigenvalue of $H(0)$ since $H(0)$ generates a positivity preserving semigroup. Since $H(\theta)$ is analytic near $\theta = 0$, we can find $\tilde f_1(\theta)$ an eigenvalue of $H(\theta)$ for $\theta \in [0, \varepsilon)$ analytic in $[0, \varepsilon)$ with $\tilde f_1(0) = E_1(0)$. Let $\varepsilon < \pi$. The only thing that can prevent one from analytically continuing past $\theta = \varepsilon$ is if $\tilde f_1(\theta) \to \infty$ as $\theta \uparrow \varepsilon$. For since $H(\theta) \geq -\|V\|_\infty$, if $\tilde f_1(\theta)$ does not approach ∞, then there is a sequence $\theta_n \to \varepsilon$ such that $\tilde f_1(\theta_n) \to \tilde E$. But then one sees that $\tilde E$ is an eigenvalue of $H(\varepsilon)$. By (c), it is a simple eigenvalue, so for $|\theta - \varepsilon| < \delta$, there is a unique eigenvalue $g(\theta)$ of $H(\theta)$ near $\tilde E$ and g is analytic for $|\theta - \varepsilon| < \delta$. In particular, for n large $g(\theta_n) = \tilde f_1(\theta_n)$ so g provides an analytic continuation for $\tilde f_1$ past ε. Thus to prove that $\tilde f_1$ can be analytically continued to all of $[0, \pi)$, we need only show that $\tilde f_1(\theta)$ remains finite as θ varies. We first show that $H(\theta)$

has no eigenvalue smaller than $\tilde{f}_1(\theta)$ if $\theta \in [0, \varepsilon)$. If it did, we could continue that back to $\theta = 0$; this continuation could not go to infinity as we decreased θ since it is always strictly less than $\tilde{f}_1(\theta)$ by the simplicity of eigenvalues and the argument above. Continuing back to $\theta = 0$, we would find an eigenvalue less than E_1. Since $\tilde{f}_1(\theta)$ is the smallest eigenvalue of $H(\theta)$ it cannot go to infinity as $\theta \to \varepsilon$. Thus, $\tilde{f}_1(\theta)$ has a continuation to $[0, \pi]$ and this continuation is the smallest eigenvalue of $H(\theta)$, i.e., it is $E_1(\theta)$.

Now look at $E_2(0)$. This may be doubly degenerate; for example it is when $V = 0$. If it is though, the degeneracy must be broken for $\theta \neq 0$ since the spectrum of $H(\theta)$, $\theta \neq 0$ is simple. By degenerate perturbation theory, the eigenvalue (or eigenvalues if $E_2(0)$ is degenerate) near $E_2(0)$ is given by analytic function(s). Let $\tilde{f}_2(\theta)$ be this function if $E_2(0)$ is simple, and the smaller of the functions if $E_2(0)$ is degenerate. Then by mimicking the argument above, $\tilde{f}_2(\theta)$ can be continued throughout $[0, \pi]$ and is the second eigenvalue $E_2(\theta)$. By repeating this argument, we can handle all the eigenvalues.

(e) This is the deepest part of the theorem, so we shall give a detailed proof. As a preliminary, we prove that $E_1(0) \leq E_1(\theta)$ for all θ. Since $e^{-tH(0)}$ is positivity improving, the eigenvector $\psi_1(0)$ associated to $E_1(0)$ is strictly positive and by the boundary condition, it has a periodic extension to all of $\mathbb{R}$. Fix k, an integer, and consider $H^{(k)}(0)$ the operator $-d^2/dx^2 + V$ on $L^2(-2\pi k, 2\pi k)$ with periodic boundary conditions. Then $\psi_1(0)$ periodically extended is a strictly positive eigenvector of $H^{(k)}(0)$ and so $E_1(0) = \inf \sigma(H^k(0))$ (see Section XIII.12). It follows that if $f \in C_0^\infty(-2\pi k, 2\pi k)$, then $(f, (-d^2/dx^2 + V)f) = (f, H^k(0)f) \geq E_1(f, f)$, and thus the operator $-d^2/dx^2 + V$ on $L^2(\mathbb{R})$ obeys $-d^2/dx^2 + V \geq E_1$. By the direct integral decomposition, $H(\theta) \geq E_1(0)$, a.e. in θ so $E_1(\theta) \geq E_1(0)$, a.e. in θ. Since $E_1(\theta)$ is continuous $E_1(\theta) \geq E_1(0)$ for all $\theta \in (0, 2\pi)$.

Now we introduce an important function, $D(E)$, associated to the differential equation

$$-u'' + Vu = Eu \tag{156}$$

Let $u_1(E, x)$ be the solution of (156) with $u_1(0) = 1$, $u_1'(0) = 0$, and let $u_2(E, x)$ be the solution of (156) with $u_2(0) = 0$, $u_2'(0) = 1$. Then $u_i(x, E)$ is analytic in E for each x by the standard theory of ordinary differential equations. Let $M(E)$ be the analytic two by two matrix

$$M(E) = \begin{bmatrix} u_1(E, 2\pi) & u_2(E, 2\pi) \\ u_1'(E, 2\pi) & u_2'(E, 2\pi) \end{bmatrix} \tag{157}$$

The **discriminant** of $-u'' + Vu$ is the function
$$D(E) \equiv \text{Tr}(M(E)) = u_1(E, 2\pi) + u_2'(E, 2\pi)$$
$M(E)$ is a natural object, for if v satisfies (156), then
$$\begin{bmatrix} v(2\pi) \\ v'(2\pi) \end{bmatrix} = M(E) \begin{bmatrix} v(0) \\ v'(0) \end{bmatrix}$$
In particular, the equation $H(\theta)\psi = E\psi$ has a nonzero solution if and only if $M(E)$ has an eigenvalue $e^{i\theta}$. Now $M(E)$ has determinant 1 since $W(x) = u_1(E, x)u_2'(E, x) - u_1'(E, x)u_2(E, x)$ is a constant. Thus its eigenvalues are λ and λ^{-1} and $D(E) = \lambda + \lambda^{-1}$. We conclude that E is an eigenvalue of $H(\theta)$ if and only if $D(E) = 2 \cos \theta$. What we will prove is that $D(E)$ has a graph somewhat like the one in Figure XIII.14.

FIGURE XIII.14 A typical discriminant.

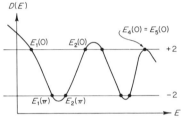

We have proven that $E_1(0) \le E_1(\theta)$ for all θ, so $D(E)$ cannot have any value in $[-2, 2]$ for $E < E_1(0)$. Now $D(E) = 2$ for $E = E_1(0)$. As θ varies from 0 to π, $D(E_1(\theta))$ varies from 2 to -2. E_1 must therefore be strictly monotone increasing since it has an inverse function Arc cos $\tfrac{1}{2} D(E_1(\theta)) = \theta$. We have $D(E_1(\pi)) = -2$. D must eventually turn around (since $H(\pi)$ has additional eigenvalues) so the next value of $D(\theta)$ in $[-2, 2]$ to occur must be -2. This occurs at $E_2(\pi)$ and then D runs from -2 to 2 as θ goes from π to 0. Thus we have the picture in Figure XIII.14. The only subtlety is that we must show that if D has a turning point at $+2$ or -2, then $H(0)$ or $H(\pi)$ has a double eigenvalue. But if $D(E)$ has a turning point at $E = E_0$ with $D(E) = +2$, then for θ near 0, $H(\theta)$ has two eigenvalues near E_0 corresponding to the fact that $D(E) = 2 \cos \theta$ has two solutions near $E = E_0$. By analytic perturbation theory E_0 must be a double eigenvalue of $H(0)$.

(f) This follows from the analytic perturbation theory of Section XII.2. ∎

The reader may have noticed that while we have been careful to avoid saying that $E_n(\theta)$ is analytic near $\theta = \pi$ or 0, it clearly is. However, if $E_n(\theta)$ is continued through $\theta = \pi$, the continuation may be $E_{n+1}(\theta)$ or $E_{n-1}(\theta)$ if $E_n(\pi)$ is a doubly degenerate eigenvalue (see Figure XIII.15). A similar phenomenon can occur at $\theta = 0$ if we identify θ and $\theta - 2\pi$.

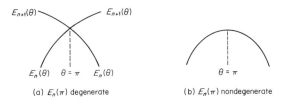

(a) $E_n(\pi)$ degenerate (b) $E_n(\pi)$ nondegenerate

FIGURE XIII.15 Crossed bands.

We can now combine Theorems XIII.85, 86, 88, and 89 to conclude:

Theorem XIII.90 Let V be a piecewise continuous function of period 2π. Let $H = -d^2/dx^2 + V$ on $L^2(\mathbb{R}, dx)$. Let $E_1(0), E_2(0), \ldots$ be the eigenvalues of the corresponding operator on $(0, 2\pi)$ with periodic boundary conditions and let $E_1(\pi), \ldots$ be the eigenvalues with antiperiodic boundary conditions. Let

$$\alpha_n = \begin{cases} E_n(0), & n \text{ odd} \\ E_n(\pi), & n \text{ even} \end{cases} \qquad \beta_n = \begin{cases} E_n(\pi), & n \text{ odd} \\ E_n(0), & n \text{ even} \end{cases}$$

Then:
(a) $\sigma(H) = \bigcup_{n=1}^{\infty} [\alpha_n, \beta_n]$.
(b) H has no eigenvalues.
(c) H has purely absolutely continuous spectrum.

Proof (a) Since the $E_n(\theta)$ are continuous, if θ_0 and ε are given, then for some δ,

$$\{\theta \,|\, |\theta - \theta_0| < \delta\} \subset \{\theta \,|\, |E_n(\theta) - E_n(\theta_0)| < \varepsilon\}$$

so by Theorem XIII.85, $\sigma(H) = \bigcup_n [\alpha_n, \beta_n]$.

(b) No function E_n is constant since the E_n are strictly monotone. Thus for each E_0, $\{\theta \,|\, E_n(\theta) = E_0\}$ is a set with at most two points. Such a set has measure zero, so by Theorem XIII.86, E_0 is not an eigenvalue.

(c) follows by Theorems XIII.86 and 89. ∎

We remark that $-d^2/dx^2 + V$ has a simple eigenfunction expansion, but since we shall give the general n-dimensional result below, we do not pause to give the details now.

The most striking feature of Theorem XIII.90 is that $\sigma(H)$ has gaps $(\beta_1, \alpha_2), \ldots, (\beta_n, \alpha_{n+1}), \ldots$. Of course, all we know is that $\beta_n \leq \alpha_{n+1}$, so that some of the "gaps" listed may be empty. In fact, if $V = 0$, then there are no gaps, so it is necessary to impose some condition on V for any given gap to be nonempty. The beautiful feature of this analysis is that the occurrence of any gap is reduced to a question about the degeneracy of some eigenvalue.

Example 1 (the Mathieu equation)   Let

$$V(x) = \mu \cos x$$

with $\mu \neq 0$. We claim that for all n, $\alpha_{n+1} \neq \beta_n$, i.e., every gap occurs. Let H_0^P (respectively, H_0^A) be $-d^2/dx^2$ on $L^2(0, 2\pi)$ with periodic (respectively, antiperiodic) boundary conditions. We need only show that $H_0^P + V$ and $H_0^A + V$ have no double eigenvalues. We give the proof for $H_0^P + V$; the proof is similar for $H_0^A + V$. Consider the functions $\varphi_n = (2\pi)^{-1/2} e^{inx}$. Then $\varphi_n \in D(H_0^P)$ and $H_0^P \varphi_n = n^2 \varphi_n$. If ψ solves $(H_0^P + V)\psi = E\psi$ and $a_n = (\varphi_n, \psi)$, then

$$(n^2 - E)a_n + \tfrac{1}{2}\mu(a_{n+1} + a_{n-1}) = 0 \tag{158a}$$

If η also solves $(H_0^P + V)\eta = E\eta$ and $b_n = (\varphi_n, \eta)$, then

$$(n^2 - E)b_n + \tfrac{1}{2}\mu(b_{n+1} + b_{n-1}) = 0 \tag{158b}$$

Eliminating the $n^2 - E$ term from (158) and using $\mu \neq 0$, we have $b_n a_{n+1} - a_n b_{n+1} = a_n b_{n-1} - b_n a_{n-1}$ so $b_n a_{n+1} - a_n b_{n+1} = c$, where c is some constant. Since $\eta, \psi \in L^2$, $\sum a_n^2 < \infty$, $\sum b_n^2 < \infty$, so $a_n \to 0$ and $b_n \to 0$ as $n \to \infty$. Therefore c must be zero, and thus

$$a_n b_{n+1} = b_n a_{n+1} \tag{159}$$

By (158a), if any two successive a_j are zero, all the a_j are zero so for any n, either $a_n \neq 0$ or $a_{n+1} \neq 0$. A similar result holds for the b_n. Now suppose that E is a doubly degenerate eigenfunction. Since $\cos x$ is even under $x \to -x$, we can choose ψ to be the even solution of $-\psi'' + V\psi = E\psi$ and η to be the odd solution since all solutions are periodic if $H_0^P + V$ has E as a degenerate eigenvalue. Since η is odd, $b_0 = \int_{-\pi}^{\pi} \eta(x) \, dx = 0$. Thus, by our remark above, $b_1 \neq 0$. Since ψ is even, $a_n = a_{-n}$ and so, in particular, by (158a)

$$-Ea_0 + \mu a_1 = 0$$

It follows that $a_0 \neq 0$ since if it were zero, a_1 would be zero, violating the remark above. Thus $a_0 b_1 \neq 0$ but $a_1 b_0 = 0$. This violates (159). We conclude that $H_0^P + V$ has no degenerate eigenvalues.

There is another example in Problem 139 where one can obtain an asymptotic formula for $\ell_n = \alpha_{n+1} - \beta_n$ as $n \to \infty$, which proves that at least for large n (where $\ell_n \neq 0$), there are lots of gaps. There are also the following general results whose proofs can be found in the references in the Notes.

Theorem XIII.91 Let V be periodic of period 2π. Then:
(a) If no gaps are present, V is a constant.
(b) If precisely one gap occurs, then V is a Weierstrass elliptic function.
(c) If all the odd gaps are absent (i.e., if $H_0^A + V$ has only degenerate eigenvalues), then V has period π. More generally, if, for fixed n, all gaps (β_k, α_{k+1}) are absent for $k \neq 2^n m$ $(m = 1, 2, \ldots)$, then V has period $2^{-n}(2\pi)$ and the converse relation is true.
(d) If only finitely many gaps are present, then V is real analytic as a function on $\mathbb{R}$.
(e) Topologize Y, the space of all C^∞ functions on $\mathbb{R}$ with period 2π, with the seminorms $\|f\|_n = \|D^n f\|_\infty$, $n = 0, 1, \ldots$. Then the set of potentials in V for which *all* gaps are nonzero is a dense G_δ set of Y. (See the discussion of "Baire almost every" in the notes to Section III.5.)

There are some general results about when two potentials V and W produce the same energy bands.

Theorem XIII.92 Let V and W be two potentials of period 2π so that $-d^2/dx^2 + V$ and $-d^2/dx^2 + W$, on $[0, 2\pi]$ with periodic boundary conditions, have the same eigenvalues. Then their energy bands are the same.

Proof We shall sketch the main ideas. Fuller details can be found in the reference in the notes. Let $D_V(E)$ and $D_W(E)$ be the respective discriminants. By the analysis used in the proof of Theorem XIII.89, it suffices to prove that the discriminants are equal. We claim that

$$|D_V(E)| + |D_W(E)| \leq C_1 \exp(C_2 |E|^{1/2}) \tag{160}$$

$$D_V(E)/2 \, \cos(2\pi\sqrt{E}) \to 1 \quad \text{as} \quad E \to i\infty \tag{161}$$

$$D_W(E)/2 \, \cos(2\pi\sqrt{E}) \to 1 \quad \text{as} \quad E \to i\infty \tag{162}$$

Deferring the proofs of (160)–(162), let us complete the proof that $D_V = D_W$. By (160) and the Hadamard factorization theorem of complex analysis,

$$2 - D_V(E) = C_V \prod_{j=1}^{\infty} (1 - E_j(V)^{-1}E)$$

$$2 - D_W(E) = C_W \prod_{j=1}^{\infty} (1 - E_j(W)^{-1}E)$$

where $E_j(V)$ are the zeros of $D_V(E) - 2$. By hypothesis, the zeros of $2 - D_V$ and $2 - D_W$ are the same. By (161) and (162), $(2 - D_V)/(2 - D_W) \to 1$ as $E \to i\infty$, so $D_V = D_W$.

(160)–(162) follow by a detailed analysis of the solutions $u_j(x, E)$. One shows that they solve integral equations, for example,

$$u_1(x, E) = \cos(x\sqrt{E}) + \frac{1}{\sqrt{E}} \int_0^x \sin((x-y)\sqrt{E}) V(y) u_1(y, E) \, dy$$

By iterating these equations, one proves (160)–(162). ∎

At first sight, one might think that there are not many pairs V, W with $D_V = D_W$. Quite the contrary: In the Notes, the reader can find references for the following two theorems:

Theorem XIII.93 Let $V(x, t)$ solve the partial differential equation

$$\frac{\partial V}{\partial t} = 3V \frac{\partial V}{\partial x} - \frac{1}{2} \frac{\partial^3 V}{\partial x^3} \tag{163}$$

with $V(x, 0)$ periodic of period 2π. Then for each fixed t, $V(x, t)$ is periodic of period 2π and has energy bands independent of t.

Equation (163) is called the **Korteweg–de Vries equation.**

Theorem XIII.94 Topologize the C^∞ functions on $\mathbb{R}$ with period 2π by the seminorms $\|D^\alpha f\|_\infty$. Fix V and suppose that the spectrum of $-d^2/dx^2 + V$ has n gaps (n may be infinite). Then $\{W \,|\, D_V = D_W\}$ is homeomorphic to an n-dimensional torus.

In the one-dimensional case one can also say quite a lot about global analytic properties of $E_n(\theta)$.

Theorem XIII.95 (Kohn's theorem) Suppose that all the gaps in a one-dimensional problem are nonzero. Then the energy band functions $E_n(\theta)$ are the branches of a single multisheeted function that has no singularities other than square root branch points on the lines Im $\theta = m\pi$, $m = 0, \pm 1, \ldots$. Explicitly, there exist positive numbers $\alpha_1, \alpha_2, \ldots$ so that the Riemann surface of $E(\theta)$ can be described as follows. E is equal to $E_n(\theta)$ on the nth sheet which is cut in $\bigcup_m [(2m + 1)\pi] \pm i(\alpha_n, \infty)$ and $\bigcup_m (2m\pi) \pm i(\alpha_{n-1}, \infty)$ for n odd; and $\bigcup_m [2m\pi] \pm i(\alpha_n, \infty)$ and $\bigcup_m [(2m + 1)\pi] \pm i(\alpha_{n-1}, \infty)$ for n even. The nth and $(n + 1)$th sheet are joined by crossing the cuts $2m\pi \pm i(\alpha_n, \infty)$ (n even) and $(2m + 1)\pi \pm i(\alpha_n, \infty)$ (n odd); see Figure XIII.16.

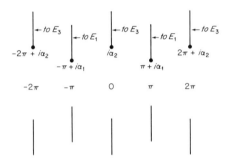

FIGURE XIII.16   The Riemann surfaces of an energy band; $n = 2$.

For a proof of this theorem, see the references in the Notes.

Now we turn to the general n-dimensional case. The direct integral decomposition in both the x-space and p-space versions will go through without significant change. The main difficulty will be in extending the analysis of the fibers of H (Theorem XIII.89). For that analysis depended critically on the simplicity of the eigenvalues of $H(\theta)$, which fails in the multidimensional case. An additional complication is that in the multivariable case, eigenvalues are not necessarily analytic (in a single-valued sense) at degeneracy points: See the example in the Notes to Section XII.1. It will turn out to be easier to analyze the fibers of the direct integral decomposition in the p-space version.

We want to allow for the possibility of local singularities in V in our discussion of the general n-dimensional case. The perturbation criteria, as stated in Section X.2, are not applicable if V is unbounded since such a periodic V cannot be in any $L^p + L^\infty$ with $p < \infty$; but if V is L^p over all bounded sets and periodic, then it will be uniformly locally L^p where:

Definition A measurable function V on $\mathbb{R}^n$ is called **uniformly locally L^p** if and only if

$$\int_C |V(x)|^p \, d^n x \leq A$$

for any unit cube C and some C-independent constant A.

The perturbation theory of Section X.2 extends to uniformly locally L^p perturbations (for suitable p) by the following localization method.

Theorem XIII.96 Let $p = 2$ if $n \leq 3$, $p > 2$ if $n = 4$ and $p > n/2$ if $n \geq 5$. Then any real-valued function on $\mathbb{R}^n$ that is uniformly locally L^p is a $-\Delta$-bounded operator with relative bound zero.

Proof Let q be such that $p^{-1} + q^{-1} = \frac{1}{2}$. We proved in Section X.2 that for any ε, there is an A_ε so that

$$\|\varphi\|_q^2 \leq \varepsilon \|\Delta\varphi\|_2^2 + A_\varepsilon \|\varphi\|_2^2 \tag{164}$$

For any cube C, let

$$\|\varphi\|_{r;C}^r \equiv \int_C |\varphi(x)|^r \, d^n x$$

Let C be a unit cube and let C' be the cube with side 3 and the same center as C. Let η be a C^∞ function with support in C' that is identically 1 on C. Now, by (164),

$$\|\varphi\|_{q,C}^2 \leq \|\eta\varphi\|_q^2$$
$$\leq \varepsilon \|\Delta(\eta\varphi)\|_2^2 + A_\varepsilon \|\eta\varphi\|_2^2$$
$$\leq 3\varepsilon \|\Delta\varphi\|_{2;C'}^2 + B\|\nabla\varphi\|_{2;C'}^2 + D\|\varphi\|_{2;C'}^2 \tag{165}$$

where we have used $\Delta(\eta\varphi) = \varphi\,\Delta\eta + \eta\,\Delta\varphi + 2\,\nabla\eta \cdot \nabla\varphi$, the triangle inequality, the fact that $(a + b + c)^2 \leq 3(a^2 + b^2 + c^2)$, and the fact that η, $\nabla\eta$, and $\Delta\eta$ are bounded functions with support in C'. Notice that since the various $\|D^\alpha \eta\|_\infty$ can be chosen independently of C, (165) holds with constants that are independent of C. For $\alpha \in \mathbb{Z}^n$, let C_α be the unit cube with center α and C'_α the corresponding C'. Then, since V is uniformly locally L^p,

$$|||V|||^2 \equiv \sup_\alpha \|V\|_{p,C_\alpha}^2 < \infty$$

Thus,

$$\|V\varphi\|_2^2 = \sum_\alpha \|V\varphi\|_{2;C_\alpha}^2$$
$$\leq \sum_\alpha \|V\|_{p;C_\alpha}^2 \|\varphi\|_{q;C_\alpha}^2$$
$$\leq \|\|V\|\|^2 \sum_\alpha (3\varepsilon\|\Delta\varphi\|_{2;C_{\alpha'}}^2 + B\|\nabla\varphi\|_{2;C_{\alpha'}}^2 + D\|\varphi\|_{2;C_{\alpha'}}^2)$$
$$= \|\|V\|\|^2 3^n (3\varepsilon\|\Delta\varphi\|_2^2 + B\|\nabla\varphi\|_2^2 + D\|\varphi\|_2^2)$$
$$\leq \|\|V\|\|^2 3^n (4\varepsilon\|\Delta\varphi\|_2^2 + (D + \tfrac{1}{4}\varepsilon^{-1}B)\|\varphi\|_2^2)$$

In the next to the last step, we have used the fact that each $x \in \mathbb{R}^n$ not on the boundary of some C_α lies in precisely 3^n of the C'_α. Then, in the last step we used,

$$\|\nabla\varphi\|_2^2 \leq \delta\|\Delta\varphi\|_2^2 + \tfrac{1}{4}\delta^{-1}\|\varphi\|_2^2$$

which via the Plancherel theorem follows by the numerical inequality $a \leq \delta a^2 + \tfrac{1}{4}\delta^{-1}$. ∎

Notice that if V is uniformly locally L^p for some $p > n/2$ then it is automatically uniformly locally $L^{n/2}$ so we stated $p = n/2$ rather than $p \geq n/2$ in the above theorem.

Given the above criterion, the following theorem has a proof that differs only in notation from the corresponding one-dimensional result (Theorem XIII.88).

Theorem XIII.97 Let $\mathbf{a}_1, \ldots, \mathbf{a}_n$ be n independent vectors in $\mathbb{R}^n$. Let V be a real-valued function on $\mathbb{R}^n$ obeying:

(i) $V(\mathbf{x} + \mathbf{a}_i) = V(\mathbf{x})$, $i = 1, \ldots, n$.
(ii) $\int_Q |V(\mathbf{x})|^p \, d^n x < \infty$ where Q is a **basic period cell**

$$Q = \left\{ \mathbf{x} \,\middle|\, \mathbf{x} = \sum_{i=1}^n t_i \mathbf{a}_i; \, 0 \leq t_i < 1 \right\}$$

and $p = 2$ if $n \leq 3$, $p > 2$ if $n = 4$; $p = n/2$ if $n \geq 5$.

For each $\boldsymbol{\theta} \in [0, 2\pi)^n$, let $H^{(0)}(\boldsymbol{\theta})$ be the operator $-\Delta$ on $L^2(Q, d^n x) \equiv \mathscr{H}$ with the boundary conditions

$$\varphi(\mathbf{x} + \mathbf{a}_j) = e^{i\theta_j}\varphi(\mathbf{x}), \qquad \frac{\partial\varphi}{\partial y_j}(\mathbf{x} + \mathbf{a}_j) = e^{i\theta_j}\frac{\partial\varphi}{\partial y_j}(\mathbf{x}) \qquad (166)$$

for all $\mathbf{x}$ with $\mathbf{x}, \mathbf{x} + \mathbf{a}_j \in \bar{Q}$ (i.e., for $\mathbf{x}$ on suitable faces of ∂Q where y_j is the coordinate given by $\mathbf{x} = \sum y_i \mathbf{a}_i$. Let

$$\mathcal{H} = \int_{[0,\,2\pi)^n}^{\oplus} \mathcal{H}' \frac{d^n\theta}{(2\pi)^n}$$

and let $U \colon L^2(\mathbb{R}^n, d^n x) \to \mathcal{H}$ by defining U on $\mathcal{S}$ by

$$(Uf)_\theta(\mathbf{x}) = \sum_{m \in \mathbb{Z}^n} e^{-i\boldsymbol{\theta} \cdot \mathbf{m}} f(\mathbf{x} + \sum m_i \mathbf{a}_i)$$

and extended as a unitary operator to L^2.
Then:

(a) For almost all $\boldsymbol{\theta} \in [0, 2\pi)^n$, V is an $H^{(0)}(\boldsymbol{\theta})$-bounded multiplication operator on $L^2(Q, d^n x)$ with relative bound zero.
(b) $U(-\Delta + V)U^{-1} = \int_{[0,\,2\pi)^n}^{\oplus} H(\boldsymbol{\theta})\, d^n\theta/(2\pi)^n$ where $H(\boldsymbol{\theta}) = H^{(0)}(\boldsymbol{\theta}) + V$.

We remark that it is possible to prove that V is everywhere, and not just almost everywhere, $H^{(0)}(\boldsymbol{\theta})$-bounded with relative bound zero. This follows, for example, from analyticity arguments.

One consequence of this theorem is the existence of an eigenfunction expansion for H in the sense of Section XI.6:

Theorem XIII.98 Each $H(\boldsymbol{\theta})$ has a complete set of eigenfunctions $\psi_m(\boldsymbol{\theta}; x)$ with eigenvalues $E_m(\boldsymbol{\theta})$. Extend $\psi_m(\boldsymbol{\theta}; x)$ to all of $\mathbb{R}^n$ by using the boundary condition (166). For $\varphi \in \mathcal{S}(\mathbb{R}^n)$, let

$$\tilde{\varphi}(m; \boldsymbol{\theta}) = \int_{\mathbb{R}^n} \overline{\psi_m(\boldsymbol{\theta}; x)}\, \varphi(x)\, d^n x$$

Then:

(a) $\quad \displaystyle \int_{\mathbb{R}^n} |\varphi(x)|^2\, d^n x = \sum_m \int_{[0,\,2\pi)^n} |\tilde{\varphi}(m; \boldsymbol{\theta})|^2\, \frac{d^n\theta}{(2\pi)^n}$

(b) $\quad \displaystyle \varphi(x) = \frac{1}{(2\pi)^n} \sum_m \int_{[0,\,2\pi)^n} \tilde{\varphi}(m; \boldsymbol{\theta}) \psi_m(\boldsymbol{\theta}; x)\, d^n\theta$

(c) Extend $\tilde{\ }$ to $L^2(\mathbb{R}^n)$ by continuity. Then $H = -\Delta + V$ obeys

$$\widetilde{H\varphi}(m; \boldsymbol{\theta}) = E_m(\boldsymbol{\theta})\tilde{\varphi}(m; \boldsymbol{\theta})$$

for all $\varphi \in D(H)$.
(d) $\tilde{\ }$ maps $L^2(\mathbb{R}^n, d^n x)$ onto $\bigoplus_m L^2([0, 2\pi)^n, d^n\theta)$.

Proof It is possible to find a complete set of eigenvectors for $H^{(0)}(\theta)$ explicitly; namely,

$$\psi_{\mathbf{k}}^{(0)}(\theta; x) = (2\pi)^{-n/2} \exp\left[i \sum_{j=1}^{n} (\theta_j + 2\pi k_j) y_j\right]$$

for $k_j \in \mathbb{Z}^n$, where y_j is defined by $\mathbf{x} = \sum_{i=1}^{n} y_i \mathbf{a}_i$. The corresponding eigenvalues tend toward infinity as $|\mathbf{k}| \to \infty$ so $H^{(0)}(\theta)$ has compact resolvent. It follows that $H(\theta)$ has compact resolvent by Theorem XIII.68. Thus it has discrete eigenvalues $\{E_m(\theta)\}_{m=1}^{\infty}$ and a corresponding complete set of eigenfunctions. By the min–max principle, one can prove that the functions $E_m(\theta)$ are measurable and that the corresponding eigenfunctions can be chosen measurably (Problem 140). Since, for θ fixed, the $\psi_m(\theta)$ are an orthonormal basis in $\mathscr{H}'$ of eigenfunctions for $H(\theta)$, we have for $\eta \in \mathscr{H} = \int_{[0, 2\pi)^n}^{\oplus} \mathscr{H}' \, d^n\theta/(2\pi)^n$,

$$(\eta, \eta)_{\mathscr{H}} = \sum_m \int |(\eta_\theta, \psi_m(\theta))_{\mathscr{H}'}|^2 \frac{d^n\theta}{(2\pi)^n}$$

$$\eta_\theta = \sum_m (\psi_m(\theta), \eta_\theta)_{\mathscr{H}'} \psi_m(\theta)$$

$$(\psi_m(\theta), (A\eta)_\theta) = E_m(\theta)(\psi_m(\theta), \eta_\theta)$$

where

$$A = \int_{[0, 2\pi)^n}^{\oplus} H(\theta) \frac{d^n\theta}{(2\pi)^n}$$

(a)–(c) now follow using the definition of U and the way we have extended $\psi_m(\theta)$. (d) has a similar proof. ∎

To give the *p*-space analysis, we need a definition that differs by a factor of 2π from the more common one.

Definition Let $\mathbf{a}_1, \ldots, \mathbf{a}_n$ be a basis for $\mathbb{R}^n$. The **dual basis** $\mathbf{K}_1, \ldots, \mathbf{K}_n$ is defined by

$$(\mathbf{K}_i, \mathbf{a}_j) = (2\pi)\delta_{ij}$$

Theorem XIII.99 Let V be a function on $\mathbb{R}^n$ with $V(\mathbf{x} + \mathbf{a}_j) = V(\mathbf{x})$ $(j = 1, \ldots, n)$ where $\{\mathbf{a}_j\}_{j=1}^{n}$ is a basis for $\mathbb{R}^n$. Let Q be the basic period cell for the basis $\{\mathbf{a}_i\}$ and let $\tilde{Q}$ be the basic period cell for the dual basis $\{\mathbf{K}_i\}$, i.e.,

$$\tilde{Q} = \left\{\sum_{i=1}^{n} t_i \mathbf{K}_i \,\bigg|\, 0 \leq t_i < 1\right\}$$

Let $\mathcal{H}' = \ell_2(\mathbb{Z}^n)$ and $\mathcal{H} = \int_{\tilde{Q}}^{\oplus} \mathcal{H}' \, d^n k$. Suppose that V is uniformly locally L^p (with $p = 2$ if $n \leq 3$, $p > 2$ if $n = 4$, $p = n/2$ if $n \geq 5$) and let $\tilde{V}_{\mathbf{m}}$ be the Fourier coefficients for V as a function on Q; i.e., for $\mathbf{m} \in \mathbb{Z}^n$,

$$\tilde{V}_{\mathbf{m}} = (\text{vol } Q)^{-1} \int_Q \exp\left(-i \sum_{j=1}^n m_j \mathbf{K}_j \cdot \mathbf{x}\right) V(\mathbf{x}) \, d^n x \qquad (167)$$

For $\mathbf{k} \in \tilde{Q}$, define the operator $H(\mathbf{k})$ on $\mathcal{H}'$ by

$$(H(\mathbf{k})g)_{\mathbf{m}} = (\mathbf{k} + \sum m_j \mathbf{K}_j)^2 g_{\mathbf{m}} + \sum_{\alpha \in \mathbb{Z}^n} \tilde{V}_\alpha g_{\mathbf{m}-\alpha} \qquad (168)$$

with domain

$$D_0 = \{g \in \mathcal{H}' \mid \sum \mathbf{m}^2 |g_{\mathbf{m}}|^2 < \infty\}$$

Finally, let $U: L^2(\mathbb{R}^n) \to \mathcal{H}$ by

$$[(Uf)(\mathbf{k})]_{\mathbf{m}} = \hat{f}(\mathbf{k} + \sum m_j \mathbf{K}_j)$$

Then U is unitary and

$$U(-\Delta + V)U^{-1} = \int_{\tilde{Q}}^{\oplus} H(\mathbf{k}) \, d^n k$$

Proof That U is unitary is just the Plancherel theorem. Moreover, it is clear that

$$[(U(-\Delta)U^{-1})g](\mathbf{k})_{\mathbf{m}} = (\mathbf{k} + \sum m_j \mathbf{K}_j)^2 g(\mathbf{k})_{\mathbf{m}}$$

since $-\widehat{\Delta f}(\ell) = \ell^2 \hat{f}(\ell)$. Because V is $-\Delta$-bounded with relative bound zero, we need only prove that

$$[(UVU^{-1}g)(\mathbf{k})]_{\mathbf{m}} = \sum_{\alpha \in \mathbb{Z}^n} \tilde{V}_\alpha g_{\mathbf{m}-\alpha}(\mathbf{k})$$

and this follows if we prove that, for $f \in \mathcal{S}(\mathbb{R}^n)$,

$$\widehat{Vf}(\mathbf{k}) = \sum_{\alpha \in \mathbb{Z}^n} \tilde{V}_\alpha \hat{f}\left(\mathbf{k} - \sum_{j=1}^n \alpha_j \mathbf{K}_j\right) \qquad (169)$$

To prove (169) we need only show that, as a tempered distribution, V has the Fourier transform

$$\hat{V}(\mathbf{k}) = (2\pi)^{n/2} \sum_{\alpha \in \mathbb{Z}^n} \tilde{V}_\alpha \, \delta\left(\mathbf{k} - \sum_{j=1}^n \alpha_j \mathbf{K}_j\right)$$

As in the one-dimensional case, this is true because the Fourier series

$$V(\mathbf{x}) = \sum_{\alpha \in \mathbb{Z}^n} \tilde{V}_\alpha \exp\left(i \sum_{j=1}^n \alpha_j \mathbf{K}_j \cdot \mathbf{x}\right)$$

is locally L^2 convergent since V is locally uniformly L^2. ∎

XIII.16 Schrödinger operators with periodic potentials

One advantage of the k-space decomposition in the form we have presented it is that the operators $H(\mathbf{k})$ have a fixed domain D_0. In fact, we can use (168) to define $H(\mathbf{k})$ for any $\mathbf{k} \in \mathbb{C}^n$ and $H(\mathbf{k})$ so defined is an entire analytic family of type (A). The easiest way of seeing this is to let

$$(\mathbf{P}g)_\mathbf{m} = \left(\sum_{j=1}^n m_j \mathbf{K}_j\right) g_\mathbf{m}$$

so that $\mathbf{P}$ is $H(0)$-bounded with relative bound zero and

$$H(\mathbf{k}) = H(0) + 2\mathbf{k} \cdot \mathbf{P} + \sum_{j=1}^n k_j^2 \tag{170}$$

$H(\mathbf{k})$ depends on n parameters; but it is useful to fix $n-1$ of them for two reasons. First, we avoid the nonanalyticity that can occur in real multiparameter eigenvalue variations. The second reason is more subtle. Let $\mathbf{a}$ and $\mathbf{b}$ be fixed vectors in $\mathbb{R}^n$. Let $z = \lambda + iy$ and define

$$E_\mathbf{m}(z) = (\mathbf{a} + z\mathbf{b} + \sum m_j \mathbf{K}_j)^2$$

Of course $H(\mathbf{k}) = H_0(\mathbf{k}) + V$ where $H_0(\mathbf{a} + z\mathbf{b})$ has a complete orthogonal set of eigenvectors with eigenvalues $E_\mathbf{m}(z)$. Now

$$\text{Im } E_\mathbf{m}(z) = 2y[\mathbf{b} \cdot (\mathbf{a} + \lambda\mathbf{b} + \sum m_j \mathbf{K}_j)]$$

is especially simple if the numbers $\mathbf{b} \cdot \mathbf{K}_j$ are rationally dependent for in that case the number in square brackets will not get arbitrarily small as $\mathbf{m}$ varies. It is thus convenient to pick $\mathbf{b}$ as the first vector in the x-space lattice, i.e.,

$$\mathbf{b} \cdot \mathbf{K}_j = 2\pi \delta_{j1} \tag{171}$$

and λ by

$$\mathbf{b} \cdot (\mathbf{a} + \lambda\mathbf{b}) = \pi \tag{172}$$

In that case, $\text{Im } E_\mathbf{m}(z) = 2\pi y(2m_1 + 1)$, so

$$|\text{Im } E_\mathbf{m}(z)| \geq \pi|y|(1 + |m_1|) \tag{173a}$$

Moreover, it is easy to see that (Problem 141a)

$$|\text{Re}(E_\mathbf{m}(z) + 1)| \geq c_1|\mathbf{m}|^2 \quad \text{if} \quad |\mathbf{m}| \geq c_2(1 + |y|) \tag{173b}$$

for suitable c_1 and c_2 (which will depend on $\mathbf{a}$ through the choice of λ in (172)). From (173), one deduces that (Problem 141):

Lemma 1 Let $n \geq 2$. Let **b** be given by (171) and λ by (172). Then

(a) If $\alpha > \frac{1}{2}n$ and $\alpha \geq n - 1$,
$$f_\alpha(y) \equiv \sum_{\mathbf{m}} |E_{\mathbf{m}}(\lambda + iy) + 1|^{-\alpha}$$
is convergent and bounded for $|y| \geq 1$.

(b) If moreover $\alpha > n - 1$, then $\lim_{y \to \pm \infty} f_\alpha(y) = 0$.

The point of Lemma 1 is that it will allow us to control $\|V(H_0(\lambda + iy) + 1)^{-1}\|$ as $y \to \infty$ for suitable V's.

Lemma 2 Let $n \geq 2$. Let V be a periodic potential whose Fourier coefficients (167) obey
$$\sum_{\mathbf{m}} |\tilde{V}_{\mathbf{m}}|^\beta < \infty \tag{174}$$
where $\beta < (n-1)/(n-2)$ if $n \geq 3$ and $\beta = 2$ if $n = 2$. Fix $\mathbf{a} \in \mathbb{R}^n$ and let **b** obey (171). Let
$$A(t) = H(\mathbf{a} + \mathbf{b}t)$$
for $t \in \mathbb{R}$. Then:

(a) Each $A(t)$ has compact resolvent.
(b) There are real analytic functions $\{E_j(t)\}$ and corresponding analytic vector-valued functions $\psi_j(t)$ so that $\psi_j(t)$ is a basis for $\mathcal{H}' = \ell^2(\mathbb{Z}^n)$ and, for each t,
$$A(t)\psi(t) = E_j(t)\psi_j(t)$$
(c) No $E_j(t)$ is constant.

Proof (a) Since $\beta \leq 2$, (174) and the Hausdorff–Young inequality imply that V is in $L^\alpha(Q)$ where $\alpha < n - 1$ if $n \geq 3$ and $\alpha = 2$ if $n = 2$. Thus V is $H_0(\mathbf{k})$ relatively bounded so it suffices to prove that $H_0(\mathbf{k})$ has compact resolvent. This follows from the fact that $H_0(\mathbf{k})$ has a complete set of eigenvectors with eigenvalues going to infinity.

(b) Let $E_j(0)$ be the eigenvalues of $A(t=0)$, ordered so that $E_1(0) \leq E_2(0) \leq \cdots$. Now, we can continue the eigenvalues and eigenvectors using, if necessary, degenerate perturbation theory. As in the one-dimensional case, we need only show that $E_j(t)$ does not go to infinity at some finite t to conclude that a continuation is possible for all t. By (170)
$$\frac{dA(t)}{dt} = 2(\mathbf{b} \cdot \mathbf{P} + \mathbf{b} \cdot (\mathbf{a} + t\mathbf{b}))$$

so, by first-order perturbation theory, and the fact that $\mathbf{P}$ is $H_0(\mathbf{k})$-bounded,

$$\left|\frac{dE_j(t)}{dt}\right| \leq c(|E_j(t)| + |t| + 1)$$

from which it follows that $E_j(t)$ cannot blow up at finite *real* t.

(c) By Lemma 1, the hypothesis on V, Hölder's inequality and Young's inequality,

$$\lim_{y \to \infty} \|(A_0(\lambda + iy) + 1)^{-1}\| = 0$$

$$\lim_{y \to \infty} \|V[A_0(\lambda + iy) + 1]^{-1}\| = 0$$

where λ is chosen so that (172) holds. From these results and standard perturbation arguments we see that for $|y| \geq Y_0$, $(A(\lambda + iy) + 1)^{-1}$ exists and

$$\lim_{y \to \infty} \|[A(\lambda + iy) + 1]^{-1}\| = 0 \tag{175}$$

Now suppose that some $E_j(t)$ is a constant C for all t. Since $A(t)$ has compact resolvent for all real t, it has a compact resolvent for all $t \in \mathbb{C}$ so that C is always an eigenvalue of $A(t)$. It follows that $(C + 1)^{-1}$ is always an eigenvalue of $(A(t) + 1)^{-1}$ so that

$$\|(A(t) + 1)^{-1}\| \geq (C + 1)^{-1}$$

This violates (175), so no $E_j(t)$ can be constant. ∎

We are now able to give a complete analysis of the spectral properties of Schrödinger operators with periodic potentials:

Theorem XIII.100 Let V be a periodic potential whose Fourier *series* coefficients are in ℓ_β where $\beta < (n-1)/(n-2)$ if $n \geq 3$, $\beta = 2$ if $n = 2, 3$. Then $-\Delta + V$ has purely absolutely continuous spectrum.

Proof $\mathbf{b}, \mathbf{K}_2, \ldots, \mathbf{K}_n$ form a basis for $\mathbb{R}^n$, so we can write the interior of $\tilde{Q}$ in terms of a decomposition $\mathbf{k} = s_1 \mathbf{b} + \cdots + s_n \mathbf{K}_n$ as $\{\langle s_1, s_\perp \rangle | s_1 \in M(s_\perp), s_\perp \in N\}$ where for every $s_\perp \in N$, $M(s_\perp)$ is an open connected set (Figure XIII.17). We can write

$$H = \int_{s_\perp \in N} \int_{s_1 \in M(s_\perp)} H(s_1 \mathbf{b} + \cdots + s_n \mathbf{K}_n)\, ds_1\, d^{n-1}s$$

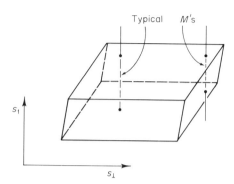

FIGURE XIII.17 The s decomposition of $\tilde{Q}$.

in a suitable direct integral decomposition. By Lemma 2 and Theorem XIII.86, the s_1 direct integral has purely absolutely continuous spectrum for each $s_\perp \in N$ and thus, by Theorem XIII.85f, H has purely absolutely continuous spectrum. ∎

We note that as in the one-dimensional case, the spectrum of H breaks up into "bands," but there are two big differences. First, due to degeneracy there may be some ambiguity in definition at singularities in the many-variable functions. Secondly, bands can "overlap" unlike the one-dimensional case.

Before discussing some of the connections between the ideas above and a simple model of solid state physics, we want to mention an arbitrariness in choice of $\mathbf{a}_1, \ldots, \mathbf{a}_n$. Given V, what is determined without any choice is

$$\mathscr{L}_V = \{\mathbf{a} \mid V(\mathbf{x} + \mathbf{a}) = V(\mathbf{x}), \text{ a.e. in } \mathbf{x}\}$$

For periodic potentials, $\mathscr{L}_V$ is always a lattice where:

Definition A **lattice** is a subset $\mathscr{L}$ of $\mathbb{R}^n$ obeying:

(i) $\mathscr{L}$ is discrete, i.e., it has no finite limit points.
(ii) $\mathscr{L}$ is a subgroup of the additive group of $\mathbb{R}^n$.
(iii) No proper vector subspace of $\mathbb{R}^n$ contains $\mathscr{L}$.

Any such $\mathscr{L}$ has a **basis**, i.e., a set $\mathbf{a}_1, \ldots, \mathbf{a}_n \in \mathscr{L}$, so that any $\mathbf{a} \in \mathscr{L}$ is uniquely of the form $\mathbf{a} = \sum_{i=1}^n m_i \mathbf{a}_i$ with $m_i \in \mathbb{Z}$. Such bases are not unique and the corresponding basic period cell Q is not unique; for example, see Figure XIII.18. What all basic cells have in common is the property that $\mathbb{R}^n = \bigcup_{\mathbf{a} \in \mathscr{L}} \tau_\mathbf{a} Q$, where $\tau_\mathbf{a} S = \{\mathbf{x} + \mathbf{a} \mid \mathbf{x} \in S\}$, with $\tau_\mathbf{a} Q^{\text{int}} \cap \tau_\mathbf{b} Q^{\text{int}} = \varnothing$, if

XIII.16 Schrödinger operators with periodic potentials

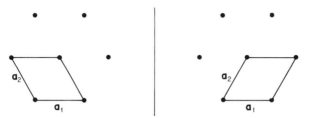

FIGURE XIII.18 Two choices of basis and basic cell.

$\mathbf{a} \neq \mathbf{b}$. There is another "basic cell," C with this property, although it is not associated to any basis. This is the **Wigner–Seitz cell for** $\mathscr{L}$ defined by

$$C = \{x \in \mathbb{R}^n \,|\, x \text{ is closer to 0 than any other point of } \mathscr{L}\}$$

Two examples of Wigner–Seitz cells are shown in Figure XIII.19. One can

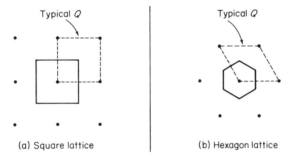

(a) Square lattice   (b) Hexagon lattice

FIGURE XIII.19 Two Wigner–Seitz cells.

show (Problem 143) that any Wigner–Seitz cell is a polyhedron, i.e., the intersection of finitely many slabs $\{x \,|\, a \leq \ell(x) \leq b\}$, ℓ a linear functional. The Wigner–Seitz cell is unique.

Similarly, the dual basis depends on the choice of basis for $\mathscr{L}_V$; but the dual lattice $\mathscr{L}'_V = \{\mathbf{k} \in \mathbb{R}^n \,|\, \mathbf{k} \cdot \mathbf{a} \in 2\pi\mathbb{Z} \text{ for all } \mathbf{a} \in \mathscr{L}\}$ and its Wigner–Seitz cell, called the **Brillouin zone** B, are independent of the basis chosen.

We mention the above terminology for the following reason. One can make an x-space direct integral decomposition with Q replaced by C and a p-space decomposition with $\tilde{Q}$ replaced by B. The study of solids in the physics literature usually begins with a construction that is a disguised form of the p-space direct integral decomposition over the Brillouin zone.

We now turn to applications of our analysis to solid state physics. To the mathematical physicist who is used to looking at atomic physics or even quantum field theory, the bewildering array of approximations known as solid state theory often appears to be more an art than a science. While there is some truth in this attitude, we would like to emphasize that the difference between atomic physics and solid state physics is really one of degree, for the "standard" purely Coulomb atomic Hamiltonian is an approximation to "real" atoms. In the first place, relativistic corrections to the kinetic energy are not included nor is spin–orbit coupling. In addition, experiments on atoms are not done with isolated atoms but with aggregates, so it is an approximation to discuss a simple atomic model and then compare it with experiment. Finally, there are the couplings to the radiation field (quantum electrodynamics) which are certainly not understood at a fundamental level. The big difference between atomic and solid state physics is that in atomic physics one model describes the most basic physical phenomena, while in solid state physics the model as described below explains qualitatively only a limited range of phenomena. Many phenomena require one to take into account lattice vibrations ("phonons") and interactions between electrons and of the electrons with the phonons. In the end, the mathematical physicist is presented with a well-defined model (or several well-defined models) to study and this is all that he or she can reasonably demand.

It is an observed phenomenon that the nuclei in a solid lie more or less in regular arrays (crystals), i.e., there is a lattice in $\mathbb{R}^n$ so that the nuclei more or less lie at the lattice points. No one has given an explanation from first principles of why crystals form; i.e., no one has proven that a large number of heavy nuclei with enough electrons to produce neutrality, interacting via Coulomb potentials, have a ground state that is approximately a crystal. We thus postulate in our model that there is a *fixed* nucleus with a number of core electrons at each site of a lattice. To obtain simplicity we replace a large solid by one filling all of $\mathbb{R}^n$. Thus if we ignore electron–electron interactions, we have electrons moving under a Hamiltonian $-\Delta + V$ with V periodic. This model is known as the **one-electron model of solids**. We want to use our analysis of periodic Schrödinger operators to describe two things:

(1) the notion of density of states and the qualitative explanation of the difference between metals and insulators;
(2) impurity scattering in the one electron model.

For simplicity of notation we suppose that space is three dimensional.

Definition Let $\tilde{Q}$ be a basic period cell in the dual lattice and let $E_n(\mathbf{k})$ be the energy levels of $H(\mathbf{k})$ (ordered by $E_1 \leq E_2 \leq \cdots$). The **density of states measure** ρ is the measure on $\mathbb{R}$ defined by

$$\rho(-\infty, E] = \frac{2}{|\tilde{Q}|} \sum_n \left| \{\mathbf{k} \in \tilde{Q} \,|\, E_n(\mathbf{k}) \leq E\} \right| \tag{176}$$

where $|\tilde{Q}|$ is the Lebesgue measure of $\tilde{Q}$ and $|\{\cdots\}|$ is the Lebesgue measure of $\{\cdots\}$.

We note that since $E_n(\mathbf{k}) \to \infty$ uniformly in $\mathbf{k}$ as $n \to \infty$ (Problem 144), the number $\rho(-\infty, E]$ is finite. Moreover, one can show easily (Problem 145) from our general analysis that ρ is absolutely continuous with respect to dE, Lebesgue measure on $\mathbb{R}$. The Radon–Nikodym derivative $d\rho/dE$ is usually called the **density of states**. To explain the importance of ρ to an analysis of solids, we introduce another notion. Let Q be the basic x-space cell and, given $m \in \mathbb{Z}$, let $Q^{(m)}$ be the set of volume $m^3 |Q|$ obtained by stacking up an $m \times m \times m$ set of Q's. Let H_m be the operator $-\Delta_P + V$ on $L^2(Q^{(m)})$ where $-\Delta_P$ denotes periodic boundary conditions. Let $P_m(\Omega)$ be the spectral projections for H_m and define

$$\rho_m(-\infty, E] = 2 \dim P_m(-\infty, E]/m^3$$

Then:

Theorem XIII.101 As $m \to \infty$, $\rho_m \to \rho$ in the sense that $\rho_m(-\infty, E] \to \rho(-\infty, E]$ for every E.

Proof We sketch the main ideas, leaving the details to the reader (Problem 147). The key point is that H_m has a direct *sum* decomposition described as follows. Parametrize $\tilde{Q}$ as $\{\sum_{i=1}^{3} t_i \mathbf{K}_i \,|\, 0 \leq t_i < 1\}$. Then

$$L^2(Q^{(m)}) \cong \bigoplus_{\alpha_1, \alpha_2, \alpha_3 = 0}^{m-1} \ell^2(\mathbb{Z}^3)$$

in such a way that H_m becomes

$$\bigoplus_{\alpha_1, \alpha_2, \alpha_3 = 0}^{m-1} H\left(\frac{\alpha_1}{m} \mathbf{K}_1 + \frac{\alpha_2}{m} \mathbf{K}_2 + \frac{\alpha_3}{m} \mathbf{K}_3\right)$$

where $H(\mathbf{k})$ are the fibers of the infinite volume operator H. The reason that this decomposition holds is that one shows that any φ periodic on $Q^{(m)}$ is a sum of φ's with $\varphi(\mathbf{x} + \mathbf{a}_j) = e^{2\pi \alpha_j i/m} \varphi(\mathbf{x})$. As a result of this decomposition,

$$\rho_m(-\infty, E] = 2m^{-3} \,\#\{n; \alpha_i \in \{0, 1, \ldots, m-1\} \,|\, E_n(m^{-1} \sum (\alpha_j \mathbf{K}_j)) \leq E\}$$

and, since $E(\cdot)$ is continuous, this expression is an approximation for $\rho(-\infty, E]$. ∎

Now we return to our model of solids. Suppose that each nucleus in free space is surrounded by ℓ electrons. Then in our model we wish to have ℓ electrons per unit cell. While we ignore interactions between the electrons, we cannot ignore the Pauli principle which asserts for noninteracting electrons that each eigenvalue of H can contain at most two electrons. How do we take this into account when H does not have eigenfunctions and when there are infinitely many electrons (in our infinite crystal lattice!)? We claim that a reasonable way of taking the Pauli principle into account is to say that in the ground state, the electrons fill up the continuum eigenstates up to that energy E where $\rho(-\infty, E] = \ell$. For if we have a large but finite $m \times m \times m$ crystal with periodic boundary conditions, there are $m^3 \ell$ electrons, and in the ground state these fill up the eigenstates of H_m to an energy E_m determined by $\rho_m(-\infty, E_m) = \ell$. The smallest number E with $\rho(-\infty, E] = \ell$ is called the **Fermi energy**, E_F. The set of $\mathbf{k} \in B$, the Brillouin zone, with $E_n(\mathbf{k}) = E_F$ for some n is called the **Fermi surface**. This picture is similar to the elementary discussion of the periodic table based on the hydrogen atom but with the complication of continuum states.

We are now in a position to explain why electron conduction is hard in some solids (insulators) and easy in others (metals). In the ground state one can use complex conjugation symmetry to prove that there is no net movement of electrons. To get flow of electrons one must excite some of the electrons. We have seen that typically a periodic Schrödinger operator has gaps in its spectrum. There is a qualitative difference if E_F occurs at the bottom of a gap or not. If E_F is at the bottom of a gap, then H has no spectrum in $(E_F, E_F + \varepsilon)$ and there is a discrete amount of energy needed to

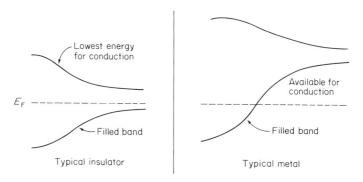

FIGURE XIII.20 Energy bands in conductors and insulators.

set up a current (see Figure XIII.20). In this case one has an insulator. If E_F is not at the bottom of a gap, one has a metal! Of course, if E_F is at the bottom of a small gap (ε small) or if E_F is not at the bottom of a gap but is fairly close to the bottom of a gap, then one has an intermediate case where the metal/insulator distinction is not sharp (semimetals, semiconductors) and in dealing with real solids one must take into account the fact that the solid is not in the ground state but rather in a finite temperature state determined by statistical mechanics. Notice that the gaps in the spectrum are crucial for this theory of insulators versus metals.

As a final topic in the one-electron theory of solids, we mention impurity scattering. Suppose that one of the lattice points has an impurity atom instead of the kind of atom at all the other sites. An electron in this crystal experiences a potential $V + W$ where V is periodic and W, which represents the difference of the potential of the impurity and what it replaces, is short range. One expects electrons in such a crystal to scatter from the impurity according to the usual scattering theory formalism.

Theorem XIII.102 Let V be a periodic potential on $\mathbb{R}^3$ that is square integrable over a basic cell. Let W be a potential in $L^1 \cap L^2(\mathbb{R}^3)$. Then

$$\Omega^\pm = \operatorname*{s-lim}_{t \to \mp \infty} \exp[+it(H_0 + V + W)] \exp[-it(H_0 + V)]$$

exist and have identical ranges and, in particular, the S matrix is unitary.

Proof Since $H_0 + V$ has purely absolutely continuous spectrum (Theorem XIII.100), we can prove the theorem by showing that $(-\Delta + V + W + c)^{-1} - (-\Delta + V + c)^{-1}$ is trace class for some c, for then we can apply the Kato–Birman theory (Theorem XI.9). Choose c so that $-\Delta + V + W \geq -c + 1$, $-\Delta + V \geq -c + 1$. Since V and W are $-\Delta$-bounded with relative bound zero,

$$(-\Delta + V + W + c)^{-1} - (-\Delta + V + c)^{-1}$$
$$= -(-\Delta + V + W + c)^{-1} W (-\Delta + V + c)^{-1}$$
$$= -[(-\Delta + V + W + c)^{-1}(-\Delta + 1)][(-\Delta + 1)^{-1} W (-\Delta + 1)^{-1}]$$
$$\times [(-\Delta + 1)(-\Delta + V + c)^{-1}]$$

The first and third factors are bounded operators by the relative boundedness and since $W \in L^1 \cap L^2$, the middle factor is trace class (see Theorem XI.20). Thus the difference of resolvents is trace class. ∎

XIII.17 An introduction to the spectral theory of non-self-adjoint operators

Thus far we have been mainly concerned with the spectral analysis of self-adjoint operators. In this final section we wish to say something about the spectral analysis of compact operators that need not be self-adjoint. In Section VI.5 we have already seen that if A is a compact operator, then $\sigma(A)$ consists of zero and a set of nonzero numbers $\{\lambda_i\}_{i=1}^{N(A)}$, where $N(A)$ is finite or countably infinite. Only zero can be an accumulation point. Moreover, each λ_i is an eigenvalue, and, by the analysis of Sections XII.1 and XII.2, each spectral projection

$$P_i = -(2\pi i)^{-1} \int_{|\lambda - \lambda_i| = \varepsilon} (A - \lambda)^{-1} \, d\lambda$$

is finite dimensional and Ran $P_i = \{\psi \,|\, (A - \lambda_i)^n \psi = 0 \text{ for some } n\}$. We call a vector ψ a **generalized eigenvector** if $(A - \lambda)^n \psi = 0$ for some $n \geq 1$. Among the natural questions are:

(1) Do the eigenvectors and generalized eigenvectors of A (vectors in the range of some P_i or solutions of $A^n \psi = 0$ for some n) span $\mathscr{H}$, i.e., are they a total set?

(2) Suppose that A is in the trace class $\mathscr{I}_1$ defined in Section VI.6. Let $\lambda_i(A)$ be a listing of the nonzero eigenvalues of A. Each eigenvalue is repeated a number of times equal to dim P_i, its **algebraic multiplicity**. Is it true that

$$\operatorname{Tr}(A) = \sum_{j=1}^{N(A)} \lambda_j(A) \tag{177}$$

(3) By the meromorphic Fredholm theorem, Theorem XIII.13, $(1 + \mu A)^{-1}$ is a meromorphic function of μ on all of $\mathbb{C}$. Can one find explicit functions, entire in μ, with simple expressions in terms of A so that $(1 + \mu A)^{-1} = F(\mu)/G(\mu)$?

We shall discuss a solution of problem 3 in the case where A is trace class and use this solution to discuss problems 1 and 2. We close the section with a discussion of "explicit" solutions to two integral equations which have arisen previously in Volume I and in this volume. Problem 2 is not as obvious as it seems. In the first place it is not a priori clear that the sum converges. Moreover, if A has only zero as eigenvalue, it is far from obvious that Tr(A), which is defined as a sum of diagonal matrix elements, is zero. Finally, as regards Problem 1, consider the following example:

Example 1 Let $\{\varphi_n\}_{n=1}^\infty$ be an orthonormal basis of $\mathscr{H}$. Let $\{\alpha_n\}_{n=1}^\infty$ be a sequence of positive numbers converging monotonically to zero. Let

$$A = \sum_{n=1}^\infty \alpha_n(\varphi_n, \cdot)\varphi_{n+1}$$

A is compact since

$$\left\| A - \sum_{n=1}^N \alpha_n(\varphi_n, \cdot)\varphi_{n+1} \right\| = \sup_{n \geq N+1} |\alpha_n| \to 0$$

Moreover

$$\|A^n\|^{1/n} = \left(\prod_{m=1}^n |\alpha_m| \right)^{1/n}$$

goes to zero as $n \to \infty$. Thus, by the spectral radius formula, Theorem VI.6, $\sigma(A) = \{0\}$. Suppose all $\alpha_n \neq 0$. Since it is then clear that $\operatorname{Ker}(A^n) = \{0\}$, A has no eigenvalues, and the span of the eigenvectors is $\{0\}$.

Example 2 Suppose that $\{\psi_n\}_{n=-\infty}^\infty$ is a basis of $\mathscr{H}$ and $\{\alpha_n\}$ is a two-sided sequence of nonzero numbers going to zero at $\pm\infty$. Let

$$A = \sum_{n=-\infty}^\infty \alpha_n(\psi_n, \cdot)\psi_{n+1}$$

Then

$$B = A^{-1} \equiv \sum_{n=-\infty}^\infty \alpha_n^{-1}(\psi_{n+1}, \cdot)\psi_n$$

is a densely defined operator on

$$D(B) = \{\eta \mid \sum \alpha_n^{-2} |(\psi_{n+1}, \eta)|^2 < \infty\}$$

Note that since $\sigma(A) = \{0\}$, $(B - \lambda)^{-1} = A(1 - \lambda A)^{-1}$. Thus, B is a closed operator with all of $\mathbb{C}$ as its resolvent set! (See also Example 5 in Section VIII.1.)

Thus, problem 1 does not have an affirmative answer in all cases. Later, we shall show that if A is a strictly m-accretive operator that is trace class, then its generalized eigenvectors span $\mathscr{H}$.

Henceforth, $\{\lambda_n(A)\}_{n=1}^{N(A)}$ will be a listing of all nonzero eigenvalues of A, a compact operator. Each eigenvalue appears a number of times equal to its algebraic multiplicity. We order them by requiring that $|\lambda_{n+1}(A)| \leq |\lambda_n(A)|$ and $\arg \lambda_{n+1}(A) \geq \arg \lambda_n(A)$ if $|\lambda_{n+1}(A)| = |\lambda_n(A)|$ where $\arg \lambda_i \in [0, 2\pi)$. As a preliminary, we shall prove that the sum $\sum_{n=1}^{N(A)} \lambda_n(A)$ converges absolutely if $A \in \mathcal{I}_1$.

Lemma 1 Let A be any compact operator. Then there exists an orthonormal set $\{e_n\}_{n=1}^{N(A)}$, so that

$$Ae_n = \lambda_n(A)e_n + \sum_{m=1}^{n-1} v_{nm} e_m \qquad (178)$$

for suitable v_{nm}. In particular,

$$(e_n, Ae_n) = \lambda_n(A) \qquad (179)$$

Proof Let P_i be the spectral projections for nonzero eigenvalues. By writing a Jordan normal form for $A \upharpoonright \operatorname{Ran} P_i$ for each i, we can find a set of algebraically independent vectors, $\{f_j\}_{j=1}^{N(A)}$ so that

$$Af_n = \lambda_n(A)f_n + \beta_n f_{n-1} \qquad (180)$$

where β_n is either 1 or 0. By applying a Gram–Schmidt procedure to $\{f_n\}_{n=1}^{N(A)}$, we find an orthonormal set $\{e_n\}_{n=1}^{N(A)}$ so that

$$e_n = \sum_{m=1}^{n} \gamma_{nm} f_m \qquad (181)$$

with $\gamma_{nn} \neq 0$. (178) follows from (180) and (181). ∎

The orthonormal set $\{e_n\}_{n=1}^{N(A)}$ is called a **Schur basis** (although it need not be a basis!).

Now let $\{\mu_n(A)\}$ be the singular values of A, i.e., the eigenvalues of $|A|$.

Theorem XIII.103 (Schur–Lalesco–Weyl theorem) For any $1 \leq p < \infty$,

$$\sum_{n=1}^{N(A)} |\lambda_n(A)|^p \leq \sum_{n=1}^{\infty} \mu_n(A)^p \qquad (182)$$

where A is compact, $\{\lambda_n(A)\}$ are its eigenvalues, and $\{\mu_n(A)\}$ its singular values. In particular, if $A \in \mathcal{I}_1$,

$$\sum_{n=1}^{N(A)} |\lambda_n(A)| < \infty$$

Proof Consider the canonical expansion for A

$$A = \sum_{n=1}^{\infty} \mu_n(A)(f_n, \cdot)g_n$$

where $\{f_n\}$ and $\{g_n\}$ are orthonormal sets (Theorem VI.17). Let $\{e_n\}_{n=1}^{N(A)}$ be a Schur basis for A. Then, by (179),

$$\lambda_m(A) = \sum_{n=1}^{\infty} \alpha_{mn} \mu_n(A)$$

where

$$\alpha_{mn} = (f_n, e_m)(e_m, g_n)$$

Now, we claim that

$$\sum_{n=1}^{\infty} |\alpha_{mn}| \leq 1 \tag{183}$$

$$\sum_{m=1}^{\infty} |\alpha_{mn}| \leq 1 \tag{184}$$

To prove (184), note that by the Schwarz inequality,

$$\sum_{m=1}^{\infty} |\alpha_{mn}| \leq \left(\sum_{m=1}^{\infty} |(f_n, e_m)|^2\right)^{1/2} \left(\sum_{m=1}^{\infty} |(g_n, e_m)|^2\right)^{1/2}$$

$$\leq \|f_n\| \|g_n\| = 1$$

where we have used Bessel's inequality in the last step together with the fact that $\{e_m\}$ is an orthonormal set. (183) follows similarly using the fact that $\{g_n\}$ and $\{f_n\}$ are orthonormal sets. Now, let q be the dual index to p. Then

$$\sum_{m=1}^{M} |\lambda_m(A)|^p \leq \sum_{m=1}^{M} |\lambda_m(A)|^{p-1} \sum_{n=1}^{\infty} |\alpha_{mn}| |\mu_n(A)|$$

$$= \sum_{n,m} |\alpha_{nm}|^{1/p} |\alpha_{nm}|^{1/q} |\lambda_m|^{p-1} |\mu_n|$$

$$\leq \left(\sum_{n,m} |\alpha_{nm}| |\mu_n|^p\right)^{1/p} \left(\sum_{n,m} |\alpha_{nm}| |\lambda_m|^p\right)^{1/q}$$

$$\leq \left(\sum_{n} |\mu_n|^p\right)^{1/p} \left(\sum_{m=1}^{M} |\lambda_m|^p\right)^{1/q}$$

In the next to the last step, we have used Hölder's inequality and the fact that $q(p-1) = p$. In the last step we used (183) and (184). (182) now follows easily. ∎

Before turning to the main topic of this section, we note one consequence of (182).

Corollary (generalized Golden–Thompson inequality) Let $-A$ and $-B$ be positive self-adjoint operators with $A + B$ essentially self-adjoint on $D(A) \cap D(B)$. Suppose that $e^{A/2} e^B e^{A/2} \in \mathscr{I}_p$. Then $e^{A+B} \in \mathscr{I}_p$ and

$$\|e^{A+B}\|_p \leq \|e^{A/2} e^B e^{A/2}\|_p$$

Proof We first prove that if C and D are bounded positive operators and $CD \in \mathscr{I}_r$, then $C^{1/2} D C^{1/2} \in \mathscr{I}_r$ and

$$\|C^{1/2} D C^{1/2}\|_r \leq \|CD\|_r$$

First note that for any bounded E, F, we have $\sigma(EF) \setminus \{0\} = \sigma(FE) \setminus \{0\}$ (Problem 166a) and if $\lambda \neq 0$ is an eigenvalue of EF it is an eigenvalue of FE with the same multiplicity (Problem 166b). Since $CD = C^{1/2}(C^{1/2} D) \in \mathscr{I}_r$, it has purely discrete spectrum away from 0, so the self-adjoint operator $C^{1/2} D C^{1/2}$ has purely discrete spectrum away from zero and thus, is a compact operator. By the above, $\lambda_n(C^{1/2} D C^{1/2}) = \lambda_n(CD)$ and since $C^{1/2} D C^{1/2}$ is self-adjoint and positive, $\lambda_n(C^{1/2} D C^{1/2}) = \mu_n(C^{1/2} D C^{1/2})$. Thus, using (182),

$$\|C^{1/2} D C^{1/2}\|_r^r = \sum_n |\lambda_n(CD)|^r$$

$$\leq \sum_n |\mu_n(CD)|^r = \|CD\|_r^r$$

Now we claim that $Q_n = (e^{A/2^{n+1}} e^{B/2^n} e^{A/2^{n+1}})^{2^n} \in \mathscr{I}_p$ with

$$\|Q_n\|_p \leq \|e^{A/2} e^B e^{A/2}\|_p$$

We prove this inductively. $n = 0$ is trivial. Let $C_n = e^{A/2^n}$, $D_n = e^{B/2^n}$ and let $r_n = 2^n p$. Then if $Q_{n-1} \in \mathscr{I}_p$, $C_n D_n \in \mathscr{I}_{r_n}$ and

$$\|Q_n\|_p = \|C_n^{1/2} D_n C_n^{1/2}\|_{r_n}^{2^n} \leq \|C_n D_n\|_{r_n}^{2^n} = \|Q_{n-1}\|_p$$

proving the bound on $\|Q_n\|_p$ inductively.

By the Trotter product formula, $Q_n \to e^{A+B}$ strongly so that (Problem 167) $e^{A+B} \in \mathscr{I}_p$ and

$$\|e^{A+B}\|_p \leq \overline{\lim} \|Q_n\|_p \leq \|e^{A/2} e^B e^{A/2}\|_p \quad \blacksquare$$

Our first goal will be a proof of (177). In proving this we shall develop a theory of infinite determinants which is useful in other contexts. These determinants will depend on our development of a little alternating algebra, i.e., the theory of antisymmetric tensor products. In Sections II.4 and VIII.10 we presented the basic definition of these objects as "Fermion Fock spaces," but we repeat the definition here with slightly different notation. Given a Hilbert space $\mathscr{H}$, $\otimes^n \mathscr{H}$ is defined as the vector space of multilinear functionals on $\mathscr{H}$. Explicitly, given $\varphi_1, \ldots, \varphi_n \in \mathscr{H}$, we define $\varphi_1 \otimes \cdots \otimes \varphi_n \in \otimes^n \mathscr{H}$ by

$$(\varphi_1 \otimes \cdots \otimes \varphi_n)(\langle \eta_1, \ldots, \eta_n \rangle) = (\varphi_1, \eta_1) \cdots (\varphi_n, \eta_n)$$

One can show (see Proposition 1 in Section II.4) that the finite span of the $\{\varphi_1 \otimes \cdots \otimes \varphi_n\}$ possesses a well-defined inner product with

$$((\varphi_1 \otimes \cdots \otimes \varphi_n), (\eta_1 \otimes \cdots \otimes \eta_n)) = (\varphi_1, \eta_1) \cdots (\varphi_n, \eta_n)$$

$\otimes^n \mathscr{H}$ is the completion of this finite span in the topology generated by this inner product. Given any $A \in \mathscr{L}(\mathscr{H})$, there is a natural operator $\Gamma_n(A)$ in $\mathscr{L}(\otimes^n \mathscr{H})$ with

$$\Gamma_n(A)(\varphi_1 \otimes \cdots \otimes \varphi_n) = A\varphi_1 \otimes \cdots \otimes A\varphi_n$$

Γ_n is a functor, that is, $\Gamma_n(AB) = \Gamma_n(A)\Gamma_n(B)$.

Let $\mathscr{P}_n$ denote the group of all permutations on n letters. Let $\varepsilon(\cdot)$ be the function on $\mathscr{P}_n$ that is $+1$ (respectively, -1) on even (respectively, odd) permutations. Define $\varphi_1 \wedge \cdots \wedge \varphi_n \in \otimes^n \mathscr{H}$ by

$$\varphi_1 \wedge \cdots \wedge \varphi_n = (n!)^{-1/2} \sum_{\pi \in \mathscr{P}_n} \varepsilon(\pi)[\varphi_{\pi(1)} \otimes \cdots \otimes \varphi_{\pi(n)}] \tag{185}$$

and define $\bigwedge^n(\mathscr{H})$ to be the subspace of $\otimes^n \mathscr{H}$ spanned by the $\{\varphi_1 \wedge \cdots \wedge \varphi_n\}$. The $(n!)^{-1/2}$ normalization factor is chosen so that if $\varphi_1, \ldots, \varphi_n$ are orthornormal, then $\varphi_1 \wedge \cdots \wedge \varphi_n$ has norm one. More generally, from (185) one can see that (Problem 149)

$$(\varphi_1 \wedge \cdots \wedge \varphi_n, \eta_1 \wedge \cdots \wedge \eta_n) = \det((\varphi_i, \eta_j)) \tag{186}$$

where $\det(a_{ij}) = \sum_{\pi \in \mathscr{P}_n} \varepsilon(\pi) a_{1\pi(1)} \cdots a_{n\pi(n)}$.

Given $A \in \mathscr{L}(\mathscr{H})$, $\Gamma_n(A)$ leaves $\bigwedge^n \mathscr{H}$ invariant, and we denote its restriction to $\bigwedge^n \mathscr{H}$ by $\bigwedge^n(A)$. Since Γ_n is a functor, so is $\bigwedge^n$.

$$\bigwedge^n(AB) = \bigwedge^n(A)\bigwedge^n(B) \tag{187}$$

When $n = 0$, we define $\bigwedge^n \mathscr{H}$ to be $\mathbb{C}$ and $\bigwedge^n(A)$ as $1 : \mathbb{C} \to \mathbb{C}$.

The connection between determinants of finite-dimensional operators and $\bigwedge^n(\cdot)$ is given by:

Lemma 2

(a) Let $\mathscr{H}$ be an n-dimensional Hilbert space. Then $\bigwedge^n \mathscr{H}$ is one dimensional.

(b) If $\mathscr{H}$ has dimension n, then $\bigwedge^n(A)$ is multiplication by the number $\det(A)$, the ordinary determinant of A.
(c) $\det(AB) = \det(A)\det(B)$.

Proof (a) We shall show more generally that $\dim \bigwedge^k \mathscr{H} = \binom{n}{k}$, the number of ways of choosing k objects from among n objects. For let $e_1, \ldots, e_n$ be an orthonormal basis for $\mathscr{H}$. We claim that

$$\{e_{i_1} \wedge \cdots \wedge e_{i_k} \mid 1 \leq i_1 < i_2 < \cdots < i_k \leq n\}$$

is an orthonormal basis for $\bigwedge^k \mathscr{H}$; this proves $\dim \bigwedge^k \mathscr{H} = \binom{n}{k}$. For the vectors in question are orthonormal and so independent, and they span $\bigwedge^k \mathscr{H}$ since $\langle \varphi_1, \ldots, \varphi_k \rangle \to \varphi_1 \wedge \cdots \wedge \varphi_k$ is multilinear and antisymmetric under interchanges.

(b) Since $\bigwedge^n(\mathscr{H})$ is one dimensional, $\bigwedge^n(A)$ must be multiplication by some number α. If $e_1, \ldots, e_n$ is an orthonormal basis, then

$$\alpha = (e_1 \wedge \cdots \wedge e_n, \bigwedge^n(A)(e_1 \wedge \cdots \wedge e_n))$$
$$= (e_1 \wedge \cdots \wedge e_n, Ae_1 \wedge \cdots \wedge Ae_n)$$
$$= \det((e_i, Ae_j))$$

by (186).

(c) follows from (b) and (187). ∎

This rather effortless proof of $\det(AB) = \det(A)\det(B)$ shows the power of alternating algebra in studying determinants.

To understand the definition that we shall take for $\det(1 + A)$ when $A \in \mathscr{I}_1$, suppose that A is an operator on a finite-dimensional space $\mathscr{H}$ with $\dim \mathscr{H} = n$. Let $\lambda_1, \ldots, \lambda_n$ be the eigenvalues of A and let $e_1, \ldots, e_n$ be a Schur basis for A. Then, it is easy to see that

$$\det(1 + A) = (e_1 \wedge \cdots \wedge e_n, (1+A)e_1 \wedge \cdots \wedge (1+A)e_n) = \prod_{j=1}^{n}(1 + \lambda_j)$$

and

$$\operatorname{Tr}(\bigwedge^k(A)) = \sum_{1 \leq i_1 < \cdots < i_k \leq n} ((e_{i_1} \wedge \cdots \wedge e_{i_k}), (Ae_{i_1} \wedge \cdots \wedge Ae_{i_k}))$$
$$= \sum_{1 \leq i_1 \cdots \leq n} \lambda_{i_1} \cdots \lambda_{i_k}$$

so that

$$\det(1 + A) = \sum_{j=0}^{n} \operatorname{Tr}(\bigwedge^j(A)) \tag{188}$$

in case dim $\mathcal{H} = n$. In the case dim $\mathcal{H} = \infty$, we shall *define* $\det(1 + A)$ by (188). To prove that the sum converges, we need the following lemma:

Lemma 3 Let A be a trace class operator with singular values $\mu_n(A)$. Then, for any k, $\bigwedge^k(A)$ is trace class and

(a) $$\|\textstyle\bigwedge^k(A)\|_1 = \sum_{i_1 < \cdots < i_k} \mu_{i_1}(A) \cdots \mu_{i_k}(A)$$

(b) $$\|\textstyle\bigwedge^k(A)\|_1 \leq \|A\|_1^k / k!$$

Proof Let $A = U|A|$ be the polar decomposition for A (Theorem VI.10). It is easy to see that $\bigwedge^k(A) = \bigwedge^k(U)\bigwedge^k(|A|)$ is the polar decomposition for $\bigwedge^k(A)$ and in particular that

$$|\textstyle\bigwedge^k(A)| = \bigwedge^k(|A|)$$

Let $e_1, \ldots$ be the orthonormal basis of eigenvectors for $|A|$. Then $e_{i_1} \wedge \cdots \wedge e_{i_k}$ are an orthonormal basis of eigenvectors for $\bigwedge^k(|A|)$. Thus

$$\begin{aligned}\operatorname{Tr}(|\textstyle\bigwedge^k(A)|) &= \sum_{i_1 < \cdots < i_k} \mu_{i_1}(A) \cdots \mu_{i_k}(A) \\ &\leq \frac{1}{k!} \sum_{i_1, \ldots, i_k} \mu_{i_1}(A) \cdots \mu_{i_k}(A) \\ &= [\operatorname{Tr}(|A|)]^k / k!\end{aligned}$$

Thus $\bigwedge^k(A)$ is trace class and (a) and (b) hold. ∎

Definition Let $A \in \mathcal{I}_1$. Then $\det(1 + A)$ is defined by

$$\det(1 + A) = \sum_{k=0}^{\infty} \operatorname{Tr}(\textstyle\bigwedge^k(A)) \tag{188}$$

Lemma 4 The sum (188) converges for each $A \in \mathcal{I}_1$. Moreover:

(a) $|\det(1 + A)| \leq \prod_{j=1}^{\infty} (1 + \mu_j(A))$.
(b) $|\det(1 + A)| \leq \exp(\|A\|_1)$.
(c) For any $A_1, \ldots, A_n \in \mathcal{I}_1$,

$$\langle z_1, \ldots, z_n \rangle \to \det\left(1 + \sum_{i=1}^{n} z_i A_i\right)$$

is an entire analytic function.

(d) For any $A, B \in \mathcal{I}_1$
$$|\det(1 + A) - \det(1 + B)| \leq \|A - B\|_1 \exp(\|A\|_1 + \|B\|_1 + 1) \quad (189)$$

Proof Since $|\text{Tr}(\bigwedge^k(A))| \leq \|\bigwedge^k(A)\|_1/k!$, the convergence of (188) and (a) and (b) follow from Lemma 3. Since we have estimates on the terms of (188) that are uniform on compact sets in $\mathbb{C}^n$, it is sufficient to prove that
$$\langle z_1, \ldots, z_n \rangle \to \text{Tr}(\bigwedge^k(z_1 A_1 + \cdots + z_k A_k))$$
is analytic. Clearly, $\bigwedge^k(z_1 A_1 + \cdots + z_k A)$ is analytic as an $\mathcal{I}_1$-valued function, so its trace is analytic for $\text{Tr}(\cdot)$ is a bounded linear functional on $\mathcal{I}_1$. This proves (c). (d) follows from (b) and (c) and the theorem immediately below. ∎

Theorem XIII.104 Let X be a complex Banach space. Let $F: X \to \mathbb{C}$ be a function with the following properties:

(i) For any $x, y \in X$, $\mu \to F(x + \mu y)$ is an entire function of μ.
(ii) For some monotone increasing function G on $[0, \infty)$,
$$|F(x)| \leq G(\|x\|)$$
for all $x \in X$.

Then
$$|F(x) - F(y)| \leq \|x - y\| G(\|x\| + \|y\| + 1) \quad (190)$$

Proof Fix x and y in X and let $f: \mathbb{C} \to \mathbb{C}$ be given by
$$f(\mu) = F(\tfrac{1}{2}(x + y) + \mu(y - x))$$
so that f is an entire function by (a). Notice that
$$|F(x) - F(y)| = |f(-\tfrac{1}{2}) - f(\tfrac{1}{2})| \leq \sup_{-1/2 \leq t \leq 1/2} |f'(t)| \quad (191)$$

Now, by the Cauchy integral formula,
$$f'(t) = (2\pi i)^{-1} \oint_{|\mu - t| = \|x - y\|^{-1}} \frac{f(\mu) \, d\mu}{(\mu - t)^2}$$
so that
$$\sup_{|t| \leq 1/2} |f'(t)| \leq \|x - y\| \sup_{|\mu| \leq 1/2 + \|x - y\|^{-1}} |f(\mu)| \quad (192)$$

For $|\mu| \leq \frac{1}{2} + \|x - y\|^{-1}$, we have that
$$\|\tfrac{1}{2}(x + y) + \mu(y - x)\| \leq \|x\| + \|y\| + 1$$
so, for such μ,
$$|f(\mu)| \leq G(\|x\| + \|y\| + 1) \tag{193}$$
by (ii). (191)–(193) imply (190). ∎

Suppose that $A \in \mathscr{I}_1$. The idea in proving that $\text{Tr}(A) = \sum \lambda_n(A)$ is the following: Suppose that we can prove that $\det(1 + zA)$ has the convergent expansion $\prod_{j=1}^{N(A)} (1 + z\lambda_j(A))$. The term linear in z in the product is clearly $\sum_{j=1}^{N(A)} \lambda_j(A)$ while the linear term in $\det(1 + zA)$ is by definition $\text{Tr}(A)$. Thus, we want to study the function $\det(1 + zA)$. We first summarize the properties of $\det(1 + A)$.

Theorem XIII.105 Let A and B be in $\mathscr{I}_1$. Then:
(a) $\det(1 + A)\det(1 + B) = \det(1 + A + B + AB)$. (194)
(b) $(1 + A)$ is invertible if and only if $\det(1 + A) \neq 0$.
(c) If $-\mu^{-1}$ is an eigenvalue of A, then $\det(1 + zA)$ has a zero of order n at $z = \mu$ where n is the algebraic multiplicity of $-\mu^{-1}$.
(d) For any ε, there is a C_ε, depending on A, so that
$$|\det(1 + zA)| \leq C_\varepsilon \exp(\varepsilon |z|)$$

Proof (a) Since the finite rank operators are dense in $\mathscr{I}_1$ (corollary to Theorem VI.21) and $\det(1 + \cdot)$ is continuous by (189), we need only verify (194) for finite rank A and B. Let V be the span of $\text{Ker}(A)^\perp$, $\text{Ker}(B)^\perp$, Ran A, and Ran B. Then V is finite dimensional, invariant under A, B, A^*, and B^*, and A and B are zero on $V^\perp$. Thus A and B leave V and $V^\perp$ invariant. Let $\tilde{A} = A \upharpoonright V$, $\tilde{B} = B \upharpoonright V$. Then $\text{Tr}(\bigwedge^k(A)) = \text{Tr}(\bigwedge^k(\tilde{A}))$, etc. Therefore,
$$\det(1 + A) = \det_V(1 + \tilde{A})$$
and similarly for B, and $A + B + AB$, where $\det_V$ is the determinant on V. (194) now follows from the finite-dimensional result, Lemma 2c.

(b) If $1 + A$ is invertible, then $(1 + A)^{-1} = 1 + B$ with $B = -A(1 + A)^{-1} \in \mathscr{I}_1$. Thus, by (194),
$$\det(1 + A)\det(1 + B) = \det(1) = 1$$
so $\det(1 + A) \neq 0$. If $1 + A$ is not invertible, then -1 is an eigenvalue of A, so $\det(1 + A) = 0$ by the result (c) that we are about to prove.

(c) Let P be the spectral projection of $-\mu^{-1}$. Let $B = AP$ and $C = A(1 - P)$. Then

$$1 + zA = (1 + zB)(1 + zC)$$

so $\det(1 + zA) = \det(1 + zB) \det(1 + zC)$. Since $1 + zC$ is invertible for z near $-\mu^{-1}$, it suffices to show that $\det(1 + zB)$ has an nth-order zero at $-\mu^{-1}$. By extending a Schur basis for B to a basis for $\mathcal{H}$, we can find an orthonormal basis, with

$$Be_i = \lambda_i e_i + \sum_{j=1}^{i-1} \alpha_{ij} e_j$$

with $\lambda_1 = \cdots = \lambda_n = -\mu^{-1}$ and $\lambda_{n+1} = \cdots = 0$. We just use the fact that Ran $B \subset P =$ span of the Schur basis. It follows easily that $\mathrm{Tr}(\bigwedge^k(B)) = \binom{n}{k}$ $(-\mu^{-1})^k$ for $k \le n$ and $\mathrm{Tr}(\bigwedge^k(B)) = 0$ if $k > n$. Thus, $\det(1 + zB) = (1 - z\mu^{-1})^n$ has an nth-order zero at $z = \mu$.

(d) Let $\mu_n(A)$ be the singular values of A. Choose N so that

$$\sum_{n > N} \mu_n(A) < \varepsilon/2$$

Then, by Lemma 4a,

$$|\det(1 + zA)| \le \prod_{j=1}^{\infty} (1 + |z|\mu_j(A))$$

$$\le \prod_{j=1}^{N} (1 + |z|\mu_j(A)) \exp(\tfrac{1}{2}\varepsilon |z|)$$

since $1 + x \le e^x$ for $x \ge 0$. Now, since $\prod_{j=1}^{N} (1 + |z|\mu_j(A))$ is a polynomial, we can find C_ε with

$$\prod_{j=1}^{N} (1 + |z|\mu_j(A)) \le C_\varepsilon \exp(\tfrac{1}{2}\varepsilon |z|) \quad \blacksquare$$

Theorem XIII.106 For any $A \in \mathcal{I}_1$,

$$\det(1 + A) = \prod_{j=1}^{N(A)} (1 + \lambda_j(A))$$

where $\{\lambda_j(A)\}_{j=1}^{N(A)}$ are the eigenvalues of A counted with algebraic multiplicity.

Proof Let $f(z) = \det(1 + zA)$. Let

$$g(z) = \prod_{j=1}^{N(A)} (1 + z\lambda_j(A))$$

If $N(A) = \infty$, the product in question converges to an analytic function by standard results (see Problem 150) because

$$\sum_{j=1}^{\infty} |\lambda_j(A)| < \infty$$

by Theorem XIII.103. We will show that $f(z) = g(z)$. By (b) and (c) of Theorem XIII.105, f and g have the same zeros including order, so f/g is an entire nonvanishing analytic function. Thus,

$$f = ge^h$$

where the ambiguity in h is determined by requiring that h be entire with $h(0) = 0$. This is possible since $f(0) = g(0) = 1$. We shall show that $h(z) = 0$ for $|z| < 1$, so h will be identically zero. For $R \geq 2$, and $|z| < R$ define

$$h_R(z) = \ln[f_R(z)]$$
$$k_R(z) = - \sum_{\{j \mid |\lambda_j|^{-1} > R\}} \ln(1 + z\lambda_j(A))$$
$$f_R(z) = f(z) / \prod_{\{j \mid |\lambda_j|^{-1} \leq R\}} (1 + z\lambda_j)$$

The ambiguities in h_R and k_R are determined by setting $h_R(0) = k_R(0) = 0$. Notice also that f_R is an entire function. Since $h = h_R + k_R$, it suffices to show that $|h_R(z)| \to 0$ and $|k_R(z)| \to 0$ as $R \to \infty$ for each z with $|z| \leq 1$.

Since $\ln(1 + x)$ vanishes at $x = 0$ and is analytic in a neighborhood of $\{x \mid |x| \leq \frac{1}{2}\}$, we have

$$|\ln(1 + x)| \leq C|x|$$

for suitable C and all x with $|x| \leq \frac{1}{2}$. Thus, for $R \geq 2$,

$$|k_R(z)| \leq |z| \sum_{\{j \mid |\lambda_j|^{-1} > R\}} |\lambda_j(A)| \to 0$$

as $R \to \infty$ since the infinite sum is convergent.

Next, consider the entire function $f_R(z)$. If $|z| = 2R$, and $|\lambda_j|^{-1} \leq R$, then $|1 + z\lambda_j| \geq 1$. Thus, if $|z| = 2R$, $|f_R(z)| \leq C_\varepsilon \exp(2\varepsilon R)$ by (d) of Theorem XIII.105. By the maximum modulus principle, this remains true if $|z| = R$, so

$$\mathrm{Re}(h_R(z)) \leq \ln C_\varepsilon + 2\varepsilon R$$

for $|z| = R$. We shall prove below a lemma that asserts that, for any function analytic in a neighborhood of $|z| \leq R$,

$$\max_{|z| \leq r} |f(z)| \leq \frac{2r}{R-r} \max_{|z|=R} [\text{Re}(f(z))] + \frac{R+r}{R-r} |f(0)|$$

So, since $h_R(0) = 0$,

$$\max_{|z| \leq 1} |h_R(z)| \leq 2(R-1)^{-1}[\ln C_\varepsilon + 2\varepsilon R]$$

Thus, for any z with $|z| \leq 1$,

$$\overline{\lim_{R \to \infty}} |h_R(z)| \leq 4\varepsilon$$

Since ε is arbitrary, $|h_R(z)| \to 0$ as $R \to \infty$. This proves the theorem. ∎

Corollary (Lidskii's theorem)   For any $A \in \mathscr{I}_1$

$$\text{Tr}(A) = \sum_{j=1}^{N(A)} \lambda_j(A) \tag{177}$$

where $\lambda_j(A)$ are the eigenvalues of A.

Proof Consider the Taylor expansion for $\det(1 + \mu A)$ about $\mu = 0$. By (188), the first term in the expansion defining det is $\text{Tr}(A)$. The first term in $\prod_{j=1}^{N(A)} (1 + \mu \lambda_j(A))$ is clearly $\sum_{j=1}^{N(A)} \lambda_j(A)$. ∎

One might expect to learn something from each term in the Taylor expansion for $\det(1 + \mu A)$, but the kth term in the series is just (177) for $\bigwedge^k(A)$, i.e.,

$$\text{Tr}(\bigwedge^k(A)) = \sum_{j=1}^{N(\bigwedge^k(A))} \lambda_j(\bigwedge^k(A)) = \sum_{1 \leq j_1 < \cdots < j_k \leq N(A)} \lambda_{j_1}(A) \cdots \lambda_{j_k}(A)$$

Because of this, one can deduce Theorem XIII.106 from (177).

In our proof of Theorem XIII.106, we used the following fact from complex analysis:

Lemma 5 (Borel–Carathéodory theorem)   Let f be analytic in a neighborhood of $|z| \leq R$. Then for any $r < R$,

$$\max_{|z| \leq r} |f(z)| = \frac{2r}{R-r} \max_{|z|=R} [\text{Re}(f(z))] + \frac{R+r}{R-r} |f(0)| \tag{195}$$

Proof We first claim that one can suppose that $f(0) = 0$. For if (195) holds for $h(z) = f(z) - f(0)$, it is easily seen to hold for f. Thus, suppose that $f(0) = 0$ and let $A = \max_{|x|=R}[\operatorname{Re} f(z)]$. Also, without loss suppose that f is not identically zero, so that $A > 0$. By the maximum modulus principle applied to e^f, $\operatorname{Re} f(z) \leq A$ for all z with $|z| \leq R$, so that

$$g(z) \equiv \frac{f(z)}{z(2A - f(z))}$$

is analytic for $|z| < R$. Writing $f = u + iv$, we see that if $|z| = R$,

$$|g(z)|^2 = \frac{1}{R^2} \frac{u^2 + v^2}{(2A - u)^2 + v^2} \leq \frac{1}{R^2}$$

since $|u| \leq |2A - u|$. Thus $|g(z)| \leq 1/R$ for all z with $|z| \leq R$ by the maximum modulus principle. Since

$$f(z) = \frac{2Azg(z)}{1 + zg(z)}$$

we see that, for $|z| = r$,

$$|f(z)| \leq \frac{2Ar/R}{1 - r/R} = \frac{2Ar}{R - r}$$

Again invoking the maximum modulus principle, we conclude that (195) holds. ∎

Next, we use the determinant theory that we have developed to investigate the third question, i.e., to find explicit functions $f(\mu)$ and $g(\mu)$ so that $(1 + \mu A)^{-1} = f(\mu)/g(\mu)$,

Theorem XIII.107 Let $A \in \mathscr{I}_1$. The function

$$F_A(\mu) = [\det(1 + \mu A)][1 + \mu A]^{-1}$$

defined on $\{\mu \mid -\mu^{-1} \notin \sigma(A)\}$ may be extended to all of $\mathbb{C}$ in such a way that F is an entire function. Moreover,

$$\|F_A(\mu)\| \leq \exp(|\mu|\|A\|_1) \tag{196}$$

and

$$\|F_A(1) - F_B(1)\| \leq \|A - B\|_1 \exp(\|A\|_1 + \|B\|_1 + 1) \tag{197}$$

Proof The function $G(\mu) = (1 + \mu A)^{-1}$ is analytic on $\{\mu \mid -\mu^{-1} \notin \sigma(A)\}$ by Theorem VI.5. If we can show that for $-\mu_0^{-1} \in \sigma(A)$, $G(\mu)$ has a pole of order k at $\mu = \mu_0$ where k is less than or equal to the algebraic multiplicity of $-\mu_0^{-1}$, then $\det(1 + \mu A)G(\mu)$ will be regular at $\mu = \mu_0$ since $\det(1 + \mu A)$ has a zero of order equal to that multiplicity. Given such a μ_0, let P be the spectral projection associated to $-\mu_0^{-1}$. Let $B = AP$ and $C = A - B$. Then

$$(1 + \mu A)^{-1} = (1 + \mu B)^{-1} P + (1 + \mu C)^{-1}(1 - P)$$

The second term on the right-hand side is nonsingular at μ_0 since $-\mu_0^{-1} \notin \sigma(C)$. Since B is a finite rank operator, we can explicitly invert $1 + \mu B$ by using the theory of finite-dimensional determinants and cofactor matrices. This realizes $(1 + \mu B)^{-1}$ as a quotient whose denominator $\det(1 + \mu B)$ is a polynomial of degree d where $d = \dim P$. Thus, $(1 + \mu B)^{-1}$ has a pole of order at most d.

It remains to prove (196) and (197). By a simple limiting argument (Problem 151), it suffices to prove (196) in the finite-dimensional case when, in addition, $1 + \mu A$ is invertible. Since $1 + \mu A = U|1 + \mu A|$ with U unitary we have

$$\|(1 + \mu A)^{-1} \det(1 + \mu A)\| = \| |1 + \mu A|^{-1} \det(|1 + \mu A|)\|$$

Let $\lambda_1, \ldots, \lambda_k$ be the eigenvalues of $|1 + \mu A|$ with $\lambda_1 \geq \cdots \geq \lambda_k$. Then

$$\| |1 + \mu A|^{-1} \det(|1 + \mu A|)\| = \lambda_k^{-1} \left[\prod_{i=1}^{k} \lambda_i\right] = \prod_{i=1}^{k-1} \lambda_i$$

Let $\alpha_1, \ldots, \alpha_k$ be the eigenvalues of $|A|$. One can show (Problem 158) that $\lambda_i \leq 1 + |\mu|\alpha_i$. Thus

$$\|(1 + \mu A)^{-1} \det(1 + \mu A)\| \leq \prod_{i=1}^{k-1} (1 + |\mu|\alpha_i)$$

$$\leq \prod_{i=1}^{k-1} \exp(|\mu|\alpha_i) \leq \exp(|\mu| \operatorname{Tr} |A|)$$

This proves (196). (197) now follows from (196) and Theorem XIII.104 applied to the functions $(\varphi, F_A(1)\psi)$ (Problem 170). ∎

One point of the estimates (196) and Lemma 4b is that they allow one to estimate errors made in truncating the power series expansions of $\det(1 + \mu A)$ and $(1 + \mu A)^{-1} \det(1 + \mu A)$. (Problem 152). It is therefore of interest to find their Taylor coefficients. We already have an expansion of $\det(1 + \mu A)$ in terms of $\bigwedge^k(A)$ and a similar expansion is possible for the function

$$D_\mu(A) \equiv A(1 + \mu A)^{-1} \det(1 + \mu A)$$

in terms of partial traces of $\bigwedge^k(A)$ (Problem 153). This expansion is essentially an abstraction of an expansion used by Fredholm in his famous paper on integral equations. Notice that $D_\mu(A)$ is defined so that

$$(1 + \mu A)^{-1} = 1 - \mu[D_\mu(A)/\det(1 + \mu A)]$$

It is not especially easy to compute $\bigwedge^k(A)$ in terms of A, so that one would like expressions for the Taylor coefficients of $\det(1 + \mu A)$ and $D_\mu(A)$ in terms of quantities like A, A^2, ... and their traces.

Lemma 6 Let A be a fixed element in $\mathcal{I}_1$. Then for $|\mu|$ small, the series $\sum_{k=1}^\infty \mu^k \operatorname{Tr}((-A)^k)/k$ converges and

$$\det(1 + \mu A) = \exp\left[-\sum_{k=1}^\infty \mu^k \operatorname{Tr}[(-A)^k]/k\right]$$

Proof Since

$$|\operatorname{Tr}[(-A)^k]| \leq \|A^k\|_1 \leq \|A\|_1 \|A\|^{k-1}$$

the series converges if $|\mu| \|A\|_{\text{op}} < 1$. Moreover, by the product expansion for det and Lidskii's theorem (Theorem XIII.106 and its corollary), we have that, for $|\mu| \max_{1 \leq j \leq N(A)} |\lambda_j(A)| < 1$,

$$\ln[\det(1 + \mu A)] = \sum_{j=1}^{N(A)} \ln(1 + \mu \lambda_j(A))$$

$$= \sum_{j=1}^{N(A)} \sum_{k=1}^\infty (-1)^{k+1} \mu^k \lambda_j(A)^k/k$$

$$= \sum_{k=1}^\infty (-1)^{k+1} \mu^k \left[\sum_{j=1}^{N(A)} \lambda_j(A)^k\right]/k$$

$$= -\sum_{k=1}^\infty \mu^k \operatorname{Tr}((-A)^k)/k$$

In the above we have used the convergent expansion for $\ln(1 + x)$ on $\{x \mid |x| < 1\}$. The interchanging of the sums is easy to justify by showing that the double sum is absolutely convergent because $\sum |\lambda_j(A)| < \infty$ and $|\mu| \max |\lambda_j(A)| < 1$. ∎

332 XIII: SPECTRAL ANALYSIS

Lemma 7 Let $f(z)$ be analytic for z small with

$$f(z) = \sum_{n=1}^{\infty} (-1)^{n+1} b_n \frac{z^n}{n}$$

Let

$$g(z) \equiv \exp(f(z)) = \sum_{m=0}^{\infty} B_m \frac{z^m}{m!}$$

Then $B_0 = 1$ and B_m is given by the $m \times m$ determinant:

$$B_m = \begin{vmatrix} b_1 & m-1 & 0 & \cdots & 0 \\ b_2 & b_1 & m-2 & \cdots & 0 \\ b_3 & b_2 & b_1 & \cdots & 0 \\ \vdots & \vdots & \vdots & \vdots & \vdots \\ b_{m-1} & b_{m-2} & b_{m-3} & \cdots & 1 \\ b_m & b_{m-1} & b_{m-2} & \cdots & b_1 \end{vmatrix} \quad (198)$$

Proof Since $g'(z) = f'(z)g(z)$, the power series are related by

$$B_n = \sum_{k=1}^{n} b_k B_{n-k} (-1)^{k+1} \left(\frac{(n-1)!}{(n-k)!} \right) \quad (199)$$

(198) clearly holds for $m = 1$; and if it holds for $B_1, \ldots, B_{m-1}$, then (199) is just the expansion in minors in the first column on the right-hand side of (198). Thus (199) holds by induction. ∎

Theorem XIII.108 (the Plemelj–Smithies formulas) Define $\alpha_m(A)$ and $\beta_m(A)$ by

$$\det(1 + \mu A) = \sum_{m=0}^{\infty} \mu^m \frac{\alpha_m(A)}{m!}$$

$$D_\mu(A) = \sum_{m=0}^{\infty} \mu^m \frac{\beta_m(A)}{m!}$$

Then $\alpha_m(A)$ is given by the $m \times m$ determinant:

$$\alpha_m(A) = \begin{vmatrix} \mathrm{Tr}(A) & m-1 & 0 & \cdots & 0 \\ \mathrm{Tr}(A^2) & \mathrm{Tr}(A) & m-2 & \cdots & 0 \\ \mathrm{Tr}(A^3) & \mathrm{Tr}(A^2) & \mathrm{Tr}(A) & \cdots & 0 \\ \vdots & \vdots & \vdots & \vdots & \vdots \\ & & & & 1 \\ \mathrm{Tr}(A^m) & \mathrm{Tr}(A^{m-1}) & \mathrm{Tr}(A^{m-2}) & \cdots & \mathrm{Tr}(A) \end{vmatrix} \quad (200)$$

and $\beta_m(A)$ is given by the $(m+1) \times (m+1)$ determinant

$$\beta_m(A) = \begin{vmatrix} A & m & 0 & \cdots & 0 \\ A^2 & \operatorname{Tr}(A) & m-1 & \cdots & 0 \\ A^3 & \operatorname{Tr}(A^2) & \operatorname{Tr}(A) & \cdots & 0 \\ \vdots & \vdots & \vdots & & \vdots \\ A^m & \operatorname{Tr}(A^{m-1}) & \operatorname{Tr}(A^{m-2}) & \cdots & 1 \\ A^{m+1} & \operatorname{Tr}(A^m) & \operatorname{Tr}(A^{m-1}) & \cdots & \operatorname{Tr}(A) \end{vmatrix} \qquad (201)$$

where (201) is to be interpreted in the sense that $(\varphi, \beta_m(A)\psi)$ is given by the numerical determinant obtained by replacing A^j by $(\varphi, A^j\psi)$ on the right-hand side of (201).

Proof (200) follows directly from Lemmas 6 and 7. For small μ, we have

$$D_\mu(A) = (A - \mu A^2 + \mu^3 A^2 - \cdots) \det(1 + \mu A)$$

on account of the geometric series expansion of $A(1 + \mu A)^{-1}$. Thus

$$\beta_m(A) = m! \left[A \frac{\alpha_m(A)}{m!} - A^2 \frac{\alpha_{m-1}(A)}{(m-1)!} + \cdots \right]$$

Expanding the right-hand side of (201) in minors in the first column and using (200), we conclude that (201) holds. ∎

As a final abstract result, we use the machinery we have developed thus far to say something about completeness of generalized eigenvectors for a special class of operators in $\mathscr{I}_1$.

Theorem XIII.109 Let A be a trace class operator that is also strictly m-accretive, that is, there is $\varepsilon > 0$ so that

$$\operatorname{Arg}[(\varphi, A\varphi)] \leq \frac{\pi}{2} - \varepsilon$$

for all $\varphi \in \mathscr{H}$. Then the generalized eigenvectors for A span $\mathscr{H}$.

Proof We shall prove that the generalized eigenvectors associated with nonzero eigenvalues span $\overline{\operatorname{Ran} A}$ and then that $\overline{\operatorname{Ran} A} + \operatorname{Ker} A = \mathscr{H}$. Let $\mathscr{M}$ be the span of the generalized eigenvectors associated to nonzero eigenvalues. Since A leaves $\mathscr{M}$ invariant, A^* leaves $\mathscr{M}^\perp$ invariant. Let B be the restriction of A^* to $\mathscr{M}^\perp$. We shall show that $B = 0$, so that $\varphi \in \mathscr{M}^\perp$ implies that $A^*\varphi = 0$, which implies that $\varphi \in (\operatorname{Ran} A)^\perp$. So $\mathscr{M}^\perp \subset (\operatorname{Ran} A)^\perp$ and thus $\overline{\operatorname{Ran} A} \subset \mathscr{M}$.

We first claim that $\sigma(B) = \{0\}$. For suppose $\lambda \in \sigma(B)$ with $\lambda \neq 0$. Then, λ is an eigenvalue of B since B is compact as the restriction of the compact operator A^*. Therefore, there is a nonzero $\varphi \in \mathcal{M}^\perp$ with $A^*\varphi = \lambda\varphi$. It follows that $\varphi \in \text{Ran}(A - \bar{\lambda})^\perp$ and that $\bar{\lambda} \in \sigma(A)$. Let P be the spectral projection for $\bar{\lambda}$. Then $\text{Ran } P \subset \mathcal{M}$ so $\varphi \in (\text{Ran } P)^\perp$. But since $\bar{\lambda} \notin \sigma(A \restriction (1 - P)\mathcal{H})$, $\text{Ran}(1 - P) \subset \text{Ran}(A - \bar{\lambda})$ so $\varphi \in \text{Ran}(1 - P)^\perp$. It follows that $\varphi = 0$ since $\text{Ran } P + \text{Ran}(1 - P) = \mathcal{H}$. Thus, B can contain only zero in its spectrum.

Let $\varphi \in \mathcal{M}^\perp$. Define the vector-valued analytic function

$$F(\mu) = (1 - \mu B)^{-1}\varphi = -\mu^{-1}(B - \mu^{-1})^{-1}\varphi$$

Since $\sigma(B) = \{0\}$, F is an entire function. Moreover, since B has no nonzero eigenvalues, $\det(1 - \mu B) = 1$ by Theorem XIII.106, and so, by (196),

$$\|F(\mu)\| \leq \exp(|\mu|\|B\|_1)\|\varphi\| \tag{202}$$

Since B is m-accretive, we know by Theorem VIII.17 that

$$\|(B - \lambda)^{-1}\| \leq [\text{dist}(\lambda, \{z \mid |\arg z| \leq \tfrac{1}{2}\pi - \varepsilon\})]^{-1}$$

Thus, for $|\arg \lambda| \geq \tfrac{1}{2}\pi - \tfrac{1}{2}\varepsilon$,

$$\|(B - \lambda)^{-1}\varphi\| \leq |\lambda|^{-1}\csc(\tfrac{1}{2}\varepsilon)\|\varphi\|$$

It follows, that for $|\arg \mu| \geq \tfrac{1}{2}\pi - \tfrac{1}{2}\varepsilon$,

$$\|F(\mu)\| \leq \alpha \tag{203}$$

where $\alpha = \|\varphi\|\csc(\tfrac{1}{2}\varepsilon)$. We claim that (203) holds for all μ. For let $\beta = \tfrac{1}{2}\pi[\tfrac{1}{2}\pi - \tfrac{1}{3}\varepsilon]^{-1}$. For any $C > 0$, the function $F_C(\mu) = F(\mu)\exp(-C\mu^\beta)$ is analytic in the set $D = \{\mu \mid |\arg \mu| \leq \tfrac{1}{2}\pi - \tfrac{1}{2}\varepsilon\}$ and goes to zero uniformly as $|\mu| \to \infty$ by (202) and the fact $\beta > 1$ and $\beta(\tfrac{1}{2}\pi - \tfrac{1}{2}\varepsilon) < \tfrac{1}{2}\pi$. Thus, by the maximum modulus principle $|F_C(\mu)|$ takes its maximum values on $\arg \mu = \pm(\tfrac{1}{2}\pi - \tfrac{1}{2}\varepsilon)$ where it is bounded by α. Thus, for any $C > 0$ and $\mu \in D$, $|F_C(\mu)| \leq \alpha$, so letting $C \downarrow 0$, we conclude (203). Since (203) holds, $F(\mu)$ is a constant by Liouville's theorem and in particular, $F'(0) = B\varphi$ is zero. Thus $B = 0$, so, by the argument already given, $\overline{\text{Ran } A} \subset \mathcal{M}$.

Now let φ be arbitrary. Let $\psi_n = n^{-1}(A + n^{-1})^{-1}\varphi$ so that $\|\psi_n\| \leq \|\varphi\|$ since A is sectorial. Let ψ be a weak limit point of the ψ_n as $n \to \infty$. We claim that $A\psi = 0$ and $\varphi - \psi_n \in \text{Ran } A$. This will show that $\varphi \in \text{Ker } A + \overline{\text{Ran } A}^w$ where ^{w} is the weak closure of $\text{Ran } A$. But, by the Hahn-Banach theorem, $\overline{\text{Ran } A}^w = \overline{\text{Ran } A}$ since $\text{Ran } A$ is a subspace. Thus, the theorem will be proven if we can show $A\psi = 0$ and $\varphi - \psi_n \in \text{Ran } A$.

Fix any η. Then

$$(\eta, A\psi) = \lim_{n\to\infty} \frac{1}{n}\left(\eta, A\left(A+\frac{1}{n}\right)^{-1}\varphi\right)$$

$$= \lim_{n\to\infty} \frac{1}{n}(\eta, \varphi) - \frac{1}{n^2}\left(\eta, \left(A+\frac{1}{n}\right)^{-1}\varphi\right)$$

$$= \lim_{n\to\infty} \frac{1}{n}(\eta, \varphi - \psi_n) = 0$$

since $\|\psi_n\| \leq 1$. This proves that $A\psi = 0$. Furthermore,

$$\varphi - \psi_n = \left[1 - \frac{1}{n}\left(A+\frac{1}{n}\right)^{-1}\right]\varphi = A\left(A+\frac{1}{n}\right)^{-1}\varphi$$

so $\varphi - \psi_n$ is in Ran A. ∎

The methods of this section can sometimes be used to say something about unbounded non-self-adjoint operators.

Corollary Let A be the generator of a holomorphic contraction semigroup and suppose that $(A+1)^{-1}$ is trace class. Then the generalized eigenvectors for A span $\mathcal{H}$.

Proof By the preceding theorem, it suffices to show that $(A+1)^{-1}$ is strictly m-accretive since the generalized eigenvectors of A are the same as those of $(A+1)^{-1}$ (Problem 159). Let $\eta \in \mathcal{H}$ and let $\varphi = (A+1)^{-1}\eta$. Then

$$(\eta, (A+1)^{-1}\eta) = ((A+1)\varphi, \varphi) = \overline{(\varphi, A\varphi)} + (\varphi, \varphi)$$

Thus

$$|\arg(\eta, (A+1)^{-1}\eta)| \leq \tfrac{1}{2}\pi - \varepsilon$$

if

$$|\arg(\varphi, A\varphi)| \leq \tfrac{1}{2}\pi - \varepsilon$$

Since A is strictly m-accretive, so is $(A+1)^{-1}$. ∎

The explicit formula for $(1+\mu A)^{-1}$ when $A \in \mathcal{I}_1$ either in the Plemelj–Smithies form which we have discussed or in the Fredholm form, which we have not discussed, are of limited value because many compact operators of interest are not trace class although they are typically in some $\mathcal{I}_p$ with $p > 1$.

Thus, the extension of the Plemelj–Smithies formula to $\mathscr{I}_p$ is of great interest. The reader is asked to make this extension in Problem 155. The resulting formulas are as simple as those for the trace class case and error estimates are also available. In the examples below we shall sometimes make use of this extended theory.

Example 3 In Section XI.6, we found the scattering amplitude for a potential $V \in R \cap L^1$ in terms of the solution φ of

$$\varphi(x, k) = |V(x)|^{1/2} e^{ikx} - \int K(x, y) \varphi(y, k) \, dy$$

where

$$K(x, y) = (4\pi |x - y|)^{-1} |V(x)|^{1/2} e^{ik|x-y|} V(y)/|V(y)|^{1/2}$$

Notice that since $V \in R$, the operator defined by the kernel K is Hilbert–Schmidt and also that K is *not* self-adjoint. The modified Fredholm theory gives explicit formulas for $\varphi(x, k)$ and $T(k, k')$. These formulas, unlike the Born series (which converges in general only when $|k|$ is large) converge for all $k \notin \mathscr{E}$, the exceptional set.

Example 4 In an example presented at the conclusion of Section VI.5 we showed how to solve the Dirichlet problem for Laplace's equation in $D \subset \mathbb{R}^3$ in terms of the solution of an integral equation with variables in ∂D. There we discussed this equation, viewing it as one on $C(\partial D)$, but one could just as well have considered the equation as one on $L^2(\partial D, dS)$. The kernel in that case is *not* in $\mathscr{I}_1$ or even $\mathscr{I}_2$, but a simple argument shows it is in $\mathscr{I}_4$ and a more subtle argument proves it is in $\mathscr{I}_3$ (or any $\mathscr{I}_{2+\varepsilon}$ with $\varepsilon > 0$) (Problem 160). Thus, one can write down "explicit" solutions for the Dirichlet problem.

Example 5 By Theorem XIII.11, the Schrödinger operator $-d^2/dx^2 + \lambda V$ has a negative eigenvalue for all small positive λ so long as V is nonpositive, in C_0^∞, and not identically zero. The determinants we have introduced in this section turn out to be the natural tool for answering two further questions one might ask about this situation: What if V is not a.e. nonpositive? Is the eigenvalue defined for small λ analytic at $\lambda = 0$? We shall suppose throughout this example that $V \in C_0^\infty$.

By the tool used in Theorem XIII.11, $E < 0$ is an eigenvalue of $p^2 + \lambda V$ if and only if 1 is an eigenvalue of $-\lambda |V|^{1/2}(p^2 - E)^{-1} V^{1/2}$ where

$V^{1/2} = |V|^{1/2}(\operatorname{sgn} V)$. Let $E = -\alpha^2$ for $\alpha > 0$. Then $(p^2 - E)^{-1}$ is an integral operator with integral kernel $(2\alpha)^{-1} \exp(-\alpha|x - y|)$. Define the operators K_α, L_α, M_α with integral kernels:

$$K_\alpha(x, y) = (2\alpha)^{-1} |V(x)|^{1/2} \exp(-\alpha|x - y|) V^{1/2}(y)$$

$$L_\alpha(x, y) = (2\alpha)^{-1} |V(x)|^{1/2} V^{1/2}(y)$$

$$M_\alpha(x, y) = K_\alpha(x, y) - L_\alpha(x, y)$$

Then a simple argument (Problem 161) shows that M_α is trace class not only for $\alpha > 0$ but for all α and is analytic in α. By the argument above, $E < 0$ is an eigenvalue of $p^2 + \lambda V$ if and only if

$$\det(1 + \lambda K_\alpha) = 0 \tag{204}$$

for $\alpha = +\sqrt{|E|}$. Now, the eigenvalue of $p^2 + \lambda V$, if it exists, goes to zero as λ goes to zero since V is a form bounded perturbation of p^2. Since M_α is continuous at $\alpha = 0$, if $\alpha(\lambda)$ is a solution of (204), then $(1 + \lambda M_{\alpha(\lambda)})$ is invertible for all small λ. Thus

$$\det(1 + \lambda K_\alpha) = \det(1 + \lambda M_\alpha) \det(1 + \lambda L_\alpha (1 + \lambda M_\alpha)^{-1})$$

so that (204) is equivalent to

$$\det(1 + \lambda L_\alpha (1 + \lambda M_\alpha)^{-1}) = 0 \tag{205}$$

for λ small. This operator is a rank one operator and for any rank-one operator B, one has (Problem 162)

$$\det(1 + B) = 1 + \operatorname{Tr}(B)$$

Thus (205) is equivalent to

$$F(\alpha, \lambda) \equiv \alpha + \tfrac{1}{2}\lambda(V^{1/2}, (1 + \lambda M_\alpha)^{-1} |V|^{1/2}) = 0 \tag{206}$$

We can now answer the two questions we posed above. F is jointly analytic near $\langle \alpha, \lambda \rangle = \langle 0, 0 \rangle$, $(\partial F/\partial \alpha)_{\lambda=0} = 1$, $F(0, 0) = 0$. Therefore, by the implicit function theorem for analytic functions, $F(\alpha, \lambda) = 0$ has a unique solution $\alpha(\lambda)$ for λ near zero with $\alpha(\lambda)$ near 0 and this solution is analytic in λ. We conclude that $p^2 + \lambda V$ has an eigenvalue for all small positive λ if and only if $\alpha(\lambda) > 0$ for λ small and positive. In that case $E(\lambda) = -\alpha(\lambda)^2$ is analytic in λ about $\lambda = 0$. When is $\alpha(\lambda) > 0$? Computing the first two terms in the Taylor series for $\alpha(\lambda)$, we see that

$$\alpha(\lambda) = -\frac{\lambda}{2} \int V(x)\, dx - \frac{\lambda^2}{4} \int V(x) |x - y| V(y)\, dx\, dy + O(\lambda^3)$$

Notice that if the $O(\lambda)$ term in this equation is zero, then the $O(\lambda^2)$ is automatically positive because $-|x - y|$ is conditionally (strictly) positive definite (Problem 163). We summarize:

Theorem XIII.110 Let V be a nonzero function in $C_0^\infty(\mathbb{R})$. Then $-d^2/dx^2 + \lambda V$ has a negative eigenvalue for *all* positive λ if and only if $\int V(x)\,dx \leq 0$, and in that case this eigenvalue is analytic in λ at $\lambda = 0$.

For further discussion of this example including the case where V is neither smooth nor has compact support and the two-dimensional case, where the eigenvalue is *never* analytic at $\lambda = 0$, see the discussion in the Notes and the reference quoted therein.

NOTES

Section XIII.1 The min–max characterization of eigenvalues was first stated as a technical lemma in E. Fischer, "Über Quadratische Formen mit reellen Koeffizienten," *Monatsh. Math. Phys.* **16** (1905), 234–49. H. Weyl in "Das asymptotische Verteilungsgesetz der Eigenwerte linearer partieller Differentialgleichungen," *Math. Ann.* **71** (1911), 441–469, used results very close in spirit to the min–max principle. However, it was R. Courant who in the 1920s first realized the far-reaching consequences of the min–max principle and its power as a tool. The first paper in his series is "Über die Eigenwerte bei den Differentialgleichungen der mathematischen Physik," *Math. Z.* **7** (1920), 1–57.

The extension of the min–max principle from operators with compact resolvent (which was the main application made by Weyl and Courant) to arbitrary operators by using the essential spectrum is something of a folk theorem. Its applications (although not its exact statement) in print go back at least to Kato's *Trans. Amer. Math. Soc.* paper quoted in the Notes to Section 3.

Section XIII.2 Many additional applications of variational methods can be found in S. H. Gould, *Variational Methods for Eigenvalue Problems*, Univ. Toronto Press, Toronto, Canada, 1957, and scattered throughout R. Courant and D. Hilbert, *Methods of Mathematical Physics*, Vols. I, II, Wiley (Interscience), New York, 1953.

The central role of the Rayleigh–Ritz method in the history of variational methods is described by R. Courant in a review talk, "Variational methods for the solution of problems of equilibrium and vibrations," Bull.*Amer. Math. Soc.*, **49** (1943), 1–23:

> Since Gauss and W. Thompson, the equivalence between boundary value problems of partial differential equations on the one hand and problems of the calculus of variations on the other hand has been a central point in analysis. At first, the theoretical interest in existence proofs dominated and only much later were practical applications envisaged by two physicists, Lord Rayleigh and Walter Ritz; they independently conceived the idea of utilizing this equivalence for numerical calculation of the solutions, by substituting for the

variational problems simpler approximating extremum problems in which but a finite number of parameters need to be determined. Rayleigh, in his classic work—*Theory of Sound*—and in other publications, was the first to use such a procedure. But only the spectacular success of Walter Ritz and its tragic circumstances caught the general interest. In two publications in 1908 and 1909, Ritz, conscious of his imminent death from consumption, gave a masterly account of the theory, and at the same time applied his method to the calculation of the nodal lines of vibrating plates, a problem of classical physics that previously had not been satisfactorily treated.

Rayleigh's *Theory of Sound* has been reprinted by Dover, New York, 1945. The two papers by W. Ritz are "Über eine neue Methode zur Lösung gewisser Variationsprobleme der mathematischen Physik.," *J. Reine Angew. Math.* **135** (1908), 1–61, and "Theorie der Transversalschwingungen einer quadratischen Platte mit freien Rändern," *Ann. Physik.* **28** (1909), 737–786. Ritz obtained for his method convergence theorems that are similar in form to Theorem XIII.4.

Temple's inequality first appeared in G. Temple, "The theory of Rayleigh's principle as applied to continuous systems," *Proc. Roy. Soc.* **119A** (1928), 276–293. Temple actually proved several inequalities for a specific problem. The wide applicability and high accuracy of the estimate of Theorem XIII.5 was first emphasized in T. Kato, "On the upper and lower bounds of eigenvalues," *J. Phys. Soc. Japan* **4** (1949), 334–339. The simple proof we give which uses some of Kato's ideas is due to T. Kinoshita in his first paper on the helium atom (see below).

Temple's inequality is a useful tool in perturbation theory. This has been emphasized by E. Harrel, II, Thesis, Princeton Univ., Princeton, New Jersey, 1976. For example, it is easy to prove using Temple's inequality that if E_0, the ground state energy of some operator H_0, is nondegenerate and discrete, and if V is a regular perturbation, then the perturbed eigenvector $\psi_n(\lambda)$ given correctly to order n, gives the perturbed energy $E(\lambda)$ to order $2n + 1$ when placed in the expression $E(\lambda) \simeq (\psi(\lambda), \psi(\lambda))^{-1}(\psi(\lambda), H(\lambda)\psi(\lambda))$ (Problem 12).

There are a wide variety of other lower bound techniques. The pre-1950 situation is summarized by Kato: "the formula given by Temple is the most precise one among them notwithstanding that it was the oldest as well as the simplest one." Three recent techniques of interest in quantum theory are the Bazley technique, the Löwdin technique, and the Thirring technique. Bazley's technique is based on a method developed by Weinstein to compute bounds on frequencies of vibration of clamped plates. This technique first appeared in A. Weinstein, "Étude des spectres des équations aux dérivées partielles de la théorie des plaques élastique," *Mémor. Sci. Math.*, No. 88, 1937. Bazley first applied the method to quantum mechanics in N. Bazley, "Lower bounds for eigenvalues with application to the helium atom," *Phys. Rev.* **120** (1958), 144–149. The method was further developed in papers of Bazley and D. Fox, "Lower bounds for eigenvalues of Schrödinger's equation," *Phys. Rev.* **124** (1961), 483–492; "A procedure for estimating eigenvalues," *J. Mathematical Phys.* **3** (1962), 469–471; and "Lower bounds for energy levels of molecular systems," *J. Mathematical Phys.* **4** (1963), 1147–1153. Löwdin's technique was developed in P.-O. Löwdin, "Studies in perturbation theory X, XI" *Phys. Rev.* **139A** (1965), 357–372; *J. Chem. Phys.* **43S** (1965), 175–185. A comparison of the two methods as applied to an x^4 oscillator is found in C. Reid, "Lower bounds for the energy levels of anharmonic oscillators," *J. Chem. Phys.* **43S** (1965), 180–189 (see especially footnote 4a).

Thirring's bounds are given in W. Thirring, *Vorlesungen über Matematische Physik*, T7, *Quantenmechanik*, Univ. Wien Lecture Notes, Section 2.9. His bounds include Temple's inequality as a special case and also the following (Problem 13): If ψ is the ground state for H_0 and $V \geq 0$, then

$$H_0 + (\psi, V^{-1}\psi)^{-1}(\psi, \cdot)\psi \leq H_0 + V$$

In particular, if the ground state energy is discrete and nondegenerate, then the ground state energy $E(\lambda)$ of $H_0 + \lambda V$ obeys

$$\lambda(\psi, V^{-1}\psi)^{-1} \leq E(\lambda) - E(0) \leq \lambda(\psi, V\psi)$$

so long as $\lambda(\psi, V^{-1}\psi)^{-1} \leq d$, the distance of $E(0)$ from the rest of the spectrum of H_0.

Except for Thirring's bounds, these methods do not provide very satisfactory bounds for systems with a large number of particles. For atoms, it is possible to use special properties of the Coulomb potential, see P. Hertel, E. Lieb, and W. Thirring, "Lower bound to the energy of complex atoms," *J. Chem. Phys.* **62** (1975), 3355–3356.

For applications of lower bound techniques to membrane and plate problems, see Weinstein's memoir, the recent book, A. Weinstein and W. Stenger, *Methods of Intermediate Problems for Eigenvalues: Theory and Ramifications*, Academic Press, New York, 1971; H. F. Weinberger, "Lower bounds for higher eigenvalues by finite difference methods;" *Pacific J. Math.* **8** (1958), 339–368; and J. Hersch, "Lower bounds for all eigenvalues by cell functions: A refined form of H. F. Weinberger's method," *Arch. Rational Mech. Anal.* **12** (1963), 361–366.

The application of Rayleigh–Ritz methods to the helium atom was first made by G. W. Kellner, "Die Ionisierungsspannung des Heliums nach der Schrödingerschen Theorie," *Z. Phys.* **44** (1927), 91–109. Systematic development of accurate test functions are due to E. Hylleraas in three fundamental papers, "Über den Grundzustand des Helium Atoms," *Z. Phys.* **48** (1928), 469–494; "Neue Berechnung der Energie das Heliums im Grundzustande, sowie des tiefsten Terms von Ortho-Helium," *Z. Phys.* **54** (1929), 347–366; and "Über den Grundterm der Zweielektronenprobleme von H^-, He, Li^+, Be^{++}, usw.," *Z. Phys.* **65** (1930), 209–225. For a delightful personal history of the problem, see E. Hylleraas, "Reminiscences from early quantum mechanics of two-electron atoms," *Rev. Modern Phys.* **35** (1963), 421–436.

A detailed discussion of relativistic corrections to the helium atom together with their history can be found in H. Bethe and E. Salpeter, *Quantum Mechanics of One and Two Electron Atoms*, Springer-Verlag, Berlin and New York, 1957.

With the availability of very accurate experimental data due to work of G. Herzberg, interest revived in the calculation of the ionization energy of helium as a test of QED. The first computations to high accuracy appeared in T. Kinoshita, "Ground state of the helium atom, I, II," *Phys. Rev.* **105** (1957), 1490–1502; **115** (1959), 366–374. More accurate calculations can be found in C. L. Pekeris, "Ground state of two-electron atoms," *Phys. Rev.* **112** (1958), 1649–1658; "1^1S and 2^3S states of helium," *Phys. Rev.* **115** (1959), 1216–1221, and "1^1S, 2^1S and 2^3S states of H^- and He," *Phys. Rev.* **126** (1962), 1470–1476. The last paper can be consulted for experimental and Lamb shift computational references.

Section XIII.3 Theorem XIII.6 (in a slightly weaker form) appears in the book by Courant and Hilbert (Notes to Section 1). Our proof is taken from B. Simon, "On the infinitude or finiteness of the number of bound states for an N-body quantum system," *Helv. Phys. Acta* **43** (1970), 607–630. The example of the three-body system for which $N(\lambda)$ alternates between finite and infinite values is also found in this paper. Additional discussion concerning Theorem XIII.6 can be found in L. Faddeev, " On the expansion of arbitrary functions in eigenfunctions of the Schrödinger operator," *Vestnik Leningrad Univ., Mat. Meh. Astronoma* **7** (1957), 164–172, and in the Birman paper quoted below in connection with the Birman–Schwinger bound.

That the model helium Hamiltonian has an infinite discrete spectrum was proven in T. Kato, "On the existence of solutions of the helium wave equation," *Trans. Amer. Math. Soc.* **70** (1951), 212–218. Zhislin's theorem (Theorem XIII.7) appeared first in G. Zhislin, "Discussion of the spectrum of the Schrödinger operator for systems of many particles," *Trudy Moskov. Mat. Obšč.* **9** (1960), 81–128. Additional discussion of this theorem, extensions to subspaces of fixed symmetry, and alternative proofs can be found in Simon's paper and in J. Uchiyama, "On the discrete eigenvalues of the many particle systems," *Publ. Res. Inst. Math. Sci.* **A2** (1966/67), 117–132; G. Zhislin and A. Sigalov, "On the spectrum of an energy operator for atoms with fixed nuclei in subspaces corresponding to irreducible representations of permutation groups," *Izv. Akad. Nauk SSSR Ser. Mat.* **29** (1965), 853–860; and E. Balslev, "Spectral theory of Schrödinger operators of many body systems with permutation and rotation symmetries," *Ann. Phys.* **73** (1972), 49–107.

There are a variety of situations where one can prove that a multiparticle system has finitely many eigenvalues in its discrete spectrum including the case of certain negatively charged ions. For example, see J. Uchiyama, "Finiteness of the number of discrete eigenvalues of the Schrödinger operator for a three particle system," *Publ. Res. Inst. Math. Sci.* **A5** (1969), 51–63; G. M. Zhislin, "On the finiteness of the discrete spectrum of the energy operator of negative atomic and molecular ions," *Teoret. Mat. Fiz.* **21** (1971), 332–341 (*Theoret. and Math. Phys.* **7** (1971), 571–578); D. R. Yafaev, "The point spectrum in the quantum-mechanical problem of many particles," *Funkcional. Anal. i Priloẑeh* **6** (1972), 103–104 (*Functional. Anal. Appl.* **6** (1972), 349–350); M. A. Antonets, G. M. Zhislin, and I. A. Shereshevskii, "On the discrete spectrum of the Hamiltonian of an N-particle quantum system," *Teoret. Mat. Fiz.* **16** (1973), 235–246 (*Theoret and Math. Phys.* **16** (1974), 800–809); I. Sigal, "On the point spectrum of the Schrödinger operators of multiparticle systems," *Commun. Math. Phys.* **48** (1976), 137–154; B. Simon, "Geometric methods in multiparticle quantum systems," CMP (1977); and R. Hill, "Proof that the H^- ion has only one bound state," *Phys. Rev. Lett.* **38** (1977), 643–646.

Theorem XIII.8 is a classical result, although its proof usually rests on a detailed analysis of the behavior of the zeros of $u_\ell(r; E)$ rather than on the min–max principle. The idea of applying Theorem XIII.8 to obtain bounds on $n_\ell(V)$ and the bound in Theorem XIII.9a appears in V. Bargmann, "On the number of bound states in a central field of forces," *Proc. Nat. Acad. Sci. U.S.A.* **38** (1952), 961–966. Bargmann's work was partially motivated by a theorem in R. Jost and A. Pias, "On the scattering of a particle by a static potential," *Phys. Rev.* **82** (1951), 840–850, who proved that $N(V) = 0$ if $\int r|V(r)|\, dr < 1$. Actually, by a general argument (Problem 23), the Jost–Pias result implies Bargmann's bound. Calogero's bound and Theorem XIII.9d first appeared in F. Calogero, "Upper and lower limits for the number of bound states in a given central potential," *Comm. Math. Phys.* **1** (1965), 80–88. For further discussion, see Calogero's book, *Variable Phase Approach to Potential Scattering*, Academic Press, New York, 1967. Theorem XIII.9c is due to V. Glaser, A. Martin, H. Grösse, and W. Thirring, "A family of optimal conditions for the absence of bound states in a potential," in *Studies in Mathematical Physics: Essays in honor of V. Bargmann* (E. Lieb, B. Simon, A. S. Wightman, eds.), Princeton Univ. Press, Princeton, New Jersey, 1976. In the same volume there is a review article by B. Simon, "On the number of bound states of two-body Schrödinger operators: A review." Theorem XIII.9e and a similar bound, $A\lambda^{3/2} < N(\lambda V) < B\lambda^{3/2}$ for large λ, are due to B. Simon, "On the growth of the number of bound states with increase in potential strength," *J. Mathematical Phys.* **10** (1969), 1123–1126. Formulas for the asymptotic behavior of $N(\lambda V)$ are discussed in Section 15. The reader should consult the Notes to Section 15 for references.

The Birman–Schwinger bound was discovered independently by M. Birman, "The spectrum of singular boundary problems," *Math. Sb.* **55** (1961), 124–174 (*Amer. Math. Soc. Trans.* **53** (1966), 23–80), and by J. Schwinger, "On the bound states of a given potential," *Proc. Nat. Acad. Sci. U.S.A.* **47** (1961), 122–129. There have been a variety of refinements of the Birman–Schwinger theorem. For references, consult the review article by Simon quoted above.

The first bounds on $N(V)$ for $n \geq 3$ having the right large coupling behavior were obtained by B. Simon in "Weak trace ideals and the number of bound states of Schrödinger operators," *Trans. Amer. Math. Soc.* **224** (1976), 367–380, who proved that for $n \geq 3$, $N(V) \leq c_{n,\varepsilon}(\|V_-\|_{n/2+\varepsilon} + \|V_-\|_{n/2-\varepsilon})^{n/2}$ for a constant $c_{n,\varepsilon}$ diverging as $\varepsilon \to 0$. Here $\|\ \|_p$ is the $L^p(\mathbb{R}^n)$

norm. Independently and by a different method, A. Martin, in "A Bound on the total number of bound states in a potential," proved that $N(V) \leq (2\pi)^{-1}(\|V_-\|_1 \|V_-\|_2^2)^{1/3}$ in three dimensions. The Cwikel–Lieb–Rosenbljum bounds which imply the Simon and Martin bounds (except for the possible size of constants) appear independently in M. Cwikel, "Weak type estimates for singular values and the number of bound states of Schrödinger operators," *Ann. Math.* **106** (1977), 93–100. E. Lieb, "The number of bound states of one-body Schrodinger operators and the Weyl problem." (to appear), and G. V. Rozenbljum, "The distribution of the discrete spectrum for singular differential operators," *Dokl. Akad. Nauk SSSR* **202** (1972), 1012–1015.

Lieb also provides new proofs of bounds on the sum of powers of the negative eigenvalues of $-\Delta + V$. Such bounds had been proven earlier by E. Lieb and W. Thirring; see Problems 31–33. The Cwikel–Lieb–Rosenbljum and Lieb–Thirring bounds are especially interesting on account of the phase-space intuition discussed in Section 15.

The proof we give of the Cwikel–Lieb–Rosenbljum bound is a simple variant of Lieb's due to him (unpublished). Cwikel's proof was based on Theorem XI.22.

The bounds of Lieb and Thirring on the sum of the negative eigenvalues of $-\Delta + V$, which are proven using the Birman–Schwinger bound on $N_E(V)$, are especially interesting because they are an important part of a simple proof of the "stability of matter"; see E. Lieb and W. Thirring, "Bound for the kinetic energy of fermions which proves the stability of matter." *Phys. Rev. Lett.* **35** (1975), 687–689.

Section XIII.4 H. Weyl in "Über beschränkte quadratische Formen, deren Differenz vollstetig ist," *Rend. Circ. Mat. Palermo* **27** (1909), 373–392, proved (what would be in modern terminology) Theorem XIII.14 in case A and B are bounded self-adjoint operators with $A - B$ compact. For non-self-adjoint operators, there are many distinct definitions of essential spectrum and a variety of forms of Weyl's theorem. For example on pp. 242–244 of Kato's *Perturbation Theory for Linear Operators*, Springer-Verlag, Berlin and New York, 1966, there is a discussion of a notion of $\sigma_{\text{ess}}(A)$ so that $\sigma_{\text{ess}}(A) = \sigma_{\text{ess}}(B)$ whenever $A - B$ is compact. If B is the operator in Example 1, according to Kato's definition $\sigma_{\text{ess}}(B)$ is the unit circle $\{z \,|\, |z| = 1\}$ while according to our definition, $\sigma_{\text{ess}}(B)$ is the unit disk $\{z \,|\, |z| \leq 1\}$. For an introduction to the literature on various kinds of essential spectra, see K. Gustafson, "Necessary and sufficient conditions for Weyl's theorem," *Michigan Math. J.* **19** (1972), 71–81.

The meromorphic Fredholm theorem in its full generality appeared in M. Ribaric and I. Vidav, "Analytic properties of the inverse $A(z)^{-1}$ of an analytic linear operator-valued function $A(z)$," *Arch. Rational Mech. Anal.* **32** (1969), 298–310. In this paper, the theorem is proven for operators on an arbitrary Banach space.

Corollary 3 to Theorem XIII.14 is a combination of results of M. Schechter, "On the essential spectrum of an arbitrary operator, I," *J. Math. Anal. Appl.* **13** (1966), 205–215, who proved the case $n = 2$ and K. Gustalfson and J. Weidmann, "On the essential spectrum," *J. Math. Anal. Appl.* **25** (1969), 121–127. An alternate "direct" proof can be found in J. Weidmann, "Spectral theory of partial differential operators," in *Spectral Theory and Differential Equations*, (W. N. Everitt, ed.), Lecture Notes in Mathematics, No. 448, Springer-Verlag, Berlin and New York, 1974.

Results related to Corollary 4 of Theorem XIII.14 and extensions of the ideas including A's that are not bounded below can be found in G. Nenciu, "Self-adjointness and invariance of the essential spectrum for Dirac operators defined as quadratic forms," *Comm. Math. Phys.* **48** (1976), 235–247.

For additional discussion of the essential spectrum of one-body Schrödinger operators see Problem 41; E. Balslev, "The singular spectrum of elliptic differential operators in $L^p(\mathbb{R}^n)$," *Math. Scand.* **19** (1966), 193–210; P. Rejto, "On the essential spectrum of the hydrogen energy and related operators," *Pacific J. Math.* **19** (1966), 109–140 and M. Schechter, *Spectra of Partial Differential Operators*, North Holland, Amsterdam, 1972.

The ideas in Example 8 are taken from J. E. Avron and I. W. Herbst, "Spectral and scattering theory of Schrödinger operators related to the Stark effect," *Comm. Math. Phys.* **52** (1977), 239–254. In the one-dimensional case, where the theory of ordinary differential equations can be exploited, one can prove that $\sigma(-d^2/dx^2 + W) = (-\infty, \infty)$ for a more general class of potentials W than those of the form $x + V$ with V short range. See, for example, E. C. Titchmarsh, *Eigenfunctions Expansions Associated with 2nd Order Differential Equations*, Oxford Univ. Press (Clarendon), London and New York, 1948; K. Kodaira, "The eigenvalue problem for ordinary differential equations of second order and Heisenberg's theory of S-matrices", *Amer. J. Math.* **71** (1949), 921–945; J. Weidmann, "Zur Spektraltheorie von Sturm–Liouville Operatoren," *Math. Z.* **98** (1967), 286–302; M. A. Neumark, *Lineare Differentialoperatoren*, Akad-Verlag, Berlin, 1960; J. Walter, "Absolute continuity of the essential spectrum of $-d^2/dx^2 + q(x)$ without monotony of q," *Math. Z.* **129** (1972), 83–94; P. Rejto, "On a theorem of Titchmarsh–Neumark–Walter concerning absolutely continuous operators, I, II," *Lett. Mat. Phys.* **1** (1975), 49–56, 57–66.

Section XIII.5 The HVZ theorem is named for W. Hunziker, C. Van Winter, and G. Zhislin for their work in the fundamental papers: W. Hunziker, "On the spectra of Schrödinger multiparticle Hamiltonians," *Helv. Phys. Acta* **39** (1966), 451–462; C. Van Winter, "Theory of finite systems of particles, I," *Mat.-Fys. Skr. Danske Vid. Selsk* 1 (8) (1964), 1–60; and G. Zhislin, "Discussion of the spectrum of the Schrödinger operator for systems of many particles," *Tr. Mosk. Mat. Obs.* **9** (1960), 81–128. Zhislin proved the result that $\sigma_{ess} = [\Sigma, \infty)$ for a restricted class of systems that included atomic Hamiltonians. His method has been further discussed by K. Jörgens, "Zur Spektraltheorie der Schrödinger Operatoren," *Math. Z.* **96** (1967), 355–372 and K. Jörgens and J. Weidmann, *Spectral Properties of Hamiltonian Operators*, Lecture Notes in Mathematics, No. 319, Springer-Verlag, Berlin and New York, 1973. Van Winter developed the Weinberg–Van Winter equations and (implicitly) proved the HVZ theorem in case $V \in L^2$ by employing the theory of integral equations with Hilbert–Schmidt kernels. Hunziker proved the $L^2 + (L^\infty)_\varepsilon$ case in work which was independent of Van Winter. The Rollnik potential case first appeared in B. Simon, *Quantum Mechanics for Hamiltonians Defined as Quadratic Forms*, Princeton Univ. Press, Princeton, New Jersey, 1972.

The Weinberg–Van Winter equations (in the form (24) rather than in the form (29)) and a generalization to N-body systems appeared in the paper of Van Winter quoted above and, independently in S. Weinberg, "Systematic solution of multiparticle scattering problems," *Phys. Rev.* **133** (1964), 232–256. Weinberg proved compactness of $I(E)$ only for three-body systems. The compactness for arbitrary N appears in Van Winter (when $V \in L^2$) and, independently, in W. Hunziker, "Proof of a conjecture of S. Weinberg," *Phys. Rev.* **135B** (1964), 800–803. Our proof of this fact (Lemma 4B) and in particular, the use of Lemma 5 are taken from Simon's monograph (see above). The symmetrized versions of the equations (29) which are needed if $V_{ij} \in R \backslash L^2$, also come from Simon's monograph.

Another set of equations for the resolvent, also with compact kernel are the Faddeev–Yakubovsky equations which appear for $N = 3$ in L. Faddeev, "Mathematical questions in the quantum theory of scattering for three particle systems," *Trudy Mat. Inst. Steklov* **69** (1963) [transl. Israel Program for Scientific Translation, 1965] and for arbitrary N in O. A. Yakubovsky, "On the integral equations in the theory of N-particle scattering," *Soviet J. Nuclear Phys.* **5** (1967), 937–942. The F-Y equations are much more complicated than the W-V equations, but they cannot have "spurious zeros" of the type discussed below. This makes them more useful in scattering theory. Still another set of equations (also without spurious zeros) appears in R. G. Newton, "Equations with connected kernels for N-particle T-operators," *J. Mathematical Phys.* **8** (1967), 851–856.

In the proof of the HVZ theorem, we saw that $\sigma_{\text{disc}}(H) \subset \{z \mid 1 - I_R(z) \text{ is not invertible}\}$. In fact, if $H\psi = E\psi$, then $I_R(E)[(H_0 - E)^{1/2}\psi] = (H_0 - E)^{1/2}\psi$. However, it can happen that $1 - I_R(E)$ is not invertible for some $E \in \rho(H)$. This was first pointed out in P. Federbush, "Existence of spurious solutions to many body Bethe–Salpeter equations," *Phys. Rev.* **148** (1966), 1551–1552, and is further discussed in R. Newton, "Spurious solutions of three particle equations," *Phys. Rev.* **153** (1967), 1502. Such values of E are called "spurious zeros." The reason for the name is that for $(1 - I_R(E))^{-1}$ to have a pole without $R(E) = (1 - I_R(E))^{-1} D_R(E)$ having a pole, $D_R(E)$ and $1 - I_R(E)$ must have compensating "zeros." Thus "zeros" of $1 - I_R(E)$ (or more precisely, of the Fredholm determinant $\det(1 - I_R(E))$) may occur at noneigenvalues.

The second proof we give of Theorem XIII.17 is a descendant of the original proof of Zhislin and a variant of a proof of V. Enss, "A Note on Hunziker's Theorem," *Comm. Math. Phys.* **52** (1977), 233–238. We closely follow the presentation of B. Simon, "Geometric methods in multiparticle quantum systems," in press. One technical simplification over the development of Zhislin, Jörgens and Weidmann, and Enss is the use of Theorem XIII.77 (Lemma 7 in this section) in place of a Weyl criterion.

Theorems of the genre of Theorem XIII.17′, dealing with the HVZ theorem when one restricts to symmetry subspaces, have been discussed in the articles of Zhislin and Sigalov, Balslev, and Simon quoted in the Notes to Section 3, in the monograph of Jörgens and Weidmann quoted above, and in the paper of Simon just quoted.

Section XIII.6 Our discussion of what we call the Aronszajn–Donoghue theory closely follows that in W. F. Donoghue, "On the perturbation of spectra," *Comm. Pure Appl. Math.* **18** (1965), 559–579, who remarks that his results are essentially contained in N. Aronszajn, "On a problem of Weyl," *Amer. J. Math.* **79** (1957), 597–610. D. Pearson has found $V \in C^\infty$, $D^j V \to 0$ at $\pm\infty$ so that $-D^2 + V$ has purely singular continuous spectrum.

Section XIII.7 The theory of smooth operators was developed by T. Kato in two remarkable papers, "Wave operators and similarity for some non-self-adjoint operators," *Math. Ann.* **162** (1966), 258–279, and "Smooth operators and commutators," *Studia Math.* **31** (1968), 535–546. In the first paper Kato defined "H-smooth" and proved Theorems XIII.22 (in a stronger form; see Problem 49), XIII.23 (implicitly), XIII.24, XIII.25, XIII.26 (in a stronger form; see Problems 53, 54 and the discussion below), and XIII.27 (in the case $N = 2$). The second paper contains a criterion for H-smoothness if H is bounded—in case H is multiplicity free, this is the criterion of Example 4 and Problem 50—and Kato's proof of the Putnam–Kato theorem (Theorem XIII.28). Our proofs of these results are patterned on Kato's arguments.

What we call Kato's smoothness theorem appears in a stronger form in his *Math. Ann.* paper. First, he proves that $H_0 + \lambda V$ and H_0 are similar, i.e., $W(H_0 + \lambda V)W^{-1} = H_0$ for an invertible bounded operator W even when λ is complex. His proof is based on a "time-independent" formulation of scattering. Second, neither H_0 or V need be self-adjoint. What is important is that $\sigma(H_0) \subset \mathbb{R}$ and $|V|^{1/2}$ be H_0-smooth and H_0^*-smooth in the sense of the basic definition (rather than the equivalent formulations in Theorem XIII.25) and that $\sup_{\mu \notin \mathbb{R}} \| |V|^{1/2}(H_0 - \mu)^{-1}|V|^{1/2} \| < \infty$.

Weak coupling results proving that $-\Delta$ and $-\Delta + V$ are unitarily equivalent for a suitable class of small V first appeared in J. Schwartz, "Some non-self adjoint operators," *Comm. Pure Appl. Math.* **13** (1960), 609–639. Schwartz dealt with the (reduced) two-body case in $\mathbb{R}^m$ with $m \geq 3$ and proved that if $\int (1 + x^2) |D^\alpha V| \, dx < \infty$ for all $\alpha \leq m - 1$, then $-\Delta$ and $-\Delta + \lambda V$ are unitarily equivalent for λ small.

Using the Dyson series (see Section X.12), R. Prosser proved a weak coupling theorem in the two-body case in "Convergent perturbation expansions for certain wave operators," *J. Mathematical Phys.* **5** (1964), 708–713. Kato proved Theorem XIII.27 in the case $N = 2$ in his *Math. Ann.* paper; in particular, (43) was used by Kato. The Prosser and Kato proofs are based on using the series

$$\Omega^- = 1 + i \int_0^\infty V_{s_1} \, ds_1 + i^2 \int_0^\infty \int_0^{s_1} V_{s_2} V_{s_1} \, ds_2 \, ds_1 + \cdots$$

where

$$V_s = e^{iH_0 s} V e^{-iH_0 s}$$

for $\Omega^- \varphi$, and the norm estimate

$$\|V_{s_1} V_{s_2} \cdots V_{s_n} \varphi\| = \| \, |V|^{1/2} \| \, \| \, |V|^{1/2} e^{i(s_2 - s_1) H_0} |V|^{1/2} \| \cdots \| \, |V|^{1/2} e^{-is_n H_0} \varphi \|$$

Small coupling results for three-body problems were proven by Hunziker in his Boulder Lectures (see the Notes to Section XI.5) who used Prosser's techniques. He remarked that the techniques could not work for N-body systems with $N \geq 4$ because $\| \, |V_{ij}|^{1/2} e^{isH_0} |V_{k\ell}|^{1/2} \|$ is constant if i, j, k, ℓ are all distinct. The idea of controlling $\int |V_{ij}|^{1/2} e^{isH_0} |V_{k\ell}|^{1/2} \, ds$ which appears under "Case (3)" in our proof of Theorem XIII.27 and thereby the N-body results if $N \geq 4$ is due to R. Iorio and M. O'Carroll, "Asymptotic completeness for multi-particle Schrödinger Hamiltonians with weak potentials," *Comm. Math. Phys.* **27** (1972), 137–145. It is an open conjecture that the conclusion of Theorem XIII.27 holds if $V_{ij} \in L^{m/2}$.

The theorem that H has purely absolutely continuous spectrum if there exists an A with $i[H, A] \geq 0$ appeared first in C. R. Putnam, *Commutation Properties of Hilbert Space Operators and Related Topics*, Springer-Verlag, Berlin and New York, 1967. The proof we give is from Kato's *Studia* paper.

Application of the Kato–Putnam theory to repulsive potentials appeared first in R. Lavine, "Absolute continuity of Hamiltonian operators with repulsive potentials," *Proc. Amer. Math. Soc.* **22** (1969), 55–60. The theory was much further developed (and in particular Theorems XIII.29 and XIII.32 appear) in R. Lavine, "Commutators and scattering theory, I. Repulsive interactions," *Comm. Math. Phys.* **20** (1971), 301–323. Theorem XIII.29 appears in a slightly weaker form (an extra condition that $r \, \partial V/\partial r \ll H_0$ is added). For potentials, repulsive, central, and only $O(r^{-1-\varepsilon})$, Lavine has proven the analogue of Theorem XIII.32 in "Completeness of the wave operators in the repulsive N-body problem," *J. Mathematical Phys.* **14** (1973), 376–379. Additional discussion of the two-body repulsive case can be found in M. Arai, "Absolute continuity of Hamiltonian operators with repulsive potentials," *Publ. Res. Inst. Math. Sci.* **7** (1971/72), 621–635. Our proof that $i[A, H_0] \geq 0$ follows Arai's paper.

Theorem XIII.31 and the notion of "H-smooth on Ω" are taken from R. Lavine, "Commutators and scattering theory, II. A class of one-body problems," *Indiana Univ. Math. J.* **21** (1972), 643–656.

The theory of smooth perturbations has been applied to the existence of propagators and scattering theory by E. B. Davies, "Time dependent scattering theory," *Math. Ann.* **210** (1974), 149–162, and extended to a Banach space setting in D. E. Evans, "Smooth perturbations in non-reflexive Banach spaces," *Math. Ann.* **221** (1976), 183–194.

Section XIII.8   Theorem XIII.33 is due to S. Agmon. He announced his results in "Spectral properties of Schrödinger operators," *Proc. Int. Cong. Math. of 1970*, Vol. 2, pp. 679–684, Gauthier-Villars, Paris, 1971. The details appear in S. Agmon, "Spectral properties of Schrödinger operators and scattering theory," *Ann. Scuola Norm. Sup. Pisa Cl. Sci.* **II**, 2 (1975), 151–218.

Our approach is based on a series of lectures by Agmon together with helpful remarks by H. Epstein, J. Ginibre, and R. Lavine. Theorem XIII.33d was proven prior to Agmon in T. Kato and S. Kuroda, "Theory of simple scattering and eigenfunctions expansions" in *Functional Analysis and Related Fields*, Springer-Verlag, Berlin and New York, 1970, 99–131. That (d) follows from Agmon's a priori estimates and the theory of local smoothness is a remark of R. Lavine in the paper quoted below.

Agmon's work represents the culmination of several lines of development. The first involves proving that $\sigma_{\text{sing}}(-\Delta + V) = \emptyset$ when V is $O(|x|^{-\mu})$ at ∞ (some of the papers quoted below require additional smoothness conditions). The earliest result was for $\mu > 2$ in the paper of Ikebe quoted in the Notes to Section XI.6. This was successively improved to $\mu > \frac{3}{2}$ by W. Jäger, "Zur Theorie der Schwingungsgleichung mit variablen Koeffizienten in Aussengebieten," *Math. Z.* **102** (1967), 62–88, to $\mu > \frac{4}{3}$ in P. Rejto, "On partly gentle perturbations, III" *J. Math. Anal. Appl.* **27** (1969), 21–67, to $\mu > \frac{5}{4}$ in T. Kato, "Some results on potential scattering," *Proc. Int. Conf. on Functional Analysis and Related Topics*, Tokyo, 1969, 206–215, and to $\mu > \frac{6}{5}$ by P. Rejto in "Some potential perturbations of the Laplacian," *Helv. Phys. Acta* **44** (1971), 708–736, and by S. Kuroda (quoted below). Rejto and Kuroda both modified their methods in response to Agmon's bootstrap argument and were able to handle all $\mu > 1$. Shortly after Agmon, and independently, the case $\mu > 1$ was handled by Y. Saito, "The principle of limiting absorption for second-order differential equations with operator-valued coefficients," *Pub. Res. Inst. Math. Sci.* **7** (1972), 581–619.

A second line of development involved the idea of proving $\sigma_{\text{sing}} = \emptyset$ by a perturbation theory of maps from X to X^* where X is a Banach space imbedded in $\mathcal{H}$. We discuss such an idea in the appendix to Section XI.6. This idea was developed by J. S. Howland, "Banach space techniques in the perturbation theory of self-adjoint operators with continuous spectra," *J. Math. Anal. Appl.* **20** (1967), 22–47, and "A perturbation-theoretic approach to eigenfunction expansions," *J. Functional Analysis* **2** (1968), 1–23, and by P. A. Rejto, "On partly gentle perturbations, I–III," *J. Math. Anal. Appl.* **17** (1967), 435–462; **20** (1967), 145–187; **27** (1969), 21–67. The use of weighted L^2 spaces was first advocated (in a slightly different context) by S. Kuroda, "On the Hölder continuity of an integral involving Bessel functions," *Quart. J. Math. Oxford Ser.* **21** (1970), 71–81.

A third line of development involved the abstract theory of eigenfunction expansions of Kuroda (see the Notes to Section XI.6) and the theory of higher order elliptic operators. In fact, Agmon's theory works for a large class of operators (see Problem 70 and below). A theory with similar results was developed (partly independently of Agmon) by S. Kuroda in "Scattering theory for differential operators, I, II," *J. Math. Soc. Japan* **25** (1973), 75–104; 222–234. Let $H_0 = \sum_{|\alpha| \leq 2m} a_\alpha(-iD)^\alpha$ where a_α is real. Suppose moreover that H_0 is **elliptic**, that is, $\sum_{|\alpha| = 2m} a_\alpha k^\alpha \neq 0$ for all $k \in \mathbb{R}^n$, $k \neq 0$. Let $P_1(k) = \sum_{|\alpha| \leq 2m} a_\alpha k^\alpha$. A point k where grad $P_1 = 0$ is called a critical point and the value of P_1 at such a point is called a **critical value**. It can be shown that H_0 has only finitely many critical values λ_i. Let $V = \sum_{|\alpha| \leq 2m} V_\alpha(x)(-iD)^\alpha$ be such that: (i) $|V_\alpha(x)| \leq C_\alpha(1 + |x|^2)^{-1/2-\varepsilon}$. (ii) V is formally self-adjoint, i.e., $(\varphi, V\varphi)$ is real for all $\varphi \in \mathscr{S}(\mathbb{R}^n)$. (iii) For each $x \in \mathbb{R}^n$ and $k \in \mathbb{R}^n \setminus \{0\}$, $\sum_{|\alpha| = 2m} (a_\alpha + V_\alpha(x))k^\alpha \neq 0$. Then Agmon and Kuroda have proven the following generalization of Theorem XIII.33:

(a) The eigenvalues of $H_0 + V$ at noncritical values of H_0 are of finite multiplicity and can have only critical values of H_0 as limit points.

(b) If $[a, b]$ is disjoint from all critical values of H_0 and all eigenvalues of $H \equiv H_0 + V$ and if $\delta > \frac{1}{2}$, then

$$\sup_{a \leq x \leq b;\, 0 < y < 1} \|(H - x - iy)^{-1}\|_{\delta, -\delta} < \infty$$

(c) $\sigma_{\text{sing}}(H_0 + V) = \varnothing$.
(d) Wave operators exist and are complete.

Extensions of the Agmon–Kuroda work to a slightly larger class of potentials can be found in S. Agmon and L. Hormander, "Asymptotic properties of solutions of differential equations with simple characteristics," *J. Anal. Math.* **30** (1976), 1–38. These authors use slightly different spaces which they consider more natural.

In Agmon's original work, he does not prove or use the existence of limiting boundary values for $(H_0 - \lambda)^{-1}$ or $(H - \lambda)^{-1}$. Lemmas 6 and 7 were replaced by a limiting argument known as the "principle of limiting absorbtion." The quadratic estimates of Lemma 6 are due to Lavine.

Theorem XIII.33 is capable of generalization in two other directions. First, V need only be form compact (see Problem 71). Secondly, one can discuss the case $V = V_1 + V_2$ where $\partial V_1/\partial r$ and V_2 are $O(r^{1-\varepsilon})$ at ∞. This is done by R. Lavine, "Absolute continuity of positive spectrum for Schrödinger operators with long range potentials," *J. Functional Analysis* **12** (1973), 30–54. See also T. Ikebe and Y. Saito, "The limiting absorbtion method and absolute continuity for the Schrödinger operator," *J. Math. Kyoto* **12** (1972), 513–542, and Y. Saito, "The principle of limiting absorbtion for the non-self-adjoint Schrödinger operator in $\mathbb{R}^n (N \neq 2)$" *Publ. Res. Inst. Math. Sci.* **9** (1974), 397–428.

Further developments of the Agmon–Kuroda theory appear in a series of papers by M. Schechter, "A unified approach to scattering," *J. Math. Pures Appl.* **53** (1974), 373–396; "Scattering theory for elliptic operators of arbitrary order," *Comm. Math. Helv.* **49** (1974), 84–113; "Scattering theory for second order elliptic operators," *Ann. Mat. Pura et Appl.* **55** (1975), 313–331, "Scattering Theory for Elliptic Systems," *J. Math. Soc. Japan* **28** (1976), 71–79; and "Nonhomogeneous elliptic systems and scattering," *Tohoku Math. J.* **27** (1975), 601–616.

The Agmon method has been extended to study $-\Delta + V + \mathbf{a} \cdot \mathbf{x}$ in I. W. Herbst, "Unitary equivalence of Stark Hamiltonians," *Math. Z.* **155** (1977), 55–71. Herbst proves that if $V = V_1 + V_2$ where $V_2 \in L^2(\mathbb{R}^3)$ has compact support, V_1 satisfies $|V_1(x)| \leq C(1 + (\mathbf{a} \cdot \mathbf{x})^2)^{-1/4-\varepsilon}$, and $-\Delta + V + \mathbf{a} \cdot \mathbf{x}$ has no eigenvalues, then $\Omega^{\pm}(-\Delta + V + \mathbf{a} \cdot \mathbf{x}, -\Delta + \mathbf{a} \cdot \mathbf{x})$ exist and are unitary. Avron and Herbst, in their paper quoted in the notes to Section 4, give criteria under which $-\Delta + V + \mathbf{a} \cdot \mathbf{x}$ has no eigenvalues.

Section XIII.9 The theory of commutative Banach algebras is discussed in Chapter XV. The elementary material used in this section may also be found in I. Gel'fand, D. Raĭkov, G. Shilov, *Commutative Normed Rings*, Chelsea, New York, 1964, or W. Rudin, *Functional Analysis*, McGraw-Hill, New York, 1973, or many other texts.

Theorem XIII.34 was proven by A. Brown and C. Pearcy in "Spectra of tensor products of operators," *Proc. Amer. Math. Soc.* **17** (1966), 162–166. M. Schechter generalized the Brown–Pearcy result in "On the spectra of operators on tensor products," *J. Functional Analysis* **4** (1969), 95–99, by proving that

$$\sigma(P(A \otimes I, I \otimes B)) = P(\sigma(A), \sigma(B))$$

where A and B are bounded operators and P is a polynomial in two variables. In particular, Schechter introduced the use of double commutant algebras. Subsequently, this spectral mapping theorem was extended to the class of rational functions in two variables analytic in a neighborhood of $\sigma(A) \times \sigma(B)$ in A. Dash and M. Schechter, "Tensor products and joint spectra," *Israel J. Math.* **8** (1970), 191–193.

Ichinose's lemma and generalizations appeared in T. Ichinose "On the spectra of tensor products of linear operators on Banach spaces," *J. Reine Angew. Math.* **244** (1970), 119–153, and "Operators on tensor products of Banach spaces," *Trans. Amer. Math. Soc.* **170** (1972),

197–219; see also "Operational calculus for tensor products of linear operators on Banach spaces," *Hokkaido Math. J.* **4** (1975), 306–334. M. Reed and B. Simon proved a spectral mapping theorem for a large class of unbounded functions of the unbounded operators $A \otimes I, I \otimes B$ in "Tensor products of closed operators on Banach spaces," *J. Functional Analysis.* **13** (1973), 107–124. Theorem XIII.35 is a special case of the Ichinose work and the Reed–Simon work. The proof given in Section 9 is new and much simpler than the proof in the Ichinose and Reed–Simon papers; but since it uses the special property that $e^{-t(A+B)} = e^{-tA} \otimes e^{-tB}$, it cannot be used to prove a spectral mapping theorem for more general functions of A and B.

Section XIII.10 There is a connection between the theory of dilation analytic potentials and the methods used in Section XI.8 to discuss analyticity of the partial wave amplitude for generalized Yukawa potentials. This later set of ideas was discussed in an N-body setting by C. Lovelace, "Three particle systems and unstable particles" in *Strong Interactions and High Energy Physics* (R. C. Moorehouse, ed.) Oliver and Boyd, London, 1964, several years before the development of dilation analytic techniques; but the connection between the two approaches was appreciated only later.

The idea of using dilations and the perturbation theory of discrete spectrum to control the singular continuous spectrum was first used by J. M. Combes in an unpublished manuscript, "An algebraic approach to quantum scattering." The idea was developed in the two-body case by J. Aguilar and J. M. Combes, "A class of analytic perturbations for one-body Schrödinger Hamiltonians," *Comm. Math. Phys.* **22** (1971), 269–279, and in the N-body case by E. Balslev and J. M. Combes, "Spectral properties of many-body Schrödinger operators with dilatation analytic interactions," *Comm. Math. Phys.* **22** (1971), 280–294. Both these papers discussed the class C_α(see Problem 73) rather than $\mathscr{F}_\alpha$ but otherwise Theorems XIII.36 and XIII.37 appear in the first and second papers (respectively).

Several years later, unaware of the Balslev–Combes work, C. Van Winter proved similar theorems by a related method in "Complex dynamical variables for multiparticle systems with analytic interactions, I, II," *J. Math. Anal. Appl.* **47** (1974), 633–670; **48** (1974), 368–399. The special role of the dilation group is not apparent in Van Winter's work, which is based on certain spaces of analytic functions.

Because Ichinose stated his theorem in terms of a bound on the resolvents of operators rather than in terms of sectorial operators or holomorphic semigroups, Balslev and Combes proved a bound on the N-body resolvents by a difficult inductive proof rather than by directly appealing to sectoriality. The class $\mathscr{F}_\alpha$ was introduced in B. Simon, "Quadratic form techniques and the Balslev–Combes theorem," *Comm. Math. Phys.* **27** (1972), 1–9. This paper also pointed out that sectorial operator methods allowed simplifications in the proofs and it removed a restriction $|\operatorname{Im} \theta| < \pi/2$ which Balslev and Combes imposed because of their direct reliance on Ichinose's lemma in its original form.

Dilation analytic techniques and, more generally, the technique of continuation in a group parameter play a role in many facets of N-body Hamiltonians and not just in eliminating the continuous singular spectrum. In particular they are of importance in the theory of resonances (Section XII.6), in exponential falloff of eigenfunctions (Section 11), and in eliminating positive eigenvalues (Section 13).

Section XIII.11 Hölder continuity of vectors in $C^\infty(H)$ for suitable N-body Hamiltonians was first proven by T. Kato in "On the eigenfunctions of many-particle systems in quantum mechanics," *Comm. Pure Appl. Math.* **10** (1957), 151–171. Kato's conditions on his potentials are slightly different from $M_\sigma^{(3)}$; namely, he supposes that $V = V_1 + V_2$ where $V_2 \in L^\infty$ and $V_1 \in L^2$ with supp V_1 compact. Kato's proof is in x space and is more complex than the proof of Theorem XIII.38, but it contains the basic idea of an L^p bootstrap argument. The idea of using p-space techniques is due to B. Simon, "Pointwise bounds on eigenfunctions and wave packets in N-body quantum systems, I," *Proc. Amer. Math. Soc.* **42** (1974), 395–401.

In the paper quoted above, Kato also proves a sharper result on the singularities of atomic wave functions ψ than is given by Theorem XIII.38. Theorem XIII.38 implies only that ψ is uniformly Hölder continuous of order θ for any $\theta < 1$. Kato shows that $\sup_{x \notin C} |\text{grad } \psi(x)| < \infty$ where C is the set of points where some coordinate difference goes to 0, and, in particular, ψ is Lipschitz continuous. Moreover, the discontinuities of grad ψ at coincident points are explicitly computed in terms of the values of ψ at such points.

The sharp results of Theorems XIII.39–XIII.42 are taken from a set of papers that appeared in preprint form in the fall of 1972. Theorems XIII.39 and XIII.40 appeared in A. O'Connor, "Exponential decay of bound state wave functions," *Commun. Math. Phys.* **32** (1973), 319–340. O'Connor's proof is based on Paley–Wiener ideas and the use of the Weinberg–Van Winter equations: It is conceptually simple but kinematically somewhat complex. The proof we give is patterned after that of J. Combes and L. Thomas, "Asymptotic behavior of eigenfunctions for multiparticle Schrödinger operators," *Commun. Math. Phys.* **34** (1973), 251–270. Combes and Thomas, who were motivated by O'Connor's work, noted the applicability of the ideas to nonthreshold eigenvalues of dilation analytic systems. Their results are stated in terms of the group generated by dilations and p-space translations.

Theorem XIII.42 appeared in the paper of Simon quoted at the start of the Notes of this section. Slightly earlier, R. Ahlrichs, "Asymptotic behavior of atomic bound state wave functions," *J. Mathematical Phys.* **14** (1973), 1860–1863, noted that uniform Hölder continuity and L^2 exponential falloff implied pointwise exponential falloff, but perhaps at a slower rate. Simon's work was motivated by the work of Ahlrichs and O'Connor.

Theorem XIII.41 has a sharp form analogous to Theorem XIII.40. For example, one can prove the following (see the Combes–Thomas paper): If each $V_{ij} \in \mathscr{F}_\alpha$ with $\alpha > \pi/4$ and if $\sum (i\pi/4)$ is the set of thresholds for $H(i\pi/4)$, then $\psi \in D(e^{ar})$ if $H\psi = E\psi$ and

$$a^2 < 2M \inf_{E_\alpha \in \Sigma(i\pi/4)} \{|E - \text{Re } E_\alpha| + |\text{Im } E_\alpha|\}$$

J. D. Morgan, III, in "The exponential decay of sub-continuum wave functions of two electron atoms," *J. Phys. A.* **10** (1977), L91, remarked that the bounds of Theorems XIII.39–42 cannot be the best possible, since one expects that $\psi(r)$ should asymptotically obey $H_0 \psi = E\psi$ as all $r_i - r_j \to \infty$, so that one would expect that $\psi(r) \sim \exp(-\sum a_i|r_i|)$ for some a_i with $\sum a_i^2/2m_i = -E$. He has made improvements in the bounds for certain three-body systems. Large classes of N-body systems are discussed, with his remark in mind, in P. Deift, W. Hunziker, and B. Simon, "Pointwise bounds on eigenfunctions and wave packets in N-body quantum systems; IV" (to appear).

Prior to the work of O'Connor, Combes, and Thomas, there was a whole series of papers proving weaker results in different cases. One line proved that discrete eigenfunctions ψ lie in $D(r^n)$ for suitable n by using operator methods. For $n = 1, 2$ and a general class of Hamiltonians, this was proven by W. Hunziker, "Space-time behavior of Schrödinger wave functions," *J. Mathematical Phys.* **7** (1966), 300–304. This was extended to all n by J. M. Combes, "Time dependent approach to multi-channel scattering," *Nuovo Cimento* **64A** (1969), 111–144. Using a similar method, in the paper quoted above Ahlrichs proved that $\psi \in D(e^{ar})$ for atomic eigenfunctions.

A second line of attack used methods from the theory of partial differential equations. I. E. Schnoll in "On the behavior of eigenfunctions of the Schrödinger equation," *Math. Sb.* **42** (1957), 273–286 (see also the discussion in I. M. Glazman, *Direct Methods in the Qualitative*

Spectral Analysis of Singular Differential Operators, Israeli Program for Scientific Translations, Jerusalem, 1965), proved that $|\psi| \leq \exp(-ar)$, but he required his potentials to be continuous and bounded from below.

A third line employed the method of integral equations. B. Simon in his *Quantum Mechanics for Hamiltonians Defined as Quadratic Forms*, Princeton Univ. Press, Princeton, New Jersey, 1971, proved that $\psi \in D(e^{ar})$ if $a < \sqrt{|E|}$ and $(-\Delta + V)\psi = E\psi$ with $E < 0$ and $V \in R$ (two-body case). Simon's analysis is based on a lemma in the theory of integral equations and the fact that ψ obeys

$$\psi(x) = -\frac{1}{4\pi} \int \frac{\exp(-\sqrt{-E}|x-y|)}{|x-y|} V(y)\psi(y)\, dy$$

For central potentials, one can use the integral equations of Section XI.8 to prove exponential falloff; see, e.g., the book by Regge and de Alfaro quoted in the Notes to Section XI.8. A final reference on the use of integral equations to prove exponential falloff is the paper of E. L. Slaggie and E. H. Wichmann, "Asymptotic properties of the wave function for a bound nonrelativistic system," *J. Mathematical Phys.* **3** (1962), 946–968, who discussed the three-body case. O'Connor's analysis is partially based on ideas from this paper.

One last line of attack uses variational methods and is described in N. W. Bazley and D. W. Fox, "Bounds for eigenfunctions of one-electron molecular systems," *Internat. J. Quant. Chem.* **3** (1969), 581–586.

Section XIII.12   The earliest results relating positivity and the nondegeneracy of an eigenvalue go back to a fundamental theorem of Perron and Frobenius: A finite matrix with strictly positive elements always has its spectral radius as an eigenvalue of multiplicity one with the corresponding eigenvector strictly positive. Notice that in the Perron–Frobenius theorem, the matrix need not be self-adjoint. The Perron–Frobenius theorem first appeared in O. Perron, "Zur theorie der Matrizen," *Math. Ann.* **64** (1907), 248–263, and F. G. Frobenius, "Über Matrizen mit positiven Elementen," *Sitzungsber. Preus. Akad. Wiss. Berlin* (1908), 471–476. The idea of extending the theorem from matrices with strictly positive elements to positivity preserving ergodic matrices appeared in F. G. Frobenius, "Über Matrizen aus nicht negativen Elementen," *Sitzungsber. Preus. Akad. Wiss. Berlin* (1912), 456–477. Further discussion of the finite matrix theorems may be found in F. Gantmacher, *Applications of the Theory of Matrices*, Wiley (Interscience), New York, 1959.

The extension of the Perron–Frobenius theorem to infinite-dimensional spaces was first accomplished by R. Jentzsch, "Über Integralgleichungen mit positiven Kernen," *J. Reine. Angew. Math.* **141** (1912), 235–244. There has been considerable development of the theory since Jentzsch's work, which dealt with a class of integral operators: See, e.g., M. Krein and M. Rutman, "Linear operators leaving invariant a cone in a Banach space," *Uspeki Mat. Nauk* **3** (1948), 3–95 [*Amer. Math. Soc. Trans.* **10** (1950), 199–235] and T. Ando, "Positive linear operators in semi-ordered linear spaces," *J. Fac. Sci. Univ. Tokyo Sect. IA Math.* **13** (1957), 214–228.

The idea of applying a theorem of Perron–Frobenius type to quantum systems is due to J. Glimm and A. Jaffe, "The $\lambda(\varphi^4)_2$ quantum field theory without cutoffs: II. The field operators and the approximate vacuum," *Ann. Math.* **91** (1970), 362–401. They used a direct proof of ergodicity for $\exp(-tH)$ where H is a spatially cutoff $P(\varphi)_2$ Hamiltonian rather than a proof going through the irreducibility of the action of $e^{-tH} \cup L^\infty(Q)$. The idea of using the irreducibility which simplifies the proof is due to I. Segal (see below). The application to nonrelativistic systems is due to B. Simon and R. Hoegh-Krohn, "Hypercontractive semi-groups and two-dimensional self-coupled Bose fields," *J. Functional Analysis* **9** (1972), 121–180. Earlier results on nonrelativistic quantum systems which do not use the Perron–Frobenius theorem are described in R. Courant and D. Hilbert, *Methods of Mathematical Physics*, Vol. I, Wiley (Interscience), 1953.

Theorem XIII.48 may be found in W. Faris and B. Simon, "Degenerate and non-degenerate ground states for Schrödinger operators." *Duke Math. J.* **42** (1975), 559–567.

The idea of using irreducibility in place of ergodicity goes back to Ando (see above); our proof of (c) ⇒ (a) in Theorem XIII.43 is taken from Simon–Hoegh-Krohn (see above). That the resolvent is positivity improving if it is ergodic is a theorem of Faris, "Invariant cones and uniqueness of the ground state for Fermion systems," *J. Mathematical Phys.* **13** (1972), 1285–1290. This paper also contains the perturbation theorem of Problems 91, 92. That an ergodic semigroup is positivity improving is a theorem of B. Simon, "Ergodic semigroups of positivity preserving self-adjoint operators," *J. Functional Analysis* **12** (1973), 335–339. The perturbation result, Theorem XIII.45, is due to I. Segal, "Construction of nonlinear local quantum processes, II," *Invent. Math.* **14** (1971), 211–241. The proof of statement (a) of Theorem XIII.49 is due to L. Gross, "Existence and uniqueness of physical ground states," *J. Functional Analysis* **10** (1972), 52–109. By a completely different method, Glimm and Jaffe (in the paper quoted above) had earlier proven the existence of an eigenvalue at the bottom of the spectrum of spatially cut-off $P(\varphi)_2$ Hamiltonians. We note that Theorem XIII.45 was extended to certain situations involving Fermion field theories in the Gross and Faris papers quoted above.

The Beurling–Deny criteria occur in A. Beurling and J. Deny, "Espaces de Dirichlet I. Le Cas Elementaire," *Acta Math.* **99** (1958), 203–224. The paper explicitly deals only with the case where the support of $d\mu$ is a discrete finite set with each point having mass 1. However, the proofs of what we have called the first and second Beurling–Deny criteria clearly extend to the general case (they are essentially the proofs we use)—moreover, the authors announced results for the general case although the detailed paper has never appeared! Additional discussion can be found in M. Fukushima, "On the Generation of Markov Processes by Symmetric Forms," Proc. Second Japan-USSR Symposium on Probability Theory, Lecture Notes in Mathematics, No. 336, Springer-Verlag, Berlin and New York,

Earlier than the Beurling–Deny paper, a condition closely related to their first criterion appeared in N. Aronszajn and K. T. Smith, "Characterization of positive reproducing kernels, applications to Green's functions," *Amer. J. Math.* **79** (1957), 611–622. Aronszajn and Smith discuss the question of when a strictly positive differential operator A has an inverse with a pointwise positive kernel. Their statements and, to some extent, their methods are restricted to differential operators. And their problem is slightly different from the one solved in the first Beurling–Deny criterion. For example $A = (-\Delta + 1)^2$ clearly has a positivity preserving inverse (since $A^{-1/2}$ is positivity preserving), but e^{-tA} is not positivity preserving for all $t > 0$ by the Levy–Khintchine formula. However, their condition is clearly related to the first Beurling–Deny criterion.

The first Beurling–Deny criterion provides a link between the proof we give of the nodelessness of the ground state, which relies on positivity of $e^{+t\Delta}$ and the proof we mentioned above of Courant and Hilbert. This latter proof exploits the inequality $\|\nabla |u|\|^2 \leq \|\nabla u\|^2$, which is just the first Beurling–Deny criterion for $e^{+t\Delta}$ to be positivity preserving! There is also a link between the positivity of $e^{+t\Delta}$ and Kato's inequality (Section X.4) provided by the first Beurling–Deny criterion; see B. Simon, "An abstract Kato's inequality for generators of positivity preserving semigroups." *Indiana Math. J.* (to appear).

With regard to Example 1 of Appendix 1, it is a classical theorem of T. J. Stieltjes, "Sur les Racines de l'Equation $X_n = 0$," *Acta Math.* **9** (1887), 385–400, that if A is a positive definite matrix with $a_{ij} \leq 0$ for $i \neq j$, then $(A^{-1})_{ij} \geq 0$ for all i, j. Example 1 provides a kind of converse to this theorem.

The notions in the second appendix were first introduced in a probability theory context involving infinitely divisible probability measures. All infinitely divisible probability measures with finite second moment were found by A. Kolmogorov, "Sulla forma generale di un processo stocastico omegeneo," *Atti. Accad. Naz. Lincei Rend. Cl. Sci. Fis. Mat. Natur.* **15** (1932), 805–808, 866–869. The general form was found by P. Lévy, "Sur les intégrales dont les éléments sont les variables aléatoires indépendantes," *Ann. Scuola Norm. Sup. Pisa Cl. Sci.* **3** (1934), 331–337; **4** (1935), 217–218. Further refinements are due to W. Feller, "On the Kolmogorov–P. Lévy formula for infinitely divisible distribution functions," *Proc. Yugo. Acad. Sci.* **82** (1937), 95–113, and A. Khintchine "Déduction nouvelle d'une formula de M. Paul Lévy," *Bull. Univ. d'Etat Moskov, Ser. Int. Sert A* **1** (1937), 1–5. The context in which infinitely divisible distributions enter is the following. n functions $f_1, \ldots, f_n$ on a probability measure space are called independent random variables if the measure on $\mathbb{R}^n$ given by $v(A) = \mu((f_1 \otimes \cdots \otimes f_n)^{-1}[A])$ is a product measure. If all the factors are the same, we say that $f_1, \ldots, f_n$ are identically distributed. If $\tilde{\mu}$ is the distribution for $f_1 + \cdots + f_n$ and μ is the one for f_i, then $\tilde{\mu} = \mu * \cdots * \mu$ (n times). If a probability distribution is a limit as $n \to \infty$ of a sum of n identically distributed random variables (the individual distribution may vary with n), then the distribution will be infinitely divisible. The classic example is that the limiting distribution be Gaussian (which corresponds to a pure $x \cdot Ax$ term in (76)). For additional discussion, see L. Breiman, *Probability*, Addison-Wesley, Reading, Massachusetts, 1968.

Conditionally positive definite functions and a theorem of the genre of Theorem XIII.52 appeared first in I. J. Schoenberg, "Metric spaces and positive definite functions," *Trans. Amer. Math. Soc.* **44** (1938), 522–536. The relation between those notions and infinitely divisible distributions has been developed by a series of Russian mathematicians: B. V. Gnedenko, "On the characteristic property of infinitely divisible distribution laws," *Bull. Mosk. Goud. Univ. Ser A.*, 1937, A. M. Yaglom and M. S. Pinsker, "Random processes with stationery nth-order increments" (in Russian), *Dokl. Akad. Nauk. SSSR* **90** (1953), 731–734; M. G. Krein, "On integral representations of Hermitian indefinite functions with a finite number of negative squares," *Dokl. Akad. Nauk. SSSR* **125** (1959), 31–34; and I. M. Gel'fand and N. Ya. Vilenkin, *Generalized Functions*, Vol. 4, Academic Press, New York, 1964. Our proof of Theorem XIII.52 follows this last presentation in part; we have corrected an error in their proof (our Theorem XIII.52; their Theorem III.4.4).

The connection between positivity preserving semigroups and the Lévy–Khintchine formula in the context of Schrödinger operator theory was made by I. Herbst and A. Sloan, "Perturbation of translation invariant positivity preserving semigroups on $L^2(\mathbb{R}^n)$," Trans. Amer. Math. Soc. (to appear), who develop some of the theory of $F(-i\nabla) + G(x)$ with F conditionally negative definite.

That e^{-tp^4} does not generate a positivity preserving semigroup has been used to construct an example illustrating the difference between different axiom schemes for Euclidean quantum field theories; see B. Simon, "Positivity of the Hamiltonian semigroup and the construction of Euclidean region fields," *Helv. Phys. Acta* **46** (1973), 686–696.

Section XIII.13 The potential constructed in Example 1 is due to J. von Neumann and E. P. Wigner, "Über merkwürdige diskrete Eigenwerte," *Z. Phys.* **30** (1929), 465–467. There is an arithmetic mistake in their paper, so that the potential they write down has $O(r^{-2})$ behavior at infinity violating Theorem XIII.58. We have corrected their arithmetic mistake. Unaware of their work, J. Weidmann in "Zur Spektraltheorie von Sturm–Liouville Operatoren," *Math. Z.* **98** (1967), 268–302 constructed an example using step functions (much in the spirit of Nelson's example discussed in Example 1 of the Appendix to Section X.1). S. Albeverio in "On bound states in the continuum of N-body systems and the Virial theorem," *Ann. Phys.* **71** (1972),

167–276, describes how one can construct potentials with bound states at any finite number of prescribed energies. See also the Atkinson paper quoted in the Notes to Section XI.8.

Theorems in the genre of Theorem XIII.56 are to some extent implicit in the work on Jost functions and solutions discussed in Section XI.8. In particular, the inverse scattering techniques discussed in the Notes to that section are critical in the Albeverio construction quoted above. Explicit theorems exploiting ordinary differential equation techniques to completely analyze the spectrum of operators with central potentials, and, in particular, results stronger than Theorem XIII.56 or the central case of Theorem XIII.58, appear in Weidmann's paper above. For additional discussion of Jost function techniques and positive energy bound states, see the appendix to Section XI.8 and its Notes.

The model for Theorem XIII.58 is a paper of F. Rellich, "Über das asymptotische Verhalten der Lösungen von $\Delta u + \lambda u = 0$ in unendlichen Gebieten," *Über. Deutsch. Math. Verein* **53** (1943), 57–65. Rellich studied solutions of $(\Delta + \lambda)u = 0$ in unbounded domains and in particular proved that there are no square integrable solutions of $(\Delta + \lambda)u = 0$ in $\mathbb{R}^n \setminus \{x \mid |x| > R\}$ if $\lambda > 0$. This implies that there are no positive eigenvalues of $-\Delta + V$ if V has compact support. This idea was developed by T. Kato, "Growth properties of solutions of the reduced wave equation with variable coefficients," *Comm. Pure Appl. Math.* **12** (1959), 403–425, who proved the case of Theorem XIII.52 with $V_2 = 0$. Results proven by similar means for potentials that are repulsive near infinity ($\partial V_2/\partial r < 0$ for r large) can be found in K. Kreith, "Differential operators with a purely continuous spectrum," *Proc. Amer. Math. Soc.* **14** (1963), 809–811, and F. Odeh, "Note on differential operators with a purely continuous spectrum," *Proc. Amer. Math. Soc.* **16** (1965), 363–366. Theorems of the form of Theorem XIII.58 (but with slightly stronger smoothness conditions) were found independently by B. Simon, "On positive eigenvalues of one-body Schrödinger operators," *Comm. Pure Appl. Math.* **22** (1969), 531–538, and S. Agmon, "Lower bounds for solutions of Schrödinger-type equations in unbounded domains," *Proceedings of the International Conference on Functional Analysis and Related Topics*, 1969, Univ. Tokyo Press, Tokyo, and "Lower bounds for solutions of Schrödinger equations," *J. Analyse Math.* **23** (1970), 1–25. Even before these later papers, J. Weidmann, "On the continuous spectrum of Schrödinger operators," *Comm. Pure Appl. Math.* **19** (1966), 107–110, used the methods from Kato's paper to show that Coulombic Hamiltonians have no positive eigenvalues (this was the first proof of this result which we discuss in Example 5 and Example 5 (revisited) using other techniques). Weidmann's methods easily extend to homogeneous potentials of degree $-\alpha$, $0 < \alpha < 2$. In the second Agmon paper, sums of potentials of the type considered in Theorem XIII.58 and these homogeneous potentials are discussed. The techniques of Theorem XIII.58 are extended to systems with magnetic fields in T. Ikebe and J. Uchiyama, "On the asymptotic behavior of eigenfunctions of second order elliptic operators," *J. Math. Kyoto Univ.* **11** (1971), 425–448. Applications of the method to assert that certain special classes of n-body operators have no eigenvalues in (E_0, ∞) for suitable $E_0 > 0$ can be found in the Simon paper and the first Agmon paper.

In the physics literature the statement of the virial theorem in quantum mechanics goes back to B. N. Finkelstein, "Über den Virialsatz in der Wellenmechanik," *Z. Phys.* **50** (1928), 293–294. Finkelstein used the formal commutator we describe in the text. While it was clear to Weidmann (who gave the first rigorous proof of the virial theorem; see below) that making Finkelstein's proof rigorous presented problems, it was Albeverio in the paper quoted above who emphasized that there were modifications of the Wigner–von Neumann example with wave functions not in the domain of the dilation generator D. Nevertheless, H. Kalf in "The quantum mechanical virial theorem and the absence of positive energy bound states of Schrödinger operators," *Israel J. Math.* **20** (1975), 57–69, did find a proof of the virial theorem along Finkelstein's lines. Essentially, the commutator argument is replaced by an explicit integration by parts. The bad domain properties imply that the boundary terms in the integration by parts do not automatically go to zero and they must be treated by a careful averaging argument.

The connection of the virial theorem and the group of dilations is a discovery of V. Fock, "Bemerkung zum Virialsatz," *Z. Phys.* **63** (1930), 855–858. The first rigorous proof of the virial theorem, which we follow in our proof, is due to J. Weidmann, "The virial theorem and its application to the spectral theory of Schrödinger operators," *Bull. Amer. Math. Soc.* **73** (1967), 452–456. Weidmann also discovered the relevance of the virial theorem to the problem of positive eigenvalues. Virial theorem techniques are further discussed in Albeverio's paper, which includes an extension to operators with magnetic fields, and Kalf's paper which includes the case of certain operators defined as sums of quadratic forms.

The application of dilation analytic techniques to recover Weidmann's theorem on the nonexistence of positive eigenvalues for Coulombic Hamiltonians was first noted in B. Simon, "Resonances in N-body quantum systems with dilation analytic potentials and the foundations of time dependent perturbation theory," *Ann. Math.* **97** (1973), 247–274. Simon also remarked that spherically symmetric dilation analytic potentials automatically have $r\,dV/dr \to 0$ at infinity so that, on the basis of Theorem XIII.83, one might expect connections between dilation analytic potentials and the lack of positive energy eigenvalues. Our discussion in Example 5 (revisited) does not follow Simon's paper but rather an unpublished remark of E. Balslev. The argument in Example 5, revisited, alternate argument, is due to W. Hunziker, "The Schrödinger Eigenvalue problem for N-particle systems," in *The Schrödinger Equation*, (W. Thirring and P. Urban, eds.) Springer, 1977, 43–72.

Carlson's theorem was proven by F. Carlson in his 1914 University of Upsala Thesis. He also proved the beautiful corollary on continuing an analytic function from its values at the positive integers (see Problem 109). Theorem XIII.61 was proven independently by B. Simon, "Absence of positive eigenvalues in a class of multiparticle quantum systems," *Math. Ann.* **207** (1974), 133–138, and E. Balslev, "Absence of Positive Eigenvalues of Schrödinger Operators," *Arch. Rational Mech. Anal.* **59**, 4 (1975), 343–357.

Unique continuation theorems were first proven for operators of second order on $\mathbb{R}^2$ by T. Carleman, "Sur un probleme d'unicité pour les systèmes d'équations aux dérivées partielles à deux variables indépendantes," *Ark. Mat.* (17) **26B** (1939), 1–9. Carleman introduced the idea, which is critical in our proof, of proving a priori estimates with constants independent of a parameter (β in our case), so that as the parameter becomes singular, the integrals in the estimates become more and more concentrated in some region of interest. Unique continuation for Schrödinger operators in n-dimensions was first proven by C. Müller, "On the behavior of the solutions of the differential equation $\Delta u = F(x, u)$ in the neighborhood of a point," *Comm. Pure Appl. Math.* **7** (1954), 505–515. This problem was further studied by E. Heinz, "Über die Eindeutigkeit beim Cauchyschen Anfangswertproblem einer elliptischen Differentialgleichung zweiter Ordnung," *Nachr. Akad. Wiss. Göttingen Math.-Phys. Kl. II* (1) (1955), 1–12, whose proof we follow in the appendix to this section. Heinz, in particular, proved a unique continuation theorem for Schrödinger operators with magnetic fields (see Problem 113). These results were extended to second-order elliptic equations with nonconstant leading coefficients independently by N. Aronszajn, "A unique continuation theorem for solutions of elliptic partial differential equations or inequalities of second order," *J. Math. Pures Appl.* **36** (1957), 235–249, and H. O. Cordes, "Über die Bestimmtheit der Lösungen elliptischer Differentialgleichungen durch Anfangsvorgaben," *Nachr. Akad. Wiss. Göttingen Math.-Phys. Kl. II* (11) (1956), 239–258.

For hyperbolic equations like the wave equation $\ddot{u} = \Delta u$, it can happen, of course, that solutions can vanish in open sets without being identically zero, but there is a uniqueness theorem in this case also: If $\ddot{u} = \Delta u$ and u vanishes in some neighborhood of the plane $\{(t, \mathbf{x}) \mid t = 0\}$, then u is identically zero. For the wave and related equations, such results go back at least to E. Holmgren, "Über Systeme von linearen partiellen Differentialgleichungen," *Öfver-*

sigt af Kongl. Vetenskaps-Akad. Förh. **58** (1901), 91–105; see also L. Nirenberg, "Uniqueness in Cauchy problems for differential equations with constant leading coefficients," *Comm. Pure Appl. Math.* **10** (1957), 89–105, and A. P. Calderón, "Uniqueness in the Cauchy problem for partial differential equations," *Amer. J. Math.* **80** (1958), 16–36. These results are discussed in detail in the book by L. Hörmander, *Linear Partial Differential Operators*, Springer-Verlag, Berlin and New York, 1964. There are examples of operators with no uniqueness theorem constructed by P. Cohen, "The non-uniqueness of the Cauchy Problem," O.N.R. Technical Report 93, Stanford, 1960 and A. Pliv, "A smooth linear elliptic differential equation without any solution in a sphere," *Comm. Pure Appl. Math.* **14** (1961), 599–617.

As emphasized by R. Lavine (unpublished), there is a serious defect in the unique continuation theorems that are available in that they require that V be bounded locally. The kind of "trapping" that occurs in Example 2 has a potential V that is not even locally L^1 and the theorem of Faris and Simon (Theorem XIII.48) suggests strongly that such non-L^1 singularities are necessary for trapping. For example, one would conjecture: If $V \in L^1_{\text{loc}}(\mathbb{R}^n)$ is positive and $H = -\Delta + V$ is defined as a sum of forms, then any eigenfunction of H vanishing on an open set is identically zero. Notice that Heinz's proof throws away a lot of information in the estimates. By carefully keeping track of this information, one should be able to prove a unique continuation theorem for certain classes of unbounded potentials.

In the application that we make of unique continuation, we only need the special case which prohibits eigenfunctions of compact support. This fact has been proven in greater generality using Theorem XIII.100 by W. O. Amrein and A. Berthier, to appear.

Section XIII.14 The basic criterion, Theorem XIII.64, and the other general results of this section (Theorems XIII.77, 78) are standard folklore. Theorem XIII.78 was first emphasized in a mathematical physics context by J. Glimm and A. Jaffe, "The $\lambda(\varphi^4)_2$ quantum field theory with cutoffs II. The field operators and the approximate vacuum," *Ann. Math.* **91** (1970), 362–40, who applied it to the spatially cutoff $P(\varphi)_2$ Hamiltonian. An application to the spatially cutoff Y_2 model and the proof we use via Theorem XIII.77 appears in their paper, "The Yukawa$_2$ Quantum Field theory without cutoffs," *J. Functional Analysis* **7** (1971), 323–357.

We have decided to call Theorem XIII.65 Rellich's criterion because the earliest related results appeared in F. Rellich, "Ein Satz über mittlere Konvergenz," *Nachr. Akad. Wiss. Göttingen Math.-Phys. Kl. II* (1930), 30–35. The authors learned the simple proof given in the text from R. Lavine.

Riesz' criterion appeared in M. Riesz, "Sur les ensembles compacts de fonctions sommable," *Acta Sci. Math. (Szeged)* **6** (1933), 136–142. Earlier related results are due to M. Fréchet, "Essai de geometrie analytique à une infinité de coordinates," *Nouvelles Ann. Math.* **8** (1903), 97–116; 289–317, and A. Kolmogoroff," Über Kompaktheit der Funktionmengen bei der Konvergenz im Mittel," *Nachr. Akad. Wiss. Göttingen Math.-Phys. Kl. II* (1931), 60–63.

Our discussion of the falloff of eigenfunctions of $-\Delta + V$ with $V \to \infty$ is taken from B. Simon, "Pointwise bounds on eigenfunctions and wave packets in N-body quantum systems, I, II, III," *Proc. Amer. Math. Soc.* **42** (1974), 395–401; **45** (1974), 454–456; *Trans. Amer. Math. Soc.* **208** (1975), 317–329. In particular, a proof of Theorem XIII.72 can be found in the paper III. Theorem XIII.70, in the $V \geq 0$ case, appeared much earlier in I. Schnol, "On the behavior of the eigenfunctions of Schrödinger's equation," *Mat. Sb.* **46 (88)** (1957), 273–286, *erratum* 259; where it was proven by different means. Our proof of Theorem XIII.70 in the $V \geq 0$ case uses a trick of E. B. Davies, "Properties of the Green's functions for some Schrödinger operators," *J. London Math. Soc.* **7** (1973), 483–491. Extensions of Theorem XIII.72b to say something about eigenfunctions other than the ground state occur in C. Bardos and M. Merigot, "Asymptotic decay of the solution of a second order elliptic equation in an unbounded domain. Applications to the

spectral properties of a Hamiltonian," *Proc. Roy. Soc. Edinburgh Sect. A* (to appear), and C. Bardos and M. Merigot, "Décroissance expontielle de la solution L^2 du problème de Dirichlet dans le complementaire d'un fermé borné," *Comp. Rend. Acad. Sci. Paris,* **281A** (1975), 561–563.
The first compact embedding theorem of the type we discuss was proved by F. Rellich in the paper quoted above. General embedding theorems for H_0^m are due to L. Gårding, "Dirichlet's problem for linear elliptic partial differential equations," *Math. Scand.* **1** (1953), 55–72. Gårding's proof is outlined in Problem 123. Rellich proved Theorem XIII.75 in the case of bounded regions with smooth boundary. The general case and the theory of regions obeying the segment property is due to S. Agmon, *Lectures on Elliptic Boundary Value Problems*, Van Nostrand, Princeton, New Jersey, 1965. Agmon also proves that if Ω has the segment property, then $\{f \mid f$ is C^∞ in a neighborhood of $\Omega\}$ is dense in H^m.
Our discussion of quantum statistical mechanics in boxes is taken from B. Simon's appendix to J. Lebowitz and E. Lieb, "The constitution of matter," *Advances in Math.* **9** (1972), 316–398, but the subject is fairly standard folklore. For further discussion of the quantum statistical mechanics of continuous systems, see J. Ginibre, "Some applications of Functional Integration in Statistical Mechanics" in *Statistical Mechanics and Quantum Field Theory, Les Houches, 1970* (C. DeWitt and R. Stora, eds.) Gordon and Breach, New York, 1971, D. Robinson, *The Thermodynamic Pressure in Quantum Statistical Mechanics*, Springer-Verlag, Berlin and New York, 1971, and D. Ruelle, *Statistical Mechanics*, Benjamin, New York, 1967.

Section XIII.15 The technique of Dirichlet–Neumann bracketing is one with a considerable antiquity; for example, Theorem XIII.78 is proven in R. Courant and D. Hilbert, *Methods of Mathematical Physics*, Vol. I, Wiley (Interscience), 1953, see especially pp. 429–431, by a method related to Dirichlet–Neumann bracketing. Weyl's original proof uses related ideas. More recent papers employing the method are E. H. Lieb, "Quantum Mechanical Extensions of the Lebowitz–Penrose Theorem on the Van Der Waals Theory," *J. Mathematical Phys.* **7** (1966), 1016–1024, and A. Martin, "Bound states in the strong coupling limit," *Helv. Phys. Acta* **45** (1972), 140–148, and the book by Robinson quoted in the Notes to the last section.
For arbitrary open regions $\Omega \subset \mathbb{R}^m$ one can choose different definitions for $-\Delta_N$ than the one we gave. For example, some authors take $Q(-\Delta_N)$ to be the closure of the form $(f, -\Delta f)$ on the set of functions in $H^1(\Omega)$ which are C^∞ in a neighborhood of Ω. For regions that obey the segment condition, the definitions agree.
The intuitive idea that one state takes up a volume h^3 in phase space entered quite early in attempts to extend Bohr's old quantum theory to gases; see, for example, O. Sackur, "Die Anwendung der Kinetischen Theorie der Gase auf chemische Probleme," *Ann. Phys.* **36** (1911), 958–980; "Die universelle Bedeutung des sog. elementaren Wirkungsquantums," *Ann. Phys.* **40** (1913), 67–86; and H. Tetrode, "Die chemische Konstante der Gase und das elementare Wirkungsquantum," *Ann. Phys.* **38** (1912), 434–442. This idea was used as a powerful tool by E. Fermi even after the development of the new quantum theory and is called the "Fermi gas picture."
Finding the number of points of $\mathbb{Z}^m$ inside the ball of radius r is of considerable number-theoretic interest. That the number is asymptotically $\tau_m r^m$ with an error at worst $O(r^{m-1})$ is a result going back at least to Gauss who used a proof which is essentially that used in Proposition 2. Surprisingly, the error is smaller than $O(r^{m-1})$ in fact, for $m = 2$ it is known to be smaller than $O(r^{2/3})$ and larger than $O(r^{1/2})$. For further discussion, see §8.7 of G. Hardy and E. Wright, *The Theory of Numbers*, Oxford Univ. Press, Oxford, 1938, and pages 183–308 of E. Landau, *Vorlesungen über Zahlen Theorie*, Vol. 2, Hirzel, Leipzig, 1927. For the analogous problem in higher dimensions, see A. Walfisz, *Weylische Exponentialsummen in der Neueren Zahlentheorie*, Deutscher Ver. Wiss., Berlin, 1963. We warn the reader that these results do not

imply that the power $m - 1$ in (116) and (117) can be improved. Quite the contrary, they imply that it cannot be improved! For the set M_{m-1} will make a contribution of order r^{m-1}, which by the above cannot be made up for by the contribution of the surface of the ball.

Theorem XIII.78 is a celebrated theorem of H. Weyl, "Das asymptotische Verteilungsgesetz der Eigenwerte linearer partieller Differentialgleichungen," *Math. Ann.* **71** (1911), 441–469, and "Über die Abhänggigkeit der Eigenschwenkgungen einer Membran von deren Begrenzung," *J. Math.* **141** (1912), 1–11. For a historical reminiscence, see H. Weyl, "Ramifications, old and new, of the eigenvalue problem," *Bull. Amer. Math. Soc.* **56** (1950), 115–139. This theorem has captured the imagination of two generations of mathematicians. Weyl's work was based on conjectures of Lorentz and Jeans who came upon the problem in the study of electromagnetic radiation theory. It shows that the spectrum of $-\Delta_\Omega$ on Ω determines the volume of Ω. This immediately raises the question of what other geometric properties of Ω are determined by the spectrum of $-\Delta_\Omega$, or more generally the Laplace–Beltrami operator on a compact Riemannian manifold $-\Delta_M$. The results up until the middle 1960s and important new directions are summarized in the beautiful article of M. Kac, "Can you hear the shape of a drum," *Amer. Math. Monthly* (Slaught Memorial Papers, No. 11) (4) **73** (1966), 1–23. More recent results are discussed in I. M. Singer's contribution to the Proceedings of the 1974 International Congress of Mathematicians entitled "Distribution of eigenfrequencies for the Laplacian." See also, R. Ballian and C. Bloch, "Eigenvalues of wave equation in a finite domain, I, II, III;" *Ann. Phys.* **60** (1970), 401; **64** (1971), 271; **69** (1972), 75. Among the things determined by the eigenvalues are the surface area of the boundary, the Euler–Poincaré characteristic, and the lengths of any closed geodesics on M. On the negative side is the existence of an example given by J. Milnor in "Eigenvalues of the Laplacian operator on certain manifolds," *Proc. Nat. Acad. Sci.* **51** (1964), 542, of two nonisometric 16-dimensional tori whose Laplace–Beltrami operators have identical spectra. These analyses of $-\Delta_\Omega$ rely on studying the integral kernel for $\exp(t\Delta_\Omega)$ rather than on the method of Dirichlet–Neumann bracketing, which is too crude to see anything but the leading term in $N(\lambda)$. A modern discussion of these Green's functions methods can be found in L. Hormander, "The spectral function of an elliptic operator," *Acta Math.* **121** (1968), 193–218.

For certain regions one can show that the error in Weyl's formula is $O(\lambda^{(n-1)/2})$; See the Hörmander paper quoted above and V. Avakumovic, "Über die Eigenfunctionen auf geschlossen Riemannschen Mannigfaltigkeiten," *Math. Z.* **65** (1956), 327–344.

For an example where a phase space picture provides a bound on a quantum-mechanical situation, see Theorem XIII.12 and Problems 31, 32, 33. We remark that while the phase space volume picture is remarkably accurate in most situations, it does fail in "borderline" situations. For example, $-\Delta - c(1 + |r|)^{-\alpha}$ on $L^2(\mathbb{R}^3)$ has an infinite number of bound states if $\alpha < 2$, $c > 0$ or $\alpha = 2$, $c > \frac{1}{4}$. The negative energy phase space volume is infinite whenever $\alpha \leq 2$, $c > 0$. Thus for $\alpha = 2$, $0 < c \leq \frac{1}{4}$, the phase space prediction is wrong.

Jordan content was an important precursor of Lebesgue measure. The basic work was done in the 1880s by Cantor, Stolz, Harnack, Peano, and Jordan. A historical discussion of the development of content can be found in I. N. Pesin, "Classical and Modern Integration Theories," Academic Press, New York, 1970. Modern treatments of Jordan content occur in L. Loomis and S. Sternberg, *Advanced Calculus*, Addison-Wesley, Reading, Massachusetts, 1968, and T. Apostol, *Mathematical Analysis*, Addison-Wesley, Reading, Massachusetts, 1960.

Theorem XIII.79 (in slightly different forms) is due to M. S. Birman and V. V. Borzov, "On the asymptotics of the discrete spectrum of some singular differential operators," *Probs. Math. Phys.* **5** (1971), 24 (*Topics Math. Phys.* **5** (1972), 19–30), A. Martin (the paper quoted above), and H. Tamura, "The asymptotic eigenvalue distribution for non-smooth elliptic operators," *Proc. Japan Acad.* **50** (1974), 19–22. Earlier, K. Chadam, in "The asymptotic behavior of the number of bound states of a given potential in the limit of large coupling, II," *Nuovo Cimento* **58A**

(1968), 191–204, had proven a similar result for one-dimensional systems and B. Simon, "On the growth of the number of bound states with increase of potential strength," *J. Mathematical Phys.* **10** (1969), 1123–1126 had proven $c\lambda^{3/2} < N(\lambda) < d\lambda^{3/2}$ for λ large; $c, d > 0$; $n = 3$. Martin uses Dirichlet–Neumann bracketing, Chadam uses Jost function methods (see Section XI.8), and Tamura uses Green's function methods. Extensions and further development may be found in a set of papers by Birman and co-workers: M. S. Birman and M. E. Solomjak, "Piece-wise polynomial approximations of functions of the classes W_p^α," *Math. Sb.* **73** (1967), 331–355 (*Math. USSR Sb.* **2** (1967), 295–313), "On the leading term in the asymptotic spectral formula for non-smooth elliptic problems," *Funckcional Anal. i Priložen* (4) **4** (1970), 1–14 [*Functional Anal. Appl.* **4** (1970), 265–275], "On the asymptotic spectrum of non-smooth elliptic equations," *Funkcional Anal. i Priložen* (1) **5** (1971), 69–70 [*Functional Anal. Appl.* **5** (1971), 56–58]; V. V. Borzov, "On some applications of piece-wise polynomial approximations of functions of anisotropic classes W_p^r," *Dokl. Akad. Nauk SSSR* **198** (1971), 499–501 [*Soviet Math. Dokl.* **12** (1971), 804–807]; G. V. Rosenbljum, "On the distribution of eigenvalues of the first boundary value problem in unbounded domains," *Dokl. Akad. Nauk SSSR* **200** (1971), 1034–1036 [*Soviet Math. Dokl.* **12** (1971), 1539–1542].

The method of proof of Theorem XIII.80 is due independently to Birman, in the paper quoted above, and to Simon in the weak trace ideal paper quoted in the Notes to Section 3.

One can also prove limit theorems on the quantities $S_\gamma(V)$ of Problem 32 and use the Lieb–Thirring bounds to extend them to arbitrary potentials in $L^{m/2+\gamma}$ (see Problems 129 and 130).

Variants of Theorem XIII.81 appear in E. C. Titchmarsh's books, *Eigenfunction Expansions, Parts I, II*, Oxford Univ. Press, London and New York, Part I, 2nd ed., 1962; Part II, 1958.

Variants of Theorem XIII.82 appear in F. H. Brownell and C. W. Clark, "Asymptotic distribution of the eigenvalues of the lower part of the Schrödinger operator spectrum," *J. Math. Mech.* **10** (1961), 31–70, and J. B. McLeod, "The distribution of eigenvalues for the hydrogen atom and similar cases," *Proc. London Math. Soc.* **11** (1961), 139–158. Generalizations to other operators appear in H. Tamura, "The asymptotic distribution of the lower part eigenvalues for elliptic operators," *Proc. Japan Acad.* **50** (1974), 185–187.

Extensions of Theorem XIII.82 to various multiparticle systems may be found in Simon's "Geometric methods" paper quoted in the Notes to Section 3.

Applications of Dirichlet–Neumann bracketing to quantum statistical mechanics can be found in D. Robinson, *The Thermodynamic Pressure in Quantum Statistical Mechanics*, Springer-Verlag, Berlin and New York, 1971; to the Fock space energy per unit volume in quantum field theory in F. Guerra, L. Rosen, and B. Simon, "Boundary conditions for the $P(\varphi)_2$ euclidean quantum field theory," *Ann. Inst. H. Poincaré* **25A** (1976), 231–334, and to the Thomas–Fermi model in E. Lieb and B. Simon, "The Thomas–Fermi model of atoms, molecules and solids," *Advances in Math.* **23** (1977), 22–116.

Theorems of the genre of Theorem XIII.82.5 go back at least to the Titchmarsh books quoted above and have been extended to more general potentials. The result is sometimes expressed in the form

$$\sqrt{E_n(V)} = n\pi + \frac{1}{2\pi} n^{-1} \int_0^1 V(x)\,dx + o(n^{-1})$$

By the remarks after the theorem, if V is sufficiently smooth (C^1 is enough), the error will be $o(n^{-2})$.

Section XIII.16 The theory of direct integrals is more general than what we discuss in that nonconstant fibers are allowed. In the case where all fibers are separable, every direct integral over M is unitarily equivalent to one of the following by a unitary which is a fibered map: Let $M = M_\infty \cup (\bigcup_{n=1}^{\infty} M_n)$ where $M_\infty, M_1, \ldots$ are disjoint measurable sets. Let

$$\mathcal{H} = \int_{M_\infty}^{\oplus} \ell_2 \oplus \left(\bigoplus_{n=1}^{\infty} \int_{M_n}^{\oplus} \mathbb{C}^n \right)$$

in terms of our constant fiber direct integrals. Direct integrals were introduced and developed by J. von Neumann, "On rings of operators: Reduction theory," *Ann. Math.* **50** (1949), 401–485. Von Neumann used his theory to decompose weakly closed algebras of operators as direct integrals of factors (see Chapter XVIII). For further discussion and references, see J. Dixmier, *Les Algèbras d'opérateurs dans l'espace Hilbertien*, Chapter II, Gauthier-Villars, Paris, 1969.

While the analysis of differential operators with periodic coefficients did not mention direct integral theory until relatively recently, the key notion that solutions should be analyzed by looking for solutions on $\mathbb{R}$ (or $\mathbb{R}^n$) which are not periodic but obey $\psi(x + a) = e^{i\theta}\psi(x)$ is an early one. It was realized in the mathematical literature by G. Floquet, and in the solid state physics literature by F. Bloch, "Über die Quantenmechanik der Elektronen in Kristallgittern," *Z. Phys.* **52** (1928), 555–600.

There is an enormous mathematical literature on ordinary differential equations with periodic coefficients, especially *Hill's equation*, $(P(x)y'(x))' + Q(x)y(x) = \lambda y(x)$ with P and Q having a common period. Books on the subject include W. Magnus and S. Winkler, *Hill's Equation*, Wiley (Interscience), 1966; M. S. P. Eastham, *The Spectral Theory of Periodic Differential Equations*, Scottish Academic Press, 1973; and parts of E. A. Coddington and N. Levinson, *Theory of Ordinary Differential Equations*, McGraw-Hill, New York, 1955; and E. C. Titchmarsh, *Eigenfunction Expansions, Parts I, II*, Oxford Univ. Press, London and New York, 1962 (2nd ed.), 1958.

The analysis of the discriminant is normally based on direct ordinary differential equation methods rather than analytic perturbation theory, see, for example, Eastham's book. The results go back to work of Liapunov, "Problème generale de la stabilité du movement," *Ann. Fac. Sci. Toulouse* (2) **9** (1907), 203–474, G. Hamel, "Über die lineare Differentialgleichungen zweiter ordnung mit periodischen Koeffizienten," *Math. Ann.* **73** (1913), 371–412; O. Haupt, "Über lineare homogene Differentialgleichungen 2. ordnung mit periodischen Koeffizienten," *Math. Ann.* **79** (1919) 278–285; and H. A. Kramers, "Das Eigenwertproblem in eindimensionalen periodischen Kraftfeldern," *Physica* **2** (1935), 483–490.

The fact that the Mathieu equation has all its gaps is due to E. L. Ince, "A proof of the impossibility of the coexistence of two Mathieu functions," *Proc. Cambridge Phil. Soc.* **21** (1922), 117–120. The example in Problem 139 has been analyzed by R. L. Krönig and W. G. Penny, "Quantum mechanics in crystal lattices," *Proc. Roy. Soc.* **130** (1931), 499–513, and is a common pedagogical model in solid state physics. Theorem XIII.91 (a) and (c) are due to G. Borg, "Eine Umkehrung der Sturm–Liouvilleschen eigenwertaufgabe Bestimmung der Differentialgleichung durch die Eigenwerte," *Acta Math.* **78** (1946), 1–96. Alternative proofs of (a) can be found in H. Hochstadt, "On the determination of a Hill's equation from its spectrum, I," *Arch. Rational Mech. Anal.* **19** (1965), 353–62; H. Hochstadt, "Function theoretic properties of the discriminant of Hill's equation," *Math. Z.* **82** (1963), 237; and P. Ungar, "Stable Hill equations," *Comm. Pure Appl. Math.* **14** (1961), 707–710. The first quoted Hochstadt paper contains a proof of (b) and the statement (d') obtained by replacing real analytic by C^∞. An alternative proof of (c) can be found in H. Hochstadt, "On the determination of a Hill's equation from its spectrum, II," *Arch. Rational Mech. Anal.* **23** (1966), 237–238. (d) is due to W. Goldberg, "On the determination of a Hill's equation from its spectrum," *Bull. Amer. Math. Soc.* **80** (1974),

1111–1112 and H. P. McKean and E. Trubowitz, "Hill's equation and hyperelliptic function theory in the presence of infinitely many branch points," *Comm. Pure Appl. Math.* **29** (1976), 143–226.

This later paper also contains (e), Theorem XIII.92, and the $n = \infty$ case of Theorem XIII.94. (e) is independently due to B. Simon, "On the genericity of non-vanishing instability intervals in Hill's equation," *Ann. Inst. H. Poincaré* **24A** (1976), 91–93. The $n < \infty$ case of Theorem XIII.94 occurs in H. P. McKean and P. van Moerbeke, "The spectrum of Hill's equation," *Invent. Math.* **30** (1975), 217–274. Theorem XIII.93 is due to C. Gardner, J. Greene, and M. Kruskal and R. Muira, "A method for solving the Korteweg de Vries equation," *Phys. Rev. Lett.* **19** (1967), 1095–1097. In comparison with (d) of Theorem XIII.91, Eastham remarked in his book that for periodic Schrödinger operators on $\mathbb{R}^m$ with $m \geq 2$, there are finitely many gaps for suitable potentials which are not even continuous.

The reader can find Hadamard's factorization theorem in E. C. Titchmarsh, *Function Theory*, Cambridge Univ. Press, London and New York, 1939. Theorem XIII.95 is due to W. Kohn, "Analytic properties of Bloch waves and Wannier functions," *Phys. Rev.* **115** (1959), 809–821. Additional discussion of this theorem and the analyticity of band functions can be found in J. Avron and B. Simon, "Analytic properties of band functions" *Ann. Phys.* (1977).

Theorem XIII.96 on uniformly local estimates is due to R. Strichartz, "Multipliers on fractional Sobolev Spaces," *J. Math. Mech.* **16** (1967), 1031–1060.

Basic papers on the spectral analysis of N-dimensional periodic Schrödinger operators are F. Odeh and J. B. Keller, "Partial differential equations with periodic coefficients and Bloch waves in crystals," *J. Mathematical Phys.* **5** (1964), 1499–1504; G. Scherf, "Das Blochsche Theorem für unendliche Systeme," *Helv. Phys. Acta* **39** (1966), 556–560; J. Avron, A. Grossman, and R. Rodrijuez, "Hamiltonians in the one-electron theory of solids," *Rep. Mathematical Phys.* **5** (1974), 113–120; L. E. Thomas, "Time dependent approach to scattering from impurities in a crystal," *Comm. Math. Phys.* **33** (1973), 335–343 and G. M. Troiunielloq, "Scattering theory for Schrödinger operators with L^∞ potentials and distorted Bloch waves," *J. Mathematical Phys.* **15** (1974), 2048–2052; F. Bentosela, "Scattering from impurities in a crystal," *Comm. Math. Phys.* **46** (1976), 153–166. Thomas' paper, in particular, contains the clever argument ((c) of the lemma to Theorem XIII.100) showing that no function $E_n(z)$ is constant and the resulting Theorem XIII.102. Thomas actually handles the case $\beta = 2$ for $n = 3$.

The eigenfunction expansion of Theorem XIII.98 is due to I. M. Gel'fand, "Expansion in series of eigenfunctions of an equation with periodic coefficients," *Dokl. Akad. Nauk SSSR* **73** (1950), 1117–1120.

There are rather subtle spectral properties of periodic Hamiltonians in electric and magnetic fields; see, e.g., A. Grossman, "Momentum-like constants of motion" in *Statistical Mechanics and Field Theory*, (R. N. Sen and C. Weil, eds.) Halsted Press, Washington, D.C., 1972.

The model of impurities that we discussed at the conclusion of this section is unrealistic in the sense that single impurity atoms do not occur but rather a low density of impurity atoms distributed randomly. There is an interesting model of this situation. Let V and W be two potentials on $[0, 1)$. Let z denote a two sided sequence $\{z_n\}$, $n = 0, \pm 1, \pm 2, \ldots$, of 0 and 1. Given z, let $V^{(z)}$ be the function on $\mathbb{R}$ that is $V(x - n)$ on $[n, n + 1)$ if $z_n = 0$ and $W(x - n)$ if $z_n = 1$. Let $H^{(z)} = -d^2/dx^2 + V^{(z)}(x)$ as an operator on $L^2(\mathbb{R})$. Let p be the density of impurities. On $\{0, 1\}^{\mathbb{Z}}$, put the product measure with $\mu(\{0\}) = 1 - p$; $\mu(\{1\}) = p$ on each factor. One now thinks of $H^{(z)}$ as a "random Hamiltonian" and asks about properties that hold for almost every $H^{(z)}$. One can prove that almost every $H^{(z)}$ has a complete set of eigenvectors! See Goldshade and Molchanov, "On the Mott's problem," *Dokl. Akad. Nauk SSSR* **230** (1976), 761–764.

Good references for elementary solid state theory are C. Kittel, *Quantum Theory of Solids*, Wiley, New York, 1963, and Ziman, *Principles of The Theory of Solids*, Cambridge Univ. Press, London and New York, 1972.

Section XIII.17   We have discussed only one aspect of the spectral theory of non-self-adjoint operators. Among the places where the reader can look for additional material are I. C. Gohberg and M. G. Krein, *Introduction to the Theory of Linear Non-self-adjoint Operators*, Amer. Math. Soc. Transl., Vol. **18**, 1969 (Russian original 1965), and *Theory and Applications of Volterra Operators in Hilbert space*, Amer. Math. Soc. Translations, Vol. **24**, 1970; I. C. Gohberg and I. A. Feldman, *Convolution Equations and Projection Methods for Their Solution*, Amer. Math. So. Transl., 1974; R. Douglas, *Banach Algebra Techniques in Operator Theory*, Academic Press, New York, 1972; N. Dunford and J. Schwartz, *Linear Operators*, Part II, Sections XI.6, 9, 10, Wiley (Interscience), 1963; B. Sz. Nagy and C. Foias, *Harmonic Analysis of Operators on Hilbert Space*, North Holland, Amsterdam, 1970, and *Proceedings of a Conference on Operator Theory*, Lecture Notes in Mathematics No. 345, Springer-Verlag, Berlin and New York, 1973; C. Pearcy (ed.), *Topics in Operator Theory*, Amer. Math. Soc., Providence, Rhode Island, 1974. In particular, the reader can consult the first Gohberg-Krein reference and the Dunford-Schwartz reference for more material on the content of this section. Our discussion of determinants is patterned after that in B. Simon, "Notes on infinite determinants of Hilbert space operators," *Adv. Math.* **24** (1977), 244–273. There are some elements in common with the treatment in J. R. Ringrose, *Compact Non-self-adjoint Operators*, Van Nostrand-Reinhold, Princeton, New Jersey, 1971.

If an operator has an eigenvector, it clearly has a nontrivial invariant subspace. The existence of compact operators with no eigenvalue (Example 1) leads to the question of whether every compact operator has a nontrivial invariant subspace. This problem was finally solved (positively) by J. von Neumann (unpublished) in the Hilbert space case and by N. Aronszajn and K. T. Smith, "Invariant subspaces of completely continuous operators," *Ann. Math.* **60** (1954), 345–350. There is a more general problem of whether every bounded operator on a Hilbert space has a nontrivial invariant subspace. This is discussed in H. Helson, *Lectures on Invariant Subspaces*, Academic Press, New York, 1964; W. B. Arveson and J. Feldman, "A note on invariant subspaces," *Michigan Math. J.* **15** (1968), 61–64; and C. Pearcy and N. Salinas, "An invariant subspace theorem," *Michigan Math. J.* **20** (1973), 21–31; and the relevant articles in the collection edited by C. Pearcy quoted above.

What we have called the Schur-Lalesco-Weyl theorem (Theorem XIII.103) has a complex history, in part, because the definition of trace class and $\mathcal{I}_p$ is so recent. The criterion for an infinite matrix to be Hilbert-Schmidt is rather simple (namely, that $\sum_{n,m} |a_{mn}|^2 < \infty$) and so this class was singled out rather early. I. Schur in "Über die charakteristischen Wurzeln einer linearen Substitution mit einer Anwendung auf die Theorie der Integralgleichung," *Math. Ann.* **66** (1909), 488–510, proved that $\sum_{n=1}^{N(A)} |\lambda_n(A)|^2 < \infty$ if A is Hilbert-Schmidt. T. Lalesco in "Une théorème sur les noyaux composés," *Bull. Soc. Sci. Acad. Roumania* **3** (1914–15), 271–272, proved that $\sum_{n=1}^{N(A)} |\lambda_n(A)| < \infty$ for a large class of operators that we would now call trace class. This was extended to the full trace class (expressed as an arbitrary product of two Hilbert-Schmidt operators!) by S. A. Gheorghiu, "Sur l'équation de Fredholm," Thèse, Paris, 1928, and E. Hille and J. D. Tamarkin, "On the characteristic values of linear integral equations," *Acta Math.* **57** (1931), 1–76. H. Weyl, in "Inequalities between the two kinds of eigenvalues of a linear transformation," *Proc. Nat. Acad. Sci. U.S.A.* **35** (1949), 408–411, proved the general relation that

$$\sum_{n=1}^{N(A)} \varphi(|\lambda_n(A)|) \leq \sum_{n=1}^{\infty} \varphi(\mu_n(A))$$

for any positive function φ with $\varphi(e^t)$ convex on $(-\infty, \infty)$ with $\varphi(0) = 0$. Weyl based his proof on general properties of convex functions and the inequality (Problem 154)

$$|\lambda_1(A) \cdots \lambda_n(A)| \leq \mu_1(A) \cdots \mu_n(A)$$

Our proof is taken from the Simon paper quoted above. Extensions of Weyl's result have been obtained by A. Ostrowski, "Sur quelques applications des fonctions convexes et concaves au sens de I. Schur," *J. Math. Pure Appl.* **9** (1952), 253–292; see also the first Gohberg and Krein book.

Using the inequality above, one can prove that

$$\mathrm{Tr}(\varphi(e^{A+B})) \le \mathrm{Tr}(\varphi(e^{A/2} e^B e^{A/2}))$$

for any φ with $t \to \varphi(e^t)$ convex. This is a result of C. Thompson, "Inequalities and partial order on matrix spaces," *Indiana Math. J.* **21** (1971), 469–480. Our proof of the corollary to Theorem XIII.103, which is similar to one of Thompson's, was suggested to us by P. Deift. The basic bound $\|C^{1/2} D C^{1/2}\|_r \le \|CD\|_r$ can also be proven using complex interpolation (Problem 168). For a discussion of the history of the Golden–Thompson inequality, see the notes to Section X.9.

Determinants of certain operators on infinite-dimensional spaces were introduced by I. Fredholm, "Sur une classe d'equations fonctionelles," *Acta Math.* **27** (1903), 365–390. He also introduced the function $D_\mu(A)$ in these cases in terms of the expansion of Problem 153. Fredholm dealt with operators A of the form $(Af)(x) = \int_a^b K(x, y) f(y)\, dy$ and wrote explicit formulas for $\mathrm{Tr}(\bigwedge^k(A))$ rather than using abstract multiltilinear algebra. For example

$$\mathrm{Tr}(\bigwedge^k(A)) = \frac{1}{n!} \int_a^b \cdots \int_a^b K\begin{pmatrix} x_1 & \cdots & x_n \\ x_1 & \cdots & x_n \end{pmatrix} dx_1 \cdots dx_n$$

where

$$K\begin{pmatrix} x_1 & \cdots & x_n \\ y_1 & \cdots & y_n \end{pmatrix} = \det(K(x_i, y_j))_{1 \le i, j \le n}$$

Fredholm's work had profound effects in the history of functional analysis leading to the abstraction of the notion of Hilbert space and to the development of the theory of compact operators. Two books describing "classical" Fredholm theory are F. Smithies, *Integral Equations*, Cambridge Univ. Press, London and New York, 1965 and W. Lovitt, *Linear Integral Equations*, Dover, New York, 1924.

Lidskii's theorem appeared in V. R. Lidskii, "Non-self-adjoint operators with a trace," *Dokl. Akad. Nauk SSSR* **125** (1959), 485–487. Our proof follows Simon's paper.

The Plemelj–Smithies formulas are due to J. Plemelj, "Zur Theorie der Fredholmschen Funktionalgleichung," *Monat. Math. Phys.* **15** (1909), 93–128; and F. Smithies, "The Fredholm theory of integral equations," *Duke Math. J.* **8** (1941), 107–130.

The theory of regularized determinants (Problem 155) for $\mathscr{I}_p$ operators is a theory with contributions made by D. Hilbert, "Grundzüge einer allgemeinen Theorie der linearen Integralgleichungen, Erste Miteilung," *Nacht. Akad. Wiss. Göttingen Math. Phys. Kl. II* (1904), 49–91; H. Poincaré, "Rémarques diverse sur l'équation de Fredholm," *Acta Math.* **33** (1910), 57–86; the papers of Plemelj and Smithies quoted above; F. Carleman, "Zur Theorie der linearen Integralgleichung," *Math. Zeit.* **9** (1921), 196–217; the paper of Hille and Tamarkin quoted in our discussion of the Schur–Lalesco–Weyl theorem; the first book of Gohberg and Krein; the book of Dunford and Schwartz; and that of H. J. Brascamp, "The Fredholm theory of integral equations for special types of compact operators on a separable Hilbert space," *Compositio Math.* **21** (1969), 59–80. This theory has turned out to be useful in studying renormalization theory in the Yukawa field theory in two space-time dimensions: See E. Seiler, "Schwinger functions for the Yukawa models in two dimensions with space-time cutoffs," *Comm.*

Math. Phys. **42** (1975), 163–182. Our treatment of $\det_n$ in Problem 155 uses some of Seiler's ideas. Theorem XIII.104, which appears in Simon's paper, is an abstraction of an argument from E. Seiler and B. Simon, "On finite mass renormalizations in the two-dimensional Yukawa model," *J. Mathematical Phys.* **16** (1975), 2289–2293.

Fredholm theory has been extended to Banach spaces by A. F. Ruston, "On the Fredholm theory of integral operators belonging to the trace class of a general Banach space," *Proc. London Math. Soc.* **53** (1951), 109–124 and "Direct products of Banach spaces and linear functional equations," *Proc. London Math. Soc.* **1** (1953), 327–384; T. Lezinski, "The Fredholm theory of linear equations in Banach spaces," *Studia Math.* **13** (1953), 244–276; and A. Grothendieck, "La Théorie de Fredholm," *Bull. Soc. Math. France* **84** (1956), 319–384. An analogue of the trace class (called nuclear operators) was developed on an arbitrary Banach space by A. Grothendieck, "*Produits tensoriel topologique et espaces nucleaires*," *Mem. Amer. Math. Soc.* **16** (1955). Grothendieck constructed an example of a nuclear operator on $L^\infty(0, 1)$ with $\sum |\lambda_n(A)|^p = \infty$ for any $p < 2$ so the Schur–Lalesco–Weyl theorem in the form $\sum |\lambda_n(A)| \leq \|A\|_1$ fails on general Banach spaces. Indeed, W. Johnson, B. Maury, H. Konig, and J. R. Retherford (in preparation) have shown that if X is a Banach space for which $\sum |\lambda_n(A)| < \infty$ for every nuclear operator, then X is topologically isomorphic to a Hilbert space! Weakened versions of the Schur–Lalesco–Weyl inequality and Lidskii's theorem on a general Banach space are discussed in the Grothendieck memoir, the paper of Johnson et al., and A. S. Marcus and V. I. Macaev, "Analogs of Weyl inequalities and the trace theorem in a Banach space," *Soviet Math. Dokl.* (1972), 299–312.

Completeness theorems for the generalized eigenfunctions of accretive operators go back to M. V. Keldys, "On the characteristic values and characteristic functions of certain classes of non-self-adjoint equations," *Dokl. Akad. Nauk SSSR* **77** (1951), 11–14. Further results of Keldys, Keldys and Lidskii, and Krein and Gohberg are discussed in detail in Chapter V of the first quoted Gohberg and Krein book. Our discussion is patterned in part on that in Dunford and Schwartz.

The discussion in Example 5 follows that in B. Simon, "The bound state of weakly coupled Schrödinger operators in one and two dimensions," *Ann. Phys.* **97** (1976), 279–288. The result that $-d^2/dx^2 + \lambda V$ has a bound state for all small λ if and only if $\int V(x)\,dx \leq 0$ is proven there for all V with $\int (|x|^2 + 1)|V(x)|\,dx < \infty$. The analyticity result of Theorem XIII.110 does not require that V have compact support but only that $\int e^{a|x|} |V(x)|\,dx < \infty$ for some $a > 0$. In two dimensions, $-\Delta + \lambda V$ has a bound state for all small λ if and only if $\int V(x)\,dx \leq 0$ so long as

$$\int |V(x)|^{1+\varepsilon}\,dx < \infty \quad \text{for some} \quad \varepsilon > 0$$

and

$$\int |V(x)|(1 + |x|^2)^\varepsilon\,dx < \infty \quad \text{for some} \quad \varepsilon > 0$$

But the eigenvalue $E(\lambda)$ is *not* analytic at $\lambda = 0$ in the two-dimensional case. In fact, $|E(\lambda)| \leq \exp(-1/a\lambda)$ for suitable $a > 0$. All these results require the use of the modified determinant, $\det_2$, for $\mathscr{I}_2$ (see Problem 155). These questions are further discussed in R. Blankenbecker, M. L. Goldberger, and B. Simon, "The bound states of weakly coupled long-range one-dimensional quantum Hamiltonians," *Ann. Physics* **108** (1977), 69–78, and K. Klaus, *Ann. Physics* (to appear).

Finally, we remark that the Jost function discussed in Section XI.8 is a Fredholm determinant. The s-wave Lippmann–Schwinger equation reads

$$\psi(\cdot, k) = \psi_0(\cdot, k) - (A_k \psi)(\cdot, k) \tag{207}$$

where A_k is the integral operator with kernel

$$k^{-1}\exp[ik(\max\{x,y\})]\sin[k(\min\{x,y\})]V(y)$$

It can be shown that $\eta(k) = \det(1 + A_{-k})$, and that the factorization $S(k) = \eta(k)/\eta(-k)$ is connected with the meromorphic solution of (207).

PROBLEMS

†1. Prove that $\mu_n(A) \leq \mu_n(B)$ for any n if $A \leq B$ (defined in Section XIII.2).

2. Use the min–max principle to prove that
$$|\mu_k(A) - \mu_k(B)| \leq \|A - B\|$$
for bounded self-adjoint operators, where μ_k is given by Theorem XIII.1.

†3. Prove Theorem XIII.4.

†4. (a) Let $p = i\,d/dx$ on $L^2(\mathbb{R})$. Prove that
$$p^2 + x^2 + \beta x^4 \leq C_1(p^2 + x^2)^2 \leq C_2(p^2 + x^2 + \beta x^4)^2$$
for any fixed $\beta > 0$.
(b) Conclude that
$$\lim_{N \to \infty} \left(\sum_{i=1}^N a_i^{(n)} \varphi_i, H\left(\sum_{i=1}^N a_i^{(n)} \varphi_i \right) \right) = \mu_n(H)$$
if $H = p^2 + x^2 + \beta x^4$, where $\{\varphi_i\}_{i=1}^N$ are the eigenfunctions of $p^2 + x^2$ and $\sum a_i^{(n)} \varphi_i$ is the expansion of the nth eigenfunction of H in terms of the φ_i.

5. Let $H = p^2 + x^2 + \beta x^{2m}$. Prove the applicability of Theorem XIII.4 using the eigenfunctions of $p^2 + x^2$ as trial functions.

6. Using φ_1 and φ_3, find an upper bound on the ground state energy of $p^2 + x^2 + 0.2x^4$. Using Temple's inequality find a lower bound. Compare with the "exact" result and perturbation theory computations in Section XII.3.

7. Using a trial vector of the form $\psi(\mathbf{r}) = c\exp(-\alpha r^2)$ and varying α, compute an upper bound for the ground state of $-\Delta - r^{-1}$ and compare with the exact answer.

8. Write a computer program to compute the lowest eigenvalue of $p^2 + x^2 + x^4$ to six-place accuracy.

9. Let $\lambda_n(c)$ denote the nth eigenvalue of $-d^2/dx^2 + x^2 + cx^4$ for $c \geq 0$. Find a constant $c_0 > 0$ so that $\lambda_2(c) - \lambda_1(c) \geq c_0$ for all c. (Hint: Compare the $\lambda_n(c)$ to the eigenvalues of $-d^2/dx^2 + x^2$ and $-d^2/dx^2 + cx^4$ and use the min-max principle.)

10. Find and prove an analog of Theorem XIII.4 for excited states.

11. (Rayleigh's theorem) Let $\mu_1, \ldots, \mu_n$ be the eigenvalues of an $n \times n$ Hermitian matrix and let $\lambda_1, \ldots, \lambda_{n-1}$ be the eigenvalues of its restriction to an $(n-1)$-dimensional subspace. Prove that $\mu_1 \leq \lambda_1 \leq \mu_2 \leq \lambda_2 \leq \cdots \leq \mu_{n-1} \leq \lambda_{n-1} \leq \mu_n$.

12. Suppose that V is a regular perturbation of H_0 and that H_0 has a discrete, nondegenerate ground state. Let $E(\lambda)$ be the ground state energy for $H_0 + \lambda V$ and $\psi(\lambda)$ the normalized ground state eigenvector. Suppose that $\varphi(\lambda)$ is given so that $\|\varphi(\lambda)\|^{-1}\varphi(\lambda) - \psi(\lambda) = O(\lambda^{n+1})$ in H_0-graph norm and $\|\varphi(\lambda)\| = 1 + O(\lambda)$ (for example, $\varphi(\lambda)$ could be the first n terms of an unnormalized Rayleigh–Schrödinger expansion).
    (a) Prove that $(H(\lambda) - E(\lambda))\varphi(\lambda) = O(\lambda^{n+1})$.
    (b) Prove that
    $$(\varphi(\lambda), H(\lambda)^2\varphi(\lambda))\|\varphi(\lambda)\|^2 - (\varphi(\lambda), H(\lambda)\varphi(\lambda))^2 = O(\lambda^{2n+2})$$
    (c) Using Temple's inequality, prove that $E(\lambda) - \|\varphi(\lambda)\|^{-2}(\varphi(\lambda), H(\lambda)\varphi(\lambda)) = O(\lambda^{2n+2})$.
    Reference: E. Harrel, II, Princeton Univ. Thesis, Princeton, New Jersey, 1976.

13. (a) Suppose that P is a projection and that $V > 0$ is a positive operator and H_0 is self-adjoint. Prove that as quadratic forms $H_0 + V \geq H_0 + V^{1/2}PV^{1/2}$.
    (b) Let A be self-adjoint with Ker $A = 0$ and let Q be a finite-dimensional projection so that Ran $Q \subset D(V^{-1/2}A)$. Let $(QAV^{-1}AQ)^{-1}$ denote the inverse of the operator $QAV^{-1}AQ$ on Ran Q. Prove that
    $$P = V^{-1/2}AQ(QAV^{-1}AQ)^{-1}QAV^{-1/2}$$
    is a projection and conclude that
    $$H_0 + V \geq H_0 + AQ(QAV^{-1}AQ)^{-1}QA$$
    (c) Prove that if ψ_0 is an eigenvector of H_0, then
    $$H_0 + (\psi_0, V^{-1}\psi_0)^{-1}(\psi_0, \cdot)\psi_0 \leq H_0 + V$$
    (Hint: Use $A = 1$, $Q = (\psi_0, \cdot)\psi_0$.)
    Reference for Part (c): W. Thirring, *Vorlesungen Über Mathematische Physik*, T7: *Quantenmechanik*, Universität Wien Lecture Notes, Section 2.9.

14. (a) Suppose that $\sigma(H) \cap (a, b) = \emptyset$ for some self-adjoint operator H. Let φ be in $D(H)$ with $\|\varphi\| = 1$. Prove that
    $$\left(\varphi, \left[H - \left(\frac{a+b}{2}\right)\right]^2 \varphi\right) \geq \left(\frac{a-b}{2}\right)^2$$
    (b) Rewrite (a) in terms of $\gamma = (\varphi, H\varphi)$ and $\eta = (\varphi, H^2\varphi) - (\varphi, H\varphi)^2$ to read $\eta \geq (\gamma - a)(b - \gamma)$.
    (c) Conclude that, if $\varphi \in D(H)$ with
    $$\eta < (\gamma - a)(b - \gamma)$$
    then $\sigma(H) \cap (a, b) \neq \emptyset$.

15. (generalized Temple's inequality) Suppose that $\sigma(H) \cap (a, b)$ consists of a single point λ. Let φ be a trial vector so that $\varphi \in D(H)$ and
    $$\eta < (\gamma - a)(b - \gamma) \tag{208}$$
    where $\gamma = (\varphi, H\varphi)$; $\eta = (\varphi, H^2\varphi) - \gamma^2$.
    (a) Prove that $\gamma \in (a, b)$.
    (b) Let $a' = \gamma - (b - \alpha)^{-1}\eta$ and $b' = \gamma + (\gamma - a)^{-1}\eta$. Prove that $a < a' < b' < b$.

(c) Let $a < a'' < a'$. Prove that $\eta < (\gamma - a'')(b - \gamma)$ and conclude that $\sigma(H) \cap (a'', b) \neq \emptyset$ (use Problem 14).

(d) Prove the generalized Temple's inequality, i.e., that under the hypothesis (208) we can conclude
$$\gamma - (b - \gamma)^{-1}\eta < \lambda < \gamma + (\gamma - a)^{-1}\eta$$

Remark See the Notes to Section XIII.2 for references to this problem.

†16. (a) Prove that
$$H_\ell = -\frac{d^2}{dr^2} + \frac{\ell(\ell + 1)}{r^2} + V(r)$$

is essentially self-adjoint on $\{u \mid u \in C_0^\infty(0, \infty)\}$ if $\ell \neq 0$ and $V \in C_0^\infty(0, \infty)$.

(b) Prove that $H_0 = -d^2/dr^2 + V(r)$ is essentially self-adjoint on $\{u \mid u \in C_0^\infty[0, \infty), u(0) = 0\}$.

†17. Let $\varphi \in C_0^\infty(\mathbb{R})$ have isolated zeros within its support. Let $H_0 = -d^2/dx^2$ on the usual domain. Prove that $|\varphi| \in Q(H_0)$.

18. (Calogero's bound) Let $V \in C_0^\infty[0, \infty)$ be negative. Let u be the solution of $u'' = Vu$; $u(0) = 0$; $u'(0) = 1$. Define $v(r)$ by $\tan v(r) = |V(r)|^{1/2}u(r)/u'(r)$.
   (a) Prove that v obeys the equations
$$v'(r) = |V(r)|^{1/2} - \tfrac{1}{2}(V'(r)/|V(r)|)\cos^2 v \tan v$$
$$v(0) = 0$$
   (b) Conclude that if V is monotone, $N(V) \leq (2/\pi) \int_0^\infty |V(r)|^{1/2} dr$.

19. (Bargmann's bound) Let u solve $u'' = [\ell(\ell + 1)r^{-2} + V(r)]u$; $u(0) = 0$.
   (a) Define $a_\ell(r)$ by
$$u'(r)[r^{\ell+1} + a_\ell(r)r^{-\ell}] = u(r)[(\ell + 1)r^\ell - \ell a_\ell(r)r^{-\ell-1}]$$
   Prove that
$$a'_\ell(r) = -(2\ell + 1)^{-1}V(r)r^{-2\ell}[r^{2\ell+1} + a_\ell(r)]^2$$
   (b) Let $b_\ell(r) = a_\ell(r)/r^{2\ell+1}$. Prove that $b_\ell(0) = 0$ and that
$$b'_\ell(r) \leq (2\ell + 1)^{-1}r|V(r)|[b_\ell(r) + 1]^2$$
   in the region where b is positive.
   (c) Prove Bargmann's bound (Theorem XIII.9a) for general ℓ.

20. (a) Use Neumann bracketing (Section 15) to prove that if V is bounded with compact support, then $-\Delta + \lambda V$ has at most one negative eigenvalue for λ small.
   (b) Use Neumann bracketing to prove that if V is a bounded function on $\mathbb{R}^n$ with compact support, then $-\Delta + V$ has only finitely many negative eigenvalues.

21. Find a potential V on $L^2(\mathbb{R})$ having compact support, with V negative on some open set, so that $-d^2/dx^2 + V$ has no negative eigenvalues.

*22. Let V be a function on $\mathbb{R}$ with $\int_{-\infty}^\infty r|V(r)| \, dr < \infty$. Prove that $-d^2/dx^2 + V$ has at most $1 + \int_{-\infty}^\infty r|V(r)| \, dr$ bound states. (Hint: Use Bargmann's bound and Dirichlet boundary conditions.)

23. (a) Fix ℓ. Let V be a central potential with $n_\ell(V) = n$. Show that $V = V_1 + \cdots + V_n$ where the V_i have disjoint support and $n_\ell(V_i) \geq 1$. (Hint: Use the zeros of $u_\ell(r; E = 0)$.)
(b) Suppose that it is known that $n_\ell(V) = 0$ if

$$\int_0^\infty f(x)g(V(x))\, dx < 1$$

for fixed f, g real and positive. Prove that

$$n_\ell(V) \leq \int_0^\infty f(x)g(V(x))\, dx$$

Reference: The Grosse, Martin, Glaser, and Thirring paper quoted in the Notes to Section 3.

24. (a) Let $n \geq 3$. Let $V \leq 0$ almost everywhere. Prove that $N(V) = 0$ if and only if $|V|^{1/2}(-\Delta)^{-1}|V|^{1/2}$ has norm less than 1.
(b) Let $n \geq 3$ and let $0 \leq \alpha \leq 1$. Show that there exists a constant $C_{\alpha, n}$ so that if $\||r^\alpha|V|^{1/2}\|_{n/1-\alpha} \leq C_{\alpha, n}$, then $N(V) = 0$. (Hint: Look at the proof of Strichartz theorem, Theorem X.21.)
(c) Let $n_0(V)$ be the number of bound states of angular momentum zero in $\mathbb{R}^3$. Prove that for $\frac{1}{2} \leq \alpha < \infty$ there exists a constant $\tilde{C}_{\alpha, 1}$ so that $n_0(V) = 0$ if

$$\int r^{\alpha/1-\alpha}|V(r)|^{1/(2-2\alpha)}\, dr \leq \tilde{C}_{\alpha, 1}$$

(d) Prove the GGMT bound except for the precise constant.
Reference for (a)–(c): The paper of Glaser et al. and the review article of Simon quoted in the Notes of Section 3.

†25. (a) Let A be a positive self-adjoint operator and let B be a form-bounded perturbation of A with relative bound 0. Suppose that $[0, \infty) \subset \sigma(A + \lambda B)$ for all $\lambda \in \mathbb{R}$. Prove that $\mu_n(A + \lambda B)$ is a monotone decreasing continuous function of $\lambda \in (0, \infty)$ by using the min–max principle.
(b) Suppose $\sigma_{\text{ess}}(A + \lambda B) = [0, \infty)$ for all $\lambda \in (0, \infty)$. Prove the result of (a) by employing the perturbation theory of Chapter XII.

†26. Fill in the details in the proof of Theorem XIII.10.

27. Prove the Ghirardi–Rimini bound

$$N(V) \leq \left(\frac{1}{4\pi}\right)^2 \int \frac{V(x)V(y)}{|x-y|^2}\, d^3x\, d^3y$$

28. (a) Let $V \in R$ and let $\{P_\Omega^\lambda\}$ be the spectral projections of $-\Delta + \lambda V$. Let $\lambda > 1$. Prove that

$$\dim(\text{Ran } P^\lambda_{(-\infty, 0)}) \geq \dim(\text{Ran } P^1_{(-\infty, 0]})$$

(b) Conclude that

$$\dim(\text{Ran } P^1_{(-\infty, 0]}) \leq (4\pi)^{-2} \int |x-y|^{-2}|V(x)||V(y)|\, d^3x\, d^3y$$

XIII: SPECTRAL ANALYSIS

†*29. Prove that for $V, W \in C_0^\infty(\mathbb{R}^n)$,

$$\int d\mu_{x,\,y;\,t} V(\omega(s))\exp\left(-\int_0^t W(\omega(s))\,ds\right)$$

is continuous in x, y, t for $t > 0$, $x, y \in \mathbb{R}^n$ where $d\mu_{x,\,y;\,t}$ is the conditional Wiener measure. (Hint: It will help to realize all integrations over the same set of paths by mapping paths ω going from 0 to 0 in time 1, to paths $\tilde\omega$ going from x to y in time t by

$$\tilde\omega(s) = \omega(st^{-1}) + x + (y-x)st^{-1}$$

†30. (a) Suppose that A is a trace class operator on $L^2(\mathbb{R}^n)$ with an integral kernel $K(x, y)$ that is pointwise positive and continuous. Prove that $\text{Tr}(A) = \int K(x, x)\,dx$. (Hint: Let P_n be a sequence of projections onto finite-dimensional spaces of functions piecewise constant on $\mathbb{R}^n$ so that $P_n \to I$ strongly. Write $\text{Tr}(A) = \lim_{n\to\infty} \text{Tr}(P_n A P_n)$ and evaluate it explicitly.)
   (b) For $W \in L^1(\mathbb{R}^n)$, prove that $W^{1/2}e^{-tH_0}$ is Hilbert–Schmidt. For $W \in L^1(\mathbb{R}^n)$ and $V \geq 0$, prove that $W^{1/2}e^{-t(H_0+V)}$ is Hilbert–Schmidt.
   (c) Prove (14).

31. (a) Let $N_\alpha(V)$ be the number of eigenvalues of $-\Delta + V$ on $L^2(\mathbb{R}^3)$ less than $-\alpha$. Use the Birman–Schwinger bound and Young's inequality to prove that

$$N_\alpha(V) \leq (4\pi)^{-1}(\sqrt{2\alpha})^{-1} \int \left|\min\left\{V(x) + \frac{\alpha}{2}, 0\right\}\right|^2 dx$$

   (b) Let $E_1(V), \ldots, E_n(V)$ be the negative eigenvalues of $-\Delta + V$. Prove that

$$\sum_i |E_i(V)| = \int_0^\infty N_\alpha(V)\,d\alpha$$

   (c) Prove the Lieb–Thirring bound

$$\sum_i |E_i(V)| \leq \frac{4}{15\pi} \int |V_-(x)|^{5/2}\,dx$$

32. Let $S_\gamma(V)$ be the sum of the numbers $|E|^\gamma$ over all negative eigenvalues of $-\Delta + V$.
   (a) Prove that $S_\gamma(V) = \gamma \int_0^\infty E^{\gamma-1} N_E(V)\,dE$, if $\gamma > 0$ and $N_E(V)$ is the number of eigenvalues less than $-E$.
   (b) Prove that for $V \in C_0^\infty$ and $W = -V_-$,

$$S_\gamma(V) \leq \gamma\Gamma(\gamma)(m+1)\int_0^\infty t^{-\gamma}\text{Tr}\left[\sum_{j=0}^m \binom{m}{j}(-1)^j e^{-(H_0+jW)t}\right]dt$$

   for any m. $\Gamma(\cdot)$ is the Euler gamma function $\Gamma(\alpha) = \int_0^\infty u^{\alpha-1}e^{-u}\,du$.
   (c) Prove that if V is a function on $\mathbb{R}^n$ and γ is any number with $\gamma \geq 0$ and $\gamma + \frac{1}{2}n > 1$, then

$$S_\gamma(V) \leq C_{n,\,\gamma} \int |V_-(x)|^{n/2+\gamma}\,d^n x$$

33. (a) Prove that for $n \geq 3$,

$$N_E(V) \leq C_n \int |(E+V)_-(x)|^{n/2}\,d^n x$$

by using the Cwikel–Lieb–Rosenbljum bound.

(b) Prove that for $V \in C_0^\infty$,

$$\gamma \int_0^\infty dE \int |(E + V)_-(x)|^{n/2} E^{\gamma - 1} \, dx = d_\gamma \int |V_-(x)|^{n/2 + \gamma} \, dx$$

where $d_\gamma = \gamma \int_0^1 (1 - E)^{n/2} E^{\gamma - 1} \, dE$.

(c) Provide an alternative proof of the bound of Problem 32c.

Remark The bounds of Problems 31, 32c, and 33c for $\gamma > 0$ are due to E. Lieb and W. Thirring, "Inequalities for the moments of the eigenvalues of the Schrödinger equation and their relation to Sobolev inequalities," in *Studies in Mathematical Physics: Essays in Honor of Valentine Bargmann* (E. Lieb, B. Simon, and A. S. Wightman, eds.), Princeton Univ. Press, Princeton, New Jersey, 1976. Their method of proof is illustrated in Problem 31.

34. (a) Let C be compact and let $A(z) = z^{-1}C$. Show that $(1 - A(z))^{-1}$ exists if and only if z is not an eigenvalue of C. Prove that the conclusion of Theorem XIII.13 may fail if $A(z)$ has compact residues instead of finite rank residues.

(b) Let $A(z) = \sum_{n=1}^\infty z^{-n} n P_n$ where $\{P_n\}$ is a family of rank-one projections with $P_n P_m = 0$ if $n \neq m$. Prove that the conclusions of Theorem XIII.13 may fail if $A(z)$ has essential singularities.

†35. Prove that $\sigma_{\text{ess}}(A)$ is a closed subset of $\mathbb{C}$ for any closed operator.

†36. Prove (b) of the strong spectral mapping theorem (Lemma 2 in Section 4).

37. Let A and B be positive self-adjoint operators such that $(A + 1)^{-n} - (B + 1)^{-n}$ is compact for some positive integer n. Prove that $\sigma_{\text{ess}}(A) = \sigma_{\text{ess}}(B)$.

38. (a) Let A and C be closed operators with $D(C) \supset D(A)$. Prove that $C(A - z)^{-1}$ is compact for some $z \in \rho(A)$ if and only if it is compact for all $z \in \rho(A)$.

(b) Prove that $C(A - z)^{-1}$ is compact if and only if C is a compact map from $\langle D(A), \|\cdot\|_A \rangle$ to $\langle \mathscr{H}, \|\cdot\| \rangle$ where $\|\psi\|_A^2 = \|\psi\|^2 + \|A\psi\|^2$.

39. Let A be self-adjoint and positive. A bounded form perturbation C of A is called **relatively form compact** if and only if C is compact from $\mathscr{H}_{+1}$ to $\mathscr{H}_{-1}$. Prove that if C is relatively form compact, then C has relative form bound zero and $\sigma_{\text{ess}}(A) = \sigma_{\text{ess}}(A + C)$.

40. Find a symmetric operator H with self-adjoint extensions A and B so that $\sigma_{\text{ess}}(A) \neq \sigma_{\text{ess}}(B)$. (*Hint:* Let h be a symmetric operator with deficiency indices $(1, 1)$ and let a and b be suitable self-adjoint extensions. Take $H = h \otimes I$, $A = a \otimes I$, and $B = b \otimes I$ where I is the identity on ℓ_2.)

41. Let $V \in L^q(\mathbb{R}^n) + L_\epsilon^\infty(\mathbb{R}^n)$ with $q \geq \max\{n/2, 2\}$ if $n \neq 4$ and $q > 2$ if $n = 4$. Prove that V is a relatively compact perturbation of $-\Delta$ so that $\sigma_{\text{ess}}(-\Delta + V) = [0, \infty)$.

†42. Using Theorem XIII.17, prove that

$$\inf\{\Sigma_D| \,\#(D) = m + 1\} \geq \inf\{\Sigma_D| \,\#(D) = m\}$$

and conclude that

$$\inf\{\Sigma_D| \,\#(D) = 2\} = \inf\{\Sigma_D| \,\#(D) \geq 2\}$$

370 XIII: SPECTRAL ANALYSIS

43. (a) Suppose that $U(t)$ is a one-parameter unitary group whose infinitesimal generator has purely absolutely continuous spectrum. Let A be compact operator satisfying $AU(t) = U(t)A$. Prove that $A = 0$ (Hint: Prove that w-$\lim_{t\to\infty} U(t) = 0$.)
    †(b) Prove that the operators associated with disconnected diagrams (according to either of the rules in Section 5) are not compact. We emphasize that the center of mass motion has been removed.

†44. (a) Given $V \in R$, show that there exist $V^{(n)} \in C_0^\infty$ so that $\|V^{(n)} - V\|_R \to 0$ as $n \to \infty$.
    (b) Given $V \in R + (L^\infty)_\varepsilon$, show that there exist $V^{(n)} \in C_0^\infty$ so that $\|(H_0 + 1)^{-1/2}(V_n - V) \times (H_0 + 1)^{-1/2}\| \to 0$ as $n \to \infty$.
    (c) Given $V_{ij} \in R + (L^\infty)_\varepsilon$, $1 \le i < j < N$, show that there exist $V_{ij}^{(n)} \in C_0^\infty$ so that $H^{(n)} = H_0 + \sum_{i<j} V_{ij}^{(n)}$ converges in norm resolvent sense to $H = H_0 + \sum_{i<j} V_{ij}$.

45. (a) Let $V \in L^2(\mathbb{R}^3) + L^\infty(\mathbb{R}^3)_\varepsilon$ and $\psi \in D(\Delta)$. Define $\psi_a(x) = \psi(x - a)$ for $a \in \mathbb{R}^3$. Prove that $\lim_{a\to\infty} \|V\psi_a\| = 0$.
    (b) Prove that $\lim_{a\to\infty} \|(-\Delta + V - \lambda)\psi_a\| = \|(-\Delta - \lambda)\psi\|$ and conclude that $[0, \infty) \subset \sigma(-\Delta + V)$.
    (c) Let H be the Hamiltonian $H_0 + \sum_{i<j} V_{ij}$ with $V_{ij} \in L^2 + (L^\infty)_\varepsilon$. Prove that $[\Sigma, \infty) \subset \sigma(H)$ without using wave operators.
    Remark This method was used by Hunziker in the paper cited in the Notes to Section 5.

46. Let $\tilde{H}_0^{(N)} = \sum_{i=1}^N (2m)^{-1}\Delta_i$ and $H^{(N)} = H_0^{(N)} + \sum_{i<j} V(r_i - r_j)$ (all V_{ij} are the same function). Suppose that $V \in R + (L^\infty)_\varepsilon$. Let $\mathcal{H}_{\text{Fermi}}^{(N)}$ be the subspace of $L^2(\mathbb{R}^{3N-3})$ of functions odd under the coordinate change induced by the interchange of any two particles. Let $H_F^{(N)} = H^{(N)} \restriction \mathcal{H}_{\text{Fermi}}^{(N)}$. Prove that

$$\sigma_{\text{ess}}(H_F^{(N)}) = [\Sigma, \infty)$$

where $\Sigma = \inf \sigma(H_F^{(N-1)})$.
Remark 1 To prove the analogue of Lemma 1 in our proof of the HVZ theorem, it is easier to use Hunziker's method (Problem 45) rather than wave operators.
Remark 2 For extensions to other permutation symmetries and symmetry groups, see the papers of Simon, Balslev, and Zhislin and Sigalov cited in the Notes to Section 5.

47. (a) Suppose the basic smoothness estimate (31) holds for a dense set of φ. Prove that for any φ, $R(\lambda \pm i\varepsilon)\varphi \in D(A)$ a.e. in λ (ε fixed, $\varepsilon \ne 0$) and that (31) holds.
    (b) Suppose that for a fixed $\varepsilon > 0$, $\int_{-\infty}^\infty \|AR(\lambda \pm i\varepsilon)\varphi\|^2 \, d\lambda < \infty$ for each $\varphi \in \mathcal{H}$. Prove that

$$\int_{-\infty}^\infty \|AR(\lambda \pm i\varepsilon)\varphi\|^2 \, d\lambda \le C\|\varphi\|^2$$

for some C and all φ. (Hint: Apply the closed graph theorem to the map $\varphi \mapsto AR(\lambda + i\varepsilon)\varphi$ of $\mathcal{H}$ into $L^2(\mathbb{R}, d\lambda; \mathcal{H})$.)
    (c) Suppose that $\sup_{\varepsilon>0} \int_{-\infty}^\infty \|AR(\lambda \pm i\varepsilon)\varphi\|^2 \, d\lambda < \infty$ for each φ. Prove that A is H-smooth. (Hint: Use (b) and the uniform boundedness principle.)

48. Let f be a bounded Borel function on $\mathbb{R}$ and suppose $f(H)$ is H-smooth for some self-adjoint operator H. Prove that $f(H) = 0$.

49. (a) Let H be self-adjoint and let A be H-smooth. Prove that A is $|H|^\alpha$-bounded for any $\alpha > \frac{1}{2}$. (Hint: Use form (3) of Theorem XIII.25 to prove that $(H^2 + 1)^{-\alpha/2}A^*$ is bounded.)

(b) Let $H = -i\,d/dx$ on $L^2(\mathbb{R})$. Prove that there exist $\varphi \in Q(H)$ that are not bounded.
(c) Find an H-smooth operator A so that A is not $|H|^{1/2}$-bounded. (Hint: Use Example 1.)

50. Let H be multiplication by x on $L^2([\alpha, \beta], dx)$ with $\alpha, \beta \in \mathbb{R}$. Suppose that A is bounded and A^*A has integral kernel K. Prove that $\|A\|_H^2 \equiv \|K\|_\infty$.

*51. Let H be multiplication by x on $L^2([\alpha, \beta], dx)$ with $\alpha, \beta \in \mathbb{R}$. Suppose that A is bounded and H-smooth. Prove that A^*A has the form (37). (Hint: First show that A^*x is in L^∞ for every x with $\|A^*x\|_\infty \leq C\|x\|_2$ and then use the Dunford–Pettis theorem (Problem 33 in Chapter V) to find a bounded measurable function F from $[\alpha, \beta]$ to $\mathscr{H}$, so that $(A^*x)(\lambda) = (F(\lambda), x)$.
Reference for Problem 51: Kato's *Studia Math.* paper (see the Notes to Section 7).

52. Let A and B be self-adjoint operators on Hilbert spaces $\mathscr{H}_1$ and $\mathscr{H}_2$. Suppose C is A-smooth and D is a bounded operator on $\mathscr{H}_2$. Prove that $C \otimes D$ is $A \otimes I + I \otimes B$-smooth.

53. Under the hypotheses of Theorem XIII.26, prove that for any $\lambda \in \mathbb{C}$ with $|\lambda| \leq 1$, $H_0 + \lambda \sum_{i=1}^n C_i$ is a strictly m-accretive form on $Q(H_0)$, and that the associated operator H has $\sigma(H) \subset \sigma(H_0)$ and obeys $\sup\| |C_i|^{1/2}(H-z)^{-1}|C_j|^{1/2}\| < \infty$ for all i, j (continued in the next problem).

*54. (continued from Problem 53) Let $R(\mu)$ be the resolvent of H_0 and $R(\mu; \lambda)$ the resolvent of $H(\lambda) = H_0 + \lambda \sum_{i=1}^n C_i$. Define $W^\pm(\lambda)$ by

$$(\varphi, W_\pm(\lambda)\psi) = (\varphi, \psi) \mp \frac{\lambda}{2\pi i} \sum_{i=1}^n \int_{-\infty}^\infty (C_i^{1/2} R(\mu \pm i0)\varphi, |C_i|^{1/2} R(\mu \mp i0, \lambda)\psi)\,d\mu$$

Prove that
(a) $W_\pm(\lambda)$ are analytic in the region $|\lambda| \leq 1$.
(b) $W_\pm(\lambda)$ are invertible and $H(\lambda) = W_\pm(\lambda) H_0 W_\pm(\lambda)^{-1}$.
(c) If λ is real, $W^\pm(\lambda) = \Omega^\pm(\lambda)$.

55. Let A and H_0 be self-adjoint operators with $\text{Ker}(A) = \{0\}$. Prove that, for any positive integers $n \neq m$, at most one of A^n and A^{-m} is H_0 smooth.

56. Let $V \in R$, the Rollnik class, with $\|V\|_R < 4\pi$. Prove that the wave operators provide unitary equivalences of $-\Delta$ and $-\Delta + V$ and in particular that scattering is complete.

57. (a) Let $H_n = H_0 + A_n^* B_n$ where B_n is H_0-smooth and A_n is H_n-smooth. Suppose that $\sup_n \|A_n\|_{H_n} < \infty$ and $\lim_{n\to\infty} \|B_n\|_{H_0} = 0$. Prove that $\Omega_n^\pm \equiv \text{s-lim}_{t \to \mp\infty} e^{itH_n} e^{-itH_0}$ converges to 1 *in norm*. In particular verify the norm continuity of $\Omega^\pm(\lambda) = \text{s-lim}_{t \to \mp\infty} e^{it(H_0 + \lambda C)} e^{-itH_0}$ for $\lambda \in (-1, 1)$, in the context of Theorem XIII.26.
(b) Let $V_n \to V$ in Rollnik norm. Prove that the corresponding S matrices converge strongly. (Hint: Write $V_n = W_n + Y_n$, $V = W + Y$, so that $Y_n \to Y$ in $L^1 \cap R$, $W_n \to W$ in R, and $\sup_n \|W_n\|_R < 4\pi$.)

58. (a) Let H_0 be the operator on $L^2[0, \infty)$ that is the closure of $-d^2/dx^2$ on $\{u \in C_0^\infty[0, \infty) \,|\, u(0) = 0\}$. Let $E \notin \sigma(H_0)$ and let

$$K_E(x, y) = E^{-1/2} \sin[\sqrt{E} \min\{x, y\}] \exp[i\sqrt{E} \max\{x, y\}]$$

where $\sqrt{E}$ is the square root with Im $\sqrt{E} > 0$. Prove that
$$[(H_0 - E)^{-1}\varphi](y) = \int_0^\infty K_E(x, y)\varphi(y)\, dy$$
(b) $|K_E(x, y)| \leq \sqrt{xy}$.
(c) Let V be a measurable function on $[0, \infty)$ with $\int_0^\infty x|V(x)|\, dx < \infty$. Then,
$$\sup_{E \in \mathbb{R}} \||V|^{1/2}(H_0 - E)^{-1}|V|^{1/2}\| < \infty$$
(d) If $\int_0^\infty x|V(x)|\, dx < 1$, then H_0 and $H_0 + V$ are unitarily equivalent and the wave operators are unitary equivalences.

59. Let A and H be bounded self-adjoint operators and let $R(\mu) = (H - \mu)^{-1}$. Prove that
$$|\varepsilon|\,|(R(\lambda + i\varepsilon)\varphi, [H, A]R(\lambda + i\varepsilon)\varphi)| \leq \|A\|\,\|\varphi\|^2$$
and use this to prove that $(i[H, A])^{1/2}$ is H-smooth if $i[H, A] \geq 0$.

60. Let A and B be bounded self-adjoint operators and c a strictly positive real number. Prove that $i[A, B] \geq cI$ is impossible by:
(a) using the theory of smooth perturbations;
(b) direct computation (look at $e^{iAt}Be^{-iAt}$).

61. Extend the Kato–Putnam theorem to the case where H is unbounded and $i[H, A] > 0$ means that $i(A\varphi, H\varphi) - i(H\varphi, A\varphi) > 0$ for all $\varphi \in D(H)$, $\varphi \neq 0$.

†62. Suppose that $f(t)$ is a Banach-space-valued uniformly continuous function on $\mathbb{R}$. Suppose also that $\int_{-\infty}^\infty \|f(t)\|^p\, dt < \infty$ for some $p < \infty$. Suppose that f is strongly differentiable with a uniformly bounded derivative. Conclude that $\lim_{t \to \infty} f(t) = 0$.

†63. Fill in the computations in the proof of Theorem XIII.29.

†64. Fill in the details of the proof of Theorem XIII.32.

65. By iterating the proof of (a) in Theorem XIII.33, prove that if $H_0\varphi = \lambda\varphi - V\varphi$ with $\lambda > 0$, then $\varphi \in L_\delta^2$ for all δ.

†66. (a) Prove that $(-\Delta - \mu)^{-1}$ is bounded from L_δ^2 to L_δ^2 for any δ and any $\mu \in \mathbb{C}\setminus\mathbb{R}$. (Hint: Prove it for $\delta > 0$ inductively in $[0, 1]$, $(1, 2]$, ….)
(b) Complete the proof of Lemma 1 in Section 8.
(c) Prove that $(-\Delta + 1)^{-1}\partial/\partial x_i$ is a bounded map from L_δ^2 to L_δ^2 for any δ.

†67. Verify the bound (62).

68. Let $\delta > n + \tfrac{1}{2}$. Prove that, for any $b > a > 0$, there is a constant C so that
$$\|(-\Delta - \lambda - i0)^{-n}\varphi\|_{-\delta} \leq C\|\varphi\|_\delta$$
for all $\varphi \in \mathscr{S}(\mathbb{R}^m)$ and all $\lambda \in [a, b]$.

69. Let $\|A\|_{\delta, -\delta}$ denote the norm of A as a map of L_δ^2 to $L_{-\delta}^2$. Let $\delta > \tfrac{1}{2}$. Prove that for any $\alpha < \min\{1, \delta - \tfrac{1}{2}\}$, $(H_0 - \mu - i0)^{-1}$ is Hölder continuous of order α as an $\mathscr{L}(L_\delta^2, L_{-\delta}^2)$-valued function of μ, i.e., for any $\mu > 0$, there is a C and an ε so that
$$\|(H_0 - \mu' - i0)^{-1} - (H_0 - \mu - i0)^{-1}\|_{\delta, -\delta} \leq C|\mu - \mu'|^\alpha$$
if $|\mu - \mu'| < \varepsilon$.

*70. Let V be an Agmon potential. Let $a_1, \ldots, a_n$ obey $\sum_{i=1}^n \partial_j a_j = 0$ (distributional sense) with $|a_j(x)| \leq C_j(1 + |x|^2)^{-1/2-\varepsilon}$. Let

$$H = -\Delta + 2i \sum_{j=1}^n a_j \, \partial_j + \sum_{j=1}^n a_j^2 + V$$

Prove that $\sigma_{\text{sing}}(H) = \emptyset$ and that $\Omega^\pm(H, H_0)$ exist and are complete. (Hint: Develop a theory paralleling Theorem XIII.33.)

*71. Suppose $V(x) = (1 + |x|^2)^{-1/2-\varepsilon} W(x)$ where W is a form relatively compact perturbation of $-\Delta$. Prove that Theorem XIII.33 remains valid.

72. Let $A \geq 0$ be a positive self-adjoint operator and let $\mathcal{H}_{+1} \subset \mathcal{H} \subset \mathcal{H}_{-1}$ be the corresponding scale of spaces. Let $U(s)$ be a unitary group on $\mathcal{H}$ with each $U(s)$ in $\mathcal{L}(\mathcal{H}_{+1}, \mathcal{H}_{+1})$. Let $V \in \mathcal{L}(\mathcal{H}_{+1}, \mathcal{H}_{-1})$ and define $V(s) = U(s) V U(s)^{-1}$ for $s \in \mathbb{R}$. Prove that if $V(s)$ can be continued to an $\mathcal{L}(\mathcal{H}_{+1}, \mathcal{H}_{-1})$-valued analytic function in a neighborhood of $s = 0$, it can be continued to an entire strip $\{s \,|\, |\text{Im } s| < \alpha\}$.

73. (a) Let H_0 be a positive self-adjoint operator and let V be a symmetric operator with $D(V) \supset D(H_0)$. Suppose that $V(H_0 + 1)^{-1}$ is a compact operator. Prove that $(H_0 + 1)^{-1/2} V (H_0 + 1)^{-1/2}$ is compact. (Hint: Use interpolation.)
   (b) Call an operator V on $L^2(\mathbb{R}^3)$ in class C_α if and only if:
      (1) V is a symmetric operator with $D(V) \supset D(H_0)$ where $H_0 = -\Delta$.
      (2) $V(H_0 + 1)^{-1}$ is compact.
      (3) The family of operators $\tilde{F}(\theta) = u(\theta) V u(\theta)^{-1} (H_0 + 1)^{-1}$ defined for $\theta \in \mathbb{R}$ has an extension to an analytic operator-valued function in the strip $\{\theta \,|\, |\text{Im } \theta| < \alpha\}$.
   Prove that $C_\alpha \subset \mathscr{F}_\alpha$.

†74. Supply the details for Example 1 in Section 10.

75. Let A be a self-adjoint operator and let $U(s) = e^{isA}$ be the generated unitary group. Let $\psi \in C^\infty(A)$ with $\sum_{n=0}^\infty t^n \|A^n \psi\|/n! < \infty$ if $|t| < \alpha$. Prove that $f(s) = e^{isA} \psi$ has an analytic continuation into the strip $\{s \,|\, |\text{Im } s| < \alpha\}$. Conversely, if $f(s)$ has such a continuation, prove that $\sum_{n=0}^\infty t^n \|A^n \psi\|/n! < \infty$ if $|t| < \alpha$.

†76. Provide the necessary connectedness argument needed in the proof of Theorem XIII.36 to show that if γ is a curve in B_α and $\lambda \notin \sigma_{\text{ess}}(H(\gamma(t)))$ for any t, then either λ is in both $\sigma_d(\gamma(0))$ and $\sigma_d(\gamma(1))$ or in neither.

†77. Fill in the details of the proof of Proposition 2 in Section 10.

†78. Under the hypotheses of Theorem XIII.37, prove that

$$\{\mu + e^{-2\theta}\lambda \,|\, \mu \in \Sigma(\theta), \lambda \in [0, \infty)\} \subset \sigma_{\text{ess}}(\theta)$$

by the following procedure:
   (a) Suppose first that each $V_{ij} \in C_\alpha$ (see Problem 73). (Hint: See Problem 45.)
   (b) Use a limiting argument if $V_{ij} \in \mathscr{F}_\alpha$.

†79. Fill in the details of the proof of Theorem XIII.37.

†80. Prove Lemma 1b of Section 11.

†81. Let $V \in M_\sigma^{(n)}$ with $\sigma > n/2$, $\sigma \geq 1$. Prove that $V < < -\Delta$.

†82. Under the hypotheses of Theorem XIII.38, prove that the embeddings $C^\infty(\tilde{H}) \subset C_\theta$ (or C_θ^1) are continuous.

XIII: SPECTRAL ANALYSIS

83. Let $(-\Delta + V)\psi = E\psi$ in distributional sense and let V be C^∞ on Ω, an open set of $\mathbb{R}^n$. Use the ideas of the elliptic regularity theorem to prove that ψ is C^∞ on Ω.

84. Find an example to show that O'Connor's lemma does not extend to the case where $P(0)$ is an infinite-dimensional projection.

85. Under the hypotheses of O'Connor's lemma, prove that $P(\alpha)$ always extends analytically to the region $\tilde{D} \equiv \{\alpha \,|\, \text{Im } \alpha = \text{Im } \alpha_0 \text{ for some } \alpha_0 \in D\}$.

†86. (a) Let $0 < a < b$. Show that there exist *unit* vectors $e_1, \ldots, e_k$ in $\mathbb{R}^n$ so that
$$\exp(a|x|) \leq \sum_{j=1}^{k} \exp(be_j \cdot x)$$
for all $x \in \mathbb{R}^n$.
(b) Verify the computation of the essential spectrum of $H(\alpha)$ in the proof of Theorem XIII.40.
(c) Show that $\sigma_{\text{ess}}(H(\alpha)) \subset \{\lambda \,|\, \text{Re } \lambda \geq \Sigma - (2M)^{-1}|\text{Im } \alpha|^2\}$.
(d) Complete the proof of Theorem XIII.40.

87. Under the hypotheses of Theorem XIII.42, prove that
$$|\psi(\zeta) - \psi(\zeta')| \leq C_\varepsilon \exp[-(a - \varepsilon)\min\{r, r'\}]|\zeta - \zeta'|^\theta$$
so long as $\theta < \min\{1, 2 - 3/\sigma\}$. Here r (respectively, r') is the radius of gyration of the configuration ζ (respectively, ζ').

†88. Fill in the missing step in the proof of Theorem XIII.42.

89. (a) Let A and B be semibounded operators on a Hilbert space with $Q(A) \subset Q(B)$. Suppose that $B - A$ is strictly positive in the sense that $(\psi, (B - A)\psi) > 0$ for all $\psi \neq 0$ with $\psi \in Q(A) \cap Q(B)$. Suppose that $\mu_1(A)$ is an isolated eigenvalue of finite multiplicity. Prove that $\mu_1(A) < \mu_1(B)$. Give an example where $\mu_1(A)$ is in the essential spectrum and $\mu_1(A) = \mu_1(B)$.
(b) Prove directly that if V is central and if $-\Delta + V$ has a ground state, then it is an s-state with no nodes.

†90. Fill in the details in Example 2 of Section XIII.12.

91. Let H_0 be an operator that generates a positivity improving semigroup. Let $-V$ be a positivity preserving bounded operator. Prove that:
(i) $(H_0 + V + \mu)^{-1}$ is positivity improving for all sufficient large μ.
(ii) $\exp(-t(H_0 + V))$ is positivity improving for all $t > 0$.

92. Let $H_0 = -\Delta$, let V be an H_0-bounded multiplication operator with relative bound less than one, and let W be a Hilbert–Schmidt operator whose kernel is negative. Prove that $H_0 + V + W$ has a nondegenerate, strictly positive ground state if it has an eigenvalue at the bottom of its spectrum.

*93. Find an operator that is positive, positivity preserving, and ergodic that is not positivity improving. (Hint: Look for a 3×3 matrix.)

94. Find a positive operator H_0 for which $(H_0 + \mu)^{-1}$ is positivity preserving for some but not all μ. (Hint: Look for a 3×3 matrix.)

95. Let $H(\lambda) = -d^2/dx^2 + x^2 + \lambda/x^4$ with $\lambda > 0$ and domain $C_{00}^\infty(\mathbb{R})$, the C_0^∞ functions with support away from the origin.
   (a) Prove that $H(\lambda)$ is essentially self-adjoint.
   (b) Prove that $H(\lambda)$ has purely discrete spectrum.
   (c) Prove that the lowest eigenvalue is degenerate.
   (d) Why does this not violate Theorem XIII.46?
   (e) Why cannot $H(\lambda)$ converge to $-d^2/dx^2$ in strong resolvent sense as $\lambda \downarrow 0$. To what does it converge?

†96. Complete the proof of the first Beurling–Deny criterion.

97. (a) Let A be a finite matrix that is not necessarily symmetric such that $a_{ij} \leq 0$ if $i \neq j$. Prove that $\exp(-tA)$ is a matrix with positive elements for all $t > 0$. (Hint: Write $A = B + D$ with D diagonal and $b_{ij} \leq 0$ for all i, j and use the Trotter product formula.)
   (b) Let A be a finite matrix that is not necessarily symmetric such that e^{-tA} is a matrix with positive elements for all $t > 0$. Prove that $a_{ij} \leq 0$ for all $i \neq j$. (Hint: Use $A = \lim_{t \downarrow 0} (1 - e^{-tA})/t$.)

98. Suppose that $H \geq 0$. Let F_n be a family of functions on $\mathbb{C}$ (respectively $\mathbb{R}$) with $|F_n(z)| \leq |z|$ and so that if $f \in Q(H)$ (respectively, is real-valued) then $F_n(f) \in Q(H)$ and $(F_n(f), HF_n(f)) \leq (f, Hf)$. Let $F_n \to F$ pointwise. Prove that for all $f \in Q(H)$, $F(f) \in Q(H)$ and $(F(f), HF(f)) \leq (f, Hf)$.

99. Let H be a Laplacian with Dirichlet or Neumann boundary conditions. Show that H satisfies the first and second Beurling–Deny criteria. (Hint: Use Problem 98 and smooth F_n approximating $|z|$ and $z \wedge 1$).

†100. Rewrite (77) in the form (76).

†101. Show that the right-hand side of (79) defines a continuous linear functional on $\{\varphi \in \mathscr{L} \mid \varphi(0) = 0 = \partial_i \varphi(0)\}$ as long as
$$\int_{0 \leq |\lambda| \leq 1} |\lambda|^2 \, d\sigma + \int_{|\lambda| \geq 1} d\sigma < \infty$$

†102. (a) Let $f \in C_0^\infty(\mathbb{R})$ with $\int f \, dx = \int xf \, dx = 0$. Prove that there is a $g \in C_0^\infty$ with $f = g''$.
   (b) Let $f \in C_0^\infty(\mathbb{R}^n)$. Write a point in $\mathbb{R}^n$ as $\langle x, y \rangle$ where $y \in \mathbb{R}$, $x \in \mathbb{R}^{n-1}$. Suppose that
$$\int f(x, y) \, dx \, dy = 0 = \int yf(x, y) \, dx \, dy$$
$$= \int x_i f(x, y) \, dx \, dy$$

Find functions $a_1, a_2 \in C_0^\infty(\mathbb{R}^{n-1})$, $\eta_1, \eta_2 \in C_0^\infty(\mathbb{R})$ so that
$$\int a_i(x) \, dx = \int x_j a_i(x) \, dx = 0, \quad i = 1, 2; \; j = 1, \ldots n-1$$

and so that
$$g(x, y) = f(x, y) - \sum_{i=1}^{2} a_i(x)\eta_i(y)$$

obeys

$$\int g(x, y) \, dy = \int y g(x, y) \, dy = 0$$

for *each* x.

(c) Prove by induction that any f of the above form can be written as a sum $\sum_{i=1}^{n} \partial^2/\partial x_i^2 \, f_i$ with $f_i \in C_0^\infty(\mathbb{R}^n)$.

(d) Prove that any $\varphi \in \mathscr{S}$ with $\varphi(0) = 0 = \partial_i \varphi(0)$ (all i) is a sum of functions of the form $\lambda_i \lambda_{i'} \psi$ with $\psi \in \mathscr{S}$.

†103. Show that the measure $d\sigma$ defined in the proof of Theorem XIII.53 satisfies $\int_{|\lambda| \geq 1} d\sigma(\lambda) < \infty$.

†104. Fill in the details of Example 2 in Section XIII.13.

†105. Prove that any distribution solution of the radial Schrödinger equation (88) on $(0, \infty)$ with $E > 0$ is a linear combination of the two Jost solutions.

†106. Fill in the details in the proof of Theorem XIII.58 by making rigorous the formal calculations on $G(m, r)$.

107. Find an explicit potential V, smooth away from $r = 0$, and an eigenfunction ψ so that ψ is not in the domain of D, the infinitesimal generator of dilatations. (Hint: Modify the Wigner–von Neumann example.)

108. Let V be the sum of a repulsive potential and a homogeneous potential of degree $-\alpha$, $0 < \alpha < 2$. Prove that $-\Delta + V$ has no positive eigenvalues.

109. Let complex numbers $a_1, a_2, \ldots$ be given. Prove that there is at most one function f analytic in $\{z \,|\, \operatorname{Re} z > 0\}$ so that $f(n) = a_n$; $n = 1, 2, \ldots$ and $|f(z)| \leq C_1 \exp(C_2 |z|)$ for some $C_2 < \pi$. (Hint: If f_1 and f_2 are two such functions, apply Carlson's theorem to $(f_2 - f_1)/\sin \pi z$.)

110. Prove that the right-hand side of (91) is finite if $\psi \in D(-\Delta)$.

111. (a) Let $c > -\tfrac{1}{4}$. Let $A_c f = -f'' + cx^{-2} f$. Show that for a suitable constant d and all $\alpha \in \mathbb{R}$, $f \in C_0^\infty(0, 1)$,

$$\int_0^1 x^\alpha |f(x)|^2 \, dx \leq d \int_0^1 x^\alpha |(A_c f)(x)|^2 \, dx$$

(b) Let $n \geq 3$. Let $h \in C_0^\infty(\mathbb{R}^n)$ with support in $\{x \,|\, 0 < |x| < 1\}$. Prove there is a constant D, independent of h, so that

$$\int |x|^\alpha |h(x)|^2 \, d^n x \leq D \int |x|^\alpha |\Delta h(x)|^2 \, d^n x$$

for all α.

(c) Prove Theorems XIII.63 and XIII.64 for any $n \geq 3$.

112. (a) Prove Theorem XIII.57 for $n = 2$ by appealing to the $n = 3$ results. (Hint: Take functions constant in one direction.)

(b) Prove Theorems XIII.63 and XIII.57 for $n = 2$ by proving that for any $h \in C_0^\infty(\mathbb{R}^2 \setminus \{0\})$ and $\alpha > 0$,

$$\int |x|^{-\alpha-2} |h(x)|^2 \, d^2x \leq \alpha^{-1} \operatorname{Re}\left[\int |x|^\alpha \overline{h(x)}(-\Delta h)(x) \, d^3x \right]$$

(Hint: Use the fact that $\int (|x|^{\alpha/2} h)[-\Delta(|x|^{\alpha/2} h)] \, d^2x \geq 0$.)

113. (a) Let $f \in C_0^\infty(0, 1)$. Let $g = -f'' + cx^{-2} f$ for some $c > -\frac{1}{4}$. Prove that for any α,

$$\int_0^1 r^\alpha |f'(r)|^2 \, dr \leq \int_0^1 r^\alpha |g(r)|^2 \, dr$$

*(b) Prove that for $n > 3$ and any h that is C^∞ and supported in $\{x \mid 0 < |x| < 1\}$,

$$\int |x|^\alpha |\nabla h|^2 \, dx \leq 2 \int |x|^\alpha |\Delta h|^2 \, dx$$

(c) Prove a unique continuation theorem for Schrödinger operators with magnetic fields. Reference: The paper of Heinz quoted in the Notes.

(d) Complete the proof Theorem XIII.62.

114. Prove that a closed subset of a metric space is compact if and only if, for any ε, it can be covered by finitely many ε-balls.

115. Let A be the self-adjoint extension with periodic boundary conditions of $i \, d/dx \upharpoonright C_0^\infty(-\pi, \pi)$ on $L^2(-\pi, \pi)$. Prove directly that the set

$$S = \{\psi \in L^2(-\pi, \pi) \mid \|\psi\| \leq 1, \|A\psi\| \leq 1\}$$

is compact.

*116. Provide an alternative proof of Theorem XIII.65 as follows:

(a) Let $H(p) = \min\{|p|, \operatorname{ess\,min}_{|q| \geq \frac{1}{2}p} \sqrt{G(q)}\}$. Prove that for $\psi \in S$, the set given in Theorem XIII.65, and $\eta \in C_0^\infty(\mathbb{R}^n)$, one has

$$\int H(p)^2 |\widehat{\eta\psi}(p)|^2 \, dp \leq (2\pi)^n 4 \left(\int (1 + 2|p|) |\hat\eta(p)| \, dp \right)^2$$

(Hint: Prove that $H(a + b) \leq 2|a| + \sqrt{G(b)}$.)

(b) Choose R and η supported in the box of side R, so that if $\psi \in S$ and $\psi_1 \equiv \eta \psi$, then $\|\psi - \psi_1\|_2 \leq \varepsilon/4$.

(c) Choose $R_0 > R$, so that for $|k_i| \leq \pi/R_0$ and $\varphi_k(x) \equiv e^{-ik \cdot x} \psi_1(x)$, $\|\varphi_k - \psi_1\| \leq \varepsilon/4$.

(d) Let $a_m(k)$ be the Fourier *series* coefficients for φ_k as a function in the box of side R_0. Prove that for some constant C_1, independent of $\psi \in S$,

$$\sum_m \int_{|k_i| \leq \pi/R_0} H\left(k + \frac{2\pi m}{R_0}\right)^2 |a_m(k)|^2 \, dk \leq C_1$$

(e) Find C_2, independent of $\psi \in S$, and some k, so that $\|\varphi_k - \psi\| \leq \varepsilon/2$ and

$$\sum_m H\left(\frac{2\pi}{R_0}(|m| - 1)\right)^2 |a_m(k)|^2 \leq C_2$$

Let $\psi_2 = \varphi_k$ for this special value of k.

(f) Find N independent of ψ so that a truncated Fourier series ψ_3 for ψ_2 obeys $\|\psi_2 - \psi_3\| \le \varepsilon/4$.
(g) Complete the proof.

117. Prove Rellich's criterion as a corollary of Reisz' criterion.

118. Let $S \subset \ell_p$, $1 \le p < \infty$. Prove that S is compact if and only if: (1) $\sup_{f \in S} \|f\|_p < \infty$; and (2) for any ε, there is an N with $\sum_{|n| \ge N} |f_n|^p \le \varepsilon$ for all $f \in S$.

119. (Strichartz theorem, form version) Let $s \ge 3$. Prove the following results for multiplication operators W on $L^2(\mathbb{R}^3)$.
(a) If $W \in L_w^s$, then
$$\|W(I - \Delta)^{-1/2}\varphi\|_2 \le C\|W\|_{s/2,\,w}\|\varphi\|_2$$
(b) If $V \in L_w^{s/2}$, then V is a form-bounded perturbation of $-\Delta$.
(c) If $V \in L^{s/2}$, then the relative bound is zero.

120. Prove Theorem XIII.68 by proving that each term in the Neumann series for the perturbed resolvent is compact.

†121. Provide the details of the proof of Theorem XIII.70.

122. Prove that if A and B are positive self-adjoint operators on a Hilbert space $\mathcal{H}$ with $B << A$, then for any closed subspace $\mathcal{H}_0$ of $\mathcal{H}$ with $\mathcal{H}_0 \cap D(A)$ dense in $\mathcal{H}_0$, $B_0 << A_0$ where A_0 and B_0 are the Friedrichs extensions of $A \upharpoonright \mathcal{H}_0$ and $B \upharpoonright \mathcal{H}_0$.

123. The purpose of this problem is to provide an alternative proof of Theorem XIII.73.
(a) Prove that it is sufficient to show that if $f_n \in C_0^\infty(\Omega)$ and $\|f_n\|_m \le 1$, then $\{f_n\}$ has a subsequence convergent in the $\|\cdot\|_j$ norm.
(b) Let $f_n \in C_0^\infty(\Omega)$ satisfy $\|f_n\|_m \le 1$. Show that $\sup_k |\hat{f}_n(k)| \le c_0$ and that $\{f_n\}$ has a subsequence $\{f_{n_i}\}$ that converges weakly in $L^2(\Omega)$.
(c) Prove that $\hat{f}_{n_i}$ converges pointwise.
(d) Show that there are constants c_1 and c_2 so that for any $r > 0$,
$$\|f_{n_i} - f_{n_\ell}\|_j^2 \le c_1 \int_{|k| \le r} |k|^{2j} |\hat{f}_{n_i}(k) - \hat{f}_{n_\ell}(k)|^2 \, dk$$
$$+ c_2 r^{2j-2m} \int_{|k| \ge r} |k|^{2m} |\hat{f}_{n_i}(k) - \hat{f}_{n_\ell}(k)|^2 \, dk$$
(e) Use the inequality in (d) to conclude the proof of Theorem XIII.73.

124. (a) Prove the first part of Corollary 2 to Theorem XIII.74 by establishing the bound
$$((\eta u), (-\Delta_D)(\eta u)) \le C[(u, (-\Delta_D)u) + (u, u)]$$
for fixed $\eta \in C_0^\infty$ identically one in a neighborhood of K.
(b) Prove the second part of Corollary 2 to Theorem XIII.74 by using the bound $x^{-2} \le C(-\Delta)$ to control $\|\eta u_n\|$.
(c) Prove Corollary 2 to Theorem XIII.75 by showing that $(\eta, (-\Delta_N)\eta)$ and (η, η) being bounded (respectively, $\|\Delta_N u_n\|$ and $(\eta, (-\Delta_N)\eta)$ being bounded) implies bounds on the H^1 norm of η (respectively, $\nabla \eta$).
Remark The proof of the second part of (c) will require estimating $\|\partial_i \partial_j \varphi\|$ by $\|-\Delta_N \varphi\|$. Such estimates are discussed in the book of Agmon quoted in the Notes to Section 14.

125. (a) Let ∂_i be the operator closure of the ith partial derivative defined on $C_0^\infty(\Omega)$, viewed as an operator on $L^2(\Omega)$. Let ∂_i^* be its adjoint. Prove that for any open set $\Omega \subset \mathbb{R}^n$, $-\Delta_D = \sum_{i=1}^n \partial_i^* \partial_i$.
   (b) For any open set $\Omega \subset \mathbb{R}^n$, prove that $-\Delta_N^\Omega = \sum_{i=1}^n \partial_i \partial_i^*$.

126. Prove Proposition 1 of Section 15 by proving the one-dimensional case and then appealing to the theory of tensor products.

127. Let C be the operator $-d^2/dx^2$ on $L^2[a, b]$ with the boundary conditions

$$\frac{df}{dx}(a) + \alpha f(a) = 0, \qquad -\frac{df}{dx}(b) + \beta f(b) = 0$$

   (a) Prove that $Q(C) = H^1(a, b)$.
   (b) Prove that for $f \in C^\infty(a, b)$,

$$(f, Cf) = \int_a^b \left|\frac{df}{dx}\right|^2 dx + \alpha |f(a)|^2 + \beta |f(b)|^2$$

128. (a) Let Ω be a contented set. Prove that Ω is Lebesgue measurable in the sense that there are Borel sets D_1, D_2 with $D_1 \subset \Omega \subset D_2$ so that $\mu(D_2 \triangle D_1) = 0$. (Hint: Let D_1 (respectively D_2) be a union (respectively, intersection) of inner approximating cubes (respectively, outer cubes).)
   (b) Find an open Lebesgue measurable set that is discontented.

129. Let A and B be semibounded self-adjoint operators and let $\tilde{N}_E(A)$ be the number of eigenvalues of A less than $-E$.
   (a) Prove that

$$\tilde{N}_E(A + B) \leq \tilde{N}_{\theta E}(A) + \tilde{N}_{(1-\theta)E}(B)$$

   for $0 \leq \theta \leq 1$.
   (b) Let $\tilde{S}_\gamma(A)$ be the sum of the γth powers of the negative eigenvalues of A. Prove that for $\gamma > 0$ and $0 < \theta < 1$,

$$\tilde{S}_\gamma(A + B) \leq \theta^{-\gamma} \tilde{S}_\gamma(A) + (1 - \theta)^{-\gamma} \tilde{S}_\gamma(B)$$

   (Hint: Prove and use the formula $\tilde{S}_\gamma(A) = \gamma \int_0^\infty E^{\gamma-1} \tilde{N}_E(A)\, dE$.)

130. Consider the formula

$$\lim_{\lambda \to \infty} S_\gamma(\lambda V)/\lambda^{\frac{1}{2}m + \gamma} = \tau_m(2\pi)^{-m/2} \int |V_-(x)|^{\frac{1}{2}m + \gamma}\, d^m x \tag{209}$$

   where $S_\gamma(W)$ is the sum of the γth powers of the negative eigenvalues of $-\Delta + W$.
   (a) Prove (209) for any $V \in C_0^\infty(\mathbb{R}^n)$ by using Dirichlet–Neumann bracketing.
   (b) Extend (209) for $n \geq 3$ to any $V \in L^{\frac{1}{2}m + \gamma}(\mathbb{R}^m)$.
   (c) Extend (209) to $n = 1, 2$ to any $V \in L^{\frac{1}{2}m + \gamma}(\mathbb{R}^m)$ if $\frac{1}{2}m + \gamma > 1$.
   (Hint: In (b) and (c) use Problem 129 and the Lieb–Thirring bounds (Problems 31–33).)

†131. Prove Theorem XIII.82.

132. Let W be such that $\dim E_{(-\infty, 0]}(-\Delta + \lambda W) < \infty$ for all λ. Suppose that

$$f(\lambda) \equiv \lim_{a \uparrow 0}[\dim E_{(-\infty, \lambda a]}(-\Delta + \lambda V)/\dim E_{(-\infty, a]}(-\Delta + V)]$$

380   XIII: SPECTRAL ANALYSIS

exists for all λ and is finite, that
$$\lim_{a \uparrow 0} E_{(-\infty, a]}(-\Delta + V) = \infty$$
and that $f(\lambda)$ is continuous at $\lambda = 1$. Prove that
$$\lim_{a \uparrow 0} [\dim E_{(-\infty, a]}(-\Delta + V + W)/\dim E_{(-\infty, a]}(-\Delta + V)] = 1$$
(Hint: First prove that if $\dim E_{(-\infty, 0]}(B) = N < \infty$, then $\dim E_{(-\infty, a]}(A + B) \leq \dim E_{(-\infty, a]}(A) + N$.)

133. Verify the properties of $A_1 \oplus A_2$ appearing before Proposition 3 in Section 15.

134. Let $f_n \in L^2(M, d\mu; \mathscr{H}') = \mathscr{H}$ where $\mathscr{H}'$ is a separable space and M is σ-finite. Suppose that $\{f_n(m)\}_{n=1}^{\infty}$ is an orthonormal basis for $\mathscr{H}'$ for almost all $m \in M$. Prove that
$$\sum_{n=1}^{N} (f_n(\cdot), f(\cdot))_{\mathscr{H}'} f_n(\cdot) \to f$$
in $\mathscr{H}$ for each $f \in \mathscr{H}$. (Hint: Use the monotone convergence theorem.)
Remark This extends Problem 12 of Chapter II.

135. Suppose that A is a self-adjoint operator. Let $U(t)$ be a unitary one-parameter group. Suppose that $\mathscr{D}$ is a core for A and that $U(t)\mathscr{D} \subset \mathscr{D}$ with $AU(t)\psi = U(t)A\psi$ for all t and all $\psi \in \mathscr{D}$. Prove that $U(t)D(A) = D(A)$ and that $U(t)e^{iAs} = e^{iAs}U(t)$.

†136. (a) Prove (b) of Theorem XIII.85.
     (b) Fill in the details in the proof of (c) of Theorem XIII.85.

137. Prove that the Fourier transform of $(2\pi)^{-1/2}(p^2 + 1)^{-1}$ is $\frac{1}{2}e^{-|x|}$. (Hint: Use contour integration.)

138. Let $f \in \mathscr{S}(\mathbb{R})$ and define
$$V(x) = \sum_{n=-\infty}^{\infty} f(x + 2\pi n)$$

(a) Using the theory of Fourier transforms of periodic distributions, (144) and (145), prove that
$$\hat{V}(k) = \sum_{n=-\infty}^{\infty} \hat{f}(n) \delta(k - n)$$

(b) Using the Fourier inversion formula for $V(0)$ prove the **Poisson summation formula**
$$\sum_{m=-\infty}^{\infty} f(2\pi m) = (2\pi)^{-1/2} \sum_{n=-\infty}^{\infty} \hat{f}(n)$$

(c) Suppose that f and $\hat{f}$ are continuous functions with $|f(x)| \leq C(1 + |x|)^{-1-\delta}$ and $|\hat{f}(k)| \leq C(1 + |k|)^{-1-\delta}$ for some $C, \delta > 0$. Prove the Poisson summation formula for f by a limiting argument.

(d) Using the known Fourier transform of $e^{-|x|}$, evaluate $\sum_{n=-\infty}^{\infty} (1 + n^2)^{-1}$.

(e) Using the known Fourier transform of $e^{-a|x|}$, evaluate $\sum_{n=1}^{\infty} n^{-2}$.

Remark As the above suggests, the Poisson summation formula is a useful tool in number theory, see for example, K. Chandrasekhar, *Introduction to Analytic Number Theory*, Springer-Verlag, Berlin and New York, 1968.

139. (The Krönig-Penny model) Let V be the potential of period 2π which is 0 on $[0, y)$ and c on $[y, 2\pi)$. Compute the discriminant explicitly and prove that as $n \to \infty$, the size of the nth gap obeys

$$\ell_n \sim \left| \frac{4c}{n} \sin\left(\frac{ny}{2}\right) \right|$$

In particular, conclude that infinitely many gaps occur.

†140. Let A be a decomposable self-adjoint operator on $\mathcal{H} = \int_M^\oplus \mathcal{H}' \, d\mu$. Suppose that each $A(m)$ is bounded from below and has compact resolvent.
  (a) Prove that $E_n(m)$, the nth eigenvalue of $A(m)$, is a measurable function. (Hint: Use the min-max principle.)
  (b) Show that $\mathcal{H}'$-valued functions $\psi_n(m)$ can be chosen which obey $A(m)\psi_n(m) = E_n(m)\psi_n(m)$ and which are an orthonormal basis of $\mathcal{H}'$ in such a way that the $\psi_n(\cdot)$ are measurable. (Hint: For given n, let $M_{k;n} = \{m \mid E_n(m) \text{ is } k\text{-fold degenerate}\}$ and choose $\psi_n(m)$ on each $M_{k;n}$.)
  (c) Suppose also that M is a topological space, that each $A(m) > 0$, and that $(A(m) + 1)^{-1}$ is norm continuous. Prove that $E_n(m)$ is continuous.
  (d) Give an example where $(A(m) + 1)^{-1}$ is weakly continuous and $E_n(m)$ is not continuous.
  (e) Explain why $\psi_n(m)$ may not be continuous even under the hypotheses of (c).

†141. (a) Verify (173b).
  (b) Let $\alpha > \frac{1}{2}n$. Verify that

$$\sum_{|m| \geq c_2(1+|y|)} |E_m(x + iy) + 1|^{-\alpha} \leq c_3(1 + |y|)^{-\alpha + \frac{1}{2}n}$$

  (c) Let $\alpha > 1$. Verify that

$$\sum_{|m| \leq c_2(1+|y|)} |E_m(x + iy) + 1|^{-\alpha} \leq c_4(1 + |y|)^{-\alpha + n - 1}$$

  (d) Conclude the proof of Lemma 1 to Theorem XIII.100.

142. Prove that any lattice in $\mathbb{R}^n$ has a basis.

143. Prove that the Wigner–Seitz cell of any lattice is a polyhedron, i.e., an intersection of finitely many half spaces. (Hint: Let $\mathbf{a}_1, \ldots, \mathbf{a}_n$ be a basis and let V be the set of 3^n points $\sum_{i=1}^n t_i \mathbf{a}_i$ with $t_i = 0, +1,$ or -1. Let $R = \max\{|x| \mid x \in V\}$ and prove that any point in C, the Wigner–Seitz cell, is in the ball of radius R. Prove that a point on ∂C is equidistant from 0 and some point in $\mathscr{L} \cap \{x \mid |x| \leq 2R\}$.)

144. Suppose that A_α is a family of self-adjoint operators, bounded from below, such that $A_\alpha \geq A$ and that A has compact resolvent. Let $E_n(\alpha)$ be the nth eigenvalue of A_α. Prove that $E_n(\alpha) \to \infty$ uniformly in α as $n \to \infty$.

145. Prove that the density of states measure given by (176) is absolutely continuous with respect to Lebesgue measure. (Hint: Prove that it is a sum of spectral measures for H.)

146. In the one-dimensional case prove that at a point E with $E_n(k) = E$ the density of states is given by
$$\frac{d\rho}{dE} = \frac{2}{|\bar{Q}|}\left(\frac{dE}{dk}\right)^{-1}$$

†147. Fill in the details of the proof of Theorem XIII.101.

*148. Prove the analogue of Theorem XIII.101 when periodic boundary conditions are replaced by Dirichlet or Neumann boundary conditions.

149. Using (185), prove (186).

150. Let $\sum_{i=1}^{\infty} |a_i| < \infty$. Let $f_N(z) = \prod_{j=1}^{N}(1 + a_j z)$. Prove that f_N converges as $N \to \infty$ uniformly on compact sets as follows:
   (a) Prove uniform bounds on $f_N(z)$ by using $|1 + x| \le e^{|x|}$.
   (b) Prove that the Taylor coefficients of f_N converge as $N \to \infty$.
   (c) Complete the proof of convergence.

151. (a) Let $A_n \to A$ in trace class norm and suppose that $\mu \notin \sigma(A)$. Prove directly (without use of Theorem XIII.107) that
$$\det(1 + \mu A_n)(1 + \mu A_n)^{-1} \to \det(1 + \mu A)(1 + \mu A)^{-1}$$
in norm.
   (b) Prove (196) knowing it is true when A is finite rank with $\mu \notin \sigma(A)$.

152. Let $d(\mu) = \det(1 + \mu A)$ and $D(\mu) = d(\mu)(1 + \mu A)^{-1} A$. Write $d(\mu) = \sum_{n=0}^{\infty} a_n \mu^n$ and $D(\mu) = \sum_{n=0}^{\infty} B_n \mu^n$.
   (a) Prove that $|a_n| \le (e/n)^n \|A\|_1^n$.
   (b) Prove that $\|B_n\|_1 \le (e/n)^n \|A\|_1^{n+1}$.

153. Define the **partial trace** Tr_{n-1} from $\mathscr{I}_1(\bigotimes^n \mathscr{H})$ to $\mathscr{I}_1(\mathscr{H})$ by $\mathrm{Tr}(C\, \mathrm{Tr}_{n-1}(A)) = \mathrm{Tr}[(C \otimes I \otimes \cdots \otimes I)A]$ for any $C \in \mathscr{L}(\mathscr{H})$.
   (a) Show that Tr_{n-1} is a well-defined contraction.
   *(b) View $\bigwedge^k (A)$ as a map on $\bigotimes^k \mathscr{H}$ by setting it equal to zero on $(\bigwedge^k \mathscr{H})^\perp$. Prove that $D(\mu) = \sum_{n=0}^{\infty} B_n \mu^n$ with
$$B_n = (n+1)\mathrm{Tr}_n(\bigwedge^{n+1}(A))$$

154. Prove that $|\lambda_1(A) \cdots \lambda_n(A)| \le \mu_1(A) \cdots \mu_n(A)$ for any compact A. (Hint: Consider the eigenvalues and norm of $\bigwedge^n (A)$.)

155. Let $A \in \mathscr{I}_p$ with $p \le n$. Define
$$R_n(A) = (1 + A)\exp\left(\sum_{k=1}^{n-1} (-A)^k/k\right) - 1$$
   (a) Prove that $R_n(A) \in \mathscr{I}_1$.
   Define $\det_n(1 + A) = \det(1 + R_n(A))$.
   (b) Prove that $1 + A$ is invertible if and only if $\det_n(1 + A) \neq 0$.
   (c) Prove that
$$\det_n(1 + A) = \prod_{j=1}^{N(A)} \left[(1 + \lambda_j(A))\exp\left(\sum_{k=1}^{n-1}(-\lambda_j(A))^k/k\right)\right]$$

Let $D_n(\mu A) = A(1 + \mu A)^{-1} \det_n(1 + \mu A)$.
*(d) Prove that
$$|\det_n(1 + A)| \le \exp(\gamma_n \|A\|_n^n)$$
$$\|D_n(\mu A)\|_n \le C_n \exp(\gamma_n \|A\|_n^n \mu^n)$$
for suitable constants C_n and γ_n.
(e) Find Plemelj–Smithies formulas for $\det_n$ and D_n.
(f) Find completeness criteria for eigenvectors of operators in $\mathcal{I}_n$.

156. Prove that the following are equivalent for $A \in \mathcal{I}_1$:
(a) A is quasi-nilpotent, i.e., $\sigma(A) = \{0\}$.
(b) $\det(1 + \mu A) - 1$.
(c) $\operatorname{Tr}(A^k) = 0$ for all $k > 0$.
(d) $\operatorname{Tr}(A^k) = 0$ for all $k > k_0$ for some k_0.

157. Prove the following are equivalent for all $A \in \mathcal{I}_1$:
(a) For all $\mu \in \mathbb{C}$, μ and $-\mu$ have identical algebraic multiplicities.
(b) $\det(1 + \mu A) = \det(1 - \mu A)$ for all $\mu \in \mathbb{C}$.
(c) $\operatorname{Tr}(A^k) = 0$ for all odd k.

†158. (a) Prove the max-min principle for singular values:
$$\mu_n(A) = \inf_{\psi_1, \ldots, \psi_{n-1}} \left(\sup_{\phi \in [\psi_1, \ldots, \psi_{n-1}]^\perp;\, \|\phi\| = 1} \|A\phi\| \right)$$
(b) Prove that $\mu_n(1 + A) \le 1 + \mu_n(A)$.
(c) Prove **Fan's inequality**
$$\mu_{n+m+1}(A + B) \le \mu_{n+1}(A) + \mu_{m+1}(B)$$

†159. Let A be a closed operator and let $\mu \notin \sigma(A)$ and P be the spectral projection for some isolated point λ in $\sigma(A)$. Prove that P is also the spectral projection associated to the operator $(A - \mu)^{-1}$ and the point $(\lambda - \mu)^{-1}$.

160. Let D be a bounded open set in $\mathbb{R}^{n+1}$ with smooth boundary ∂D. Let $\mathcal{H} = L^2(\partial D, dS)$. Let $K(x, y)$ the kernel of an integral operator A on $\mathcal{H}$ and suppose that
$$|K(x, y)| \le C|x - y|^{-\alpha}$$
with $\alpha < n$. Suppose that L is the kernel of an integral operator B, with
$$|L(x, y)| \le D|x - y|^{-\beta}$$
(a) If $\alpha + \beta < n$, prove that AB has a bounded kernel.
(b) If $\alpha + \beta > n$, prove that AB has a kernel M with
$$|M(x, y)| \le E(x - y)^{-\gamma}$$
$\gamma = \alpha + \beta - n$.
(c) Let $\alpha < n[1 - (2k)^{-1}]$, for k equal a positive integer. Prove that $A \in \mathcal{I}_{2k}$.
*(d) Let $2 \le p < \infty$. Let $\alpha < n(1 - p^{-1})$. Prove that $A \in \mathcal{I}_p$. (Hint: Use complex interpolation.)

161. (a) Let $f \in L^2(\mathbb{R})$ have an L^2 derivative. Prove that there are $g, h \in L^2$ with $f = g * h$.
    (b) Let $f \in L^2(\mathbb{R})$ have an L^2 derivative; $V, W \in L^2(\mathbb{R})$. Prove that
    $$V(x)f(x-y)W(y)$$
    is the kernel of a trace class operator on $L^2(\mathbb{R})$.
    (c) Let $V, W \in L^2(\mathbb{R})$ have *compact support*. Let f be *locally L^2* with a *locally L^2* derivative. Prove that
    $$V(x)f(x-y)W(y)$$
    is the kernel of a trace class operator.

162. Let B be a rank one operator. Prove that $\det(1 + B) = 1 + \text{Tr}(B)$. (Hint: Replace B by λB and prove it for small λ.)

163. (a) Prove that the Fourier transform of the distribution $|x|$ on $\mathbb{R}$ is a distribution equal to the function $-(2/\pi)^{1/2} k^{-2}$, away from $k = 0$.
    (b) Prove that $-|x|$ is conditionally strictly positive definite; i.e., if $(1 + |x|)V \neq 0$ is in L^1 and $\int V(x)\, dx = 0$, then
    $$-\int V(x)|x-y|V(y)\, dx\, dy > 0$$

164. Suppose that A and B are trace class operators with $A \geq 0$.
    (a) Prove that $|B|^{1/2}(1 + \mu A)^{-1/2}$ converges to zero in Hilbert–Schmidt norm as $\mu \to \infty$. (Hint: Compute $\text{Tr}((1 + \mu A)^{-1/2}|B|(1 + \mu A)^{-1/2})$ in an eigenbasis for A.)
    (b) Prove that $B(1 + \mu A)^{-1}$ converges to zero in trace class norm as $\mu \to \infty$.
    (c) Prove that $\lim_{\mu \to \infty} [\det(1 + B + \mu A)/\det(1 + \mu A)] = 1$; where det is the Fredholm determinant.

165. Consider the one-dimensional Schrödinger operator $-d^2/dx^2 + V(x)$ with V positive and periodic with period 1. Let $D(E)$ be the discriminant for this equation and $D_0(E) = 2\cos\sqrt{E}$, the discriminant for $V = 0$. Let G_0 be the inverse of $(h_0 + 1)^{-1}$ with h_0 the operator $-d^2/dx^2$ on $L^2(0, 1)$ with periodic boundary conditions. Prove that
    $$D(E) - 2 = C[\det(1 + G_0 V - (E + 1)G_0)]$$
    where det is the Fredholm determinant and
    $$C = D_0(-1) - 2$$
    (Hint: First prove that $D(E) - 2 = C_V[\det(1 + G_0 V - (E + 1)G_0)]$ by showing that both sides are entire functions given as infinite products and that they have the same zeros. Then, using Problem 164, prove that $C_V = C_0$ by showing that $(D(E) - 2)/(D_0(E) - 2)$ and $\det(1 + G_0 V - (E + 1)G_0)/\det(1 - (E + 1)G_0)$ approach one as $E \to -\infty$. Find C_0 by taking $E = -1$.)

166. Let E and F be two bounded operators.
    (a) Suppose that $0 \neq \lambda \notin \sigma(EF)$. Prove that $-(\lambda)^{-1}[1 - F(EF - \lambda)^{-1}E]$ is a two-sided inverse for $FE - \lambda$ and conclude that $\lambda \notin \sigma(FE)$.
    (b) Let $\lambda \neq 0$, and $\mathcal{M} = \{\varphi \mid EF\varphi = \lambda\varphi\}$; $\mathcal{N} = \{\varphi \mid FE\varphi = \lambda\varphi\}$. Prove that F is a bijective map of $\mathcal{M}$ onto $\mathcal{N}$ and conclude that $\dim \mathcal{M} = \dim \mathcal{N}$.

167. Let $p \neq \infty$. Suppose that $A_n \in \mathcal{I}_p$ with $\sup_n \|A_n\|_p < \infty$ and let A be a bounded operator with $A_n \to A$ weakly. Prove that $A \in \mathcal{I}_p$ and $\|A\|_p \leq \varlimsup \|A_n\|_p$. (Hint: Let q be the dual index to p. Prove that $|\text{Tr}(FA)| \leq \|F\|_q \varlimsup \|A_n\|_p$ for any finite rank operator F.)

168. Prove the bound $\|C^{1/2}DC^{1/2}\|_r \leq \|CD\|_r$, used in the proof of the generalized Golden–Thompson inequality by using complex interpolation.

169. This problem asks the reader to prove Hölder's inequality for matrices (Proposition 5 of the appendix to Section IX.4) from Hölder's inequality for sums without recourse to complex interpolation.
    (a) Let $\{a_{ij}\}_{1 \leq i, j \leq n}$ be a matrix with $\sum_i |a_{ij}| \leq 1$ for all j and $\sum_j |a_{ij}| \leq 1$ for all i. Prove that
    $$\left|\sum a_{ij}\mu_i v_j\right| \leq \left(\sum_i |\mu_i|^p\right)^{1/p} \left(\sum_j |v_j|^q\right)^{1/q}$$
    for $p^{-1} + q^{-1} = 1$. (Hint: Look at the proof of Theorem XIII.103.)
    (b) Let $A = \sum \mu_i(\varphi_i, \cdot)\psi_i$ and $B = \sum v_j(\eta_j, \cdot)\gamma_j$ be the canonical expressions for A and B. Prove that
    $$|\text{Tr}(AB)| \leq \sum |a_{ij}|\mu_i v_j$$
    where $\sum_i |a_{ij}| \leq 1$, $\sum_j |a_{ij}| \leq 1$.
    (c) Conclude that $|\text{Tr}(AB)| \leq \text{Tr}(|A|^p)^{1/p} \text{Tr}(|B|^q)^{1/q}$.

170. Let $F_A(\mu)$ be the function of Theorem XIII.107. The purpose of this problem is to show that for fixed $A, B \in \mathcal{I}_1$, the function $\lambda \mapsto F_{A+\lambda B}(1)$ is entire in λ. This is needed to prove (197).
    (a) Fix μ with $-\mu^{-1} \notin \sigma(A)$. Show that $F_{A+\lambda B}(\mu) = F_C(\lambda)F_A(\mu)$ where $C = \mu(1 + \mu A)^{-1}B$ and conclude that $\lambda \mapsto F_{A+\lambda B}(\mu)$ is entire for such μ.
    (b) Let $\mu_n \to \mu_\infty$ and let $f_n(\lambda) = F_{A+\lambda B}(\mu_n)$. Using the first part of the proof of Theorem XIII.107, show that
    $$\sup_n \sup_{\lambda \in R} |f_n(\lambda)| < \infty$$
    for any bounded region R and that $f_n(\lambda) \to f_\infty(\lambda)$ for each λ. Conclude that $F_{A+\lambda B}(\mu)$ is entire in λ for fixed μ.

List of Symbols

A foolish consistency is the hobgoblin of little minds.

Ralph Waldo Emerson

Superscript 1 and 2 indicate that these are pages in Volumes I and II, respectively.

$\mathbb{C}$		the complex numbers	$H_0(\theta)$		183
$C(X)$		102^1	H_D		121
$C_0^\infty(\mathbb{R}^n)$		145^1	$H_0^m(\Lambda)$	(Sobolev spaces)	253
C_α		373	$H^m(\Lambda)$	(Sobolev spaces)	253
$C_\theta(\mathbb{R}^n)$		191	iDm		130
$C_\theta^1(\mathbb{R}^n)$		192	$\sim iDm$		130
$d\Gamma(A)$		$302^1, 208^2$	$\mathcal{I}_p$		$207^1, 208^1, 41^2$
$D(\cdot)$	(domain)	249^1	$I(E)$		125
$\mathcal{D}(\mathbb{R}^n), \mathcal{D}(\Omega)$		147^1	Ker		185^1
$\mathcal{D}'(\mathbb{R}^n), \mathcal{D}'(\Omega)$		148^1	KLMN		167^2
D^α		2^2	ℓ_p		69^1
$\det(\cdot)$		323	$L^p(X, d\mu)$		68^1
$\det_n(\cdot)$		382	$L^2(X, d\mu; \mathcal{H}')$		40^1
$D(E)$		125	L^p_{loc}	(functions locally L^p)	353^2
$D \triangleright D'$		130	L^2_δ	(weighted L^2)	168
$\mathcal{E}$		(Section XI.6)	L^p_w	(weak L^p)	30^2
$f(A)$	(continuous functional		$L^r + L^s$		165^2
	calculus)	222^1	$\mathcal{L}(\mathcal{H})$		182^1
$\hat{f}, \mathcal{F}$	(Fourier transform)	1^2	$\mathcal{L}(X, Y)$		69^1
$\check{f}, \mathcal{F}^{-1}$	(inverse Fourier		$M_\sigma^{(n)}$		193
	transform)	1^2	$n_\ell(V)$		91
$\mathcal{F}(\mathcal{H}), \mathcal{F}_a(\mathcal{H}), \mathcal{F}_s(\mathcal{H})$			$N_\ell(E; V)$		91
	(Fock spaces)	$53^1, 54^1$	$N(V)$		91
$\mathcal{F}_\alpha, \overline{\mathcal{F}}_\alpha$		184, 235	$P_\Omega(A), P_\Omega^A$		234^1
$\mathcal{H}$		39^1	$p(D)$		45^2
$\mathcal{H}_{pp}, \mathcal{H}_{ac}, \mathcal{H}_{sing}$		230^1	$\mathcal{P}_n$		321
H_0	(free Hamiltonian)	55^2	$\mathbb{R}$		the real numbers

387

388   LIST OF SYMBOLS

R	(Rollnik class)	170^2
R	(center of mass)	197
r	(radius of gyration)	197
Ran		185^1
$R(\lambda + i\mu), R_\lambda(T)$ (resolvent)		188^1
$\mathscr{R}(A)$		177
supp		$139^1, 17^2$
$\mathscr{S}(\mathbb{R}^n)$		133^1
$\mathscr{S}'(\mathbb{R}^n)$		134^1
$\mathrm{tr}(\cdot)$		$207^1, 208^1$
$u(\theta)$		183
$U(\theta)$		56
$W_m(\Omega)$	(Sobolev spaces)	$50^2, 253$
x^2		2^2

$\Gamma(T)$	(operator graph)	250^1
$\Gamma(A)$		$309^1, 208^2$
$d\Gamma(A)$		$302^1, 208^2$
$\Gamma_n(A)$		321
Γ	(resonance width)	54
Δ	(symmetric set difference)	388
Δ	(Laplacian on $\mathbb{R}^n$)	388
Δ_D^Ω		263
Δ_N^Ω		263
$\mu_n(H)$		76
μ_ψ		225^1
$\rho(T)$		188^1
$\sigma(T)$		118^1
$\sigma_{pp}, \sigma_{cont}, \sigma_{ac}, \sigma_{sing}$		231^1
σ_{disc}		$236^1, 13$
σ_{ess}		236^1
σ_{ap}, σ_r		178
χ_A		2^1

$\|\cdot\|_p$	(functions)	68^1
$\|\cdot\|_p$	(operators)	41^2
$\|\cdot\|_R$	(Rollnik norm)	170^2
$\|\cdot\|_\infty$	(functions)	67^1
$\|\cdot\|_\infty$	(operators)	81^2
$\|f\|_\delta$	(weighted-L^2)	168
$\|f\|_{(\theta)}$		192
$\|f\|_{(\theta), 1}$		192

$\oplus$		$40^1, 78^1$
$\otimes$	(measures)	26^1
$\otimes$	(Hilbert spaces)	49^1
$\otimes$	(functions)	141^1
$\otimes$	(operators)	299^1

$\leq$	(operators)	75, 85, 269
$<<$	infinitesimally small (operators)	162^2
$\prec\prec$	infinitesimally small (forms)	168^2
$\overline{}$	(closure)	92^1
$\overset{\circ}{,}$ int	(interior)	92^1
$*$	(adjoint)	187^1
$*$	(dual space)	72^1
$*$	convolution	$6^2, 7^2$
$\to \infty$		247
$\overset{\|\cdot\|}{\to}, \overset{w}{\to}, \overset{s}{\to}$		$182^1, 183^1$
$\lvert\cdot\rvert$	(absolute value of an operator)	196^1
$\perp$	(orthogonal complement)	41^1
$\setminus$	(set difference)	1^1
$/$	(quotient)	$78^1, 79^1$
$\upharpoonright$	(restriction)	2^1
$\langle \cdot, \cdot \rangle$	(ordered pair)	1^1
$(\cdot, \cdot)$	(inner product)	36^1
$\{\cdot, \cdot\}$	(Poisson bracket)	314^2

Index

Superscripts 1 and 2 indicate that these are pages in Volumes I and II, respectively.

A

Absolute value of an operator, 196^1
Absolutely continuous subspace, 230^1
Accretive operator, 240^2
Adjoint, Banach space, 185^1
Adjoint, Hilbert space, 186^1
Adjoint, unbounded operator, 252^1
Agmon–Kato–Kuroda theorem, 169
Agmon potential, 169
Algebraic multiplicity, 9
Analytic family
  in the sense of Kato, 14
  type (A), 16
  type (B), 20
  type (b), 20
Analytic Fredholm theorem, 201^1
Analytic function, vector-valued, 189–190^1
Analytic vector, 201^2
Angular momentum, 90
Anharmonic oscillator, 175^2, 184^2, 206^2, 266^2, 270^2, 20, 32, 41, 64, 84

Antilinear operator, 69^1
Approximate identity, 251^1, 9^2
Approximate point spectrum, 178
Aronszajn–Donoghue theory, 140
Ascoli's theorem, 30^1
Associated string, 129
Asymptotic series, 26
  order k strong, 43
  strong, 40
Atomic model, 304^1
Auger states, 52
Autoionizing states, 52
Axiomatic quantum field theory, 62^2

B

Baire measure, 105^1, 110^1
Balslev–Simon theorem, 237
Banach space, 67^1
Barely connected, 125
Bargmann's bound, 94, 366

389

390 INDEX

Basic period cell, 303
Basis of a lattice, 310
Bessel's inequality, 38^1
Beurling–Deny criteria, 209
Birman–Schwinger bound, 98
B.L.T. theorem, 9^1
Bochner integral, 119^1
Bochner–Schwartz theorem, 14^2
Borel–Carathéodory theorem, 328
Borel set, 14^1, 105^1
Borel summability method, 44
Borel transform, 44
Boson Fock space, 53^1
Bound state energies, 79
Bound states, 79
Bounded holomorphic semigroup, 248^2
Bounded operator, 8^1
Breit–Wigner resonance shape, 53
Brillouin zone, 311

C

C^∞-vector, 201^2
Calogero's bound, 94, 366
Canonical commutation relations, 274^1, 218^2, 232^2
Canonical form for compact operators, 203^1
Carleman's theorem, 39
Carlson's theorem, 236
Cauchy sequence, 5^1
Center of mass, 197
Central potential, 90
Characteristic equation, 2
Characteristic function, 2^1, 221
Classical Weyl theorem, 117
Closed graph theorem, 83^1
Closed operator, 250^1
Closed quadratic form, 277^1
Cluster decomposition, 129
Commuting (unbounded) operators, 271^1–272^1
Compact operator, 199^1, see also Relatively compact; Relatively form compact
  determinant, 323, 382
  general theory, 198^1–206^1, 316–338
  Hilbert–Schmidt, 210^1

ideal theory, 41^2–44^2
resolvent, 244–260
trace class, 206^1–213^1
Compact sets, 97^1, 244
Compact support
  distributions, 139^1, 178^1, 17^2
  functions, 111^1
Complex scaling, see Dilation analytic potentials
Complex thresholds, 191
Conditionally negative definite, 215
Conditionally positive definite, 214
Contented set, 271
Continuous functional calculus, 222^1
Contraction semigroup, 235^2
Convex function, 104
Convex set, 109^1
Convolution
  distributions, 7^2
  functions, 6^2
Core, 256^1
Coupling constant, 11
Critical value, 346
Cwikel–Lieb–Rosenbljum bound, 101

D

D-thresholds, 189
Darwin correction, 81
Decomposable operator, 281
Deficiency indices, 138^2
Degenerate eigenvalue, 2
Density of states, 313
Density of states measure, 312
Determinant, 323, 382
Diagram
  barely connected, 125
  disconnected, 125
  k-connected, 129
Dilation analytic potentials, 184
Dilation operators, 56, 183
Dirac operator, 337^2
Direct integral, 280, 358
Direct sum
  of Banach spaces, 78^1
  of Hilbert spaces, 40^1
Dirichlet Laplacian, 263

Dirichlet–Neumann bracketing, 262
Dirichlet problem, 204[1]
Disconnected part, 125
Disconnected string, 129
Discrete spectrum, 236[1], 13
Discriminant, 296
Distribution of eigenvalues
 Dirichlet Laplacian, 260
 Schrödinger operators, 86
Distributions
 compact support, 139[1], 178[1], 17[2]
 tempered, 134[1]
Domain
 analytic function, 189[1]
 form, 276[1]
 function, 2[1]
 unbounded operator, 249[1]
Dominated convergence theorem, 17[1], 24[1]
Double well potential, 34
Dual basis, 305
Dual space, 43[1], 72[1]
Dunford functional calculus, 245[1]
Dyson expansion, 282[2]

E

Eigenvalue, 188[1]
 degenerate, 2
 discrete, 13
 multiplicity, 9, 316
 nondegenerate, 9, 13
 pseudo-, 48
 simple, 9
 stable, 29
Eigenvector, 188[1]
 generalized, 316
 pseudo-, 48
Elliptic operator, 112[2], 346
Elliptic regularity, 49[2]
Energy levels, 11, 79
Energy operator, see Hamiltonian
Equicontinuous function, 28[1]–30[1].
Ergodic operator, 202
Essentially self-adjoint, 256[1]
Exceptional set (§ XI.6, Vol. III)
Extension of an operator, 250[1]

F

Fan's inequality, 383
Fermi energy, 314
Fermi gas picture, 356
Fermi golden rule, 52, 59
Fermi statistics, 117[2]
Fermi surface, 314
Fermion Fock space, 54[1]
Feynman–Kac formula, 279[2]
Fibers, 280–281
Finite partition, 208
First resolvent formula, 191[1]
Fock space, 53[1]
Form, see also Quadratic forms
 core, 277[1]
 domain, 276[1]
 of operator, 277[1]
Formal series, 73
Fourier inversion theorem, 3[2]
Fourier transform, 1[2]
Friedrichs' extension, 177[2]
Functional calculus, 222[1], 225[1], 245[1], 263[1], 286[1]–287[1]
Functions of rapid decrease, 133[1]

G

Generalized eigenspace, 9
Generalized eigenvector, 316
Generalized function, 148[1]
Generator
 group, 268[1]
 semigroup, 237[2], 246[2]
Geometric eigenspace, 9
Geometric multiplicity, 9
GGMT bound, 94
Ghirardi–Rimini bound, 367
Golden–Thompson inequality 333[2], 320
Graph, 83[1], 250[1]
Graph limits, 293[1], 268[2]
Green's functions, 59[2], 263
Ground state, 201
Ground state energy, 201
Gyration, radius of, 197

H

Hadamard's three line theorem, 33^2
Hahn–Banach theorem, 75^1–77^1
Hamiltonian, 303^1
 free, 55^2, 220^2
 time dependent, 109^2
Hausdorff–Young inequality, 11^2
Helium Hamiltonian, 80
Hermite functions, 142^1
 completeness, 121^2
Hermitian operator, see Symmetric operator
Hilbert–Schmidt operators, 210^1
Hilbert–Schmidt theorem, 203^1
Hilbert space, 36^1
Hölder continuous, 81^2, 191
Hölder's inequality, 68^1, 34^2
 matrices, 385
 operators, 41^2
Holomorphic semigroup, 252^2
 bounded, 248^2
H-smooth on Ω, 163
H-smooth operator, 142
Hughes–Eckart term, 80
Hunziker's theorem, see HVZ theorem
HVZ theorem, 121
Hypercontractive semigroup, 258^2
Hyperfine structure in hydrogen, 19

I

Ichinose's lemma, 183
Index of a string, 129
Infinitely divisible, 221
Infinitesimal generator
 group, 268^1
 semigroup, 237^2, 246^2
Infinitesimally small form, 168^2
Infinitesimally small operator, 162^2
Interpolation theorems, 32^2
Inverse Fourier transform, 1^2
Irreducible, 232^2
Iorio–O'Carroll theorem, 154

J

Jensen's inequality (17), 104
Jordan content, 271
Jordan normal form, 10
Jost solution, (§XI.8, Vol. III)

K

k-connected diagram, 129
Kato–Agmon–Simon theorem, 226
Kato–Rellich theorems, 162^2, 15
Kato–Simon theorem, 193
Kato's inequality, 183^2, 351
Kato's projection theorem, 22
Kato's smoothness theorem, 152
Kato's theorem, 166^2
KLMN theorem, 167^2
Kohn's theorem, 301
Korteweg–de Vries equation, 300
Krönig–Penny model, 381

L

L^p inequalities, 32
Lamb shift, 81, 86
Laplacian
 Dirichlet, 263
 Neumann, 263
 on $\mathbb{R}^n$, 54^2
Lattice
 ordered sense, 310^1
 subgroup sense, 310
Lavine's theorem, 159
Levy–Khintchine formula, 215
Levy–Khintchine theorem, 221
Lidskii's theorem, 328
Lieb–Thirring bound, 368
Lipschitz, 154
Local Sobolev spaces, 51^2, 253
Local smoothness (operators), 163

M

m-accretive, 168^1, 240^2
Mathieu equation, 320^2, 298
Measurable, weakly and strongly, 64^1, 115^1–117^1
Measurable functions, sets, 14^1–16^1, 19^1–24^1, 104^1–111^1
Measurable unbounded operator-valued function, 283
Measure, 19^1–25^1, 104^1–111^1
Meromorphic Fredholm theorem, 107
Min–max principle, 76
Monotone convergence theorem, 17^1, 24^1
  for nets, 106^1
Multiplicity of an eigenvalue, 9
Multiplicity theorem, 231^1

N

Neumann Laplacian, 263
Norm, 8^1
  operator, 9^1
Norm resolvent sense, convergence in, 284^1–291^1
Normal coordinates, 197
Normal operator, 246^1

O

O'Connor–Combes–Thomas theorem, 198
O'Connor's lemma, 196
One-electron model of solids, 312
One-parameter unitary group, 265^1
Operator
  adjoint
    Banach space, 185^1
    Hilbert space, 186^1
    unbounded, 252^1
  bounded, 8^1
  compact, 199^1
  compact resolvent, 244
  contraction, 151^1, 235^2
  decomposable, 281
  dilation, 183
  elliptic, 112^2, 346
  energy (Hamiltonian), 303^1
  general theory, 81^1–84^1, 182^1–216^1
  Hamiltonian, 303^1
  Hermetian, see Operator, symmetric
  Hilbert–Schmidt, 210^1
  ideals, 41^1–44^2
  infinitesimally form small, 168^2
  infinitesimally small, 162^2
  locally smooth, 163
  momentum, 304^1, 63^2
  non-self-adjoint, 316–338
  norm, 9^1
  normal, 246^1
  positive, 195
  positivity improving, 201
  positivity preserving, 201
  relatively bounded, 162^2
  relatively compact, 113
  relatively form bounded, 168^2
  relatively form compact, 369
  resolvent, 188^1, 253^1
  Schrödinger, 79
  self-adjoint, 187^1, 255^1
  smooth, 142
  symmetric, 255^1
  tensor product, 299^1
  topologies, 182^1–185^1, 283^1–295^1, 41^2–44^2, 268^2–273^2
  trace class, 207^1–210^1
  unbounded, general theory, 249^1–312^1
Order k strong asymptotic series, 43

P

Paley–Wiener theorems, 16^2, 17^2, 18^2, 23^2, 109^2
Parseval's relation, 45^1–46^1
Part of the spectrum in X, 46
Partial trace, 382
Period cell, 303
Periodic potentials, 279
Perron–Frobenius theorems, 202, 350
Perturbation series, 2
  asymptotic, 26
Puiseux, 4

Raleigh–Schrödinger, 1
  strong asymptotic, 40
  wildly divergent, 27
Phase space, 313^2
Phragmen–Lindelöf principle, 236
Plancherel theorem, 10^2
Plemelj–Smithies formulas, 332
Poisson bracket, 314^2
Poisson summation formula, 380
Polar decomposition, 197^1, 297^1
Polarization identity, 63^1
Positive definite, see also Positive type
  conditionally, 214
Positive distribution, 182^2
Positive function, 201
Positive operator, 195^1
Positive type
  distribution, 14^2
  function, 12^2
Positivity improving operator, 201
Positivity preserving operator, 186^2, 201
Principle of uniform boundedness, 81^1, 132^1
Product formula, 295^1–297^1, 245^2
Projection, 187^1
Projection valued measure (p.v.m.), 234^1–235^1, 262^1–263^1
Pseudo-eigenvalue, 48
Pseudo-eigenvector, 48
Puiseux series, 4
Putnam–Kato theorem, 157

Q

Quadratic forms, 276^1
  closed, 277^1
  form core, 277^1
  form domain, 276^1
  Friedrichs' extension, 177^2
  infinitesimally bounded, 168^2
  positive, 276^1
  relatively bounded, 168^2
  relatively compact, 369
  Riesz lemma, 43^1
  semi-bounded, 276^1
  strictly m-accretive, 281^1
  strictly m-sectorial, 282^1
  symmetric, 276^1

R

Radius of gyration, 197
Radon–Nikodym theorem, 25^1
Rayleigh's theorem, 364
Rayleigh–Ritz technique, 82
Rayleigh–Schrödinger coefficients, 5–8
Rayleigh–Schrödinger series, 1, 5
Reduced disconnected part, 131
Reduced resolvent, 130
Relatively bounded, 162^2
Relatively compact, 113
Relatively form bounded, 168^2
Relatively form compact, 369
Rellich's criterion, 247
Rellich's theorem, 4
Resolvent, 188^1, 253^1
Resolvent set, 188^1, 253^1
Resonance eigenvalues, 191
Resonance pole, 55
Resonance thresholds, 191
Retardation term, 81
Riccati equation, 96
Riemann–Lebesgue lemma, 10^2
Riesz's criterion, 248
Riesz's lemma, 43^1
Rollnik potential, 170^2

S

Scale of spaces, 278^1, 44^2
Schrödinger equation, 303^1
Schrödinger operators, 79
Schur basis, 318
Schur–Lalesco–Weyl theorem, 318
Schwartz space, 133^1
Schwarz inequality, 38^1
Secular equation, 2
Segment property, 256
Self-adjoint operator
  bounded, 187^1
  unbounded, 255^1
Semibounded operator, 137^2
Semibounded quadratic form, 276^1
Semigroup
  contraction, 235^2
  holomorphic, 248^2, 252^2

hypercontractive, 258^2
infinitesimal generator, 237^2, 246^2, 248^2
L^p-contractive, 255^2
strongly continuous, 235^2
Singular support, 88^2
Singular value of a compact operator, 203^1
Smooth operator, 142
Sobolev inequality, 31^2, 113^2
Sobolev lemma, 52^2
Sobolev spaces, 50^2, 253
Solid state theory, 311
Sommerfeld correction, 80
Spectral concentration, 45–50
Spectral mapping theorem, 222^1, 109, 181–183
Spectral measures, 228^1
  associated with a vector, 225^1
Spectral projections, 234^1
Spectral theorem
  functional calculus form, 225^1, 263^1
  multiplication operator form, 227^1, 260^1
  p.v.m. form, 235^1, 263^1–264^1
Spectrum, 118
  absolutely continuous, 231^1
  approximate point, 178
  asymptotically in a set, 46
  continuous, 231^1
  continuous singular, 231^1
  discrete, 236^1, 13
  essential, 236^1
  point, 188^1, 231^1
  residual, 188^1
Spin–orbit correction, 80
Spin–spin interaction, 81
Stable eigenvalue, 29
Standard 2^{-n} cube, 271
Stark effect, 200^2, 50, 118
Stone's formula, 237^1
Stone's theorem, 265^1–267^1
Strichartz's theorem, 171^2, 378
Strictly m-accretive form, 281^1
Strictly m-accretive operator, 281^1
Strictly m-sectorial form, 282^1
Strictly positive function, 201
String, 129
Strong asymptotic series, 40
Strong operator topology, 182^1
Strong resolvent sense, convergence in, 284^1–291^1
Strongly continuous semigroup, 235
Strongly continuous unitary group, 265
Strongly measurable, 64^1, 115^1–117^1
Sturm oscillation theorem, 92
Summability methods
  Borel, 44
  Padé, 63
  regular, 63
Support of a distribution, 139^1, 17^2
  singular, 88^2
Symmetric operator, 255^1
Symmetric quadratic form, 276^1

T

Tempered distributions, 134^1
Temple's inequality, 84
  generalized, 365
Tensor product
  Hilbert spaces, 49^1
  operators, 299^1
  spectrum of, 177
Time-dependent perturbation theory, 51–60, 66–68
Total mass, 197
Trace class, 207–210^1
Trotter–Kato theorem, 288^1
Trotter product formula, 295^1–297^1, 245^2

U

Uncertainty principle lemma, 169^2
Uniform boundedness principle, *see* Principle of uniform boundedness
Uniform operator topology, 182^1
Uniformly Hölder continuous, 191
Uniformly locally L^p, 302
Unique continuation, 226, 239
Unitary operator, 39^1

V

Virial theorem, 231
von Neumann's theorem, 268^1, 143^2, 180^2
von Neumann's uniqueness theorem, 275^1

W

Watson's theorem, 44
Weak derivative, 138¹
Weak graph limit, 294¹
Weak-L^p, 30²
  inequalities, 32²
Weak topology, 93¹, 111¹
Weakly measurable (vector-valued) function, 114¹
Weighted L_2 space, 76², 168
Weinberg kernel, 125
  symmetrized, 131
Weinberg–Van Winter equation, 126, 132
Weyl relations, 275¹, 231²
Weyl's criterion, 237¹, 152²
Weyl's lemma, 53²
Weyl's theorem
  classical, 117
  essential spectrum, 112
Width of a resonance, 55
Wiener measure, 278²
  conditional, 102
Wigner–Seitz cell, 311
Wigner–von Neumann potential, 223
Wronskian, 150²

Y

Young's inequality, 28²

Z

Zhislin's theorem, 89

ISBN 0-12-585004-2